SERVICE AVAILABILITY

SERVICE AVAILABILITY
PRINCIPLES AND PRACTICE

Editors

Maria Toeroe
Ericsson, Canada

Francis Tam
Nokia Research Center, Finland

A John Wiley & Sons, Ltd., Publication

Library of Congress Cataloging-in-Publication Data

Service availability : principles and practice / editors, Maria Toeroe,
Francis Tam.
        p. cm.
    Includes bibliographical references and index.
    ISBN 978-1-119-95408-8 (cloth)
    1. Reliability (Engineering)   I. Toeroe, Maria.   II. Tam, Francis.
    TA169.S465 2012
    620$'$.00452–dc23

                                            2011047219

A catalogue record for this book is available from the British Library.

Print ISBN: 9781119954088

Typeset in 9/11pt Times by Laserwords Private Limited, Chennai, India
Printed and bound in Malaysia by Vivar Printing Sdn Bhd

1   2012

Contents

Part II   THE SA FORUM SYSTEM: SERVICES AND FRAMEWORKS

3   Overview of the Service Availability Architecture   33
        Dave Penkler

**4   The SA Forum Information Model: The Heart of Control
        and Monitoring   63**
        Maria Toeroe

5   Consistent and High Level Platform View   85
        Maria Toeroe

List of Contributors

Mario Angelic, Ericsson, Stockholm, Sweden

Robert Hyerle, Hewlett-Packard, Grenoble, France

Jens Jensen, Ericsson, Stockholm, Sweden

Ali Kanso, Concordia University, Montreal, Quebec, Canada

Ferhat Khendek, Concordia University, Montreal, Quebec, Canada

Ulrich Kleber, Huawei Technologies, Munich, Germany

Anik Mishra, Ericsson, Town of Mount Royal, Quebec, Canada

Dave Penkler, Hewlett-Packard, Grenoble, France

Sayandeb Saha, RedHat Inc., Westford, Massachusetts, USA

Francis Tam, Nokia Research Center, Helsinki, Finland

Maria Toeroe, Ericsson, Town of Mount Royal, Quebec, Canada

Foreword

The need to keep systems and networks running 24 hours a day, seven days a week has never been greater, as these systems form some of the essential fabric of society ranging from business to social media. Keeping these systems running in the presence of hardware and software failures is defined as service availability. In some areas of networking, such as telecommunications, it has formed an essential requirement for almost 100 years; it is part of why traditional plain old telephone service (POTS) would still be available when power went out. With the advent of the Internet, service availability requirements are increasingly being demanded in the marketplace, not necessarily due to regulatory requirements, as was the case with telephone networks, but due to business requirements and pressures from the marketplace. Of course, it's not just communications where service availability is important, many other industries such as aerospace and defense have similar requirements. Imagine the impact of a loss of control during missile flight, for example.

After the Internet bubble of the late 1990s, and an almost global deregulation of the telecommunications market, it was increasingly recognized that the high cost of development for proprietary hardware and software systems was no longer viable. The future would increasingly be based on commercial off-the-shelf (COTS) systems, where time to market for new services, outweighs the elegance of proprietary hardware and software systems. High availability middleware, which forms a core aspect of delivering service availability, was one of these complex components. Traditionally viewed as high value and differentiating, in this new environment of time to market service emphasis, where rapid application development, adaptation, and integration are key, proprietary middleware is both time consuming to develop and costly to maintain.

The Service Availability Forum (SA Forum) was established in 2001 to help realize the vision of accelerating the implementation and deployment of service available systems, through establishing a set of open specifications which would define the boundaries between hardware and middleware and between the middleware and the application layer. At the time, concepts which are generally accepted today, such as a layered approach to building systems, the use of off-the-shelf hardware and software, and defacto standards developed through open source, were in their relative infancy.

The Founders of the SA Forum, Force Computers, GoAhead Software, HP, IBM, Intel, Motorola, Nokia, and Radisys all recognized that in 2001 the world was changing. They understood that redundancy and service availability would spread downstream from the traditional high end applications, such as telecommunications and that the key to success was a robust ecosystem built around a set of open specifications for service availability. This would allow applications to run on multiple platforms, with different hardware and operating systems, and enable rapid and easy integration of multiple applications onto a single platform, realizing the vision of rapid development to meet the demands of new services in the marketplace. None of what was envisioned precluded the continued development of proprietary systems, but the concepts were clearly aimed at the increased use of COTS hardware and software with a view accelerating the interoperation between components.

Although it has changed over time, as the organization and the market has evolved, the current mission statement of the SA Forum characterizes the objectives set out in 2001.

> The Service Availability Forum enables the creation and deployment of highly available,
> mission critical services by promoting the development of an ecosystem and publishing
> a comprehensive set of open specifications. A consortium of industry-leading companies,
> the SA Forum maintains 'There is no Upside to Downtime.'

It is always a challenge to create an industry organization when so much investment in proprietary technology already exists. On the one hand, there needs to be a willingness to bring some of this expertise and possibly intellectual property to the table, to serve as a basis for creating the specifications. This has to be tempered with the fear that someone will contribute intellectual property and later aggressively seek to assert patent rights. To avoid issues in this area, the SA Forum was established as a not-for-profit organization and a key aspect of the bylaws was that all members agreed to license any intellectual property to any other members on fair and reasonable terms. Since the SA Forum was dealing primarily in software application programming interfaces around an underlying conceptual architecture, the assertion of patents is quite difficult, but in any event, the Forum has always operated on a cooperative model, with everyone seeking to promote the common good and to address differences within the technical working groups. To further control the objective of a common goal, the SA Forum established three levels of membership, promoters, contributors, and adopters. An academic (associate) membership level was added at later date, and the status of adopter was conferred on anyone with an implementation and use of the specifications in a product.

Promoters were the highest level, and only promoters could be on the board of directors. They were the founders of the organization, and hence the main initial contributors. To avoid predatory actions by other companies, additional promoters could be added only by a unanimous vote of all the promoters. While this may seem overly restrictive, it has worked well in practice, and companies who have demonstrated commitment and who have contributed to the Forum have been offered promoter status.

In order to participate in SA Forum work groups and contribute to the specifications, companies had to be contributor members. This proved to be the workhorse membership level for the organization and many valuable contributions came from this group of members.

The adopter members have generally been companies with interest in supporting the SA Forum's work, or who have developed products that have incorporated some aspect of the SA Forum's specifications.

The cooperative nature of the SA Forum has led to the development of a robust set of specifications for service availability. Indeed, that is what this book is all about, the concepts and use of the SA Forum specifications.

The first tentative steps after the formation in 2001 were white papers on the then new concepts of service availability and a layered architecture approach. These were followed by the initial specifications focused on the hardware platform interface (HPI), which has gone through a number of revisions and enhancements. The most recent release of the HPI specification includes provisions for firmware upgrades and hardware diagnostics.

Work on the more challenging application interface specification (AIS), which address the interfaces to applications, management layers, and overall control of the availability aspects of a distributed system. Early work focused on what has come to be known as the utility services, the fundamental services necessary to create a service available system, cluster concepts, checkpointing, messaging, and so on. By the 2005–2006 timeframe, the Forum was ready to address overall system concepts, such as defining the framework and policy models for managing availability. This resulted in the Availability Management Framework (AMF) and the Information Model Management (IMM). These critical services provide both the flexibility to architect a system to meet application requirements, but also a common mechanism for managing availability, with extensibility to manage applications themselves if desired. This complex work really created the core of the SA Forum AIS and it is in many ways a remarkable piece of work. More recent developments have included the Software Management Framework (SMF) to enable seamless upgrading (and downgrade if necessary) campaigns for systems, demonstrating the true idea of service availability, and platform management (PLM), which enables a coherent abstraction of a system. This encompasses complex hardware designs with computer boards equipped with mezzanine cards, which are themselves compute engines, and enables modern virtual machine architectures to be embraced by the SA Forum system model. This in turn enables the SA Forum specifications to become an essential part of cloud computing concepts.

The SA Forum itself has been responsible for the genesis of other industry organizations. It was recognized that the scope of the SA Forum was insufficient to meet the objective of the wide-spread adoption of off-the-shelf technology and the cooperation between the component layers of the solution. By its very charter, the SA Forum was focused on service availability and middleware. An outgrowth of the Forum was the creation in 2007 of the SCOPE Alliance.

The SCOPE Alliance was founded by Alcatel-Lucent, Ericsson, Motorola, NEC, Nokia, and Siemens. It is a telecom driven initiative which now includes many leading network equipment providers, hardware, and software companies, with the mission to enable and promote a vibrant carrier grade base platform (CGBP) ecosystem for use in telecom network element product development. The SCOPE members believe that a rich ecosystem of COTS and free open source software (FOSS) communities provide building blocks for the network equipment manufacturers to adopt, accelerating their time to market and better serving the service provider marketplace.

To accomplish these goals, SCOPE has created a reference architecture which has been used to publish profiles that define how off-the-shelf technologies can be adopted for various application and platform requirements. These profiles also identify where gaps exist between the various layers of CGBP technology. A core component of the CGBP is service availability middleware, based on SA Forum specifications.

Creating specifications is a complex and intellectually challenging task. This is an accomplishment in and of itself. However, the success of the SA Forum and its specifications is really measured by their adoption in the marketplace and their use in systems in the field. Over the years, there have been a number of implementations of the specifications. When the Forum was founded, and the use of open source software was in its infancy, it was foreseen that the specifications would enable multiple implementations and the portability would be accomplished at the application programming interface (API) layer. From 2006 onwards, the Forum had various initiatives aimed at demonstrating portability. Multiple companies did indeed implement some or part of the specifications to varying degrees. These implementations ranged from selected services to complete implementations of the specifications.

On the hardware side, most major hardware vendors have adopted the HPI specification. There are both proprietary, commercial implementations and an open source solution, OpenHPI, available in the marketplace. With the broad adoption of HPI, this can be very much considered a success in the marketplace.

AIS is much more complex and a range of proprietary and open source solutions have appeared in the marketplace since the mid-2000s. These have had various levels of implementation relative to the specifications discussed in this book, and they have included internal development by network equipment manufacturers, proprietary commercial products, and open source solutions. OpenAIS is an open source solution dating from around 2005 and it has been used extensively for clustering in the Linux community. The most complete implementation of the AIS is the OpenSAF project, this is a focus for many adopters of the SA Forum AIS moving forward, with rollout commitments from major equipment manufacturers and a vibrant ecosystem.

Many people, from a wide variety of companies, have contributed to the SA Forum specifications, and their effort and foresight have led to a framework that is now being implemented, adopted, and deployed. The current focus is on expanding the use cases for the SA Forum specifications and demonstrating that they address a broad range of applications. This goes beyond the traditional five and six '9's' of the telecom world and the mission critical requirements of aerospace and defense, to the realms of the Enterprise and the emerging cloud computing environment.

Timo Jokiaho
Chairman of the SCOPE Alliance, 2011, President of the SA Forum, 2003

John Fryer
President of the SA Forum, 2011

Preface

How This Book Came About

Maria's Story

I joined the Service Availability (SA) Forum in 2005 with the mandate of representing Ericsson in the efforts of the SA Forum Technical Working Group (TWG) to define the Software Management Framework. This is where I met Francis and the representatives of other companies working on the different specifications. The standardization has been going on already for several years and I had a lot to learn and catch up with. Unfortunately there was very little documentation available besides the specifications themselves, which of course were not the easiest introduction to the subject.

Throughout the discussions it became even more obvious that there was an enormous 'tribal knowledge' – as someone termed it – at the base of the specifications. This knowledge was not written anywhere, not documented in any form. One could pick it up gradually once he or she started to decipher the acronym ridden discussions flying high in the room and on the email reflectors. There were usually only a handful who could keep up with these conversations at the intensity that was typical at these discussions. For newcomers they were intimidating to say the least. This was an issue for the SA Forum from the beginning and for the years to come even though there was an Educational Working Group with the mandate to prepare training materials. Many TWG members felt that it would be good to write a book on the subject, but with everyone focusing on the specifications themselves there was little bandwidth to spare for such undertake.

Gradually I picked up most of the tribal knowledge and was able to participate in those discussions, but preparing educational materials or writing a book still did not come to my mind until Ericsson started a research collaboration with Concordia University. Suddenly I had to enlighten my students about the mysteries of the SA Forum specifications. These specifications are based on the years of experience of telecom and information technology companies in high-availability cluster computing. These systems evolved behind closed doors in those companies as highly guarded secrets and accordingly very little if any information was available about them in the public domain. This also meant that the materials were not taught at universities nor were books readily available to which I could refer my students. Soon the project meetings turned into an ad-hoc course where we went through the different details, the intricacies of the specifications and the reasoning behind the solutions proposed. These solutions were steeped in practice and brewed for production. They reflected what has worked for the industry as opposed to theoretical models and proofs more familiar to the academia. This does not mean that they lack theoretical basis. It just means that their development was driven by practice.

Understanding all these details was necessary before being able to embark on any kind of research with the students and their professors. These discussions of course helped the students but at the same time they helped me as well to distill the knowledge and find the best way to present it. Again it would have been nice to have a book, but there was none, only the specifications and the knowledge I gathered in the TWG discussions.

A few years later OpenSAF, the open source implementation of the SA Forum specifications reached the stage when people started looking at it from the perspective of deployment. They started to look for documentation, for resources that they could use to understand the system. OpenSAF uses mostly the SA Forum specifications themselves as documentation for the services compliant to these specifications.

These people faced the same issue I had experienced coming to the world of the SA Forum. I was getting requests to give an introduction, a tutorial presentation so that colleagues can get an idea what they are dealing with, how to approach the system, where to start. After such presentations I would regularly get the comment that 'you should really write a book on this subject.' At this time I saw the suggestion of writing a book more realistic and also with the increasing demand for these presentations it made a lot of sense.

In a discussion with my manager I mentioned the requests I was getting to introduce the SA Forum specifications and the suggestions about the book. He immediately encouraged me to make a proposal. This turn of events transformed the idea I have toyed with for some time into a plan and the journey has begun. I have approached Francis and others I knew from the SA Forum to enroll them in the book project. This book is the realization of this plan, the end of this journey. It is a technical book with a rather complex subject that we, the authors and editors tried to present in a digestible way.

Francis' Story

My contribution related to the SA Forum specifications in this book was based on the project titled 'High Availability Services: Standardization and Technology Investigation' that I worked on during 2001–2006 in Nokia Research Center. The project was funded by Strategy and Technology, the then Nokia Networks (now part of Nokia Siemens Networks), with the objective to support the company's standardization effort in the SA Forum and contribute to a consistent carrier-grade base platform architecture for the then Nokia Networks' business. I became one of the Nokia representatives to the SA Forum and took part in the development of the first release of the Availability Management Framework specification with other member companies' representatives. Subsequently, I took up the role of co-chairing with Maria the Software Management specification development group. Regrettably I had to stop my participation in the SA Forum at the end of 2006 before the Software Management Framework was published.

Parallel to my full-time employments over the years, I have been giving a 12-hour seminar course on highly available systems to the fifth (final) year Master of Engineering students in Computer Science at INSA Lyon (Institut National des Sciences Appliquées de Lyon) in France almost every year since 1993. It has been widely recognized in the academic community that there is a lack of suitable books for teaching the principles and a more pragmatic approach to designing dependable computer systems. Very often such materials have to be gathered from various sources such as conference proceedings, technical reports, journal articles, and the like, and put together specifically for the courses in question. On a number of occasions, the thought of writing such a book came to my mind but it left rather quickly, probably due to my senses were warning me that such an undertaking would have been too much.

I remember it was a few years ago when Maria asked me if I could recommend a book in this area for her teaching. After explaining to her about the general situation with regard to books in this subject area, I half-jokingly suggested to her that we could write one together. She left it like that but only returned in January 2010 and asked if I would be interested in a book project. As they say, the rest is history.

The Goal of the Book

Our story of how the book came about has outlined the need that has built up and which it was time to address with a book. It was clear that the approach to the subject should not be too theoretical, but rather an explanation of the abstractions used in the SA Forum specifications that would help practitioners in mapping those abstractions to reality; it also needed to make the knowledge tangible, to show how to build real systems with real applications using the implementations of the SA Forum specifications. The time was right as these implementations were reaching maturity fast.

At the same time we did not want to write a programmers' guide. First of all a significant portion of the specifications themselves is devoted to the description of the different application programming interface (API) functions. But there is so much reasoning in these systems and the beauty of their logic cannot be delivered just by discussing the APIs, which are like the scrambled puzzle pieces do not reflect the complete picture, the interconnection and interdependencies until they are put together piece by piece. They give little information on the reasoning which animates the picture and fills in even missing puzzle pieces.

The specifications may not be perfect at this time yet but they bring to the light this technology that has been used and proved itself in practice to provide the magic five-nine figures of in service performance, but has been hidden from the public eye. At this time they already come with open source implementations meaning that they are available for anyone to experiment with or to use for deployment, and also to evolve and improve.

The concepts used in these specifications teach a lot about how to think about systems that need to provide their services continuously 24/7 in the presence of failures. Moreover they are designed to evolve respecting these same conditions, that is, these systems and their services develop without being taken out for planned maintenance, they evolve causing minimal service outage. They are ideal for anyone who needs to meet stringent service level agreements or SLAs.

The concepts presented in this book remain valid whether they are used in the framework of the SA Forum specifications or transpired to cloud computing or any other paradigm that may come. The SA Forum specifications provide an excellent basis to elaborate and present the concepts and the reasoning. They also set the terminology allowing for a common language of discussion, which was missing for the area.

We set out to explain these concepts and their manifestation in the specifications and demonstrate their application through use cases.

So who would benefit from this book? The obvious answer is that applications and systems designers who intend to use the SA Forum middleware. However since we look at the specifications more as one possible manifestation of the concepts, ultimately the book benefits anyone who needs to design systems and applications for guaranteed service availability, or who would like to learn about such systems and applications. We see this book as a basis for an advanced course on high service availability systems in graduate studies or in continuous education.

The Structure of the Book

The book is divided into three main parts:

Part One introduces the area of service availability, its basic concepts, definitions, and principles that set the stage for the subsequent discussions. It also delivers the basic premise that makes the subject timely. Namely that in our society the demand for continuous services is increasing in terms of the number and variety of services as well as the number of customers. To meet this demand it is essential to make the enabling technologies widely available by standardizing the service APIs so that commercial off the shelf components can be developed. Enabling such an ecosystem was the mission of the SA Forum, whose coming about is presented also in this part.

Part Two of the book focuses on the specifications produced by the SA Forum to achieve its mission. The intention was to provide an alternative view of the specifications, a view that incorporates that 'tribal knowledge' not documented anywhere else and which provides some insight to the specifications, to the choices that were made at their design.

We start out with the architectural overview of the SA Forum middleware and its information model.

The subsequent chapters elaborate on the different services defined by the SA Forum Architecture. Among them the Availability Management Framework and the Software Management Framework each has their own dedicated chapter while the other services are presented as functional groups: the Platform services, the Utility services, and the Management Infrastructure services.

Rather than discussing all the SA Forum services at a high level we selected a subset on which we go into deeper discussions so that the principles become clear. We do not cover the Security service in our discussions as it is a subject crosscutting all the services and easily filling a book on its own.

The presentation of the different services and frameworks follow more or less the same pattern:

First the goals and the challenges addressed by the particular service are discussed, which are followed by an overview of the service including the service model and architecture supporting the proposed solution.

Rather than presenting the gory details of each of the API functions like it would be in a programmer's guide we decided to explain the usage through the functionality that can be achieved by using the APIs. This approach reveals better the complete picture behind the puzzle pieces of the API functions. We mention the actual API functions only occasionally when it makes it easier to clarify the overall functionality.

Whenever it is applicable we also present the administrative perspective of the different services. The goal of these sections is to outline what a system administrator may expect to observe in a running system and what control he or she can obtain through configuration and administrative operations according to the specification. Sometimes these details could be overwhelming, so the anticipation is that different implementations of the standard services may restrict this access while other vendors may build management applications that enhance the experience by assisting the administrator in different ways.

Subsequently the service interactions are presented inserting the service discussed thus far in isolation into the environment it is expected to operate. Since the specifications themselves are written in a somewhat isolated way, these sections collect information that are not readily available, which require the understanding of the overall picture.

Finally the open issues and recommendations conclude each of the service overviews.

Particularly the open issues deserve some explanation here: even though the SA Forum specifications are based on the best practice developed in the industry over the years, the specifications themselves are not the reflection of a single working implementation. Rather they are based on the combined knowledge derived by the participants from different working implementations. So at the time of the writing of the different specifications the SA Forum system existed only in the heads of the members of the SA Forum TWG. It was this common vision that was scrutinized in the process of the standardization that obviously reshaped and adjusted the vision.

As the work progressed and people started to implement the different specifications the results were fed back to the standardization process. In case of the simpler services most of the issues found through these implementations have been resolved by the time of the writing of this book. But for the more complex services there are still remaining open issues.

There are also a few cases where the TWG deliberately left the issues open so that the implementations have the freedom to resolve them in a way most suitable for the particular implementation; for example, the system bootstrapping was left implementation specific. These are usually cases that do not impact applications using the services, but for which service implementers would like to have an answer (but typically not the one the specification would offer).

Part Three of the book looks at the SA Forum middleware in action, that is, at the different aspects of the practical use of the specifications presented in Part Two.

It starts with the overview of the programming model used throughout the definition of the different service APIs. There is a system in the API definitions of the different specifications and Chapters 11 and 12 serve as Ariadne's thread in what seem to be a labyrinth. This is followed by a bird's-eye view at the two most important open source implementations of the SA Forum specifications: OpenSAF and OpenHPI.

To help integrators and application developers to use these middleware implementations in Chapter 14 we discuss different levels of integration of the VideoLAN Client (VLC) application originally not developed for high availability. This exercise demonstrates in practice how an application can take advantage of the SA Forum Availability Management Framework even without using any of its APIs. Of course better integration and better user experience can be achieved using the APIs and additional services, which is also demonstrated.

After this 'hands on' exercise the problem of migrating large scale legacy applications is discussed. This chapter gives an excellent insight not only for those considering such migration, but also to designers and developers of new applications. It demonstrates the flexibility of the SA Forum specifications which people usually realize only after developing an intimate relationship with them. The mapping of the abstractions defined by the specifications is not written in stone and it is moldable to meet the needs of the situation. This is demonstrated on the example of two different database integrations with the SA Forum middleware depending on the functionality inherent in the database.

The final chapter of Part Three takes yet again a different perspective. It discusses the issues complementary to the specifications but necessary for the operation of the SA Forum middleware. It introduces the use of formal models and techniques to generate system configurations and upgrade campaigns necessary for the Availability and the Software Management Frameworks to perform their tasks. This approach was part of the vision of the SA Forum specifications as they defined the concepts enabling such technology opening the playground for tool vendors.

We could have continued exploring the subject with many exciting applications, but we had to put an end as we reached our page limit as well as the deadline for delivering the manuscript. So we leave the rest of the journey to the reader who we hope will be well equipped after reading our book to start out with their own experimentations.

Acknowledgments

The group of people that were essential for the creation of this book are the Service Availability (SA) Forum's Technical Working Group representatives of the different member companies; who concocted the specifications and provided a challenging yet inspiring environment for learning and growing in the field. We cannot possibly list all the participants without missing a few, so we will not do so. There were however a few outstanding:

We had extremely constructive and rewarding discussions with the SA Forum Software Management Working Group when we were creating the Software Management Framework, for which we would like to thank Peter Frejek, Shyam Penubolu, and Kannan Kasturi. We probably should not forget about another regular participant of our marathon-length conference calls: the Dog whose comments broke the seriousness of the discussions.

We would like to thank Fred Herrmann, who left his fingerprints over most if not all SA Forum service specifications, and for the numerous stimulating discussions and debates which made the experience so much more exciting. And in the debates it was a pleasure to have the calming wisdom of Dave Penkler. Dave was also instrumental in the writing and reviewing of this book. We are grateful to him for graciously stepping up and helping out with key chapters when we were under pressure of time and short of a pair of fresh eyes.

We are deeply obliged to our co-authors for helping us create this book. For most of them this meant the sacrifice of their spare time – stealing it from their families and friends to deliver the chapters and with that make the book so much more interesting.

Finally we would like to thank Wiley and in particular Sophia Travis for recognizing the vision in our book proposal and helping us through the stress of the first book with such an ease that it truly felt like a breeze.

From Maria

First and foremost I would like to thank the generosity of Ericsson and within that of my managers Magnus Buhrgard and Denis Monette for allotting me the time to work on this book and their continuous support and trust that it would be completed. Not that I ever had a doubt, but it definitely took more time and efforts than I anticipated. Their support made the whole project possible.

I am also grateful to the MAGIC team of Concordia University. The professors: Ferhat Khendek, Rachida Dssouli, and Abdelwahab Hamou-Lhadj, the students Ali Kanso, Setareh Kohzadi, Anik Mishra, Ulf Schwekendiek, Pejman Salehi, and the post-docs: Pietro Colombo and Abdelouahed Gherbi. They provided me with a completely different learning experience. All of them had their own approach to the problem and in the discussions I had to learn to investigate the subject from many different sometimes unconventional angles and answer questions that within industry were taken for granted. These discussions and working together on the problems led me to a fresh look and a deeper understanding of the subject all facilitating (at least in my belief) a better delivery.

Finally I would like to thank my colleagues in Montreal and across the sea in Stockholm who were the initiators of this project with their requests and suggestions, who joined my family and friends, in supporting and encouraging me in my writing from the beginning.

A heartfelt thank to all of you.

Maria Toeroe
September, 2011

From Francis

The undertaking to write a book is a daunting commitment even in the best of times, having to do it in my spare time after the day job was rather demanding. My contribution to this book would not have been possible if it was not for the thoughtful understanding and unreserved support from my wife Riikka, who has the shared belief that this book project was good for me. She deserves a medal for putting up with my long evenings and weekends of writing.

As if my lack of time were not enough, I went through one round of company reorganization and was under the threat of lay-off for some weeks – a slightly different kind of redundancy I originally planned to think about. My warm thank you goes to Minna Uimonen, who has always encouraged me and reminded me of the Finnish *sisu* during this difficult time. I am grateful to all my friends for their kind wishes and understanding of my short disappearance. I look forward to re-integrating with the community and do what I do best – as a highly available 'Chief Entertainment Officer.'

Francis Tam
September, 2011

List of Abbreviations

3G	3rd generation
3PP	Third Party Product
ABI	Application Binary Interface
AIS	The SA Forum Application Interface Specification
AMF	The SA Forum Availability Management Framework
AMM	Availability Management Middleware
ANSI	American National Standards Institute
API	Application Programming Interface
ARP	Address Resolution Protocol
ASN.1	Abstract Syntax Notation One
ATCA	Advanced Telecommunication Computing Architecture
ATL	ATLAS Transformation Language
BASH	Born Again Shell
CASE	Computer-Aided Software Engineering
CCB	Configuration Change Bundle
CGBP	Carrier Grade Base Platform
CIM	Common Information Model
CIMOM	Common Information Model Object Manager
CKPT	the SA Forum Checkpoint Service
CLC-CLI	component life-cycle command line interface
CLI	command line interface
CLM	the SA Forum Cluster Membership Service
CORBA	Common Object Request Broker Architecture
COTS	commercial-off-the-shelf
CPU	central processing unit
CSI	component service instance
CST	component service type
DAM	dependability analysis modeling
DAT	domain alarm table
DBMS	database management system
DET	domain entity tree
DIMI	Diagnostics Initiator Management Instrument
DMTF	Distributed Management Task Force

DN	distinguished name
DNS	domain name server
DRT	domain reference table
(E)AM	external active monitoring
EE	execution environment
ETF	entity types file
EVT	the SA Forum Event Service
FAR	Federal Acquisition Regulation
FOSS	free open source software
FRU	field replaceable unit
FT	fault tolerant
ftp	file transfer protocol
FUMI	Firmware Upgrade Management Instrument
GUI	graphical user interface
HA	high availability
HE	hardware element
HP	Hewlett-Packard
HPI	the SA Forum Hardware Platform Interface
HTTP	hypertext transmission protocol
HW	hardware
IBM	International Business Machines
ID	identifier
IDR	Inventory Data Record
IEEE	Institute of Electrical and Electronics Engineers
IETF	Internet Engineering Task Force
IFIP	International Federation for Information Processing
iLO2	HP Integrated Lights-Out 2
IMM	the SA Forum Information Model Management Service
I/O	input/output
IP	the Internet Protocol
IPMB	intelligent platform management bus
IPMI	intelligent platform management interface
ISP	in-service performance
ISV	independent software vendor
IT	information technology
ITU	International Telecommunication Union
Java EE	Java Enterprise Edition (formerly J2EE)
JCP	Java Community Process
JMX	Java Management eXtenstions
JSR	Java specification request
JVM	Java virtual machine
LDAP	lightweight directory access protocol
LOG	the SA Forum Log service
MAGIC	Modeling and Automatic Generation of Information for Configuration and upgrade campaigns for service availability
MARTE	OMG's Modeling and Analysis of Real-Time Embedded systems
MDA	model driven architecture
MDE	model driven engineering
MDS	message distribution service
MIB	management information base

MOF	meta object facility
MSG	the SA Forum Message Service
MTBF	mean time between failures
MTTF	mean time to failure
MTTR	mean time to repair
NAM	the SA Forum Naming Service
NEC	Nippon Electric Company, Limited
NETCONF	network configuration protocol
NIO	New I/O
NTF	the SA Forum Notification Service
NTP	network time protocol
OA	onboard administrator
OAM	operations, administration, and maintenance (or management)
OCL	object constraint language
OI	object implementer
OI-API	the object implementer API of the IMM service
OM	object manager
OM-API	the object management API of the IMM service
OMG	Object Management Group
OS	operating system
NP-hard	nondeterministic polynomial-time hard
PCI	peripheral component interconnect
PICMG	PCI Industrial Computer Manufacturers Group
PLM	the SA Forum Platform Management Service
POSIX	Portable Operating System Interface for Unix
POTS	plain old telephone service
RIBCL	remote insight board command language
RDN	Relative Distinguished Name
RDR	resource data records
RPM	RPM Package Manager or Redhat Package Manager
RPT	resource presence table
RSA	remote supervisor adapter
RTAS	run-time abstraction services
RTP	real-time transport protocol
RTSP	real-time streaming protocol
SA	service availability
SAF	Service Availability Forum
SAI	Service Availability Interface
SAN	Storage Area Network
SEC	the SA Forum Security Service
SI	service instance
SG	service group
SMF	the SA Forum Software Management Framework
SMIv2	Structure of Management Information Version 2
SNMP	simple network management protocol
SOAP	simple object access protocol
SPNP	stochastic Petri net package
ssh	secure shell
SSL	secure socket layer
SU	service unit

SW	software
TCP	transmission control protocol
TID	HP Telecom Infrastructure Division
TIPC	transparent inter-process communication
TMR	the SA Forum Timer Service
TWG	the SA Forum Technical Working Group
UCMI	universal chassis management interface
UML	unified modeling language
UmL	User-mode Linux
UCS	upgrade campaign specification
URI	uniform resource identifier
VLC	VideoLAN Client
VLM	VideoLAN Manager
VM	virtual machine
VMM	virtual machine monitor
VoD	Video on Demand
WBEM	Web-Based Enterprise Management
WG	working group
XMI	XML metadata interchange format
XML	eXtensible Markup Language
xTCA	ATCA and MicroTCA

Part One

Introduction to Service Availability

1

Definitions, Concepts, and Principles

Francis Tam
Nokia Research Center, Helsinki, Finland

1.1 Introduction

As our society increasingly depends on computer-based systems, the need for making sure that services are provided to end-users continuously has become more urgent. In order to build such a computer system upon which people can depend, a system designer must first of all have a clear idea of all the potential causes that may bring down a system. One should have an understanding of the possible solutions to counter the causes of a system failure. In particular, the costs of candidate solutions in terms of their resource requirements must also be known. Finally, the limits of the eventual system solution that is put in place must be well understood.

Dependability can be defined as the quality of service provided by a system. This definition encompasses different concepts, such as reliability and availability, as attributes of the service provided by a system. Each of these attribute can therefore be used to quantify aspects of the dependability of the overall system. For example, reliability is a measure of the time to failure from an initial reference instant, whereas availability is the probability of obtaining a service at an instant of time. Complex computer systems such as those deployed in telecommunications infrastructure today require a high level of availability, typically 99.999% (five nines) of the time, which amounts to just over five minutes of downtime over a year of continuous operation. This poses a significant challenge for those who need to develop an already complex system with the added expectation that services must be available even in the presence of some failures in the underlying system.

In this chapter, we focus on the definitions, concepts, principles, and means to achieving service availability. We also explain all the conceptual underpinning needed by the readers in understanding the remaining parts of this book.

1.2   Why Service Availability?

In this section, we examine why the study on service availability is important. It begins with a dossier on unavailability of services and discusses the consequences when the expected services are not available. The issues and challenges related to service availability are then introduced.

1.2.1   Dossier on Unavailability of Service

Service availability – what is it? Before we delve into all the details, perhaps we could step back and ask why service availability is important. The answer lies readily from the consequences when the desired services are not available. A dossier on the unavailability of services aims to illustrate this point.

Imagine you were one of the one million mobile phone users in Finland, who was affected by a widespread disturbance of a mobile telephone service [1] and had problems receiving your incoming calls and text messages. The interrupt of service, reportedly caused by a data overload in the network, lasted for about seven hours during the day. You could also picture yourself as one of the four million mobile phone subscribers in Sweden when a fault, although not specified, had caused the network to fail and unable to provide you with mobile phones services [2]. The disruption lasted for about twelve hours, which began in the afternoon and continued until around midnight.

Although the reported number of people affected in both cases does not seem to be that high at first glance, one has to put them in the context of their populations. The two countries have respectively 5 and 9 millions of people so the proportion of the affected were considerable.

These two examples have given a somewhat narrow illustration of the consequences when services are unavailable in the mobile communication domain. There are many others and they touch on different kinds of services, and therefore different consequences as a result. One case in point was the financial sector reported that a software glitch, apparently caused by a new system upgrade, had resulted in a 5.5 hour delay in shares trading across the Nordic region including Stockholm, Copenhagen, Helsinki, as well as the Baltic and Icelandic stock exchanges [3]. The consequence was significantly high in terms of the projected financial loss due to the delayed opening of the stock market trading.

Another high-profile and high-impact computer system failure was at the Amazon Web Services [4] for providing web hosting services by means of its cloud infrastructure to many web sites. The failure was reportedly caused by an upgrade of network capacity and lasted for almost four days before the last affected consumer data were recovered [5], although 0.07% of the affected data could not be restored. The consequence of this failure was the unavailability of services to the end customers of the web sites using the hosting services. Amazon had also paid 10-day service credits to those affected customers.

A nonexhaustive list of failures and downtime incidents collected by researchers [6] gives further examples of causes and consequences, which includes categories of data center failures, upgrade-related failures, e-commerce system failures, and mission-critical system failures. Practitioners in the field also maintain a list of service outage examples [7]. These descriptions further demonstrate the relationships between the cause and consequence of failures to providing services. Although some of the causes may be of a similar nature to have made the service unavailable in the first place, the consequences are very much dependent on what the computer system is used for. As described in the list of failure incidents, this could range from the inconvenience of not having the service immediately available, financial loss, to the most serious result of endangering human lives.

It is important to note that all the consequences in the dossier above are viewed from the end-users' perspective, for example, mobile phone users, stockbrokers trading in the financial market and users of web site hosting services. *Service availability* is measured by an end-user in order to gauge the level of a provided service in terms of the proportion of time it is operational and ready to deliver. This is a user experience of how ready the provided service is. Service availability is a product of the availability of all the elements involved in delivering the service. In the example case of a mobile phone user above, the elements include all the underlying hardware, software, and networks of the mobile network infrastructure.

1.2.2 Issues and Challenges

Lack of a common terminology and complexity have been identified as the issues and challenges related to service availability. They are introduced in this section.

1.2.2.1 Lack of a Common Terminology

Studies on dependability have long been carried out by the hardware as well as software communities. Because of the different characteristics and as a result a different perspective on the subject, dissimilar terminologies have been developed independently by many groups. The infamous observation of 'one man's error is another man's fault' is often cited as an example of confusing and sometimes contradictory terms used in the dependability community. The IFIP (International Federation for Information Processing) Working Group WG10.4 on Dependable Computing and Fault Tolerance [8] has long been working on unifying the concepts and terminologies used in the dependability community. The first taxonomy of dependability concepts and terms was published in 1985 [9]. Since then, a revised version was published in [10]. This taxonomy is widely used and referenced by researchers, practitioners, and the like in the field. In this book, we adopt this conceptual framework by following the defined concepts and terms in the taxonomy. On the general computing side, where appropriate, we also use the Institute of Electrical and Electronics Engineers (IEEE) standard glossary of software engineering terminology [11]. The remainder of this chapter presents all the needed definitions, concepts, and principles for a reader to understand the remaining parts of the book.

1.2.2.2 Complexity and Large-Scale Development

Dependable systems are inherently complex. The issues to be dealt with are usually closely intertwined because they have to deal with the normal functional requirements as well as the nonfunctional requirements such as service availability within a single system. Also, these systems tend to be large, such as mobile phone or cloud computing infrastructures as discussed in the earlier examples. The challenge is to manage the sheer scale of development and at the same time, ensure that the delivered service is available at an acceptable level most of the time. On the other hand, there is clearly a common element of service availability implementation across all these wide-ranging application systems. If we can extract the essence of service availability and turn it into some form of general application support, it can then be reused as ready-made template for service availability components. The principle behind this idea is not new. Over almost two decades ago, the use of commercial-off-the-shelf (COTS) components had been advocated as a way of reducing development and maintenance costs by buying instead of building everything from scratch. Since then, many government and business programs have mandated the use of COTS. For example, the United States Department of Defense has included this term into the Federal Acquisition Regulation (FAR) [12].

Following a similar consideration in [13] to combine the complementary notions of COTS and open systems, the Service Availability Forum was established and it developed the first open standards on service availability. Open standards is an important vehicle to ensure that different parts are working together in an ecosystem through well-defined interfaces. The additional benefit of open standards is the reduction of risks in a vendor lock-in for supplying COTS. In the next chapter, the background and motivations behind the creation of the Service Availability Forum and the service availability standards are described. A thorough discussion on the standards' services and frameworks, including the application programming and system administrator and management interfaces, are contained in Part Two of the book.

1.3 Service Availability Fundamentals

This section explains the basic definitions, concepts, and principles involving service availability without going into a specific type of computer system. This is deemed appropriate as the consequences of system failures are application dependent; it is therefore important to understand the fundamentals instead of going into every conceivable scenario. The section provides definitions of system, behavior, and service. It gives an overview of the dependable computing taxonomy and discusses the appropriate concepts.

1.3.1 System, Behavior, and Service

A *system* can be generically viewed as an entity that intends to perform some *functions*. Such entity interacts with other systems, which may be hardware, software, or the physical world. Relative to a given system, the other entities with which it interacts are considered as its *environment*. The system *boundary* defines the limit of a system and marks the place where the system and its environment interact.

Figure 1.1 shows the interaction between a given system and its environment over the system boundary. A system is structurally composed of a set of *components* bound together. Each component is another system and this recursive definition stops when a component is regarded as atomic, where further decomposition is not of interest. For the sake of simplicity, the remaining discussions in this chapter related to the properties, characteristics, and design approaches of a system are applicable to a component as well.

The functions of a system are what the system intends to do. They are described in a *specification*, together with other properties such as the specific qualities (for example, performance) that these

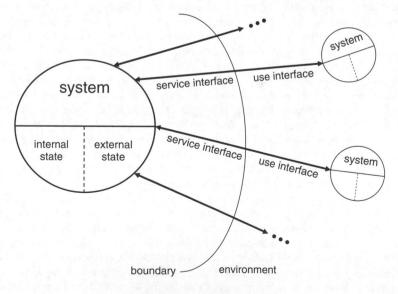

Figure 1.1 System interaction.

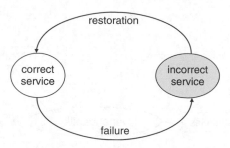

Figure 1.2 Service state transitions.

functions are expected to deliver. What the system does to implement these functions is regarded as its *behavior*. It is represented by a sequence of *states*, some of which are internal to the system while some others are externally visible from other systems over the system boundary.

The *service* provided by a system is the observed behavior at the system boundary between the providing system and its environment. This means that a service user sees a sequence of the provider's external states. A *correct service* is delivered when the observed behavior matches those of the corresponding function as described in the specification. A *service failure* is said to have occurred when the observed behavior deviates from those of the corresponding function as stated in the specification, resulting in the system delivering an incorrect service. Figure 1.2 presents the transition from a correct service to service failure and vice versa. The duration of a system delivering an incorrect service is known as a service outage. After the restoration of the incorrect service, the system continues to provide a correct service.

Take a car as an example system. At the highest level, it is an entity to provide a transport service. It primarily interacts with the driver in its environment. A car system is composed of many smaller components: engine, body, tires, to name just a few. An engine can be further broken into smaller components such as cylinders, spark plugs, valves, pistons, and so on. Each of these smaller components is connected and interacts with other components of systems.

As an example, an automatic climate control system provides the drivers with a service to maintain a user-selected interior temperature inside the car. This service is usually implemented by picking the proper combination of air conditioning, heating, and ventilation in order to keep the interior temperature at the same level. The climate control system must therefore have functions to detect the current temperature, turn on or off the heater and air conditioning, and open or close air vents. These functions are described in the functional specification of the climate control system, with clear specifications of other properties such as performance and operating conditions.

Assuming that the current interior temperature is 18 °C and the user-selected temperature is 20 °C, the expected behavior of the automatic climate control system is to find out the current temperature and then turn on the heater until the desired temperature is reached. During these steps, the system goes through a sequence of states in order to achieve its goal. However, not all the states are visible to the driver. For example, the state of the automatic climate control system with which the heater interacts is a matter of implementation. Indeed whether the system uses the heater or air conditioning to reach the user-selected temperature is of no interest to the user. On the other hand, the state showing the current interior temperature is of interest to a user. This gives some assurance that the temperature is changing in the right direction. This generally offers the confidence that the system is providing the correct service. If for some reason the heater component breaks down, the same sequence of steps does not raise the interior temperature to the desired 20 °C as a result. In this case, the system has a service failure because the observed behavior differs from the specified function of maintaining a user-selected temperature in the car. The service outage can be thought of as the period of time when the heater breaks down until it is repaired, possibly in a garage by qualified personnel and potentially takes days.

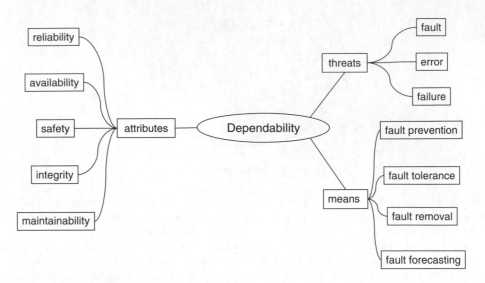

Figure 1.3 Classifications of dependability concepts.

1.3.2 Dependable Computing Concepts

As discussed in the introduction, availability is one part of the bigger dependability concept. The term dependability has long been regarded as an integrating concept covering the qualities of a system such as availability, reliability, safety, integrity, and maintainability. A widely agreed definition of *dependability* [10] is 'the ability to deliver service that can justifiably be trusted.' The alternative definition, 'the ability to avoid service failures that are more frequent and severe than is acceptable' is very often served as a criterion to decide if a system is dependable or not.

Figure 1.3 shows the organization of the classifications. At the heart is the main concept of dependability, which is comprised of three subconcepts: threats, attributes, and means. It must be pointed out that the concept of security has been taken out due to the subject being outside the scope of this book. A *threat* is a kind of impairment that can prevent a system from delivering the intended service to a user. Failures, errors, and faults are the kinds of threats that can be found in a system. Since dependability is an integrating concept, it includes various qualities that are known as *attributes*. These include availability, reliability, safety, integrity, and maintainability of the intended service. The *means* are the ways of achieving the dependability goal of a service. To this end, four major groups of methods have been developed over the years, namely, fault prevention, fault tolerance, fault removal, and fault forecasting.

1.3.2.1 Threats

In order to understand the consequences of a threat to a service, it is important to differentiate the different types of threats and their relationship. The fault–error–failure model expresses that a fault, a physical defect found in a system, causes an error to the internal state of a system, and in turn finally causes a failure to a system, which can be detected externally by users. Faults are physical defects and that means they could be wiring problems, aging of components, and in software an incorrect design. The existence of a fault does not mean that it immediately causes an error and then a failure. This is because the part of the system that is affected by the fault may not be running all the time. A fault is said to be in a dormant state until it becomes active when the part of the system affected is exercised.

The activation of a fault brings about an error, which is a deviation from the correct behavior as described in the specification. Since a system is made up of a set of interacting components, a failure does not occur as long as the error caused by a fault in the component's service state is not part of the external service state of the system.

1.3.2.2 Attributes

- **Reliability**
  This is defined as the ability of a system to perform a specified function correctly under the stated conditions for a defined period of time.

- **Availability**
  This is defined as the proportion of time when a system is in a condition that is ready to perform the specified functions.

- **Safety**
  This is defined as the absence of the risk of endangering human lives and of causing catastrophic consequences to the environment.

- **Integrity**
  This is defined as the absence of unauthorized and incorrect system modifications to its data and system states.

- **Maintainability**
  This is defined as a measure of how easy it is for a system to undergo modifications after its delivery in order to correct faults, prevent problems from causing system failure, improve performance, or adapt to a changed environment.

1.3.2.3 Means

- **Fault prevention**
  This is defined as ensuring that an implemented system does not contain any faults. The aim is to avoid or reduce the likelihood of introducing faults into a system in the first place. Various fault prevention techniques are usually carried out at different stages of the development process. Using an example from software development, the use of formal methods in the specification stage helps avoid incomplete or ambiguous specifications. By using well-established practices such as information hiding and strongly typed programming languages, the chances of introducing faults in the design stage are reduced. During the production stage, different types of quality control are employed to verify that the final product is up to the expected standard. In short, these are the accepted good practices of software engineering used in software development. It is important to note that in spite of using fault prevention, faults may still be introduced into a system. Therefore, it does not guarantee a failure-free system. When such a fault activates during operational time, this may cause a system failure.

- **Fault tolerance**
  This is defined as enabling a system to continue its normal operation in the presence of faults. Very often, this is carried out without any human intervention. The approach consists of the error detection and system recovery phases. Error detection is about identifying the situation where the internal state of a system is different from that of a correct one. By using either error handling or fault handling in the recovery phase, a system can perform correct operations from this point onwards. Error handling changes a system state that contains errors into a state without any detected errors. In this case, this action does not necessarily correct the fault that causes the errors. On the other hand, a system using fault handling in the recovery phase essentially repairs the fault that causes the errors. The workings of fault tolerance are presented in Section 1.4 in more details.

- **Fault removal**
  This achieves the dependability goal by following the three steps of verification, diagnosis, and correction. Removal of a fault can be carried out during development time or operational time. During the development phase, this could be done by validating the specification; verifying the implementation by analyzing the system, or exercising the system through testing. During the operational phase, fault removal is typically carried out as part of maintenance, which first of all isolates the fault before removing it. Corrective maintenance removes reported faults while preventive maintenance attempts to uncover dormant faults and then removes them afterwards. In general, maintenance is a manual operation and it is likely to be performed while the system is taken out of service. A fault-tolerant system, on the other hand, may be able to remove a fault without disrupting service delivery.

- **Fault forecasting**
  This is concerned with evaluating the system behavior against likely failures. This involves identifying and classifying the failure modes and assessing how well a system deals with anticipated faults in terms of probabilities. The assessment is usually carried out by two complementary quantitative evaluation approaches: modeling and operational testing. A behavioral model of the system is first created and then processed to estimate the dependability measure of the system. The data used in processing the model can either be obtained by operational testing, which is performed in the system's operational environment, or based on data from past system failures. It is important to point out that fault forecasting only provides a prediction of how well the current system copes with anticipated faults. If appropriate actions are taken, for example, by using fault removal, the quality of the system would improve over time. However, there are situations when the identified faults are not removed from the system. This is usually due to economic reasons such as the costs of removing a fault outweighing its benefit, especially when the probability of the fault occurring is low. If a fault is elusive and difficult to pin down, then fault tolerance could be an alternative to fault removal.

It would be naïve to believe that any one of the four approaches can be used on its own to develop systems with high dependability. In practice, all or a combination of the methods are used at each design and implementation stage of developing dependable systems. For example, it is common to have extensive fault prevention and fault removals throughout the various system development stages in a fault-tolerant system. After all, one would not want any faults to be introduced into the critical phases of error detection and system recovery. Once a system is operational, live data are used in fault forecasting in order to feed in the improvements and/or corrections for the next version of the system.

1.3.3   The Meaning of Availability

Due to the fact that studies of dependable systems have been carried out by a diverse group of communities, there is a constant source of confusion over the use of some terms. Specifically, some definitions associated with the measures of failures must be explained. Before we discuss the meaning of availability, we first give the following definitions and then discuss their relationships in the context of expressing a system's availability.

- **Mean time to failure (MTTF).** This is defined as the expected time that a system takes to fail. It is the same as uptime, which can be described as the duration of time when a system is operational and providing its service.
- **Mean time to repair (MTTR).** This is defined as the expected time required to repair a system such that it can return to its normal, correct operation. Assuming that all the overheads are accounted for in arriving at the repair time estimate, it should be the same as the downtime. MTTR is known

to be difficult to estimate. One possible approach is to inject faults into a system and then determine experimentally by measuring the time to fix them.

- **Mean time between failures (MTBF).** This is defined as the expected time between two successive failures in a system. This measure is meaningful only if a system can be repaired, that is, a repairable system, because a system fails the second time only if it has been repaired after the first failure.

Figure 1.4 illustrates the relationship among the different failure measures in a system. It shows the status of a system, which can be either up or down, against the time of operation. At time T_{start}, the system starts its operation and continues until time $T_{failure-1}$, when it encounters the first failure. After the duration of MTTR, which is the downtime, the system is repaired and continues its normal operation at time $T_{repaired-1}$. It continues until the second failure hits the system at time $T_{failure-2}$.

As shown in Figure 1.4, the relationship among the failure measures of a system can be expressed as:

$$MTBF = MTTF + MTTR$$

Another way of expressing the MTBF of a system is the sum of the downtime and uptime during that period of time. This is essentially linked to a system's dependability attribute of availability.

Availability is defined as the degree to which a system is functioning and is accessible to deliver its services during a given time interval. It is often expressed as a probability representing the proportion of the time when a system is in a condition that is ready to perform the specified functions. Note that the time interval in question must also include the time of repairing the system after a failure. As a result, the measure of availability can be expressed as:

$$Availability = MTTF/MTBF = MTTF/(MTTF + MTTR)$$

The availability measure is presented as a percentage of time a system is able to provide its services readily during an interval of time. For example, an availability of 100% means that a system has no downtime at all. This expression also highlights the fact that the availability of a system depends on how frequently it is down and how quickly it can be repaired.

It is generally accepted that a highly available system, such as those used in telecommunications, must have at least 99.999% availability, the so-called systems with 5–9s availability requirements. Table 1.1 shows the maximum allowable downtime of a system against the different number of 9s availability required under various operating intervals. As shown, a 5–9s system allows for just 5 minutes 15 seconds of downtime in 1 year's continuous operation.

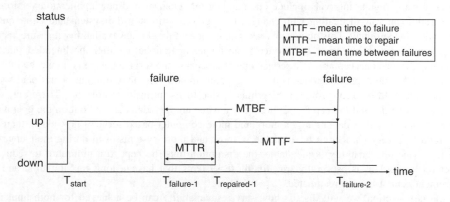

Figure 1.4 Different failure measures.

Table 1.1 Maximum allowable downtime for different availability levels

Years of continuous operations	1	2	3
Availability	Maximum allowable downtime		
99.0000% (2–9s)	3 d 15 h 36 min 0 s	7 d 7 h 12 min 0 s	10 d 22 h 48 min 0 s
99.9000% (3–9s)	8 h 45 min 15 s	17 h 31 min 12 s	1 d 2 h 16 min 48 s
99.9900% (4–9s)	52 min 34 s	1 h 45 min 7 s	2 h 37 min 41 s
99.9990% (5–9s)	5 min 15 s	10 min 31 s	15 min 46 s
99.9999% (6–9s)	32 s	1 min 3 s	1 min 35 s

In order to relate to the service users of a system, the availability expression above can be rewritten for a service in terms of service uptime and service outage as:

$$\text{Service availability} = \text{service uptime}/(\text{service uptime} + \text{service outage})$$

As stated in the definition in Section 1.2, a measure of service availability is the product of the availability of all the elements involved in implementing the service. This turns out to be more complicated than just looking at the plainly stated downtime for a single element. In reality, a system implementing end-user services is typically composed of many subsystems. The availability calculation must therefore take into account the availability of all the constituent subsystems. In general, subsystems must achieve a high level of availability in order to meet the expected service availability requirement. In practice, however, the availability requirements for each subsystem are more fine-grained and are usually attached to individual functions, instead of a single figure for the entire subsystem. Nevertheless, a whole range of possible failures such as hardware, operating system, application, operator errors, and environmental problems may still contribute to the piling up of downtimes in the availability calculation equation.

While the main source of impacting service availability has been the failure of some of the underlying elements, the need to modify a system after it has been put into operation is increasingly becoming a vital factor for system designers to consider. All systems need changes, be it an upgrade or replacement of hardware components, new features or simply bug fixes. In order to deal with an increase in service capacity, for instance, more service requests per second, additional hardware is typically put into the system to increase its storage and/or processing capability. In the case of repairing a faulty hardware component, it is normally taken away and replaced by a new and functional one. Other required changes at some regular intervals include operating system changes and new application versions.

The traditional way of dealing with changes or upgrades is to stop the system and carry out the modifications accordingly. This obviously makes an impact on the service availability measure because from the end-users' perspective, the service is no longer operational to offer the intended function. This is also deemed unacceptable to a service operator because the service outage is translated into loss of revenues. A mobile phone service is a case in point. In addition, upgrading an operational system has been known to be error-prone [3, 5] primarily due to its intrinsic complexity. Therefore, some methods must be devised to keep the services going while an upgrade is carried out on the operational system. As the society is depending on more and more computer-based services nowadays, there is a clear trend to expect services to be available in more and more systems even during their upgrades. Even if service unavailability is inevitable, the disruption must be kept to a minimum to reduce its impact on users. This is especially true for those systems that have a long mission time and this requirement cannot be underestimated.

In the next section, we will discuss how service availability can be achieved for both unplanned and planned events, primarily caused by failures and upgrades respectively.

1.4 Achieving Service Availability

This section focuses on how the notion of service availability can be realized in systems. It shows how fault tolerance can be used as the basis for delivering service availability to systems. It explains various forms of protective redundancy and the workings of fault tolerance. It also highlights the interrelationship between upgrade without loss of service and protective redundancy in the context of providing service availability in a system.

1.4.1 Following the Framework of Fault Tolerance

Fault tolerance aims to be a failure avoidance approach. It attempts to intervene and defend against faults from causing failures. The general assumption is that in spite of all the fault prevention methods employed, there is still a chance that faults can creep into a system and cause a failure. The basic principle behind fault tolerance is to ensure that a system is sufficiently designed to withstand such possibility by means of error detection and system recovery. The approach tries to make sure that if a fault is activated, the caused error can be detected and then handled accordingly before it causes a failure.

Fault tolerance is considered to be equally applicable to implementing service availability for unplanned events such as those caused by failures of the underlying system. The four phases of fault tolerance [14] are outlined below to show how the principles can be applied to delivering service availability:

1. **Error detection.**
   Before any action can be taken to address a service failure, the presence of an error must first be identified.

2. **Damage confinement and assessment.**
   The level of damage caused by a fault is evaluated and if possible, the effect of the error should be restrained as much as possible. It is worth pointing out that an error can be passed as data, which may be in the form of either user data over the service interface or as system state information. This phase basically tries to limit the scope and propagation of an error from one part of the system to another.

3. **Error recovery.**
   Error recovery is the process of transforming an erroneous system state into one that has no detected errors. Therefore, the possibility of the activated fault causing a service failure can be eliminated.

4. **Fault treatment and service continuation.**
   If the fault that causes the detected error to occur can be identified and corrected, it could prevent the same fault from being activated again. Even without this fault handling, a system can continue to provide its intended service because the error condition has now been cleared. The key concern in this phase is the attempt to create a perception to service users that the intended service continues to be available as if nothing has happened.

In the ideal case, a service continues operating regardless of what kinds of faults it encounters. Due to the application requirements and/or resource constraints, a service implementation does not necessarily have all of the above fault tolerance phases. As always, there is a trade-off between the resources needed and the protection it offers. There are also the influences of the application requirements. For example, some application scenarios have strict requirements on the types of faults to be tolerated, while other application scenarios have less stringent requirements. Some applications may even allow for service failures provided that certain safety conditions are fulfilled. Some applications are in between, where a partial service failure is acceptable.

In a traditional fault-tolerant system, the failure response defines how such a system reacts when a failure is encountered. As a result, the response can be viewed as the level of fault tolerance provided by a system. The same definitions are also relevant to services:

- **Fail-operational**
  A service continues to operate in the presence of errors with no loss of functionality or performance. This is the highest level of service availability as the users do not even notice any significant differences in terms of their expectations. Needless to say, the required additional resources to cover all the conceivable failure scenarios are usually prohibitively high for most applications.

- **Fail-soft (graceful degradation)**
  A service continues to operate in the presence of errors with a degradation of functionality and/or performance. This is perhaps the most common level of application requirements. The idea behind this is to at least keep some of the functions going with the current, limited resources available. The system can subsequently be taken off for repair before full service is restored. The choices of which functions are kept naturally depend upon the types of error the system is facing. They also depend on how an application is designed. The most intuitive line of thinking would be placing more protection on critical functions than their less essential counterparts. Relaxing nonfunctional requirements such as performance at the time the system is in trouble is another option available to a system designer. For example, reducing serving capacity would put less strain on the demand of limited resources.

- **Fail-safe**
  A system maintains its integrity and halts its intended operation. It must be pointed out that halting the intended operation does not necessarily mean stopping the system. For example, a traffic light reverting to flashing reds in all directions [11] is not the intended operation for directing traffic. Instead, it is a warning signal to all road users that there is something wrong with the traffic lights. This is a safety measure to help ensure that the possibility of lights erroneously going into greens in all directions when there is a failure is excluded. The primary concern in this type of application is safety. However, the precise meaning of what is a safe condition is entirely application dependent: a service being stopped and not doing anything is a possible and sometimes applicable scenario.

- **Fail-stop**
  A service halts immediately upon a detected malfunction. In some literature, this property is also a synonym for *fail-fast*, which adds the connotation that the stopping is done very quickly. It is a useful and desirable property if a system cooperates with other systems, especially when it prevents the erroneous system from spreading incorrect information to others.

1.4.2 Redundancy is a Requisite

The key to a fault tolerance approach is to have protective redundancy built into a system. Protective redundancy is an additional resource that would be unnecessary if a system operates correctly. That is why the term redundant. Protective redundancy, or redundancy for short, is usually a replica of a resource. In the event that a fault causes an error to occur, the replicated resource is used instead of the erroneous one. Thus, the replica protects a system against failures, giving the impression that services are uninterrupted even when there are failures in the underlying system. It is precisely this property that makes fault tolerance stand out as a candidate approach (discussed in Section 1.3.2) that can be used as a basis for achieving service availability. There are two main aspects related to redundancy that require consideration: what should be replicated and how the redundancy is structured in order to achieve the desired service availability.

There are many forms of resources that can be replicated in order to provide redundancy in a system, for example, hardware, software, communications, information, and even time. Duplicating a piece of hardware is perhaps the most common form of redundancy to deal with hardware faults. Running a copy of a piece of software on different hardware can tolerate hardware faults. It must be stressed that if the fault lies in the software, which is basically a design fault, a copy of the same software in the exact same state with the exact same data would only duplicate the same fault. A whole range of solutions based on diversity can be found in the additional reading [15]. Having more than one communications path among interconnected systems is a way to handle failures in connectivity. Maintaining multiple copies of critical information or data has long been used as a mechanism for dealing with failures in general. This inevitably brings in the issues of consistency among replicas. Simply by repeating the execution of a task has been known to be able to correct transient hardware faults. This is also referred to as temporal redundancy as time is the additional resource in this context. In practice, a combination of these forms of redundancy is used according to the application requirements and resources available.

Since there are replicated resources in the system, it is necessary to differentiate what roles each of these resources should take. A resource is considered to have an *active* role if it is used primarily to deliver the intended service. The redundant resource is regarded as taking a *standby* role if it is ready to take over the active role and continue delivering the service when the current active element fails. It must be noted that a standby element usually needs to follow what the current active element is up to in order to be successful in taking over. This typically requires that a standby element has up-to-date state information of the current active element, implying that there are communications between active and standby elements. The end result is that users are given the perception that a service continues as if there was no failure.

The most basic structure for a system employing protective redundancy is to have one active element handling all the service requests, while a standby element tracks the current state of its active counterpart. If the active element fails, the standby element takes over the active role and continues to deliver the intended service. The manner in which resources are structured in terms of their roles in a system is known as the redundancy model.

Figure 1.5 illustrates a frequently used redundancy model known as active-standby involving both the active and standby roles. As shown in (a), an active role is taken to provide service S by an application A running on computer node X under normal operation. The application is replicated on node Y and assumes the standby role of providing service S. During this time, the necessary information for the standby application to take over the active role must be obtained and maintained. If there is a failure on node X rendering the application in the active role to fail as well, as depicted in (b), the application with the standby role takes over and continues to provide service S. At this point, however, there is no redundancy to protect further failures. If the failed node X is not repaired, a further failure on node Y will cause service outage.

There are many ways of structuring redundancy in a system, involving different numbers of active or standby roles, each of which has resource utilization and response time implications. For example, it is possible to have one standby element protecting a number of active elements in order to increase the overall resource utilization. With only a single redundant element, the resulting system will not be protected against further failures if the faulty element is not repaired. The trade-off of how many active and standby elements and in what way they are structured must be weighed against the application

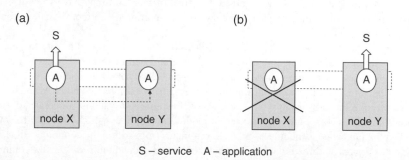

S – service   A – application

Figure 1.5 Active-standby redundancy model (a) Active-standby: normal operation; (b) Active-standby: node X failed.

requirements. It is worth noting that with the coordination of the active and standby roles, an application becomes distributed in nature and that makes it more complex. In Chapter 6, some of the commonly used redundancy models in the context of managing service availability are discussed in more details.

1.4.3 Dealing with Failures

Fault tolerance is performed by means of error detection and then system recovery. In this section, we look at the mechanics of fault tolerance in more detail during the four phases, in particular, how service availability can be accomplished by following the framework of a fault tolerance approach.

1.4.3.1 Error Detection

Error detection aims to identify the occurrence when the state in a system deviates from its expected behavior as specified. This can be done by either the system itself or an external entity monitoring the system. There are many ways to detect an error and the precise mechanism used is obviously application dependent. In general, these techniques fall into two main types: concurrent detection and pre-emptive detection. The main difference is that concurrent detection is carried out during normal service delivery while pre-emptive detection is performed when normal service delivery is suspended. Pre-emptive detection aims at finding latent errors and dormant faults. Although the service is unavailable in the case of the pre-emptive detection type, it is worth mentioning that its use may still be appropriate for some applications provided that the anticipated service outage is within the allowable limit.

1.4.3.2 Damage Confinement and Assessment

The main purpose of damage confinement and assessment is to contain the detected error and limit its effect from spreading to other parts of a system. At design time, a commonly used approach is to hierarchically decompose a system into smaller and manageable parts. Not only is each part a unit of service, but also a unit of failure [16]. If such a unit fails, it is replaced by a similar, functional unit. The notion of a field replaceable unit (FRU) has long been used in hardware. The equivalent in software is usually a software module. However, defining a unit of failure for software is more complicated because software tends to be more complex due to its structure, and that different pieces of software usually cooperate to achieve some functions. As pointed out in a previous section, errors can be spread via incorrect data among these cooperating software entities. Therefore, the boundary of a unit of failure changes depending on what the fault is.

A *fault zone* is used for fault isolation and repair. It defines the boundary of a unit of failure for a specific fault and can be viewed as a collection of entities that may be damaged by this fault. This can therefore be used to determine how widespread the effect of this fault may have when it is activated. A fault zone is associated with a recovery strategy so that appropriate corrective actions can be taken in the recovery phase.

Figure 1.6 shows the structure of a simple application and is used to illustrate the concept of a fault zone in isolating faults. The application, which is composed of three subsystems, interacts with its environment. Each subsystem, which consists of a number of software modules, has some defined functions to perform. In the communication subsystem, for example, modules A and B cooperate in order to provide the subsystem's functions. Altogether, the subsystems cooperate to provide the application's functions. If a fault originates from module A and does not cause any error outside the module, then the fault zone is module A. If a fault originates from within the communication subsystem and may cause errors within the subsystem but not outside, the fault zone is the communication subsystem. If a fault may cause errors all over the application, the corresponding fault zone is the application. The corresponding recovery action of a fault zone will be described in the next section on error recovery under recovery escalation.

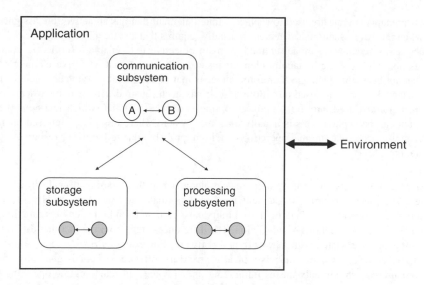

Figure 1.6 Example application.

1.4.3.3 Error Recovery

Error recovery is essentially a way of correcting the error condition, which results in masking the effect of an error. After a successful recovery action, a system continues to deliver a seemingly error-free service. In general, error recovery does not normally repair the fault that caused the error in the first place. In some cases though, it has the combined effect of error recovery and fault repair. There are many techniques available and they are not mutually exclusive. That is, if an error persists, another method can be attempted next according to some predefined recovery strategy for a system.

- **Rollback**
  Rollback, or backward error recovery, attempts to restore the system to a correct or error-free state so that it can continue to provide the intended service from this point onwards. It assumes that the state just before an error appears is correct. A rollback mechanism requires the process known as checkpointing to record prior system states at various predetermined recovery points. When an error is detected, the system state is restored to the last saved, presumably error-free state. Rollback does not require the knowledge of the error it deals with and therefore provides a general recovery scheme. However, it takes additional resources for the checkpointing operation and time redundancy to perform.

- **Roll-forward**
  Roll-forward, or forward error recovery, tries to move the system into a new and correct state so that it can continue from there to provide the intended service. It requires that a new and correct state be found. Roll-forward is commonly used to gracefully degrade a system by moving the system to a state where reduced capability is delivered to its users. It is generally regarded as being quick but it requires application-specific knowledge. In the context of transaction processing, roll-forward is typically performed by restoring a database from a backup and then applying a separately recorded changes to the database to bring it up to date.

- **Failover**
  A *failover* action corrects an error condition by using an element that previously has the standby role assigned instead of the one currently assigned the active role to deliver the service. This change of

role assignments is typically used to replace a failed element that has an active role assigned at the time when the error condition is detected. It usually requires that a system has protective redundancy in place and a redundancy model defined. A main objective is to enable a service to continue as soon as possible by using a standby element instead. It assumes that an active element's service state has not been corrupted and therefore neither is that of the standby. Typically the element that has a detected error is not used any more until it has been repaired. At the point when a repaired element is reintroduced into the system, an option known as *fail-back* which reassigns the active role again can be applied. This is usually used in a system where there is a preferred element for delivering services under normal operating condition, probably due to the richer resources available in this element.

- **Restart**
  A *restart* action clears an error by stopping an element and then starting it again in a controlled manner. The idea is to bring an element back to a normal state, which is usually found at the beginning of its operation. This term is used in its most generic form in this context and the concept encompasses all the variations of this action. The differences stem from the fact that different terms are traditionally associated with hardware and software. For example, *reset* is commonly used in referring to hardware, while *restart* is typically linked to software. *Reboot* is generally connected to a computer which normally means initializing its hardware and reloading the operating system. There are also differences in terms of how a restart operation is carried out, for example, whether an element is stopped in an abrupt or orderly manner; how many changes to the element's settings are retained when its operation starts again; and if power is taken off from the element or not are just some of these variations. Caution is therefore required when this term is interpreted with reference to the precise actions taken.

It must be noted that a recovery action can normally mask an error condition only at the place where the error was detected. However, if the fault does not originate from the affected element and it has not been repaired, the same error condition is likely to return, in addition to which more errors may be detected on other elements due to the spreading of damage caused by the fault. In a system where fault zone is used for isolating faults, *recovery escalation* can be used to raise the error condition to its containing fault zone, that is, the next higher level of fault zone. The recovery action associated with this fault zone is carried out on all the entities in the same fault zone. If the recovery action in the new fault zone still cannot remove this persistent error condition, this process continues to the next higher level until the condition is cleared or there are no more levels to go up to. It is important to remember that due to the hierarchical structure, the higher the level of fault zone goes, the more entities are impacted during the corresponding recovery action. If services of the affected elements are disrupted during a recovery action, this may potentially have an impact on the overall service availability of the system.

In calculating the service availability measure for a system, recovery time is an important factor. The smaller the recovery time, the higher the service availability in a system. The recovery times of various elements in a system, for example, hardware, operating systems, and software must also be taken into account.

1.4.3.4 Fault Treatment and Service Continuation

Fault treatment in this phase aims to stop faults from activating again in a system, thus preventing faults from causing errors that may lead to a system failure. Treating faults is usually accomplished by corrective maintenance, which aims to remove faults from a system altogether while a system is in its operational phase. Handling a fault first of all requires the identification of the fault in terms of its location and type. This is followed by excluding the faulty elements from taking part in delivering the service, before a healthy and correctly functioning element is put into the system to replace the faulty one. A new element may be a spare part for hardware, or a new version of program with bug fixes for software. In order to provide service continuation, the new element must be put into the system without stopping it. For hardware, technologies such as hot swap have been successfully used to replace components without the need for taking the power off a system. As for software, it typically requires a restart and therefore it may render services unavailable for a period of time during this operation. However, with protective redundancy deployed in a system to achieve service availability, it is possible to avoid or minimize a service outage if an upgrade is carefully planned.

1.4.4 Upgrade Matters

We have seen in the previous section how unplanned events such as failures are dealt with by following the framework of fault tolerance. In this section, we turn our attention to planned events – the issues of upgrading a system with minimum or no service outage.

In order to reduce the chances of losing availability of a service during an upgrade, an intuitive way is to decrease the scope of affected elements in a system to the smallest possible. In a system where protective redundancy is already put in place for dealing with anticipated failures, it can also be used to keep a service functioning while some parts of the system are being upgraded. Rolling upgrade is one such popular method that attempts to limit the scope of impact by upgrading one part of a system at a time. If the operation is successful, it continues to another part of the system until all the required elements have been upgraded. During this upgrade process, when an element to be upgraded must be taken out of service, the protective redundancy is there to help provide service continuation.

Figure 1.7 shows an example of rolling upgrade and the basic principle behind the method. The objective of the upgrade in the example is to update the currently running service from version 1 (S1) to version 2 (S2). The initial state of the system is shown in (a), where nodes X and Y are configured in an active-standby redundancy mode in which node X assumes the active role whereas node Y is assigned the standby role. When the rolling upgrade begins in (b), the standby role of node Y is taken away when the application is being updated to version 2. Although service S1 is still available at this time, a failure would incur a service outage because the protective redundancy has been removed. Step (c) shows that after a successful upgrade of the application on node Y, it is the turn for node X. At this point, node Y with the newly updated service takes over the active role to provide the service (S2) while node X is taken off for the upgrade. Similar to the condition in step (b), a failure at this point would incur a service outage. Finally in step (d), it shows that node X has been successfully upgraded and therefore can be assigned the standby role, returning the system to its initial state of having an active-standby redundancy model to protect it from service outage.

In practice, there are variations to the basic steps as illustrated. For example, if node X is the preferred provider of the service in an application scenario then it must be reverted back to the active role after a successful upgrade. It must be added that if we want service continuity during an upgrade, S1 needs to behave in the same way as S2. When the system transitions from step (b) to (c), A2 on node Y must synchronize with A1 on node X. Indeed there are many other upgrade methods, including some custom schemes that are needed to meet the requirements of a specific site in terms of its capacity planning and service level agreement. The key issue here is to provide applications with a flexible way for applying different upgrade methods, with the aim of eliminating or reducing the window of vulnerability of service outage (as shown in steps (b) and (c) above), if any.

In case the new version does not work, a plan is also needed to ensure that the system can be restored to the state prior to the upgrade such that the service, albeit the previous version, continues to operate. Therefore, the monitoring and control of the upgrade, together with measures for error recovery of the upgrade process, are essential. Since an upgrade is a planned event, these extra

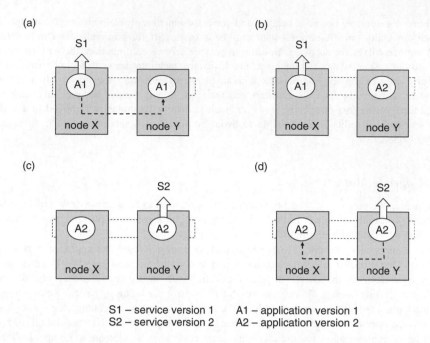

S1 – service version 1   A1 – application version 1
S2 – service version 2   A2 – application version 2

Figure 1.7 Rolling upgrade example (a) Initial state: current versions; (b) Application A on node Y is being upgraded; (c) Application A on node X is being upgraded; (d) Upgraded system: new versions.

actions can be taken into consideration and designed into an upgrade operation. A related issue is that error-prone manual processes are typically used in upgrades. Experience has shown that human mistakes account for a considerable number of cases of system crashes during an upgrade, especially in large-scale systems. The upgrade process thus needs to be automated as much as possible. Chapter 9 has an in-depth treatment on the subject of upgrade.

1.5 Conclusion

We have presented the definitions, basic concepts, and principles of dependable computer systems in general, service availability in particular, at the level appropriate for the remaining of the book. By showing the consequences when expected services were not available, we examined why the study on service availability was important. Issues and challenges related to service availability were introduced. We presented how fault tolerance can be used as the basis for delivering service availability to systems. Various forms of protective redundancy have been explained, together with its interrelationship with upgrade without loss of service in the context of delivering service availability to a system.

The subject area of dependable computer systems is a broad one. It is therefore impossible to cover all aspects and the wide-ranging issues in a single chapter. Interested readers are referred to the additional reading list. In [17] an in-depth treatment of designing dependable systems is presented along the development life cycle that includes the specification, design, production, and operation stages. The fundamentals of reliable distributed systems, web services and applications of reliability techniques are thoroughly covered in [18]. Pullum [19] gives an account of the techniques that are based on design and data diversity, which are applicable solutions for tolerating software design faults. A closely related subject of disaster recovery is covered in full in [20]. A document [21] published by the High Availability Forum, a predecessor of the Service Availability Forum, provides an insight into the background, motivation, and the approach to developing the specifications discussed in the rest of this book.

In the next chapter, we will examine the benefits of adopting open standards of service availability in general for carrier-grade systems, and the Service Availability Forum specifications in particular. It includes the rationale behind the development of the Service Availability Forum from both the technical and business perspectives.

2

The Birth of the Service Availability Forum

Francis Tam
Nokia Research Center, Helsinki, Finland

2.1 Introduction

On 4 December 2001, a group of leading communications and computing companies announced an industry-wide coalition to create and promote an open standard for service availability (SA) [22, 23]. The announcing companies included the traditional network equipment providers, IT companies, hardware suppliers, and software vendors. A series of questions ensued. What was going on? Why were these companies working together? Why did they need an open standard for SA? What were they going to standardize?

These were the typical questions one heard after the SA Forum launch events. In order to answer these questions, one has to take a step back and take a look at what was happening at around the time of the late 1990s and early 2000. For example, the rapid technological advancements in the communications sector, the incorporation of mobility in traditional services and the considerably increased use of the Internet had all contributed to the convergence of the information and communications industries. As a result, the landscape of the overall business environment had changed, requiring the necessary matching actions to be taken in developing these new products and services.

This chapter first examines the technology developments at around that time in the information and communications technology industry. It then discusses the changes from the business perspective as a result of these technological developments. We then delve into the rationale behind the forming of the SA Forum and the benefits that it intended to provide to the users of the standards. We elaborate on the scope and the approach taken by the SA Forum middleware, together with the justification, and explain how it enables a SA ecosystem for the telecommunications industry.

2.2 Technology Environment

In the first chapter, we introduced the concept, principle, and fundamental steps for achieving SA. We have also highlighted a few reported cases of service outage and their corresponding consequences.

Service Availability: Principles and Practice, First Edition. Edited by Maria Toeroe and Francis Tam.
© 2012 John Wiley & Sons, Ltd. Published 2012 by John Wiley & Sons, Ltd.

The described unavailability of mobile communications services perhaps had the most impact in terms of the number of people that were directly affected. It is therefore not surprising to observe that a very high level of SA has been a well-known characteristic expected of a communications service, be it a traditional landline or the more contemporary mobile communications system. Indeed, the term carrier-grade has long been used to refer to a class of systems used in public telecommunications network that deliver up to five nines or six nines (99.999 or 99.9999%) availability. Its origin comes from the fact that a telecommunications company provides the public for hire with communications transmission services is known as a carrier [24]. The equipment associated with providing these highly available services has traditionally been dubbed 'carrier-grade.' This reinforces our association of communications services with high level of availability.

Ever since the introduction of the public telephone service in the late nineteenth century, which was a voice-grade telephone service for homes and offices, very little in terms of service features had changed over a long period of time, and the expectation of new telephony services was not very high either. During this time the development was more on the quality of the provided services such as intelligibility of transmitted voice and issues with reliability. Decades later, the introduction of digital electronics had led to its use in telephony services and systems. By the 1960s, digital networks had more or less replaced the analog ones. The benefits of doing so were the lower costs, higher capacity and more flexible to introduce new services.

The technology for communications services took a fork along the path of mobility, with the first commercial analog mobile phones and networks were brought out in 1983. The digital counterparts appeared as commercial offerings in 1991. In 2001, the first 3G (3rd Generation) networks were launched in Japan. 3G was all about communicating via a combination of text, voice, images, and video while a user was on the move. In just a mere two decades, we saw the tremendous pace of advancement in communications technologies as well as an exponential growth in the telecommunications market.

In parallel to the developments of 3G, another trend around this time was the considerably increased use of the Internet – 6 August 1991 is considered to be the public debut of the World Wide Web [25]. This trend turned out to be one of the biggest technological drivers for the paradigm shift in the information and communications industry. If we look at the Internet today, it is clear that it has become the convergence layer for different technologies, regardless of whether the information is data or voice, passive or interactive, and stored or real-time. At the time of these changes, information technology has also found its way into the deeper part of communications systems such as the infrastructure. A significant influence of this trend in the communications networks was apparent, based on the observation of the emerging of all Internet protocol (IP) core networks for mobile communications and services at the time. This transitioning from the traditional way of implementing and offering communications services using closed, proprietary technologies into the then new era had substantial effect on the way these services should be developed.

From a user's perspective, however, all these new and exciting technological developments were primarily translated into new features they could now experience. The expectation of the high level of SA was nevertheless untouched because this perception had been around since the beginning, certainly for the majority who had enjoyed the highly available communications services.

2.3 Business Environment

While the pace of technological developments in the information and communications industries were remarkable during just a short period of two decades, it had brought with it some significant changes to the landscape of the environment in which businesses were conducted. On the one hand, the technological advances had considerably broadened the scope and increased business opportunities for a company. On the other hand, a company had to change and adapt to this new environment. In the following subsections, we will speculate about the reasons why the seemingly disparate companies decided to work together, and the need for an open standard for SA.

2.3.1 Ecosystem

An important aspect of the changes in the industry was that transferring bits over communications lines was no longer that profitable. New services based on innovations during this time of convergence with a clear customer value were needed. Due to the increased competition from previously different industry segments, most communications companies had the urgency to roll-out these new services as soon as possible. This was easier said than done when the sheer scale and complexity of the expected products and services were daunting to say the least. Many companies had started to look for new ways of developing these products and services, with a clear objective to increase productivity by reducing the development effort and costs. These included model-driven architecture with automatic code generation, and reusing software assets. The former attempts to produce code as fast as possible at the push of a button, while the latter tends to avoid duplication of effort by using the same code over and over again.

It should be noted that assets for reuse can be either developed in-house or bought from outside. These kinds of ready-made components were generally referred to as *Commercially-Off-The-Shelf* (*COTS*) components. They were used as a building block and incorporated into products or services. It is important to point out that a COTS component may come in many different forms. The broad notion of a COTS component is that it is available to the general public and can be obtained through a variety of manners including buying, leasing, or licensing. It is worth mentioning that open source software is therefore considered to be COTS components. In some cases, the open-source implementations have become so dominantly accepted that they are even considered to be de facto standards.

At about this time, the communications industry was going through a period of transitioning from building everything itself in a proprietary way to adopting solutions from the information technology world. For example, the use of COTS components for hardware such as processors and boards; for system software such as operating systems and protocol stacks. One key question was what sorts of items should be made a commodity from a business perspective. Since SA was a natural common function across most communications products and services, support in a middleware was deemed appropriate in this new hybrid architecture.

As a sidenote, middleware was originally developed to provide application developers with assistance to deal with the problems of diverse hardware and complexity of distribution in a distributed application [26]. It positions between an application and its underlying platform, hence the name middleware. One such example is the Object Management Group's Common Object Request Broker Architecture (CORBA) [27], which is a standard enabling software components written in different programming languages and running on different computers to interoperate.

The term platform is used here to collectively refer to its operating system, processor, and the associated low-level services. By providing a set of common functions for an application that resides on different platforms, middleware relieves an application developer's burden of handling interactions among these different application instances across a network. Early use of middleware in database systems and transaction monitors had proved to be effective. By incorporating the SA support functions into a middleware in a similar fashion, it was generally believed that comparable benefit could be gained.

Many communications companies had gone down this path and started to develop their own SA middleware for their own internal reuse. Some had even gone further by releasing them as products, for example, Ericsson's Telecom Server Platform [28]. Other IT and software vendors had also worked with their partners in the communications sector to develop similar products, for example, Sun's Netra Carrier-Grade Servers [29] and Go Ahead's Self Reliant [30]. There were many more at the time. Some companies – and, in some cases, products – are no longer in the market.

While these solutions addressed product efficiency internally within each of the companies, they did not offer the same interface to external application developers. Therefore the same applications still could not be offered across platforms without laborious porting, adaptation, and integration work. At the same time the number of applications that have been implemented in these systems and offering the same functionality was steadily growing yet still lagging behind the (anticipated) demand for new common services. This was the early indication of an ecosystem in which new cooperation among different companies could be beneficial to all parties involved. This explains why a diverse range of companies such as network equipment providers, IT companies, hardware suppliers, and software vendors were interested in working together, as pointed out at the beginning of this chapter.

2.3.2   COTS and Open Systems

These trends underlie the point argued in [13] that COTS and open systems are related concepts but they are not the same. An open system has the characteristics of having interfaces defined to satisfy some stated needs; group consensus is used to develop and maintain the interface specifications; and the specifications are available to the public. The last two qualifying criteria of an open system have essentially made the interface specifications open standards. An implementation of such a standard specification can therefore be made available as a COTS component, although this is not required by an open system. While the different SA middleware solutions could be considered as COTS components, the systems they were part of remained closed.

There was a potential that by opening up what was a closed and proprietary solution on SA support could create an environment in which new cooperation among different companies would become beneficial to all parties involved. However this had to be a collective effort in the tradition of tele- and data communications standardization.

The reasons behind the companies' joining forces together were far from obvious. There were many speculations and observations. My co-editor has theorized the IT bubble [31] and its burst could also be a contributing factor. She has suggested that the World Wide Web brought about the IT bubble in which the trends were as we have described them in this chapter. Initially when the companies had money for development, everyone was doing it on their own and hiring many people to keep up with the demand and the competition. Being different was considered to be a benefit as it locked-in customers for the coming future.

Then in the year 2000 when the bubble burst the profit went down. Existing systems boosted up during the bubble were still working fine and therefore they did not need to be replaced. The main opportunity to increase profit was to address the appetite of the customers which was still growing for the new services and applications. Eventually companies started to look inward and reorganize their processes to save money, and at the same time trying to meet the demand so that they could keep or even grow market share. Standardization was a way of accomplishing this goal.

Her conclusion was that the burst of the IT bubble was a contributor if not really the trigger for the companies' collaboration.

Regardless of whatever the reasons were that caused the companies joining forces and working in a standardization body, adopting a standards-based approach is considered to be a sound risk management strategy. In addition, compatibility of products delivered by different vendors can be ensured. A key role played by standards is to divide a large system into smaller pieces with well-defined boundaries. As a result, an ecosystem is created with different suppliers contributing to different parts. The standardization process ensures that the stakeholders are involved in the development and agreement of the outcome, resulting in conforming products being compatible and interoperable across interface boundaries. This is particularly important to those businesses involving many vendors: the only way to ensure that the system as a whole works is to have standardized interfaces.

Standard COTS components have the added advantage of having a wider choice of vendors and one can normally take the best solution available. By standardizing the design of commonly used high availability (HA) components and techniques, it opens up a competitive environment for improving product features, lowering costs and increasing performance of HA components. The application developers can concentrate on using their core competence during application development, leaving the SA support to the middleware. As a result, a wider variety of application components can be developed simultaneously in shorter development time, thus addressing exactly the trend of the demands we were observing. Being a standard also enables the portability of highly available applications between hardware, operating systems, and middleware. This to a certain extent has answered the question of 'Why did they need an open standard for SA?'

2.4   The Service Availability Forum Era

The predecessor of the SA Forum was an industry group called the HA Forum. Its goal was to standardize the interfaces and capabilities of the building blocks of HA systems. However, it only went as far as publishing a document [14] that attempted to describe the best-known methods and a guide to a common vocabulary for HA systems. The SA Forum subsequently took over this initiative and developed the standards for SA.

The focus of the SA Forum is to build the foundation for on-demand, uninterrupted landline, and mobile network services. In addition, the goal is to come up with a solution that is independent of the underlying implementation technology. This has been achieved by first of all identifying a set of building blocks in the context of the application domain, followed by defining their interfaces and finally, obtaining a majority consensus among the member companies.

So 'What is in the SA Forum middleware then?' A short answer is the essential, common functions we need to place in the middleware in order to support the applications to provide highly available services. This was carried out by extracting the common SA functions that were not only applicable to telecommunications systems, but also to the upcoming new technology and applications areas. The basis of the standardization was drawn from the experiences brought into the Forum by the member companies. It is important to note that they contributed the technical know-how [28–30] to the specifications.

A high-level view of such a SA middleware is shown in Figure 2.1, illustrating an overall architecture and areas of functions. The diagram also shows the relationship between a SA middleware with its application, platform, and cooperating nodes over a system interconnect. Each of these nodes is connected, forming a group to deliver services in a distributed system environment. According to the

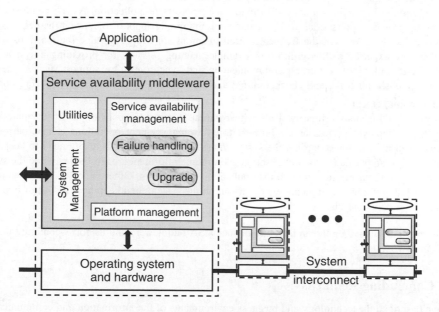

Figure 2.1   High-level view of a service availability middleware.

functions required to support SA, the identified functions are put into four functional groups. They are platform management, SA management, common utilities for developing distributed applications, and system management.

Although the functions have been partitioned into different groups, ultimately all of them must work together in order to deliver SA support to applications. We will go through each functional group and describe what it does; why it is needed; and what kinds of functions are expected.

- **Platform management**
  This group of functions manages the underlying platform, which consists of the operating system, hardware and associated low level services. If any of these underlying resources are not operating correctly, the problem must be detected and handled accordingly. Since we are in a distributed environment, and the separate instances of the middleware must cooperate, there is a need to maintain information regarding which node is healthy and operating and which is not at any point in time. It must be mentioned that this kind of group membership information is distributed and dynamic in nature, and must be reliably maintained. The expected functions in this group include the monitoring, control, configuring, and reconfiguring capabilities of platform resources; and a reliable facility for maintaining membership information for a group of nodes over a system interconnect.

- **SA management**
  This group of functions is essentially the core for providing SA support to applications. As discussed in Chapter 1, both unplanned and planned events may impact the level of SA. Therefore, the functions are further split into two subgroups to deal with failures handling and upgrade respectively. The expected functions in failures handling include support for error detection, system recovery, and redundancy models. For the upgrade, the expected functions include a flexible way to apply upgrades with minimum or no service outage; monitoring, control, and error recovery of upgrades. Although there are two subgroups in this area, it must be pointed out that the functions need to collaborate closely because protective redundancy is used in both.

- **Common utilities for developing distributed applications**
  One of the original goals of developing a piece of middleware was to hide the complexity caused by the required interactions among distributed instances of an application. As the name suggests, this group provides applications with distributed programming support to conceal the behavior of a distributed system. Most of the frequently used functions for supporting distribution transparency are therefore expected in this group. For example, naming services for providing a system-wide registry and look-up of resources; communication and synchronization support such as message passing, publish/subscribe, and checkpointing; and coordinating distributed accesses to resources.

- **System management**
  This group of functions primarily deals with external systems or applications for the purpose of managing a deployed system as a whole. Before any management operations can be performed, management information regarding the system must be present. There is also a need to keep track of the configuration data for the initial set-up and runtime information of resources while the system is operating, and having functions to manipulate this information. Other expected functions in this group include informing significant events arising in the system; and keeping a record of important events for further analysis.

A thorough discussion on the architecture of such a SA middleware and its corresponding functions are in Part Two.

2.5   Concluding Remarks

We have described the technology and business environments of the information and communications industries around the period of the late 1990s to early 2000, when there was a tremendous pace of advancement in communications technologies. Against this backdrop of what was going on, we explained how the business environment had changed, and how the companies had reacted by joining forces and cooperating to build a viable ecosystem. We distinguished between COTS and open systems, and explained the benefits of having an open standard for a SA middleware. We have also given a high-level view of a SA Forum system and described the intended functions of the middleware.

There you have it: the answers to the questions we raised about the announcement in the beginning of the chapter. These were the circumstances and motivations behind the founding of the SA Forum dated back to the year of 2001 – the SA Forum was born!

In the years that followed 2001, the SA Forum, backed by a pool of experienced and talented representatives from its member companies and with the desire to realize the Forum's vision, had diligently worked to develop the necessary specifications. There have also been separate open source implementations to deliver what were written in the specifications, instead of just a pile of papers containing the descriptions of the interfaces.

In the next part of the book we will be looking at the results of these developments. We will run through the reasoning and design decisions behind the development of these specifications. We will also give hints on the best way to use the various SA Forum services and frameworks, and the pitfalls to avoid where appropriate.

Part Two

The SA Forum System: Services and Frameworks

3

Overview of the Service Availability Architecture

Dave Penkler
Hewlett-Packard, Grenoble, France

3.1 Introduction

The Service Availability (SA) Forum architecture is presented as a logical software architecture that is comprised of a set of interface abstractions, operational models, and information models. The architecture is not of itself prescriptive but is intended to provide a categorization of the service interfaces and a view of how the various services fit together in the context of a system. We begin by examining the context surrounding the architecture behind the SA Forum service availability specifications. This includes some historical background, the requirements and assumptions regarding the scope and physical systems that were used in selecting the functionality to be included in the specifications. We then look at the problem of software architecture in general and specifically in the context of interface standards.

3.1.1 Background and Business Context

As the explosive growth of the Internet in the late 1990s started to blur the boundaries between traditional telecommunications and information technology (IT) services a joint need emerged between IT and telecommunication equipment manufacturers to reduce the development and maintenance cost of their software-based infrastructure products while accelerating the time to market of new products and services. On the one hand IT manufacturers were maintaining a number of different hardware and software stacks for their respective equipment manufacturer customers. The differences were mainly due to specific legacy requirements but also a number of custom functional details. The commercial IT hardware or cluster based fault tolerant solutions were only applicable to a small subset of the types of network elements used in modern networks. On the other hand in order for the equipment manufacturers to address the cost issue they were seeking to reuse standard components in a base platform that would be able to support a broad variety of network elements. They were also keen to move off their monolithic proprietary technologies to be in a position to benefit from

Service Availability: Principles and Practice, First Edition. Edited by Maria Toeroe and Francis Tam.
© 2012 John Wiley & Sons, Ltd. Published 2012 by John Wiley & Sons, Ltd.

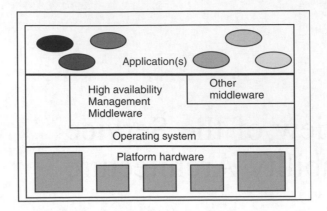

Figure 3.1 Open HA individual system architecture.

the rapidly improving price to performance ratio of commercial of the shelf (COTS) products such as microprocessors, processor boards, storage, and system interconnect technologies.

Under the initiative of a number of companies, comprised of hardware and software suppliers, an industry group called the high availability (HA) Forum (see Chapter 2) was formed to address the complexity of integrating COTS products to build highly available systems. They produced a document 'Providing Open Architecture High Availability Solutions' [21] which collected the best practices and capabilities needed for building highly available systems. It was intended as a guide to a common vocabulary and potentially applicable HA functions from which it was expected that the system designer would select the appropriate functions for each system based on the specific HA requirements, design complexity, and cost. The basic structure of the open HA architecture described in this document was used as a starting point for the SA Forum reference architecture. The architecture for a single constituent COTS system from [21] is depicted above in Figure 3.1.

The SA Forum took this initiative forward by developing a set of open standard application programming interface (API) specifications for the services needed to enable the delivery of highly available carrier-grade systems based on cost-effective COTS components. Wide adoption of these standard interfaces for HA middleware would enable reuse and portability, thereby reducing cost and effort for developers, system integrators, and operators. The specifications were shaped by group of highly talented and experienced system architects from major IT and network equipment manufacturers.

The goals and requirements that guided the process for creating the specifications are outlined in the next section.

3.1.2 Goals and Requirements

During the specification development process at the SA Forum a set of requirements and goals were used to assess the proposals made by various member companies and elaborate the final specification. These were derived from the business considerations and technical experience of the members. We describe these in some detail here to give insight into the background and motivation of the specifications. Among the architectural and business driven objectives the following considerations were borne in mind:

● Separation of hardware from software concepts to allow for the independent evolution of hardware, middleware, and application technologies and products.
● Providing a common set of abstractions for hardware and software resources for the purposes of HA management.

- Enabling the open integration of COTS components into a highly available system by factoring out common availability functions into a modular set of frameworks and services.
- Specifying interfaces rather than protocols to enable the development of common APIs that will protect application investment. It also allows for innovation and differentiation by middleware based on nonfunctional aspects such as performance, latency, and scalability. This also allows implementations to choose the protocols that best suit their particular application requirements.

A number of so called nonfunctional requirements were also taken into account. It was not feasible to directly assess the extent to which the proposals made by the various members for the candidate interface specifications complied with these requirements. The SA Forum relied on the implementation experience of the evaluation committees to make the appropriate judgments.

- **Usability:** Providing a simple programming model that covers a broad spectrum of systems and applications. Specifications should use a common consistent naming and programming style to facilitate learning and adoption. The interfaces should be described in directly usable programming interface for which there are established tools and integrated development environments.
- **Portability:** Dependencies on unique operating system or hardware specific features should be avoided and exposed system functions should be wrapped in abstractions that can be readily and efficiently mapped to different platforms.
- **Upgradeability:** Interfaces should be constructed such that they facilitate backward compatibility with newer releases of the specification.
- **Deployment:** Interfaces should be able to be implemented as one or more profiles representing subset-solutions for resource constrained platforms.
- **Integration:** Interfaces should be designed to facilitate integration with other interfaces and in particular with the configuration and management models.
- **Performance:** The nature of Service Availability Interface (SAI) specifications can to a certain extent influence performance trade-offs that need to be made when implementing and using the specification. For example, different replication styles impose varying trade-offs on central processing units (CPUs) and I/O overhead against fail-over and recovery times. Flexibility to use or configure low overhead operation variants, possibly with concomitant loss of functionality, is important for soft real-time systems such as communication controller network elements.
- **Reliability:** While not a property of interfaces per se; ensuring simplicity and clear functional separation between operations defined in the interface tends to lead to more robust implementations and applications.
- **Scalability:** As with performance there is a trade-off between imposing strong consistency semantics on the implementations and ensuring that systems will scale linearly and incrementally when adding resources, such as processors, memory, I/O interfaces, and nodes.
- **Security:** Interfaces should be designed to facilitate the application of the principle of least privilege policies in a system. This requires that separate functions are defined for operations requiring different privilege levels.

Whereas interoperability is important at mechanical and electrical interfaces as well as for low-level HW management (e.g., automatic bus enumeration, etc.) it was decided not to make it a fixed requirement to provide for interoperability of HA middleware services provided by different vendors on the same system.

Openness and choice of implementation technologies were identified as the key requirements which put the emphasis on defining interface specifications that could be implemented on a reasonably rich modern operating system with portability of the applications being the primary goal.

3.1.3   Service Availability Architecture Scope and Presentation

In working out the specifications, certain assumptions concerning the type of the systems being targeted had to be made. Before describing the conceptual model of the SA architecture we first describe the scope of the systems which the architecture is expected to address.

3.1.3.1   Scope

The architecture is intended to provide a structured view of the SA functions defined in the specifications, how they relate to each other and how they relate to the hardware and other software in the system. A system in this context is considered to be a configured set of hardware, software, and applications which are managed by or use SA functions. The system is further constrained by the following assumptions:

- The physical computational, networking, and storage resources of a system are assumed to be co-located in a single environment (such as a data-center or central office) under the control of a single operator. This implies that the physical security of all the resources is maintained by a single trusted organization. There can however be applications running on the system from more than one provider. Furthermore, no assumptions are made about the number of administrative entities managing their respective applications on the system.
- The computational resources are represented by a set of compute elements that are interconnected by a (possibly redundant) physical network in a logical mesh topology. A computing element is a unit comprised of a set of hardware resources capable of hosting or running applications and SA management software.
- Another implication of the co-location assumption is that the failure rates due to environmental conditions (e.g., air-conditioning or power failures), acts of nature, or vandalism are the same for all components of the system.
- The physical, software, and data resources of the system are assumed to be adequately protected against intrusion from external elements.
- Apart from the local or network storage required for compute elements to load their operating systems, middleware, and application software, no further assumptions are made on persistent storage.

The specified SA functions apply to the installation, operation, and upgrade lifecycle phases of the hardware, software, and applications that use them. While the architecture provides some guidance for system design phase, the development and test phases are currently not covered.

We note however, that implementations of the SA architecture and services could be built that perform adequately beyond this envisaged scope.

3.1.3.2   Architecture Description Approach

There is no common agreed definition of software architecture but for our purposes the following definition from Bass *et al.* [32] is appropriate:

> The software architecture of a system is the structure or structures of the system, which comprise the software components, the externally visible properties of those components, and the relationships among them.

The SA Forum architecture itself is comprised of a comprehensive set of services that can be used in the construction of software systems capable of delivering the high levels of SA and continuity.

Thus in our case the components of the architecture to be described are in fact the services provided to the applications as opposed to the functional components that constitute an actual final system. Also for reasons of modularity most of the services are fairly self-contained such that there is little or no coupling between them. The architecture is presented in terms of the common and unique attributes among the services on the one hand, and the relationships between the services and other elements of the system on the other. In our description of the architecture we will adopt some common views used in practice to describe software architectures:

- Interface abstraction – functional view;
- Operational model – dynamic view;
- Information model – static view;
- Use case view.

The published interface specifications of the various services define their externally visible properties which include their functional interfaces, information model, operational model, and administrative interface where appropriate. Some of the service specifications also provide use case views. In Section 3.3 we will look at the different services and how they relate to one another from a functional and dependency perspective. Brief details on the functional, operational, and information model view for each of the services as they relate to system and application availability are given. Part of the requirements for the specifications called for the cohabitation of the specified services with custom or legacy services within the scope of a SA system. For brevity and clarity we do not cover the interactions with these types of services here.

In the remainder of this section we make some general remarks concerning the first three views as regards the architecture and specifications.

3.1.3.3 Service Interface Abstraction Models

A compromise was sought between a high level of abstraction and a very implementation oriented specification. With a high level of abstraction broad applicability could be achieved but it was considered that this would leave too much room for interpretation in implementations which in turn would lead to divergence. An implementation detail oriented specification would need to make a choice between currently available technologies which would limit the addressable designs with the risk of not being future-proof. In order to satisfy the requirements for openness a number of interface abstraction models were eliminated:

- Fault-tolerant operating system abstraction;
- Protocol specific fault tolerance middleware (e.g., FT-CORBA (fault-tolerant common object request broker architecture) [33];
- Fault-tolerant programming languages or fault-tolerance extensions for programming languages.

The primary interface abstraction model adopted in the SAI specifications for the HA middleware is that of a concrete API specified in ANSI (American National Standards Institute) 'C' syntax that can be implemented and delivered as binary or source level interface libraries. A specification that describes the Java bindings for the interfaces has also been released. The specification for each service provides the signature and associated data-types for the functions provided by the service.

One of the consequences of the COTS hardware requirement is that the implementations of the specified HA middleware services should assume a 'shared nothing' platform model. This implies that the different elements of an HA service implementation can only communicate with one another via message passing. The SAI specifications only standardize the APIs leaving the design of the distributed algorithms and message formats to the implementers of the HA middleware services. Experience had shown that there is not a 'one-size-fits-all' distributed computing paradigm which led to the trade-off of sacrificing interoperability between different HA middleware implementations for application portability and system flexibility. In this context, to promote usability while meeting the scalability and reliability requirements, it was necessary that the interface specifications hide the distributed nature of the underlying platform from the service user. Since many functions would require the exchange of messages within the system, which would cause the function call to block for a duration proportional to the message latency and number of messages sent, many functions provide both synchronous and asynchronous invocation styles. Furthermore the programming model ensures that the application has complete control of the computing resources. When any HA middleware service needs to communicate with its user it is done via a callback mechanism. The programming model allows the application to fully control the dispatching of the callback invocations. Further details on the C/C++ programming model and the Java mappings are provided in Chapters 11 and 12 respectively.

3.1.3.4 Operational Model

Each service is specified to follow the same life-cycle in terms of initiating and terminating the service on behalf of a service user. Services also follow the same model for initiating and handling asynchronous operations as well as for tracking state changes in one or more of their defined entities. This uniformity in the programming model meets our usability requirement. Services also expose an internal operational model on the entities they define. These models are described in the specifications themselves. The complexity of the models varies significantly between the different services.

3.1.3.5 Information Model

There is no single encompassing information model for the architecture; rather the information model is the aggregation of the entities defined by the various services. As for the functional view this was driven by the need for modularity so that designers and implementers could adopt only those services of interest for their needs.

Let us now consider a SA system as consisting of a set of resources. Some of these are physical resources such as CPU boards, network interfaces and the like, others are software resources like operating systems and yet others could be system services that entities in the system can invoke such as network name to address translation. Depending on the set of HA middleware services configured in the system and their coverage in the system, a greater or smaller number of the resources of the system will be reflected in aggregated information model. Each HA middleware service represents the resources within its scope by the logical entities that are defined in its specification. Some of these entities can be acted on programmatically through the APIs of the respective services.

The services also expose some of their logical entities as managed objects in the management information model. Figure 3.2 illustrates how certain hardware and software resources are represented as logical entities in the system model and how some logical entities and resources are represented as managed objects in the management information model. This managed object model is managed by the Information Model Management service (IMM), which is intended to be used by all HA middleware services. Applications can also expose locally defined logical entities in the managed information model. These application defined entities may represent resources internal to the application or resources that are external to it, even resources that are external to the system. As part of the specifications the SA Forum provides the object definitions of all the managed objects defined by the various services as the 'SA Forum Information Model' file in XML Metadata Interchange (XMI) format [34]. Note that this only contains the managed logical entities of the different services exposing their configuration attributes, runtime attributes and administrative operations through the IMM.

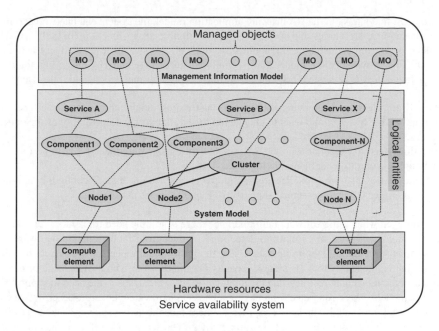

Figure 3.2 Service availability system and information models.

In summary we may regard the architectural information model as the aggregation of the managed and unmanaged logical entities defined by the set of HA middleware services published in the specifications.

Chapter 4 further elaborates the SA Forum Information Model.

3.2 HA Concepts Applied

In the previous section we presented the scope of the architecture as a system with certain properties and constraints. We also described the overall structure of the architecture and how it is to be presented. Before examining the services and their interrelations we review the HA concepts that strongly influenced the specifications and how they can be applied on an abstract level.

3.2.1 To Be or Not to Be High Availability Aware

To deliver SA with cost effective COTS components requires the use of software fault tolerance techniques. For some high value transaction applications such as financial services or airline reservation systems the use of expensive hardware fault tolerant platforms may be appropriate. However hardware fault tolerance is not cost effective for modern internet information, communication and entertainment infrastructure systems where the total cost per transaction can only be a small fraction of a cent and is dropping. Nonetheless HW fault tolerant nodes may sometimes be used to provide certain critical functions such as system control in a largely COTS-based system.

In general systems based on software fault tolerance can only achieve the highest levels of SA with the active involvement of the applications. This is the so called HA aware application model. In this model some of the HA decisions are delegated to the application when a failure occurs allowing the application to continue to provide service albeit in degraded mode while recovery and repair actions are in progress. In contrast, in the HA unaware application model, the system can make no assumptions about what subset of resources and services an application needs to function correctly. As a result it will delay restarting or resuming the application until all the system level recovery actions have completed. This is typically the model that is employed in commercial HA clustering solutions. In order to achieve very high levels of SA all parts from the system must cooperate. For example:

- The hardware elements comprising the system must expose status and control interfaces for the system to be able to detect and react to changes in the availability and operability of the underlying resources.
- Applications need to cooperate with the SA middleware by signaling their ability to perform certain tasks while reacting to reconfiguration requests.
- All layers, that is, platform, middleware, and applications need to expose configuration and management interfaces to the operational support or IT management systems.

However, even when using software fault tolerance there is still the need for some level of hardware redundancy and excess system capacity to detect and recover from faults, tolerate hardware failures or reduce down time during system upgrades. The flexibility requirements on system scalability and reliability, that is, the ability to scale system capacity up and down as well as increasing or decreasing the availability level without restructuring the application software, leads to the need to support various redundancy models. These can range from having a back-up hardware element for each active element to having no back-up at all. In the latter case if an element fails the active software load is redistributed across the remaining elements potentially resulting in degraded service level if the remaining resources are insufficient to meet the total workload.

The function that is responsible for receiving the hardware and software failure notifications and redistributing the workload on the available and correctly functioning system resources is the availability management. It is in a way the core of the HA middleware.

3.2.1.1 Availability Management

Service Availability (SA) is ensured by the availability management functions at the platform, middleware, and application levels. In particular it is important for these functions to be able to cooperate and coordinate their actions. We will refer to the subset of functions in the HA middleware that are related to availability management as the *availability management middleware* (*AMM*). To ensure SA subject to varying scalability and reliability constraints over the lifetime of the system it was necessary that services be explicitly represented as entities in the system model of the AMM. One can think of an instance of a service that needs to be delivered by an application as *job* that needs to be done. Further, jobs are done by workers in a workplace and in our analogy workers are software resources and workplaces are hardware resources. In order for software to do its job it needs operational hardware to perform it on. So in a SA system we have a set of jobs to perform and a set of hardware and software resources to perform those jobs with. The aim of the AMM is to ensure that all the jobs are being performed properly as a function of the availability of the hardware and software resources. At a high level of abstraction software resources can be modeled as components. Components can be thought of as the workers that receive assignments for the jobs they need to perform by interacting with the AMM. In the system model a *component* is the basic unit of software from which applications are built. Essentially a component is comprised of a set of software functions that can be managed as a unit from the perspective of the AMM. In the design of our architecture a component is conceived as being instantiated by one or more processes (or tasks) executing on the same operating system on a hardware element with a single physical memory address space. Processes (or tasks) correspond to the basic software abstraction as provided in modern operating systems. The rationale for basing the component on process boundaries include the following:

- A process (or task) is the natural software unit that can be started and stopped by the operating system.
- Most current operating systems use hardware memory management functions to ensure isolation between processes by allocating a separate logical address space for each of them.
- A process can stop itself as a unit allowing well-written components that detect their internal errors to implement fail-stop semantics.
- The software of a process can also be replaced as a unit for upgrade or downgrade purposes.

We will call a hardware element that is capable of running an operating system a *computing element* to distinguish it from other hardware elements such as power supplies, and so on. Note that a computing element may host multiple operating systems concurrently as is the case with virtual machines. Thus from an architectural perspective we have three nested natural failure zones within a compute element:

1. The computing element itself;
2. The operating system running on the computing element;
3. The processes running in an operating system constituting a component.

To maintain SA in the event of a failure in one of the failure zones it is necessary to identify the set of components that are capable of taking over the jobs that were being performed by the failed component(s) (i.e., the worker(s) who are unable to continue to perform their jobs because they have died, become sick, or their workplace has disappeared). Note that if a computing element or OS goes down all the components housed by that computing element or OS are also considered as having failed.

Any form of automatic fault tolerant electronic data processing system must perform the following availability management functions:

- Fault handling
  - Detect that a fault that has occurred;
  - Identify the failed hardware or software element;
  - Generate an error report to trigger recovery and repair;
  - Contain or limit the propagation of the effects of the fault in the system.
- Recovery and repair
  - Continue processing on the remaining healthy resources;
  - Initiate repair operations on the faulty resources;
  - Integrate repaired resources back into the system.

3.2.1.2   Fault Handling

With software-based fault tolerance systems some hardware and most software failures are detected by software monitoring and auditing. We also note in passing that the majority of failures in modern systems are in fact caused by software defects. In our system, hardware and software faults are monitored by various specialized HA services as described in Section 3.3. Note that whereas the AMM receives notifications of computing element, OS and process failures in the system from its internal monitoring functions, for HA-aware applications it relies on components to actively report any failures occurring within them that require its intervention.

3.2.1.3   Recovery and Repair

Some transient hardware faults are detected and corrected by the hardware, for example, error correcting memory. A component may also be able to recover from a failure occurring with its scope without the aid of the AMM. For example, when a process fails in a multi-process component the component may simply restart the failed process. When the AMM receives a failure notification in the form of an error report from a component, it will initiate recovery actions to maintain or restore the system functionality within the constraints of the available operational resources. Some of the recovery actions it has recourse to are the following:

- Restart the computing element on which the failure occurred;
- Restart the operating system in which the failure occurred;
- Restart the component in situ;
- Restart the component on a different OS/hardware element;
- Reassign the jobs of the failed components to other components;
- Restart the whole system.

As the granularity of the component, which is determined by the natural isolation boundaries provided by the operating systems, is fairly coarse we require the ability for a single component to be able to take on multiple jobs of the same or different types.

The flexibility afforded by the redundancy models and the ability of assigning single or multiple jobs to a component leads to a certain degree of complexity in the system model of the AMM. Despite this complexity it is still possible to accommodate simple single service HA unaware or basic active-standby HA-aware application models as well as the more sophisticated multi-service load balancing configurations.

3.2.2 HA Aware Application Perspective

For an application that is HA unaware it suffices that it be configured in the system model of the AMM. An HA-aware application, on the other hand, will have one or more components that will interact with the AMM. Here we will refer to the middleware service with which applications interact for availability management as the *availability manager*. Within the SA Forum system architecture it is called the Availability Management Framework (AMF) which is described in Chapter 6. Components are expected to detect faults within their scope, as mentioned above and to apply appropriate local recovery actions. If they require system level recovery actions to be taken they send an error report to the availability manager. A component may influence the choice of the recovery action that will be taken by accompanying the error report with a recommended recovery action to reduce service downtime.

The availability manager interacts with components by assigning jobs to or taking jobs away from them. Each job assigned to a component has a role associated with it which is determined by the availability manager. The role associated with a job tells the component whether it should play an active or standby role for that job. The way the availability manager maintains SA is by assigning and removing jobs from components and controlling the roles of the assigned jobs. For example, in the case where there is a corresponding standby job assigned for each failed active job, the availability manager performs a 'fail-over' by simply changing the roles of the standby jobs to active. The main responsibilities of the components with regard to the availability manager are to report detected errors and respond to the assignment of jobs and to changes in the role of their assigned jobs.

Some applications maintain an internal state that needs to be preserved between interactions with the external environment. In order to mask failures occurring in these applications some form of replication of their internal state across failure zones is required. Whereas hardware based fault tolerance systems transparently replicate all application state, in a software fault tolerant system it is up to the application or middleware to replicate sufficient state to be able to recover from a failure while incurring little or no loss in service. From a component perspective: If a component has an active job assigned, it must provide the associated service.

When a component has a standby job assigned, it must be ready to rapidly provide the associated service in the event that the availability manager sets the availability state of the job to active. In order to react swiftly to the change from standby to active availability state for one of its assigned jobs a component may require access to internal state from the component that was previously providing active service. To protect this internal state against the failure of a component or the node it is running on the state must be replicated to another component or node.

The methods used to replicate state between two or more components that are protecting the same job fall into two categories. The first is so called passive replication. In this case the critical state maintained by the component that has the active job is replicated or 'check-pointed' to an alternate location at regular intervals or when 'significant' state changes have occurred. The second category is the active replication method. With active replication each component sees the exact same sequence of service requests from the external environment where for the job assignments in the active state they generate outputs or responses to the external environment whereas for the standby job assignments they do not.

If the components (by design) start off with the exact same initial state with respect to the protected job under consideration, when a 'fail-over' does occur the component taking over the active state for the job will have an up-to-date copy of the application state. Active replication does, however, require that all choices in the execution streams of the participating components be deterministic to avoid the possibility that their copies of the state information diverge. Functions that support or provide active or passive replication mechanisms are generally provided as part of the HA high availability middleware of a software fault tolerant solution. Even so, in some cases components may resort to synchronizing their state using application specific mechanisms in order to take advantage of 'piggy-backing' replication data in application information flows.

While the availability manager consolidates the error reports from all HA-aware components it is often necessary for software components to reliably communicate with one another about changes in application or external state or synchronize access to resources. These features are also generally provided by the HA middleware to ensure consistent cluster semantics across the components and the availability manager.

In this section we have looked at HA high availability concepts and how they apply in the context of providing flexible levels of SA in a software fault tolerant environment. Before describing the different services in more detail we will introduce the architecture of the SA Forum Service Availability Interface Specification.

3.3 Architecture

3.3.1 Basic Architectural Model

As mentioned in the requirements of Section 3.1.2 a fundamental separation of concerns that needed to be respected was the decoupling of software management from hardware management to facilitate hardware evolution while protecting software investment. Accordingly two sets of specifications were produced: The Hardware Platform Interface (HPI) [35] and the AIS. This separation of concerns is based on the nature of the logical entities that are represented and manipulated in the interfaces. For the HPI the primary area of concern is to model hardware resources and to discover, monitor, and control them in a platform independent manner. The AIS focus is to facilitate the implementation and management of applications providing high levels of SA and continuity. Whereas the HPI functions are generally implemented in hardware or processors embedded in the hardware, the AIS services are implemented as middleware that typically runs on general purpose operating systems. The requirement for HPI implementations to operate in environments with limited processing and storage capabilities led to the difference in programming models adopted between the HPI and AIS. However, in order to provide a convenient, complete, and consistent programmatic view of both software and hardware in the AIS model a service dedicated to platform management was added to the AIS.

3.3.1.1 The Service Availability System

The SA system as it is exposed to an application is the set of computing resources across which the AIS services are available to it. In general terms it can be seen as a cluster of interconnected nodes on which the services and applications run. The nodes represent the actual computing resources which have corresponding entries in the configuration. Prior to virtualization being supported on general purpose COTS computing elements there was a convenient one-to-one correspondence between compute elements and nodes. Now however we need to make the distinction between a node as an operating system instance and a computing element as a single hardware element capable of hosting one or more than one operating system instance. In the SA system each node of a cluster corresponds to a single operating system instance running on a computing element.

As shown in Figure 3.3 the computing elements in the system are attached to a possibly redundant system interconnect. The AIS software running on the nodes of the computing elements communicates over the system interconnect. From the platform perspective the system consists of the set of compute elements that have been specified in the platform configuration data. The AIS can use the HPI services to discover which of the computing elements are physically present and match them to its configured resources.

The HPI services operate independently of the configured system in that it may 'see' computing elements that are not part of the AIS system configuration. In typical implementations the HPI uses a separate out-of-band communication mechanism to discover and control the various hardware resources within its domains. Thus the topology and resource coverage of this out-of-band communication facility can be different from those of the system interconnect. There may also be separate communication paths to other computational, storage, and network resources that are not part of the computing elements or the system interconnect. In an ideal system all the hardware resources are discovered by

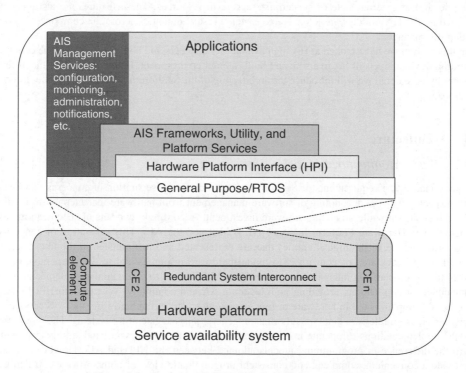

Figure 3.3 Service availability high level system architecture.

the HPI and exposed in the management information model which identifies the set of configured computing elements present among them from the platform configuration data. The configuration data also specifies which operating systems should be loaded on the configured computing elements. Each operating system instance, referred to as a *node*, also has a unique *node name* assigned to it in the cluster configuration. The list of these *node names* in the cluster configuration data defines the system at the cluster level. At the highest level the availability manager configuration defines a set of nodes, some of which map onto cluster nodes. For applications managed by the availability manager the SA system is the set of these nodes that map to the healthy cluster nodes. In summary then the SA system exists at three levels: Platform level, Cluster level, and Availability management level. There is a one-to-one mapping of the availability management node to the cluster node and the cluster node to the platform operating system instance running on a compute element. For each level the configured system may differ from the actual physical system which may be different to the set of currently active and healthy nodes. The three level architecture of the SA system is depicted in Figure 3.4 below.

3.3.1.2   The Hardware Platform Interface (HPI)

The primary purpose of the HPI is to expose the monitoring and control functions of the hardware through a set of standard, platform independent APIs to applications, and SA management software. It includes functions to discover and access the inventory of hardware entities, sensors, and controls within the scope of the addressable hardware domain controllers. It also supports the fault handling, recovery, and repair mechanisms of the system through monitoring, control, and event management functions. The level of abstraction of the API provides ample room for hardware manufacturers to innovate and differentiate themselves by allowing certain fault handling, recovery, and repair actions to be performed automatically at the hardware platform level. For example, on failure of a fan the hardware controller might adjust the speeds of the remaining fans or resetting or power cycling a computing element via a watchdog timer.

Further details on HPI are discussed in Chapter 5.

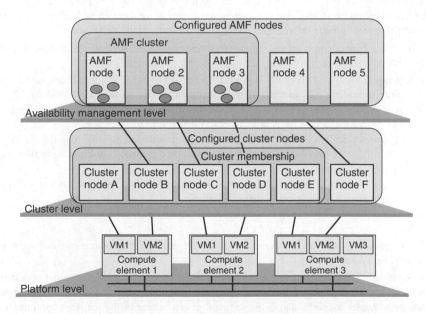

Figure 3.4   Service availability system architecture.

3.3.1.3   The Application Interface Specification (AIS)

The AIS is split into a set of relatively independent services where each service covers a specific set of related functions. The service lifecycle of each AIS service used within an application is independent from that of the other services being used by the application. For convenience of presentation, related services are further categorized into four functional groups. We briefly present these functional groups before describing their constituent services in further detail.

- The *AIS Platform Services* are responsible for handling all the platform related functions and providing the necessary abstractions and controls for the other services. It is divided into two services:
  - The Platform Management service (PLM) [36] which provides the abstractions of the hardware and low-level system software.
  - The Cluster Membership service (CLM) [37] provides a cluster level abstraction of the system in terms of the current set operational nodes in the system. It supports the node level software availability management functions of other services and applications.
- The AIS Management Services group together the essential management services that are commonly required by software that provides management interfaces to external systems. In the SA Forum system these management services are the following:
  - The IMM [38] which provides a consistent view of the configuration, administration, and management of all the hardware, software, and system resources that are exposed in the management information model.
  - The Notification service (NTF) [39] hat provides a consistent way for applications and services to notify external entities about alarms and changes in the system. There is a separate error reporting mechanism used for the internal availability management of the system that is provided by the AMF.
  - The Log service (LOG) [40] providing a system wide log management facility where alarms and other notifications including application specific notifications can be recorded and consulted.
  - The Security Service (SEC) [41] which provides a means for the HPI and different AIS services to control access to their different functions from authenticated applications.
- The *AIS Utility Services* is a set of support services commonly used in distributed HA software applications.
  - The Checkpoint service (CKPT) [42] provides a set of programming interfaces used by applications to replicate their state in the system to maintain service in the case of failures.
  - The Event service (EVT) [43] is a publish-subscribe event distribution mechanism providing a many-to-many, anonymous, asynchronous communication facility.
  - The Message service (MSG) [44] is a mailbox style distributed message passing facility.
  - The Lock service (LCK) [45] is a distributed resource access coordination facility based on system wide lock objects.
  - The Naming service (NAM) [46] is a simple distributed facility used for binding system object references to names and retrieving them by name.
  - The Timer service (TMR) [47] is a node local facility for setting timers and receiving expiry events.
- The *AIS Frameworks* are special services that directly manage the applications and software configured in their system models.
  - The AMF [48] provides the availability management functions to its configured applications.
  - The Software Management Framework (SMF) [49] maintains SA during system or application software upgrades by coordinating the application of changes in the software configuration with the AMF.

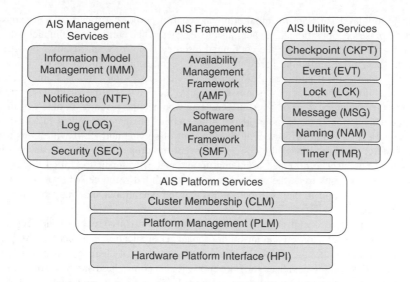

Figure 3.5   Functional grouping of the HA middleware services.

The AMM we introduced in Section 3.2.2 maps to the group of availability management functions of the HPI, PLM, CLM, and AMF. A high level functional representation of the HPI and AIS service functional groups is provided in Figure 3.3. The functionality in each box depicted above the operating system exposes APIs to those above it and uses the APIs of those below. Each compute element in the system hosts one or more operating system instances. The software functionality depicted above the operating system layer is distributed across all the computing elements.

In this section we have introduced the various services constituting the HA middleware specified by the SA Forum. The functional grouping of the services is summarized in Figure 3.5.

3.3.2   The AIS Services and Frameworks Architecture

Now that we have defined the SA system and introduced the services we will briefly describe the roles and scope of the various AIS services and frameworks in providing SA support capabilities. In particular we will examine how their fault handling and recovery-repair functions interact with each other and the applications using them.

3.3.2.1   AIS Platform Services

The combined AIS Platform Services provide the interfaces and abstractions for the AMF to monitor and control platform resources. Applications and other availability management software running on the system can also avail itself of these interfaces to manage hardware resources attached to the computing elements that are not covered by the implementation of AIS platform service being used. These will typically be resources used to provide application specific services such as specialized digital signal processing capabilities in digital radio communications systems.

Platform Management Service
PLM provides a model in which to represent the desired configuration of the hardware and low level software (hypervisor and operating system) resources. It also provides a view of the actual

hardware and low level software for monitoring and control purposes. In particular it exposes information on the presence, operational, and administrative states of the entities in the model. For example, the presence state indicates whether the configured resource is not-present, inactive, or active. When the presence state for a configured hardware resource becomes active it can be taken into account for availability management purposes. PLM allows service users to track this and other state changes affecting availability via its readiness status tracking APIs. For the purposes of this exposition we consider that the *readiness state* of a resource reflects whether the resource is able to function correctly or not. This notion is made more precise in the actual specification. When PLM detects that a resource has failed, its readiness state will be set to out-of-service and PLM will attempt to isolate the failed resource from the rest of the system to prevent the propagation of errored state. The PLM implementation may also be designed to attempt to automatically repair the failed resource. Should any of the AMM functions track the readiness state of the failed resource, when a successful repair action has been effected on it, they will be notified that the operational state has now transitioned to the in-service state.

Through administrative operations on a hardware resource or low-level software entity (e.g., an operating system or hypervisor) an administrator or manager application can shut them down or bring them back into service without regard to the actual implementation of the mechanisms used to perform the specific operations on the device or software entity. It also allows hardware and software resources modeled by PLM to be isolated from the system even if they are considered operational at the PLM level.

PLM also allows for the modeling of dependency relationships between the entities in its information model. This allows the reflected state of an entity be determined, in addition to the entity's own status, by the status of the other entities it depends on greatly simplifying the view that needs to be exposed to the PLM users.

The tracking interface provided by the service further simplifies the view for AMM functions and other users by allowing the tracking of groups of entities in a single request. The tracking interface also provides for fine grained coordination of managed operations. For example, let us suppose a blade computing element, currently actively providing service, is to be extracted for maintenance purposes. When the extraction lever of the blade is opened the HPI informs PLM of the intent to remove the blade. PLM will check whether the operation may proceed by asking its interested users such as the AMM to validate the operation. The AMM checks to see if there are sufficient resources to maintain service and if so it allows PLM to proceed with the operation. PLM now notifies users tracking this resource or the entities that depend on it, in particular the nodes or operating system running on it, that a termination operation has been started and waits for their responses. The AMF for its part receives its notification via CLM for each affected node and will accordingly migrate any active services off of the affected nodes by assigning the active jobs to components running on other blades. Once all the users have responded to the start of termination notification PLM terminates the entity and sends an operation completed notification to all tracking users. At the same time, when all software entities running on the blade have been terminated, PLM instructs the HPI to put the resource into the inactive state and sets the presence state to inactive. This toggles the state of a light-emitting diode (LED) on the blade informing the service technician that the board can now safely be extracted.

Reports of state changes of each of the entities in the PLM's information model are also sent to the AIS NTF where, depending on the configuration, they will also be logged. Operations support systems can automatically generate trouble tickets from these reports. The logs can be a valuable resource for operations, the system architect and reliability engineering staff.

3.3.2.2   Cluster Membership Service

The Cluster Membership service (CLM) is responsible for maintaining a consistent view of the set of currently operational nodes in the cluster. It uses PLM to track the state of the various hardware resources that need to be monitored to assess whether a node is functioning or able to function correctly. Depending on the implementation CLM may need to have recourse to other mechanisms such as node to node heartbeating over the system interconnect to determine the state of health of the configured nodes in the cluster. A healthy node must be able to provide all the services required of the node by the AIS services of the system as well as the application components configured to run on that node. In particular it must be able to communicate with the other nodes of the cluster.

The precise semantics of what constitutes a healthy node is not prescribed by the specification and is left up to the implementation. In a tightly coupled system a healthy node must be able to communicate with every other healthy node of the cluster. Effectively ensuring this tight semantics on a system with a large number of nodes can be very difficult. In another more loosely coupled model it may suffice for a node to be able to communicate with a small set of 'controller' nodes in order to be considered healthy.

One of the primary roles of CLM in the SA architecture is to notify the AIS services (and applications) using it of the addition and removal of nodes from the current set of operational nodes in the cluster. For example, when CLM informs the AMF that a new node has become operational, the AMF will start up the configured components on the node and assign the appropriate jobs to them and the role they should perform for these assigned jobs. As we saw from the board extraction example with the PLM in section 'Platform Management Service' on page 47, CLM notifies the AMF not only of node removals but also of pending node removals, allowing the AMF to take the appropriate action before the removal actually takes place. When a node that is operational at the PLM level, is not considered healthy at the CLM level (possibly due to connectivity issues) CLM can use PLM functions to isolate and optionally repair the node by possibly rebooting it. In the SA system CLM is the ultimate authority about which nodes constitute the membership of healthy nodes in the system at any given time. The coherency of the AIS services in the system depends on the extent to which all services in the system rely on CLM for information on the membership of the healthy nodes.

In summary the Platform services provide fault detection, isolation, error notification, and repair on the entities within their domain of control in so far as these are handled by the implementation. PLM deals with hardware and low-level system software while CLM deals with the cluster and cluster nodes. CLM uses PLM for the fault handling and recovery functions but adds the necessary fault detection mechanisms in order to measure all the criteria needed for determining the state of health of a cluster node. Finally the Platform services provide interfaces for administrative management.

Chapter 5 presents the Platform services in more detail.

3.3.2.3 AIS Management Services

There are a number of common management functions that are needed in distributed systems providing highly available services. These common functions include configuration and runtime management; notifying applications, and external management systems of significant events occurring in the system; saving a record of significant events for further analysis; controlling access to critical system functionality. In this section we examine only how the AIS management services providing these functions contribute to maintaining the availability of services and applications. For more details see Chapter 8.

Information Model Management Service
In the AIS it is IMM that provides the interface for the configuration and runtime management functions of the manageable objects in the system. Each service in the AIS exposes their configuration and runtime information as well as their administrative interfaces through IMM. In the following we will describe examples of these functions as used by the AIS platform services.

As mentioned in Section 3.3.1.1 the SA system consists of the configured computing resources that correspond to the nodes of the cluster. This cluster configuration data is stored and managed by IMM. In the architectural model of the AIS the desired state of the system is described by the configuration and administration information held in the system management information model. Until now we have used the term system management information model to distinguish it from the logical information model which may contain programmatic entities that are not exposed in the management information model. The information model of IMM is the system management information model.

When the AIS services start they initialize themselves according to their configuration data. Once running they are notified of any changes to their configuration by IMM and are expected to apply the changes to the system. For example, the system configuration can have more computing elements configured than are actually physically present on the system interconnect. This allows the cluster to be scaled up to add capacity and/or redundancy by simply adding one or more physical computing elements that match the existing entries in IMM configuration. When one such computing element is added it is detected by HPI and reported to PLM which starts the configured operating system(s) on it. Once the operating system has started the node is detected by the cluster management service which adds it to the cluster membership and informs all interested parties which in turn then take the appropriate actions as determined by their configuration.

The condition where there are unconfigured physical compute elements on the cluster interconnect can also occur. These will only be taken into account if configuration data in the information model is added to reflect them. In the case of the cluster only those nodes having an entry in the cluster configuration data can become members of the cluster and therefore be used by the AIS services and the applications. To permanently remove a configured node (say) from the cluster it is simply removed from the configuration data. In order to remove its configuration data, the node would first have to be administratively locked and all references to the node in other parts of the information model would also have to be removed.

As we mentioned, when the configuration data for a particular entity is changed the service responsible for the entity will be informed by IMM of the change. In order to coordinate change of configuration that affects multiple entities, as in our example of removing a node's configuration, IMM provides a lightweight transaction mechanism whereby a set of related changes can be grouped into a configuration change bundle or CCB which will succeed or fail as a unit. If any operation or configuration change in the bundle is rejected none of the changes in the bundle is applied. This allows services to maintain availability by being able to enforce dependency or state change constraints that are not explicitly modeled in the information model where a sufficiently comprehensive implementation of IMM could take them into account.

IMM also plays a critical role in the case of a cold start or cluster restart in that it must provide for persistence of all the configuration data and persistent state of the system (such as the administrative state of the various entities). To protect the configuration against storage or other failures during system start-up before the other AIS services are available IMM must implement its own specific HA mechanisms. These might include replicating the configuration data onto different nonvolatile storage devices and being able to probe and test these devices to ensure the system has the latest valid configuration.

Notification Service

The AIS management service that provides the standard way for a service or application to notify other entities of significant events occurring in the system is the Notification service (NTF). The service defines various types of notifications specifically supported by the APIs including alarms, state change, security alarms, and object-lifecycle changes. Here we will examine how the general and alarm specific notification information contributes to the SA architecture. Further details on NTF can be found in Chapter 8.

NTF provides three interfaces corresponding to the role of the entity using them. These are the *producer* API for creating and sending notifications and the consumer APIs consisting of a *subscriber* API for receiving notifications as they are sent and the *reader* interface for retrieving stored notifications. In order to reduce the complexity of notification handlers, users of the consumer APIs can restrict the types of notifications they will receive or retrieve by applying filters on the values of some general and notification specific fields. For example, the general event type field for an alarm notification broadly classifies it according to the affected system area: communications, quality of service, software, hardware, or environment.

Other general fields include a timestamp, a cluster unique notification identifier, a reference to the object in the system information model emitting the notification as well as a reference to the object that caused the alarm. The unique notification identifier of one or more previous notifications may also appear in the general correlation field which together with the other fields mentioned allows the sequence of events relating to an alarm to be reconstructed for on-line fault management or reliability engineering purposes.

In addition to the general fields the alarm notification has a number of specific fields relating to the cause, perceived severity, trend, and proposed repair actions. Particular care should be taken in applications and services when setting the perceived severity field as the severity is often context dependent and can be interpreted by various consumers. Six perceived severity levels are provided for in the specification: alarm-cleared, undetermined, warning, minor, major, and critical. For alarms and security alarms the NTF implementation must as far as possible guarantee delivery as the availability of the system may depend on the notification reaching the appropriate fault handling function external to the AMM.

The architecture envisages that all AIS services and their user applications exclusively use NTF for alarm reporting to ensure consistent fault handling in the system. As the HPI is not an AIS service user it is up to PLM to produce HPI-related notifications. Within the AMM, for example, the AMF includes a separate error reporting function that allows components to inform the framework of errors to trigger recovery and repair actions; therefore no alarms are needed. The AMF produces NTF notifications based on the component error reports and sets the type and severity fields appropriately. This separate error reporting function provides for a tight coupling between the AMF and high-availability aware application components whereas the notification service's producers and consumers are loosely coupled and mostly independent of one another.

For consumers to reliably retrieve alarms even after a cluster restart NTF must ensure the persistence of alarm notifications.

In our architecture the NTF implementation is supposed to provide guaranteed delivery.

Log Service

The Log service (LOG) of the AIS provides a standard interface for writing log records and standard log record format rules which can be used to allow custom log analysis applications to be designed and implemented. Four types of log streams are defined: alarm log, notification log, system log, and application specific logs. The SA architecture envisions NTF as being the exclusive writer on the alarm and notification log streams.

From an availability management perspective LOG provides persistence for certain log streams and a log filtering mechanism to reduce the load LOG imposes on the system during high service load conditions. As part of the application specific log stream configuration data LOG provides a HA flag. When set LOG ensures that the log files associated with the log stream are persistent across cluster restarts.

LOG in and of itself is not critical to the functioning of the system and as such it should not unnecessarily consume system resources during times of overload. For operational, governance, business, or system engineering reasons some log records must be kept under all circumstances whereas other may be discarded without incurring severe consequences. To this end LOG defines a log filtering mechanism based on a log record severity attribute. Log record filtering applies only to the system log stream and application defined log streams because the alarm and notification log streams are considered to be essential in the context of the SA architecture. When writing a log record its log severity level can be set to one of seven values defined by the specification. In decreasing order of severity these are: emergency, alert, critical, error, warning, notice, and informational.

LOG provides an administrative interface to control the logging of records according to their log severity. Upon the onset of high load conditions a system-wide overload control facility can set the log filter on the log stream to block records with selected severities. When the load lightens previously filtered severities can be unblocked. At the time of writing there are no provisions for signaling specific load conditions in the AIS.

From an implementation perspective LOG must provide persistence of the highly available log files. LOG implementations should also specify how log file names in the configuration map to the node level path name of the valid copy of the associated log if node level replication is used.

Security Service

The Security service (SEC) is primarily concerned with preventing denial of service situations from occurring due to the excessive use of HPI or AIS services by unauthorized applications. Essentially it provides a framework for the authentication and authorization of it client service users. The enforcement of the authorization is delegated to the actual service in each case. The system designer must ensure that processes protecting a given service execute with same security privileges in order to avoid them being denied access to AIS services during a fail-over.

3.3.2.4 AIS Frameworks

What distinguishes the AIS frameworks from the other services is that they explicitly model application software artifacts in their system and management information models. For in depth introductions to these frameworks see Chapters 6 and 9. Here we will limit ourselves to discussing how these frameworks fit into the overall SA architecture.

Availability Management Framework

The AMF specifies the APIs that can be used by HA aware applications for them to actively participate in the fault handling and recovery actions of the system. The APIs allow the application components to be managed by the framework in accordance with the resources, services, and policies configured in the information model. Interactions between the AMF and the components of HA-aware applications follow the same model as the other AIS APIs, that is, AMF controls are received by components via callbacks. For a discussion of the AIS programming model see Chapter 11. HA unaware applications that are managed by the framework do not use the API but only appear in the AMF configuration.

As mentioned in Section 3.3.1.1 the AMF defines its own view of the system as a cluster of AMF configured nodes which map onto the CLM configured cluster nodes. This indirection or 'late binding' of AMF nodes to CLM nodes allows an AMF managed application to be configured independently from the clusters on which it will be deployed. The system model of the AMF defines a number of manageable entities to facilitate the configuration and administration of applications. Here we will limit ourselves to applications and the entities exposed at the AMF API, that is, components and jobs (which are formally called *component service instances*). Thus for our purposes we will consider that an application is composed of components and the jobs that those components are to perform.

The AMF application configuration specifies how the AMF should allocate the components to the AMF nodes and how to assign the jobs to those components. This includes the configuration of the redundancy relationship between components. Typically components are in a redundancy relationship if they can be assigned the same jobs in active and standby roles. The AMF defines a number of different redundancy models corresponding to the level of protection for the jobs that is to be provided by the components. In general higher levels of protection require higher levels of redundancy. The additional redundancy is manifested in the configuration by a greater number of components and AMF nodes.

When mapping AMF nodes to the CLM nodes in virtualized environments the administrator must ensure that the components in a redundancy relationship are not allocated to nodes that are hosted by the same computing element, if they are to protect their services against hardware failures. This can be verified programmatically by following the mappings from the AMF to CLM to PLM and then checking the PLM containment dependency relations in the PLM information model and its mapping to HPI entities.

The AMF manages fault detection for applications by relying on the CLM for node failure detection and implementation specific operating system mechanisms to detect the failure of the processes of the component that it knows about. It also relies on components to report component failures on themselves and on other components using an error report API. An HA-aware component reports faults it has detected to the AMF to trigger recovery and repair actions.

The AMF also provides a configuration and administrative command to start an external process to monitor the health of a component. This may entail having the external process send dummy service requests to the component or have it inspect information about the component. When the external process detects a problem with the component it also uses the error report API to notify the AMF.

An HA-aware component can also use the AMF health-check API to trigger periodic health checks and have the AMF monitor for solicited or unsolicited health-check responses. Should the component not respond within a configured time interval the AMF will assume the component has failed and take the appropriate recovery actions.

Components and their related administrative entities have a comprehensive set of states (e.g., administrative, operational, presence, etc.). Of particular interest with regard to availability management are the readiness and the HA states of the jobs assigned to a component. When the AMF has a job to assign, it evaluates the candidate components based on their readiness state. As we already mentioned, when the AMF assigns a job to a component it tells the component what role (or HA state) it should assume for that job, that is, active or standby. Thus a component has an HA state for each job assigned to it. The interpretation of this active or standby role is application specific in each case. AMF only coordinates the appropriate number of assignments in each role for each job.

The AMF also provides a job dependencies configuration parameter for each job. Unless all the jobs listed in the dependencies parameter of a particular job already have an active assignment the AMF will not attempt to assign that particular job to a component with the active HA state. In other words this is a way to inform that AMF that a given job can only be actively performed if the jobs it depends on are already being actively performed.

In this subsection we have briefly summarized the AMF in terms of the functions, states, and configuration only as they pertain to AMF nodes, applications, components, and jobs (aka *component service instances*) in order to provide the reader with a simple overview of its architecture. For further details the reader is referred to Chapter 6.

Software Management Framework

Modern software systems are continually evolving, whether it is to correct defects or to introduce new functionality on the software side or add, remove or replace resources on the hardware side. The purpose of the Software Management Framework (SMF) is to provide a well defined and ordered process to move the software and hardware configuration of a system from the current state to the new desired state while minimizing service impact. The sequence of changes and actions that describe how the system is to be moved to the new state is defined in a structured machine readable eXtensible Markup Language (XML) file called an *upgrade campaign specification* (*UCS*). The execution of such an UCS is managed through a state-model that the SMF maintains in the IMM.

The campaign is divided up into phases, procedures, and steps. As the campaign progresses through the steps it can roll-back to retry a failed step while not undoing all the work already done in previous steps. In the case of an unrecoverable situation it can roll-back all the executed steps, or when even that is not feasible fallback to the original configuration that was backed up in the initialization phase.

The upgrade campaign designer must take into account the dependency and compatibility constraints present in the system but that are not modeled in the system information model where the AMF automatically would take them into account. For example, when a component is being upgraded it may change the representation of the state information it replicates using CKPT. In this case the new component would not be compatible with the old component meaning that without application coordination just mechanically upgrading the components will not work if no outage can be tolerated. SMF exposes an API for such coordination.

The SMF relies on the AMF to maintain SA. It executes the upgrade campaign by applying CCBs to the IMM configuration of the AMF managed entities. It also performs the software installations and removals and interacts with AMF to lock/unlock, stop, and start the AMF entities. For details of the SMF the reader is referred to Chapter 9.

3.3.2.5 AIS Utility Services

These services cover the common functions typically required when developing highly available distributed applications.

Checkpoint Service

The Checkpoint service (CKPT) is designed to allow applications to replicate their state outside their address space of their components such that their state can be preserved across failures. In order to facilitate the implementation of highly efficient replication mechanisms no strong ordering requirements are imposed on consistency and the outcomes of concurrent updates to the same area of a checkpoint by different processes. On the other hand ordering of writes by a single writer must be preserved and inconsistencies signaled to the applications.

Redundancy is achieved by having multiple copies of a checkpoint on different nodes. A copy of a checkpoint maintained by the service is called a *replica*. The propagation of updates between replicas is determined at creation time by the application. At any one time there is a single *active* replica to which the updates are being applied with the updates being propagated to the remaining replicas by CKPT. If an application has chosen synchronous replication, write operations block until all replicas have been updated. In the case of asynchronous replication, write operations return as soon as the active replica has been written to while the propagation of updates to the other replicas occurs asynchronously.

Interestingly, CKPT has no configuration data in the system information model: the management of checkpoints and their replicas across the nodes of the cluster is handled automatically between the service and its users. Applications can control the number and placement of replicas by creating a so-called collocated checkpoint. A replica is created on each node on which an application opens a previously created collocated checkpoint. Collocated checkpoints can only be used with asynchronous replication. In order to ensure the highest performance for writing checkpoints the service provides a mechanism whereby an application can request the active replica be collocated with it. In this way when the AMF assigns a job in the active role to a component the component can set its local replica active to ensure that the state information needing to be replicated for the job can be written with low overhead while the service asynchronously propagates the information to the other replicas which were opened by the components that were assigned the same job in the standby role. If subsequently the AMF were to change the state of one of the standby assignments to active, when the component accordingly requests its local replica to be active the service will ensure that the replica is up-to-date before completing the request.

Although CKPT does not have any configuration data it does expose runtime information and statistics about checkpoints and their replicas through the IMM.

Chapter 7 takes a deeper look at CKPT.

Event Service

The Event service (EVT) is intended to provide a cluster wide event distribution service whereby applications can publish information to anonymous subscribers on named cluster wide event channels. EVT is typically used by application designers wishing to distribute state information between applications in a loosely coupled scalable manner.

As with CKPT, EVT does not define any configuration in the system information model. Event channels are created programmatically by the applications. An application may create or open an event channel taking on the role of publisher, subscriber, or both at once. Events are in fact just messages with a common header allowing subscribers to request the service to only forward those events to them that match their specific filter criteria on the header.

The specification calls for best effort and at most once delivery of published events to subscribers. Publishers set a priority level for each event sent. In the case of overflow due to resource limitations lower priority events may be lost. Should events be lost the service informs affected subscribers of the fact. Each subscriber sees events in published order within a given priority level and for a given publishing process. In other words the specification does not require strict global ordering with regard to multiple receivers and multiple senders.

In order to further reduce coupling between publishers and subscribers the service allows a retention time to be specified on events. This allows a subscriber that starts some time after an event has been published to still be able to receive it. EVT does expose runtime information about each event channel in the system information model. It includes information such as the number of subscribers, publishers, retained and lost events for each channel.

More details on the EVT can be found in Chapter 7.

Message Service

The Message service (MSG) was conceived to provide a flexible, cluster wide distributed message passing capability. The main use case is to facilitate the implementation of reliable scalable cluster internal client-server applications. Here typically the client in the cluster receives service requests from the external environment and distributes them to a set of servers. Additionally it was required that the service masks failures in a receiver from senders and to support load sharing between receivers.

In order to allow one receiver to take over from another receiver without the involvement of the sender the abstraction of a message queue or mailbox was introduced. A message queue acts as a buffer for messages sent to the queue where they are held until retrieved by the receiver. For example, assume we associate a single message queue, to which service requests are being sent, with a job that the AMF assigns to a pair of components. The job assignment for the first component is set to the active state and that of the other to standby. The first component, with the active state for the job, has opened the queue and is retrieving and processing the messages from the queue. In the meantime the second component simply waits for instructions from the AMF. Now assume that the first component suddenly crashes. When the AMF detects the failure of the first component it sets the state of the second component's job to active. The second component then opens the same queue and proceeds to receive and process the service requests from it. Any processes sending services requests to the queue will be unaware that there has been a failover. Any messages sent to the queue after the component crashed will have been buffered in the queue until received by the second component. Ensuring that sent messages are preserved in the face of random failures is a challenging task for MSG implementers.

The specification does not require that messages stored in message queues be preserved across node failures or cluster restarts. This allows message queues to be memory based for performance reasons. Sophisticated implementations may however provide protection against node failures without compromising performance. Various message sending models are specified: send and forget, notify sender on delivery to queue, and request-reply from the receiving process.

A message queue can have any number of processes sending messages to it but can only be opened by one process at a time for receiving messages. This avoids the complexity of having to implement a mechanism for notifying multiple processes for each message sent to the queue and providing ordering guarantees between them for receiving the messages.

In order to support load sharing the concept of message queue groups was defined. A message queue group has a name just like a single queue. Processes sending a message specify its destination by providing a name which can be either that of a single queue or that of a message queue group. The specification currently defines four load sharing policies for message queue groups of which round-robin is the only mandatory one. To optimize for latency an optional local round-robin policy is defined which behaves the same way as standard round-robin except that if there is a queue opened by a process that is co-located on the same node with the sender that queue will be preferred. A variant on this last policy is also defined which selects the local queue with the largest amount of available buffer space. These are all so called uni-cast policies in which a message sent to the group will only be delivered to a single member queue of the group.

The other policy is the broadcast or multi-cast policy which delivers messages that are sent to the group to each queue of the group that has sufficient room to buffer it. Again to simplify and ensure robust implementations atomic multicast semantics are not required.

MSG does not specify any configuration objects in the system information model but does expose detailed runtime information about queues and message queue groups.

Further discussion on the service is presented in Chapter 7.

Lock Service

The Lock service (LCK) specifies a mechanism to perform simple cluster wide distributed resource access coordination. The service defines an abstract entity called a lock resource which applications associate to actual resources in the system to which access by different components needs to be coordinated.

Both shared and exclusive locking of the lock resource are supported. Hierarchical or composite and recursive locks are not catered for at the time of writing. Any number of processes can have a resource locked in shared mode when no process has it locked exclusively. Only one process can have a resource locked exclusively at any one time. The principal use cases being multiple concurrent readers or a single process that can both read and write to the resource associated with the lock.

Locks held by a process are automatically released by the service if the process terminates or the node on which it is running leaves the cluster membership. The service specifies two optional features: dead-lock detection and orphaned locks. The orphaned locks feature allows the automatic releasing of locks by the service to be disabled in order to allow clean-up operations to be undertaken by other parts of the application. The application must explicitly purge the orphaned locks.

LCK maintains useful runtime information about lock resources in the system information model.

Naming Service

The Naming service (NAM) is intended to provide both node local and cluster wide name registry and look-up functions. Names are simple strings that can be bound to object references or other resource references which typically would include communication or service access point addresses.

The naming conventions used are a local matter for applications. In order to avoid namespace conflicts the service defines the notion of a context in which names are bound to references. Two contexts are defined by the service itself: a cluster wide context and a node local context. Applications can define their own contexts with cluster wide or node local scope. All contexts represent mutually disjoint namespaces irrespective of their scope. User created contexts with node local scope including the bindings they contain do not persist across departures of the local node from the cluster membership.

Apart from creating, looking-up, updating, and deleting name-reference bindings the specification also provides a mechanism for an application to monitor for binding creation, updates, and delete actions related to a name within a context. This can be used for an application to learn when a service becomes available or when it needs to rebind to a new service access point. For example, suppose an application needs to access some service called 'S1' on an interface that is determined dynamically at run time. The application will monitor for changes to the binding of the name 'S1' in the application's context by registering a call back function with NAM. Initially, when looking up the 'S1' it finds that no associated binding exists. When the interface is created and an address is assigned to it, a configuration management process binds the name 'S1' to an object reference representing the interface (possibly the distinguished name of a runtime object in the system information model). At this point NAM will invoke the call back function that was previously registered by the application to notify it of the new binding, obviating the need for the application to have to poll for changes or doing a subsequent lookup.

Another example is where a service endpoint is implemented as a socket interface with an Internet protocol (IP) address and port number as configuration parameters. When the job that provides service at that socket interface fails-over, the new active service endpoint may have a different IP address and port number. The process invoking services on the endpoint can be notified of the changed address by using the same monitoring technique as in the previous example. The component whose job end-point HA state is set to active updates the name binding. When this occurs, the service user is informed by NAM of the binding update and the service user connects to the new IP address and port number. This is in contrast to a failover of a service endpoint using a MSG queue instead of a socket address. Using the MSG queue avoids the need of the invoking process to be involved at all in the failover procedure.

NAM defines a configuration object class for the service defined default cluster wide and node local naming contexts in the system information model. This class allows the system administrator to set the configuration parameters of these contexts. It also exposes run-time information for them. A separate runtime object is exposed for each programmatically created naming context. The life-cycle of the configured default cluster wide and local naming contexts is coupled to the life-cycles of the cluster and the cluster nodes respectively. Applications, however, are responsible for removing any unused user created naming contexts and cleaning up all stale or unused bindings.

Timer Service

The Timer service (TMR) is a node local service specifies an operating system independent interface for common timer functions such as setting timers and receiving timer expiration notifications. TMR supports both single event (one-shot) and periodic (repetitive) timers. The expiry time can be specified as an absolute time or as a duration relative to the time at which the timer is started. Any timers started by a process are only visible and usable by that process. They are cancelled (i.e., destroyed) upon termination of the process.

Timers are a frequently used mechanism in HA applications to monitor for state changes or responses to service requests which need to occur within specific time bounds. Should the change or arrival of the response not occur within the time interval some recovery action such as retry or cleanup must be invoked. The specification does not require timers to survive process restarts or failures. A key design decision was to allow for implementations to support a very large number of timers per process. Very large numbers of operating system based system timers can generate considerable system overhead as it cannot make any assumptions about them. Implementations of the TMR must assume that the vast majority of timers will be cancelled before expiry requiring them to be extremely 'lightweight.' Timer values are specified with the standard time type which has a resolution of nanoseconds. However to promote portability and broad platform support for implementations the specification exceptionally imposes a nonfunctional requirement which is that the timer resolution should be no longer than 10 ms.

Unlike the other AIS utility services TMR is not a distributed service. It continues to provide service even on nodes that are removed from the cluster membership. One of the main reasons that TMR is not a distributed service is that it is complex and resource intensive to provide cluster wide scope for a large number of timer instances. Another is that it would require highly reliable fine grained time synchronization between the nodes in the cluster membership further exacerbating complexity and increasing the number of platform failure modes. The additional burden on applications to replicate timestamps of timer start times with their associated service requests and restarting them as part of application recovery action is relatively small. Note, however, that applications must also deal with the implications of time variations between different nodes on the cluster.

TMR has no administrative interface and exposes neither configuration data nor runtime information in the system information model.

3.3.3 Service Dependencies

As mentioned in Section 3.1.3.2 the different services were specified such that each interface is self contained with minimal coupling to other services. For example, the HPI and AIS specifications do not have any formal dependencies upon one another. In other words they can each be used separately or together. In the absence of an HPI implementation the PLM could be implemented directly on proprietary interfaces or another server system management standard such as Systems Management Architecture for Server Hardware (SMASH) [50]. A CLM implementation could even dispense with both HPI and PLM and simply use native operating system functions directly and perform the system network probing over the system interconnect to discover the set of operational nodes.

The SA architecture design does however impose two essential dependencies between the services: dependencies on CLM and dependencies on the AIS management services. In order to provide a coherent view of the scope of the operational system resources to applications all AIS distributed services depend on CLM. The two AIS services that do not depend on CLM are the PLM and TMR services. PLM was designed to facilitate the implementation of CLM. TMR is a node local only service as discussed in section 'Timer Service' on page 57 and has no dependencies but all other AIS service could use TMR.

The dependency of all AIS services except TMR on the AIS management services is intended to provide a common administrative view on all services. This includes management of configuration and runtime information, administrative interfaces, notification and logging of system events and changes.

In the preceding presentation of the various services we have already discussed some of the interactions they have with other services. Figure 3.6 depicts the ideal dependency relations between the various services as envisaged in the SA architecture from a service implementation perspective. These dependency relations are not normative and may vary in actual implementations.

Here we limit ourselves to summarize the ideal and assumed dependency relationships.

The PLM was designed to work with HPI and to provide the bridge between HPI and AIS. The relationships of AMF depending on CLM depending on PLM for availability management were illustrated in Section 3.3.2.1. Although there are no direct dependencies of other services on the AMF, service implementers can choose to model the service implementation as an AMF application. The SMF clearly relies on the AMF in order to manage its activation and de-activation units (for details please see Chapter 9). However, it does not directly interact with the AMF by calling the AMF APIs. It drives the AMF by applying changes to the AMF managed entities in the IMM and using the administration interface of the AMF.

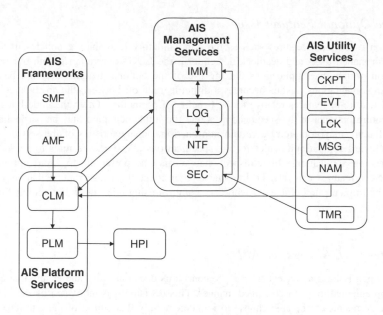

Figure 3.6 Ideal dependency relationships between services.

The AIS Frameworks, Platform services, and all the Utility services except for TMR depend on IMM and NTF for management and administration purposes where appropriate.

All distributed services except PLM depend on CLM which depends on the AIS Management services. Thus there is a mutual dependency between CLM and the Management services. Resolving the mutual dependency between CLM and IMM during system start-up requires special care on the part of the service implementers.

NTF ideally depends on LOG for persisting alarm and notification events and LOG depends on NTF for sending alarms and log stream life-cycle and change notifications. All services that send alarms or notifications indirectly depend on the LOG. For simplicity in the diagram of Figure 3.6 we thus group LOG and NTF together.

Note that IMM depends on NTF for sending miscellaneous notifications and NTF depends on IMM for its filter configuration settings. The use of the SEC for some of the AIS services has not yet been defined but it is planned for all services to use SEC. SEC for its part does not have a direct dependency on NTF as its service users are expected to send notifications and security alarms as part of their authorization procedures.

In addition implementations may use one or more of the AIS utility services in the implementation of other services. For example, TMR can be used in most other services. MSG can be used by the AIS management services and the other utility services.

3.4 Open Issues

Over the past decade much has transpired in the industry and much experience has been gained by the SA Forum members and the ecosystem of implementers and adopters of the specifications. In this section we will look at some of the open issues that have emerged over time and some additional services that still remain on the drawing board.

3.4.1   The Optional Features Issue

The definition of optional features such as the nonmandatory load-sharing policies in the MSG or support for orphaned locks and deadlock detection in the LOG is at variance with our primary goal of application portability. Applications making use of the optional features may not be portable to other AIS implementations. It thus becomes a difficult trade-off between the advantages realized by the optional feature or features relative to the benefits of portability. Often the underlying reason for making a feature optional in the specification is that the implementation may be particularly onerous or that the feature may compromise certain nonfunctional characteristics of the service. We should add that backward compatibility rules for the specifications allow new features to be added without affecting existing applications. In retrospect then it may be preferable to avoid optional features altogether. New features can then be introduced after it has been ascertained that they are useful and able to be implemented efficiently while not compromising other desirable nonfunctional service characteristics.

3.4.2   Integrated AIS Service API

The second open issue is a consequence of the conscious decision to try and ensure that each service could be implemented in a self-contained manner. This decision was motivated by the business need for software companies to be able choose to implement only that subset of the interfaces they were interested in and for which they had sufficient development resources without depending on services provided by other suppliers. As a result each service has its own independent life-cycle as seen from the application perspective. While the consistency of the life-cycle and programming model required by the architecture of the AIS services simplifies the use of many different services in an application, it still creates a substantial housekeeping burden on the application to maintain the separate handles and contexts of the different services. Thus there is the temptation for a project that will be using many of the services in its programs is to wrap the life-cycle, selection and dispatch routines of the different services into a single common proprietary library, which is then used across the project. Now that high-quality, open-source implementations of most of the services exist there is no longer a pressing need for independent services. Thus in retrospect it might have been better to define a single life-cycle and interaction API framework into which the services could have been plugged.

3.4.3   Common Low Level Communication Facility Interface

The third issue is also related to the integration of independent services. When integrating different AIS services, that have been implemented by separate independent suppliers into a single coherent system there is the inevitable issue that there will be duplicated low level functions. In particular all distributed services require some form of underlying cluster wide communication facility. Whereas some services can use the MSG in their implementation, others cannot. The duplicate implementations may have different behaviors when encountering a common fault affecting both facilities. Furthermore when isolating a node during fault handling, multiple communication facilities need to be notified and participate in the isolation. Of particular concern is the issue when the reachable set of nodes is different over the various facilities. In short the duplication of communication facilities entails increased configuration complexity as well as testing burden for the integrator. A common low-level communication abstraction interface specification with well-defined semantics for use by all service implementations would greatly facilitate the integration task.

3.4.4 Common Distributed Process Management Interface

Another missing common function needed by all service implementations is that of distributed process (or task) management. While it was a stated objective to remain operating system neutral it would nonetheless have been possible to define an abstract process management service that would support the instantiation, monitoring, and termination of processes in the cluster. Users of this service could include the AMF, service implementations not managed by the AMF and multi-process components.

3.4.5 System Trace Service

LOG is designed to maintain a record of system and application related events in support of the system operator and application service provider. Distributed systems are notoriously difficult to test and debug and there is a need to record runtime system implementation related information in support of the developer and system integrator. The difference in requirements between tracing and logging is sufficiently great to warrant a separate service. Firstly the nature of the information recorded should allow the identification of the functional area in the source code emitting the trace record. The service must be extremely efficient since it must support very high rates of trace message recording in live systems without affecting service levels. Furthermore it should support the selective enabling of global trace levels and tracing of specific functional areas without requiring specific explicit trace configuration data for each process in the system information model. Finally it requires relatively fine grained cluster-wide consistent time stamps to enable post hoc failure cause analysis across a set of node local trace logs. This service has not yet been specified.

3.4.6 Diagnostics Framework

Another service that has been on the drawing board for some time is the diagnostics framework. As part of the repair procedures following a failure it is often desirable to test the potentially still faulty or repaired component in situ without its administrative or operational state changes being taken into account by the AMF. Similarly it should be possible to schedule routine diagnostics on different parts of the system and have the results in a standard format to facilitate automated analysis.

3.4.7 Overload Control Framework

Although disastrous total system failures are relatively rare, when they do occur, most often the cause is inadequate overload control. While there are some discrete mechanisms defined such as the HA readiness state for service assignments, node, and service capacities in the AMF, log levels or message and event priorities that can be used to limit the use of resources when overload conditions occur there are as yet no consolidated runtime usage states and throttling mechanisms defined in the specifications to signal and control overload conditions. The runtime state might include usage states for cluster resources and component job assignments. Controls may take the form of service priorities and service request admission rate policies. Like security, one of the challenges with specifying overload detection, prevention, and control services is that they affect most of the existing specifications. Making the changes to accommodate overload control represents a considerable effort requiring great care to maintain backward compatibility. Service related resource usage information for an overload control framework could also be used to define additional load-sharing mechanisms in MSG. Adding this feature may be a first step toward a full overload control framework.

3.5   Conclusion

In this chapter we have presented the SA architecture with its constituent services. We began by giving some background on the specification creation process and presenting the architecture description approach. We then reviewed some basic HA concepts and outlined how they are applied in software based fault tolerant systems. In the presentation of the architecture and its services we examined the interactions between the services that illustrate how they work together to provide high levels of SA and continuity. Throughout we have tried to describe the design considerations and trade-offs that were made in selecting the functionality that was included in the specification. Designing and implementing reliable, flexible, and high performance distributed systems is a complex and difficult task. The choices of features and functionality in the specifications were strongly influenced by the many lessons learned in previous proprietary systems. It is hoped that this exposition will help the reader to obtain a better understanding of the specifications and to encourage the brave to tackle some of the remaining open issues.

4

The SA Forum Information Model: The Heart of Control and Monitoring

Maria Toeroe

Ericsson, Town of Mount Royal, Quebec, Canada

4.1   Introduction

This chapter introduces the Service Availability (SA) Forum information model, which was created to answer management needs and in particular that of the Availability Management Framework (AMF) to be discussed in Chapter 6. Subsequently the concepts of the AMF information model were applied to the management information of other Application Interface Specification (AIS) services; thus, the SA Forum information model was born.

The first question is why AMF needs an information model and what type of model it requires.

From Part I of the book one may see that SA management is all about coordination of the available resources so that at any moment in time there is at least one resource available in the system which is able to provide any given service that is required to be highly available. To pull off this attraction the availability manager needs to know or have a view of:

- the resources with their availability status and the services they are able provide;
- the services that the system needs to provide; and
- the relationship between these two sides.

What does all this mean when applied to some 'random' software application we want to be managed? To begin with someone needs to tell to our availability manager at least part of this information, that is, what the resources are and what services they should provide. They need to be given in a way that suits this 'random' software as well as the availability manager. In other words a system administrator needs to configure for the availability manager the application so it can interpret this information. Thus, we need an information model that includes at least this configuration information.

Service Availability: Principles and Practice, First Edition. Edited by Maria Toeroe and Francis Tam.
© 2012 John Wiley & Sons, Ltd. Published 2012 by John Wiley & Sons, Ltd.

The information needs to be provided in an abstract and uniform way applicable to the wide variety of applications that one may want to be managed by the availability manager.

The information model needs to provide enough information for the management functionality to fulfill its task. In our case for the availability management the information needs to be enough so that it can map the software resources executing in the system at runtime and their services into this model representation and use this model representation to control those software resources with their services.

Yet the information needs to be simple enough so that an administrator will actually be capable of providing it, that is, it will be able to compose as well as to interpret the information and find its way within the model.

Besides the configuration information the availability manager needs to maintain the availability status and the current relationship between the resources capable of providing the services and the to-be-provided services. This is a piece of information that the administrator may also be interested in. For him or her knowing this information may provide an explanation why some required services are not provided by the system at a particular moment in time or that the risk is high for such an outage.

The status information is typically runtime information that is collected from the system and provided to the administrator to facilitate the monitoring and to support the administrative control. The exposure of this information is not necessary for the management functionality (e.g., availability management) itself, but there could be parties interested in the information even beyond the system administration.

In any case we can say that in a typical system a management facility managing some physical resources – whether they be software or hardware resources – abstracts these managed resources into some logical entities that characterize the resources from the perspective of that management functionality.

In turn some or all of these logical entities can be reflected in an information model as managed objects that may expose the status of the represented resources from the perspective of the management functionality and/or allow the system administrator to configure and control the management functionality and its resources. The collection of these managed objects composes an information model, which serves as an interface between the system and the system administration.

Considering different management functionalities in a system where each of them exposes its own information model, it is possible that the combined system information model exposes the same physical resource through multiple managed objects, each representing a particular management aspect of that resource. For example, even within the availability management we have already distinguished the service provider aspect of a resource from the services it provides.

In this chapter we take a look at the main concepts used in SA Forum systems when it comes to management information. This allows newcomers to have an easier way to understand the application of these concepts to the different SA Forum services when they try to interpret their information models. In turn this information also provides guidelines for application designers wanting or needing to extend the SA Forum system information model for their applications.

4.2   Background

4.2.1   Management Models Out There

Many systems have faced in the past, and will face in the future, this problem of how to provide an adequate management view of the system; and there are numerous solutions addressing it. Some of these solutions are more suitable for the management functionality itself or a particular class of them; in other cases they favor the administration functions. It is almost always a trade-off between providing a comprehensive view and set of controls of the various system functions and resources on the one hand and a simple and secure view for the system administrator on the other. The solution is further flavored by the particular area of technology for which it was developed.

Among the best known existing solutions we need to mention the Internet Engineering Task Force (IETF) regulated Management Information Base (MIB), which is a 'virtual information store' – essentially a database of management information for a particular type of system. Initially the target system was the Internet, but it could be and was applied to other systems as well including the SA Forum system. To provide a management interface the SA Forum started out with defining MIBs for its services.

The initial version of MIB-I [51] quickly evolved into MIB-II [52], which has become ubiquitous on the Internet and it is used in conjunction with the Simple Network Management Protocol (SNMP) [53]. The managed objects of the MIB are defined using a subset of the Abstract Syntax Notation One (ASN.1) [54]. IETF specifies the used subset in the Structure of Management Information Version 2 (SMIv2) [55]. Multiple related objects are organized into tables, which together with the scalar objects (i.e., defining a single object) are organized into a tree hierarchy. From the management perspective, an object is a data structure that characterizes some resource (e.g., device, interface) in the system that an administrator would like to control. To this end it uses an SNMP 'agent,' which is capable of receiving the instructions of the administrator in reference to such an object, interpreting the data structure, and applying it to the resource represented by the object.

The SA Forum applied the MIB approach to the hardware platform interface for which it was well suited but it turned out to be less suited to the needs of the AMF [48] – the flagship specification of the SA Forum. In particular it was cumbersome to accommodate the dynamic nature of the relationships of the AMF model with the numerical enumeration of the naming hierarchy of the MIB. Similar conclusions lead IETF to develop the Network Configuration Protocol (NETCONF) [56] and its associated data modeling language YANG [57]. NETCONF provides mechanisms to install, manipulate, and delete the configuration of network devices, for which the configuration and state data are modeled in YANG. The primary target area of NETCONF and YANG – not surprisingly – remained the Internet as it was for the MIB and SNMP. YANG would have been more suitable for the SA Forum needs, but it came late. By the time it has been approved by IETF the SA Forum had its model defined using the Unified Modeling Language (UML) [59].

A competing and similarly recent standard defining management information is the Common Information Model (CIM) [58] standardized by the Distributed Management Task Force (DMTF) for the IT environments in general. It is an object-oriented approach to represent the management information in a vendor independent way for a wide variety of managed elements such as computer systems, operating systems, networks, middleware, devices, and so on. In the Core Model DMTF defines the concepts common for all targeted areas. Although extendable, this part is expected to stay stable over time. In addition for each of the target areas DMTF defines a still technology independent Common Model. The Core and Common Models provide a starting point to analyze and describe any managed system. The Common Models are extendable in Extension Schemas to capture the different technology specifics.

DMTF publishes the models as schemas. These are supplemented by specifications that define the infrastructure, interchange format, and compliance requirements. CIM is a UML-based [59] technology.

The initial attempts to derive the information models for the various SA Forum AIS services (and AMF in particular) from the CIM led to overly complex object hierarchies without providing an intuitive mapping onto SA Forum defined entities and relationships.

4.2.2 The SA Forum Needs

At the same time as the IETF and DMTF put efforts to define YANG and CIM respectively, the SA Forum has also been working on an appropriate representation of the management data of SA Forum compliant systems. The result falls somewhere between these two solutions. Let us examine why.

First of all there is a necessity of system configuration. For example, the availability management function needs to know all services it needs to maintain and all the resources it can use for this purpose. In a complex system – and SA Forum compliant systems are complex – this means a detailed representation showing each component with its services and state. While such a detailed representation is necessary for the availability management functions it is overwhelming for an administrator. Therefore there is a need for an organization that simplifies this view such as higher level aggregation and abstraction even when there is no physical manifestation of such compound entities in the system or if the entities are quite different from the function manager's perspective, but they are similar for the administrator.

High-availability systems are relatively autonomous in the sense that once they have been configured and deployed they operate 24/7 without continuous administrative control as the systems themselves implement mechanisms to cope with emergency situations within the defined limits. (We will see this in details in the discussion of the AMF [48] in Chapter 6.) This also means that these systems are able to handle many other workload related (e.g., increase of traffic) issues and changes dynamically.

In particular the SA Forum utility services allow the creation and deletion of their service entities dynamically through application programming interface (API) calls. The originator of these operations may or may not be part the system.

For example, new checkpoints may be created in response to increased traffic, increased number of open sessions toward the system or due to a new application that has been added to the system configuration. In either case the checkpoint service (CKPT) creates these checkpoints at runtime in response to the requests it receives from its user processes via the service API and not through configuration.

The information model needs to accommodate this feature. More specifically, the creation and deletion of these utility service resources at runtime via the service API is very similar to the creation and removal of configuration resources by an administrator from the perspective of handling; that is, the life-cycle of such an entity is controlled by the user and its representation should remain in the model until the API user requests the removal of the entity and therefore its representation. On the one hand, they have a similar 'prescriptive power' as the configuration has. On the other hand, they need to be distinguished from the configuration information as they are created by the service management functionality on behalf of the user and they should not be configured by the administrator.

Finally the information model also needs to be able to satisfy the needs of system monitoring and system discovery. System administrators need to be able to find out the system state, easily find their way around in the model and interpret the information even when they know very little about the functionality a particular application. In other words, in spite of the wide range of application functionality deployed on such systems and the variety of configurations, the model needs to express clearly the system's organization and state.

This requirement is related again to organizational aspects of the model, but also the differentiation of the information depending on its source.

There could be different solutions for these requirements depending on the preferences or the target behavior. The primary goal was to provide a management interface to external management systems to manage and configure a system based on the SA Forum specifications and a number of suitable open standard solutions could have been used. It was the need to provide runtime management access to the service implementations themselves that drove the decision to specify the interfaces to the UML model based on an adapted Lightweight Directory Access Protocol (LDAP) [60] model. However the administrative interface exposed by the resulting Information Model Management service (IMM) [38] is well adapted to be exploited by external management systems using agents based on the existing open management standards. We describe the SA Forum definition of the different object class categories catering to these two requirements further in this chapter.

Satisfying the above requirements were the primary drivers in the definition of the SA Forum information model while also drawing on the concepts and techniques and therefore aligning it with existing standards and developmental solutions such the MIB [52] and LDAP [60] from IETF, CIM [58] from DMTF and UML [59] from Object Management Group (OMG).

4.3 The SA Forum Information Model

4.3.1 Overview of the SA Forum Solution

4.3.1.1 The Managed Object Concept

The SA Forum system consists of many different *resources* – software and hardware alike [61] (Figure 4.1). These include the hardware nodes composing the system platform that run different operating system instances within or without virtual machines providing execution environments for different processes in the system. The processes may implement some SA Forum defined system functionality or some application functionality the availability or other aspects of which are still managed by the SA Forum services.

For the purpose of the management of these different resources an SA Forum service defines some logical concepts that abstract the aspect of the resources, which is managed by the given service. We refer to these logical concepts as *SA Forum entities*. The service semantics and functionality are defined for these logical entities. For example, the Platform Management service (PLM) abstracts most of the details of any hardware resource and represents all of them as hardware elements. Similarly, operating systems, virtual machines, and virtual machine monitors are all summarized as execution environments. Then the PLM specification describes the operation, the managed states, and other semantics of the PLM in terms of these two logical entities, that is, the hardware element and the execution environment.

In addition, a service may also define logical entities that cannot be mapped directly to any single resource present in the system, but which reflects some organizational aspect of those entities that do have a physical manifestation. Typical examples would be the AMF's service unit and service group concepts. An AMF component may manifest in the system as a process managed by AMF. The service unit, however, is only a grouping of such components and known only to AMF. If we look at the reasons why it was defined in AMF we see that one of the reasons behind the service unit is the fact that tightly collaborating components tend to fail together; it is the reflection of this coupling.

There was however a second reason behind the definition of the service unit. It simplifies the management view for administrators by adding a level of hierarchy and as a result the administrator

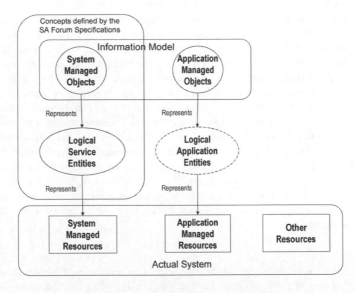

Figure 4.1 The SA Forum managed object concept. (Based on [61].)

does not need to deal with individual components, but can control them simultaneously as a service unit. Yet the option of observing each individual component still remains possible whenever it is needed.

This brings us to our subject, the SA Forum information model: these different logical entities defined in the different SA Forum services are presented to the system administration in the form of *managed objects* of the SA Forum information model (Figure 4.1). For each of the SA Forum services that exposes any of its logical entities to the system administration the classes of the managed objects are defined in the service specification using UML class diagrams. These diagrams define:

- the class name and category;
- the different attributes with their types, categories, multiplicities, and if applicable IMM [38] representation; and finally
- the administrative operations applicable to objects of the class.

Objects of these classes represent instances of the respective logical entities that expose the status of these logical entities in the system as appropriate and also allow the administrator to control these entities and through them the behavior of the represented logical entities and their physical resources as appropriate. Thus, the information model provides a management view of the SA Forum system.

The SA Forum IMM [38] was specified with the purpose of maintaining and managing the SA Forum information model. As part of this it provides the management interface to the SA Forum system as well as an interface for the services and applications that handle the logical (service) entities and the resources they represent. We will take a closer look at the details of IMM in Chapter 8.

4.3.1.2   Object and Object Attribute Categories

In Section 4.2.2 we pointed out that in an SA Forum system a service entity may manifest as a result of an administrator configuring it or a service user requesting the service through the API to create it.

Configuration Objects

The administrator may configure the service to instantiate the entity regardless of whether it has an actual physical manifestation or only the logical service entity exists defining some service aspects for the service implementation itself. In this case, the administrator configures the managed object in the information model via the management interface exposed by IMM. In turn IMM informs the service implementation responsible for the particular portion of the SA Forum information model so that it can deploy the configuration changes intended by the administrator. Because of this configuration aspect these objects and their classes belong to the configuration object class category.

Configuration objects are prescriptive in nature, that is, they express the intention of the system administration about the organization and the behavior of the system. The physical resources associated with such objects are controlled and manipulated by the service implementation to which the objects belong so that their status and behavior matches that of the intention of the administrator.

For example, the information model of the already mentioned AMF is primarily composed of configuration objects. Since AMF manages the life-cycle of the application components it controls, it requires first the configuration of the application: The components it is built from, their organization and the different policies applicable to them that guide among others the recoveries should an error occur. An AMF implementation cannot figure out all this information by itself; it cannot know the intention of the administrator.

Runtime Objects

Alternatively, service entities may be created by the SA Forum service itself as a result of user requests received via the service API. For example, the application configured for AMF may use checkpoints defined in the SA Forum CKPT to synchronize the execution state between the component actively providing a service and the one standing by to protect the service against any failure. In this case it is the application, the processes represented by the AMF components that instruct the CKPT to create, read, write, and remove the checkpoint as necessary. The CKPT handles in the system the physical representations, the replicas of such a checkpoint according to the requests received through its API. It also exposes the checkpoint and its status in the information model. Again the service functionality is defined in terms of logical service entities, which then can be implemented in different ways, for example, some database records, or files depending on the given service implementation.

The important part is that the user processes refer to the logical service entity (e.g., checkpoint) and the operations the service needs to perform to fulfill their needs via the API. The system administration is not involved, yet, these entities need to be present as long as the user process requires it. So they also have some prescriptive nature toward the service implementing them. Nevertheless from the perspective of the information flow its direction is opposite to that of the configuration objects. It is the service implementation which inserts the managed object into the information model in response to the API request creating the logical entity. The system administrator is primarily an observer of the life-cycle of these objects that as a result reflect the runtime status of the system and of the particular service implementing these objects. Accordingly these classes of objects are called runtime objects.

Object Life-Cycle

Considering the life-cycle of objects of these categories, configuration objects are created and deleted by an administrator. In case of runtime objects, they are created as a consequence of a service user request. Depending on the service their removal may be explicitly the result of an API request, but more often since they may be accessed by several service user processes simultaneously they are removed as the consequence of inactivity. The definition of inactivity varies from service to service, it may be declared after the expiration of some period without any process accessing the associated service entity; immediately after the last user process indicates that it stopped using the entity; or when such situations imply, for example, due to a process failure.

Since we are dealing with fault tolerant systems we need to consider different error situations: The service user process may fail, the service implementation process may fail and/or the IMM implementation process may fail.

The assumption is that if the service implementation process (including IMM) fails then there is a standby process, which will take over in providing the service, that is, the service as a whole does not fail. This means that even though a user session with the initial service process may terminate the entities created as a result of that session remain in the system and so do the runtime objects representing them.

If the user process fails then any session it has opened toward a given service is implicitly terminated. The SA Forum services should detect the termination of the application processes even if it was abrupt. The service entities created through this session, however, remain in the system as long as they are active; that is, there are still users that may access them or their timer has not expired.

Considering a checkpoint, it will remain in the system in all these cases and its representation will remain in the information model as well.

The story changes when we consider termination or restart operations.

For example, if the application is restarted whether the service entities created by the application and the runtime objects representing them remain in the system depends on different factors: If the restart is due to a failure the application components are terminated abruptly, so the same applies as in case of the process failure. If it is a graceful termination then depending on the service the application processes may clean up their logical entities, which results in the cleanup of the information model. For example, the application processes have the possibility to close the checkpoints they use and that means that all the associated resources will be freed by the CKPT. It will also remove the associated checkpoint objects from the information model.

If the IMM is terminated or restarted, it has no obligation to preserve and restore runtime objects in the information model. It is required, however, to persist and restore configuration objects. Once IMM becomes available again it is the responsibility of the various SA Forum services to restore their runtime objects in the information model. Picking up our example of the checkpoints, IMM will not restore the runtime objects representing the checkpoints in the information model, but the CKPT will keep and maintain the checkpoint replicas themselves regardless IMM. It will recreate the runtime objects in the model when the IMM becomes available again.

On the other hand, if the service implementation is terminated or restarted this implies that the operation also terminates all its logical service entities. That is, the checkpoint replicas go away together with the CKPT since it cannot maintain them any more. But the IMM – since it is up and running – continues to maintain all the runtime objects representing them (e.g., the checkpoint objects) unless they are explicitly removed by the quitting service implementation, which it may only do if it was terminated gracefully.

If the entire cluster is terminated or restarted all the logical service entities and their model representations are removed. But again the IMM is required to preserve and restore configuration objects. That is, all checkpoint replicas in the system and the objects representing them in the information model go away, but the objects representing the AMF components configured by the administrator will be kept by IMM and restored in the model as soon as IMM is running. Using this information AMF can start the appropriate application processes in the system, which in turn will need to open their checkpoints again triggering the recreation of the associated runtime objects.

In-between: Persistent Runtime Objects

This difference in the handling of runtime and configuration objects may not be appropriate in all the cases since as we pointed out runtime objects do have a prescriptive nature even if it is less emphasized than for configuration objects. To close the gap and also enable the preservation of runtime objects the category of persistent runtime objects was introduced in the SA Forum information model. That is, the IMM is required to persist and restore this object category in a similar manner to what it does for configuration objects even though they are runtime objects. From our discussion it also follows that for the service entities represented by this object category the service needs to define an API for explicit life-cycle handling. Note that none of the defined SA Forum services currently uses this object category, but it could be used by an application that wishes to extend the SA Forum information model with its own classes. We could assume an application, which enhances the Checkpoint by declaring its checkpoint object class as persistent runtime. In this case the objects representing the checkpoint will remain in the information model whether IMM or the entire cluster is restarted so that this application may use this information to restore the checkpoint replicas in the system accordingly.

One may ask the question why not just use configuration objects instead? This would require that the application process accesses the information model the same way management processes and the administration do, which may not be desirable. First of all it complicates the application process as now it needs to use an additional service and not a simple one for that matter: configuration changes are heavy operations because they may have wide impact; they also need to be validated at runtime as they may have been initiated by a human operator, who may not be aware of all the consequences.

The alternative of the service (e.g., CKPT) inserting the configuration object is not viable due to the information flow, as the same service initiating the change is the receiver of this model change which creates a potential deadlock.

Attribute Categories

Specific individual object attributes of a configuration object may in fact be assigned to the runtime category. This was done to avoid having to model a single logical entity with separate configuration and runtime classes. Accordingly IMM preserves the values of configuration and persistent runtime attributes, while it takes no such responsibility for runtime attributes.

Configuration objects may have attributes of any of these categories. In their case the runtime attributes may serve as a feedback mechanism through which the service or application implementing the logical and physical entity represented by the object can inform the administrator to what extent it was able to align the service entity with the requirements of the configuration attributes. That is, these runtime attributes reflect entity states.

Runtime object classes, on the other hand, may only have runtime attributes. If any of the runtime attributes is declared as persistent that makes the object class itself persistent as well. To be able to preserve the object with its persistent runtime attribute the name attribute of the object also needs to be declared persistent.

This leads us to the next topic we need to discuss about the SA Forum information model: its organization and naming conventions.

4.3.1.3   LDAP DNs and RDNs versus UML

The SA Forum adopted the basic naming conventions from the LDAP [60].

LDAP organizes the directory objects into a tree structure and the objects' name reflects this structure. Namely, an object which is the child of another object in the tree will have a name which distinguishes it from any other child object of this same parent object – it will have a Relative Distinguished Name or (RDN). Since this is true for the parent object and its ancestors as well, we can generate a unique name for any object within the tree just by concatenating the RDNs of the objects we need to go through to reach the root of the tree. Such a name is called the Distinguished Name, or the (DN) of the object.

The root of the tree has no name. The specificity of LDAP and therefore the SA Forum DNs is that the DN of a child object is constructed from the child object's RDN followed by the parent object's DN. So the RDN works exactly as the first name: it distinguishes the kids in the family indicated by the last name. Unfortunately people's full name is not generated through the ancestry any more, so it does not provide the uniqueness at this point in time; however, we are pretty sure the idea was the same – at least for a given locality. In comparison addresses still work in this manner in many countries, that is, the addressee's name is followed by the house number, then the street, the municipality and if applicable by the province or state, and finally the country. The same logic applies to the generation of the DN of an object in the LDAP tree: it leads from the object to the root in case of left-to-right reading. To find the object from the root, the DN needs to be read right-to-left. In this direction it can be interpreted as a containment hierarchy and this is often the implied interpretation used in the SA Forum model.

The SA Forum also adopted the method of typing of relative names, but it did so with a slight twist. LDAP defines an RDN as an `attribute_type=value` pair, where the `attribute_type` is a standard naming attribute and the `value` is the relative name of the part. SA Forum managed objects are named in a similar manner. The difference is that the naming attribute does not have a fixed naming type but the type can be determined by the object class. In this way structurally important attributes can be used as the link in the naming (containment) hierarchy without imposing a type on them.

The different SA Forum specifications when they define an object class the name of the first attribute of this class defines the `attribute_type` portion of the RDN for the objects of the class and it is identified as the RDN attribute. The value of this attribute for each given object will provide the `value` portion of the object's RDN.

For example, the AMF specification defines the configuration classes `SaAmfApplication`, `SaAmfSG`, and `SaAmfSI` to represent respectively applications, service groups, and service instances in the information model for which we only show the RDN attributes in Figure 4.2. They have their RDN attributes named respectively: `safApp`, `safSg`, and `safSi`. Note that these naming attributes

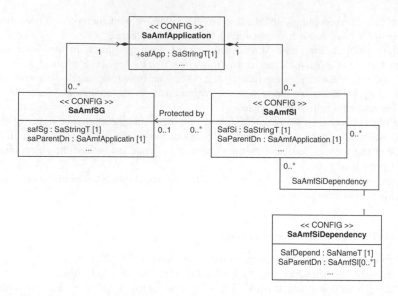

Figure 4.2 Example of the RDN attribute definitions in AMF configuration object classes.

all start with the 'saf' prefix as opposed to all other attributes of the SA Forum defined classes that start with 'sa'.

The <CONFIG> stereotype in the diagram is used to indicate that the object class belongs to the configuration object class category. Using these classes we can define an object model of an application myApp providing two service instances myService1 and myService2, each of which is protected by a respective service group which could also be named myService1 and myService2 even though they are also children of this same myApp application object as the actual providers of those service instance. The reason is visible from our Figure 4.3, which shows the RDN attribute with its value for each of these objects as well as in the saParentDn attribute the DN of the object which is the parent of the given object in the SA Forum information model.

Accordingly the objects mentioned so far have the following DNs:

- 'safApp=myApp' – the object representing the application;
- 'safSi=myService1,safApp=myApp' – the object representing one of the service instances;
- 'safSi=myService2,safApp=myApp' – the object representing the other service instance;
- 'safSg=myService1,safApp=myApp' – the object representing the service group providing the first service instances;
- 'safSg=myService2,safApp=myApp' – the object representing the service group providing the second service instances.

We can see that the RDN attribute name used as the RDN tag distinguishes the service groups from the service units and therefore make their DNs unique.

As we mentioned earlier the SA Forum specifications define the object classes using UML diagrams. The complete model is also published as an XMI file [62].

Both figures are UML diagrams: Figure 4.2 is a class diagram as it is used in some specifications, while Figure 4.3 is an object or instance diagram one would use to represent the information model instance describing the SA Forum system.

On these UML diagrams we also specify relations between classes and respectively their object instances such as the already mentioned parent child relation reflected by the DNs. In some cases the

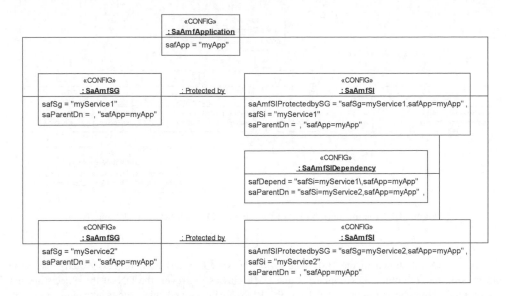

Figure 4.3   Objects of some AMF configuration object classes with all their links.

relationship between some object classes is expressed through a UML association class. This is the case for the SaAmfSIDependency class in Figure 4.2, which implies appropriate objects in the object diagram and also in the SA Forum information model. Unfortunately these relations do not readily convert into a tree organization required by the LDAP naming conventions presented so far. The solution used in the SA Forum specifications is the following:

One of the object classes in such an association is selected as the parent object class for the association class objects in the information tree. Since usually such associations have a direction it is the source object class, which is selected for the parent role. Among the two objects that participate in such a relationship the object of this class will provide the parent DN for the DN of the association class object. To link the association class object with the second, the target object of the association relation, the DN of this second object is used as RDN value of the association class object.

In our example of Figure 4.3, we may have a dependency between the two service instances indicating that the second service instance 'safSi=myService2,safApp=myApp' cannot be provided unless 'safSi= myService1,safApp=myApp' is provided. This we could show in UML as a simple directed association. However, since there could be a short period while 'safSi=myService2,safApp=myApp' can tolerate the outage of 'safSi=myService1,safApp=myApp' we need to define this period in an association class object of the SaAmfSIDependency object class of Figure 4.2. This object in Figure 4.3 will then have the DN: 'safDepend=safSi=myService1\,safApp=myApp,safSi=myService2, safApp=myApp'. Note that the 'safSi= myService1\,safApp=myApp' is an RDN value and in this situation we need to use an escape character ('\') so that the following comma is not processed as an RDN delimiter. As this DN implies and the saParentDn attribute shows in the SA Forum information model, this association class object is the child of the 'safSi=myService2,safApp=myApp' object representing the second, the dependent service instance, which references through the RDN attribute value the first service instance on which it depends and which is represented by the object 'safSi=myService1,safApp=myApp'.

Figure 4.4 shows again the same AMF objects organized based on their DNs into a tree as it is used in the SA Forum information model. Note that the saParentDn attribute in the UML model

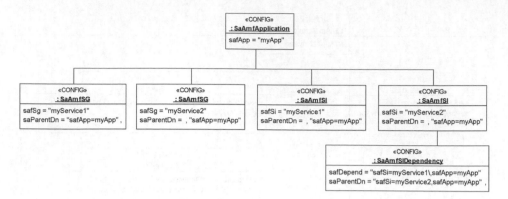

Figure 4.4   The tree organization of the sample AMF information model objects.

is private and not shown in the UML classes of the specifications as it reflects exactly the same information that the object DN implies. In the UML classes, however, the DN of the objects are not represented. They can be constructed from the RDN attributes and this private attribute (which in UML terms is the association end of the association representing the parent–child relation).

Other relations are expressed explicitly through attributes (or association ends in UML terms). For example, the `saAmfSIProtectedbySG` attribute, which indicates the name of the service group protecting the service instance.

But – and there is a big but here – the early versions of the model would choose a descriptive name for these attributes and reference simply the `SaNameT` as the type of the attribute. This approach did not reflect the class at the other end of this association. Essentially the `SaNameT` type works like a generic pointer that can point to any class it can hold, the DN of any object. Since it is important to see what classes are associated through which relation therefore the newer model will indicate the class of the object at the other end for these associations whenever it is known. The `SaNameT` type is used when the class is not known and the associated object may not even be present in the information model.

One of the consequences of this change is that now the `saParentDn` attribute (by having the appropriate object class as the attribute type) also reflects the way the information model tree can be built, which was only implied before. Previously the DN formats guiding these relations were only defined in the specification as a table.

The DN of an object cannot be changed without destroying the object. In other words both the RDN and the `saParentDn` attributes are read only.

4.3.1.4   Consideration for Designing an Information Model

Now that we understood the tree structure of the information model, we may want to define information models for our applications. As mentioned earlier the different services define the object classes they use for their portion of the information model. In addition to the classes the specifications also define the DN format. For some classes this definition is very restrictive, for others it allows for different organizations.

For the AMF classes we used in our example the specification defines that objects representing applications should be at the root of the SA Forum information model, that is, their DN has the format of `'safApp=...'`. Service groups have the definition of `'safSg=...,safApp=...'` and service instances the `'safSi=...,safApp=...'` indicating that they need to be children of an application object, and so on. In these cases the information model also has the appropriate class indication for saParentDN attribute.

As opposed to this the CKPT [42] allows the placement of a checkpoint object anywhere in the SA Forum information model based on the DN format: 'safCkpt=...,*'. It is also a runtime object class, which means that it is created by the CKPT upon the request of a user. In fact the user provides the DN of the checkpoint object when it creates the checkpoint and the CKPT uses this information to fit the object into the model. This assumes that the software developer has a very good idea of the organization of the related portion of the information model at the development time.

To this end, it is important to consider the object's and its ancestors' life-cycle and their category. According to the specification defining the C programming model [63] and the IMM specification [38] no object should parent an object of stronger persistency. However the precedence of object categories with respect to persistency is not defined nor is it completely relevant to the life-cycle of the objects.

The idea is that objects should not be removed from the information model while they still have children. If only configuration objects are involved, this becomes only a matter of ordering as one can remove the object starting from the leaf-objects proceeding toward the root.

The story is different with runtime objects as they are ideally controlled only by the service or application that created them. For successful creation of such an object it is necessary that all its expected ancestry already exists: There is a path in the information model matching the object's DN given by the service user.

Appending an object to an existing tree, however, locks in that branch of the tree, as in this case it is not allowed to delete an ancestor before all its children have been deleted. These two aspects imply an ordering at the creation as well as at the deletion of these objects. This ordering is implied by the information model, its organization and not by the services owning the relevant portions of the information model. As we have mentioned the service entities may explicitly be deleted by an API user or cease to exist due to inactivity making the control of this ordering at the time of deletion virtually impossible in many cases.

The restart of IMM or the cluster – which are both extremely drastic measures in this context – will purge all the 'simple' runtime objects from the information model, but preserve the persistent runtime objects.

Considering persistent runtime objects the IMM is obliged to maintain them even across cluster restarts. One cannot expect IMM to figure out when such objects become obsolete in the information model. They can only be removed by the service user explicitly. We emphasize this as none of the SA Forum services defines persistent runtime objects and this category is expected to be used by applications mainly.

After restart IMM may restore an object only if all ancestor objects are also persistent (belong to a persistent runtime or a configuration object class). That is, if a class is runtime class at least its RDN attribute needs to be defined as persistent.

For all the classes in the ancestry of a persistent runtime object, the user also needs to be able to indicate the removal of the entities these object classes represent, at which time the application needs to remove the object.

This also means that, since there is no other way to remove these objects but wait for their owner to do so, an IMM implementation needs to be able to resolve the case if the owner application of such a preserved persistent runtime object never comes back after it crashed for example. There is a need for some type of garbage collection.

We assume that if the application is removed from the system it cleans up its part of the information model, but the more open the system for third party applications the less assumptions one may make. Or, conversely, the system needs to protect itself from faulty implementations, which may behave in an unexpected way and the integrity of the information model is essential in this respect.

In any case, we see that the organization of the information model requires careful consideration in cases where the specification provides different levels of freedom.

4.3.1.5 More Conventions

In this section we go into further details on how to read the information model to provide some clarification on the specifications. These details are not necessary for the general understanding of the information model and may be overwhelming so if that is the case we suggest to skip to Section 4.3.2 and return to this section later as it may become necessary.

Issues with Inheritance

Probably the biggest surprise to a seasoned UML user when looking at the SA Forum information model specification is still the fact that it barely uses inheritance and when it does so it may be the reflection of the semantics rather than the class inheritance of object-oriented technology.

Nevertheless classic object-oriented inheritance is used in the definitions of relations. For example, in the PLM the abstract class PLM entity is defined which is specialized in two concrete classes for hardware elements and execution environments. The reason for the abstract class is the dependency relationship that can be defined within and among the two concrete classes. In all these cases the parent class is an abstract class and therefore the system information model will never include an object of this class.

Inheritance is indicated in other cases too, but they do not act in the classic way as in case of PLM. To clarify this statement let us consider the class defining the component base type of the AMF, which is specialized as component type class. The only attribute defined in the component base type is the name, that is, the RDN attribute. However the component type class has its own RDN attribute as required by the LDAP roots of the model. So in the way the object-oriented technology uses inheritance there is no inheritance here. On the other hand, the semantics of the component base type is that it collects the component types that have common features that are different versions of the same application software, for example. So semantically the component base type is the generalization of the component type. The relation between these classes does represent inheritance at this different angle of abstraction.

There was another relation where inheritance could be considered, and that was the relation between entities and their types, for example, the components of this component type is question.

The role of the component type is to provide the default values for the attributes of its components. It can do so because the type object semantically represents an implementation, the software; while the entity object represents an instantiation of this implementation such as a process running the software.

In the information model there will be at least two objects present: one for the type and one for the entity of the type, which will reference the type object. In these relations there is always only one type object, which provides a single point of control of the configuration for the one or more entities represented by the objects referencing this type object. That is, in this relation the type locks in not only the set of attributes, but also the values for the objects of the entity class.

Instead of inheritance the relation was defined as a realization reflecting that the type object provides the specification, the blueprint for the entity objects. We will look further in section 'Multiplicities and Defaults' at the use of this relation as single point of control. Before doing so we need to explain first the naming conventions.

With respect to inheritance the conclusion is that the SA Forum information model does not use inheritance for other than documentation purposes, therefore IMM does not need to deal with inheritance per se.

Naming Conventions

We have mentioned that the RDN attributes defined by the SA Forum all start with the 'saf' prefix. This is the only attribute in an object class that has such a prefix. All other attributes have the 'sa' prefix followed by an <area> tag that identifies the service specification to which the attribute belongs. If the service defines more than one class then typically the next tag identifies the service entity that the object class represents. For example, in the `saAmfSIProtectedbySG` attribute seen in Figure 4.3 `sa` indicates that this is an attribute defined by the SA Forum. `Amf` indicates the AMF specification as the area and `SI` indicates that the service entity, an aspect of which is characterized by the attribute, is the service instance. The rest of the name describes this aspect, that is, in this particular case this attribute links together the service instance (SI) with the service group (SG) that protects it.

The same approach is used in naming the object classes, except that they use the `Sa` prefix. Accordingly the class defined for AMF service instances is named `SaAmfSI`.

Well-Known Names

In addition to these naming conventions of the classes and their attributes the programming model specification [63] also defines some well-known names for objects representing the different SA Forum services. Each SA Forum service is represented in the SA Forum information model by an AMF application object with a standard name. This facilitates the exposure of the status and the administrative control of the middleware services in a similar way as it is done for the applications running in the cluster. A middleware implementation may follow through completely this alignment or remain at the level described here. That is, it may only provide the application level well-known objects so that users can identify from the information model the middleware services deployed in the platform and their versions.

These names follow the format `safApp=saf<area>Service`, where the `<area>` is the standard three or four letter abbreviation of the particular SA Forum service, for example, the object representing the AMF implementation in the system has the DN `safApp=safAmfService`. Through the use of a colon symbol the DN may include the particular implementation name, that is, `safApp=safAmfService:OpenSAF` for the implementation of the OpenSAF Foundation [64].

Since one may also be interested in finding out the version of the specification, the implementation is compliant to as well as the implementation version itself. For this the already mentioned type concept is used. That is, the specification [63] also defines the application base type and the application version type objects that the above mentioned application object should reference. The application base type has the same RDN value as the application object, but with a different RDN tag `safAppType=saf<Area>Service[:<vendImplRef>]`. The application type object indicates the specification release and optionally the implementation version. `safVersion=<specRel>[:<vendVersion>]`.

Accordingly the object representing the mentioned OpenSAF implementation may reference in its type attribute an application type object with the DN of `safVersion=B.04.01:4.0, safAppType=safAmfService:OpenSAF`. In particular the version B.04.01 indicates the specification release which defines the classes used in the information model in order to allow management applications correctly handle the content of the information model. No lower level granularity (e.g., class level) of versioning is available in the information model.

Multiplicities and Defaults

Another notoriously questioned issue is the multiplicity used in the specifications. To understand the problem we need to discuss a little bit of the background.

As discussed, one of the goals of the information model is to provide a management interface, for example, for human operators. To simplify the information model the specifications define single classes for entities that may exist in the system in a number of varieties. These classes represent the basic concepts of these logical entities important for an administrator rather than all the sophistication required for the management functionality implementing them. For example, the basic concept of the mentioned AMF service group is the protection of service instances and this is what is important for an operator at a high abstraction level he or she deals with; and who may not know about all the

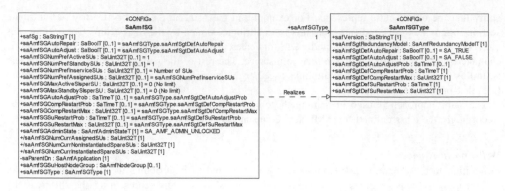

Figure 4.5 The service group and the service group type configuration object classes.

application details regarding different redundancy models and their implied attributes. (Just like we are ignorant at this point of all those details – we will get into them soon enough.) This allows for this higher level system view and an application (functionality) independent navigation within the information model.

This intention is opposed by the original purpose of the information model to provide the configuration information for the AMF, which does require all the low level details of the configuration as well, such as the redundancy model of the service group and all the appropriate attributes. The problem in this case is that each redundancy model has its own set of configuration attributes. This nonapplicability cannot be reflected in a single UML class, while specialization (e.g., defining different classes for each redundancy model) would have complicated the administrative view which was a definite no-no in this context.

The resolution – the trade-off – between these opposing tendencies was made by creating the object class attributes for representing the service groups as the union of those different sets of attributes required by the different redundancy models. The result is shown in Figure 4.5. In the UML object class the attributes that may be absent for some redundancy models are defined as optional attributes having the multiplicity [0..1], while those that are required for all service group objects were defined as mandatory with multiplicity [1].

The twist is that the specification defines default values for most of the attributes regardless whether they are mandatory or optional and these default values and their applicability may depend on the actual redundancy model. For example the saAmfSGNumPrefStandbySUs attribute is only needed for the N+M redundancy model while it is applicable to the 2N as well. It is not applicable for the other redundancy models even though it has a default value. This default is defined so it satisfies the redundancy models the attribute is applicable.

Unfortunately this may create some confusion as optional attributes are also used with defaults (e.g., saAmfSgtDefAutoRepair with the default SA_TRUE) to indicate that the value does not need to be provided at the creation of an object of the class as in case of absence it will take the default value.

The possibility of not providing some values was perceived as simplification of the configuration process and therefore embraced.

The approach presented so far was applied in general at the definition of all object classes of the information model. That is, only one object class was defined for each service entity category. If service entities of a given category needed differently sets of attributes depending on their subcategories, then rather than defining different object classes the union of the attributes of the different the subcategories was used to form the object class that captures all the possibilities. In this class if needed additional attributes were added to reflect the subcategories.

The attributes required only in some subcategories were defined optional while those required in all subcategories were defined as mandatory.

To reduce the configuration information whenever it was possible default values were defined for configuration attributes and the attributes themselves were made optional. So whether a configuration attribute is optional or not reflects whether the attribute value needs to be provided at object creation and not its applicability. If it is not provided then the defined default may be used, but it is up to the service or application using the configuration information to decide whether this default value needs to be used.

Attributes may have also multiple values. These are typically attributes that are optional in its true meaning and therefore have no defaults.

There are different ways to define the default value of an attribute:

- as a literal (e.g., `saAmfSgtDefAutoRepair` the default value is `SA_TRUE`);
- as a variable depending on some feature of the model (e.g., `saAmfSGNumPrefInserviceSUs` has a default of the number of service units in the service group);
- as a variable based on another attribute in the same object (e.g., the default for the `saAmf-SGNumPrefAssignedSUs` attribute is the value of the `saAmfSGNumPrefInserviceSUs` attribute of the same object);
- as a variable based on another attribute in a different object of the model (e.g., the default of the `saAmfSGAutoRepair` in the service group object is defined based on the `saAmfSgtDefAu-toRepair` attribute of the related service group type object).

In cases when the default value is defined based on another attribute's value that attribute's name is given for the default value.

Whenever this referenced attribute is in another object class if it is possible the path to the attribute is given (e.g., `saAmfSGType.saAmfSgtDefAutoRepair`). This is usually formed as two attribute names: the first one is attribute name in the referencing object class that holds the reference to the object of the object class (e.g., `saAmfSGType` the attribute which is the association end that points to the service group type class); the second name is the name of the attribute in the referenced class which provides the default value (e.g., `saAmfSgtDefAutoRepair`).

The first attribute name can be omitted if the referenced object class is singleton as long as the attribute name is distinguishable from all the attributes in the referencing class.

The UML model indicates these referenced defaults to document where the applicable default values originate from. However, in the information model instance the default values are not set. They are fetched as needed by whoever needs it from the referenced attributes.

This is the mechanism through which an object class representing a type can be used as a single point of control for the configuration of all its entities that we mentioned in section 'Issues with Inheritance.'

As long as the linked attribute does not have a value in the referencing object, the applicable value is fetched from the referenced attribute whenever it is needed. When such an attribute is set to a given value, for example in an object representing an entity of a given type, this value overrides the value provided by the attribute of the referenced object representing the entity's type. It is used as set without looking into the referenced attribute.

For example, as long as the `saAmfSGAutoRepair` attribute in the service group object is not set AMF will check the value in the `saAmfSgtDefAutoRepair` attribute in the service group type reference by the `saAmfSGType` of the service group object. If there are 100 service group objects in the model all referencing this same service group type then for all of them the service group type object will determine the applicable value and this will change as this attribute is changed in the service group type object. If in any of this 100 service group objects we set the value to directly in the `saAmfSGAutoRepair` attribute to `SA_TRUE`, this attribute of the service group object becomes decoupled from the `saAmfSGType.saAmfSgtDefAutoRepair` attribute and will only change if it is modified directly.

As we mentioned the expectation is that it is the service or application using the attribute that interprets this link.

Attribute Types and Constraints

The C programming model specification defines a set of types that are used by the different SA Forum specifications. These types are also defined for the SA Forum information model.

However the service specifications or applications may extend this common set of types in different ways, some of which may show up in the information model of the particular service or application. The most common cases are:

- The runtime attribute representing a state of the service entity for which the values are defined by the related service specification as an enumeration.
- Attributes reflecting associations between classes of a service or an application as these classes are defined by the given services or applications and they are not known in advance.

Of course the IMM managing the information model cannot be prepared for all these different extensions in advance, so instead in the class definitions such an attribute indicates two types:

- As the attribute type it indicates the type defined in the owner service specification or application (e.g., `SaAmfSGType` for the `saAmfSGType` attribute of the service group class shown in Figure 4.5).
- As the IMM representation type given as a constraint of the attribute it indicates the IMM type to which the attribute type is mapped to represent the attribute value in IMM (e.g., `SA_IMM_ATTR_SANAMET` which indicates the `SaNameT` C type). The definition of the mapping between these types is the responsibility of the service specification or the application.

It is important to note that when an attribute's type is a class, it is always interpreted as a reference to an object of this class, which has a proper DN. This DN is used as the reference to the object. This implies that the class of the referenced object includes an RDN attribute and there is at least an implied DN format indicating the positioning of these objects in the tree of the information model. In addition the `saParentDn` attribute may also reflect this information in the class definition.

Besides the IMM representation type, the attribute categories discussed in Section 4.3.1.2 and other features are also defined as constraints. We already mentioned the configuration, runtime, and the persistency categories and RDN role of an attribute, all of which are indicated as constraints.

In addition for configuration attributes the writability, for runtime attributes the caching are given as constraints.

Class Operations

Since the information model is concerned of the management view, the only operations defined in the UML object classes are the administrative operations applicable to the objects of the given class.

4.3.2 Administrative and Management Aspects

As we have noted the SA Forum Information Model is managed by the IMM, which exposes it to administrators and object manager (OM) applications through a 'northbound' API referred to as the IMM Object Manager or OM-API. In turn IMM delivers the administrative operations and changes applied to the information model to the AIS services owning (or responsible for) the objects being manipulated to carry out the changes in the system and return any result, which then returned to the initiator again using the OM-API. The IMM and the use of its APIs are covered in details in Section 8.4.

However, IMM provides only a general framework for managing the information model. The actual management objects, their meaning and the applicable operations are defined by the different AIS services and we present them as we go through each discussing their information model and how they map into the different service entities the services manipulate.

4.3.3 Application Information Models

Applications may also have their own information model and they may want to expose it to the system administration. It makes sense, and it is actually desirable for the system administration that the applications expose their model in the same way as the middleware, the SA Forum system, does.

As we have hinted already in our discussion (Figure 4.1), the SA Forum information model is extensible for applications as long as they follow the concepts and principles of the system model. In particular, the model needs to be organized into a tree and use the LDAP naming schema.

The object classes representing the application entities can be added to the SA Forum model using the IMM. The application implementation is expected to be the owner (or implementer) of these object classes. The classes may be of any of the categories defined for the services. This means that this owner application should not create or delete the objects or manipulate the configuration attributes of the configuration object classes. On the other hand it is responsible for maintaining the runtime attributes and objects of the runtime object classes.

Note that certain aspects of the SA Forum information model are not mapped into its representation maintained by the IMM. These include the different associations, the administrative operations, and the referenced default values.

4.3.4 Open Issues and Recommendations

The core portion of the SA Forum information model is the information model of the AMF, the concepts of which were extended to the other SA Forum services as well as to applications designed for SA Forum compliant systems. This increasing scope also reflects the state of maturity of the model. It has been used mostly for AMF and the SA Forum services, while for applications the use is just starting as the first mature implementation (i.e., OpenSAF [64]) of the SA Forum AIS is taking off.

When looking at the open issues we need to distinguish the issues related to the information model itself and those related to the IMM maintaining the representation of the information model within an SA Forum middleware implementation. Here we will focus only on the issues of the information model itself.

Probably the biggest issue with the information model is that even though it uses UML it does not follow completely the UML concepts. It combines UML with different more traditional solutions which occasionally result in cumbersome solutions. We have seen this in the discussions of the LDAP naming which reflects a tree structure that does not fit easily the UML association class concept used in the model simultaneously.

UML was initially used mostly for documentation purposes. As a result, it was not so much the UML semantics as the semantics of the domain and targeted use that were the guiding principles. On the one hand this may put off people with UML expertise. On the other hand the same feature may be embraced by people who use the model in day-to-day system management and who are less comfortable with the abstractions of UML. For example, configuration management does not require the use of inheritance and in the information model in many cases it is not used in an object-oriented (i.e., UML) way.

The SA Forum model is designed to be as simple as possible, so when in a class definition another class is given as an attribute type it is always interpreted as a reference to another object. In other words, the information model does not support complex types as attributes which is also an issue that some may question.

Nevertheless at this stage these issues are not open for discussion. The SA Forum made its choices and defined the information model accordingly.

The two biggest issues that still not settled are:

- the rules ensuring model consistency; and
- the IMM representation of the information model.

With respect to model consistency the issue which is not completely resolved is the ordering of the creation and the deletion operations of different objects in the model. The only defined requirement in the specifications (noted in both [38] and [63]) is that objects of weaker persistency cannot parent objects of stronger persistency. Accordingly a runtime object may not be the parent of a configuration or a persistent runtime object.

This expectation was defined from the perspective of failures and restarts of the IMM and the entire cluster due to the fact that IMM cannot restore a persistent object if its ancestor is a runtime object, which is flushed at such an event. But in a system designed to run continuously these events are rare and the information model content is preserved for a long time even in the case of runtime objects. So the question is raised: how to combine the objects' life cycle with their category as well as semantics?

For example, the configuration objects of an application that require regular reconfiguration may have a shorter life-cycle than some runtime objects representing the platform itself.

A related issue is the power of the system administrator with respect to runtime objects. The SA Forum services have been defined with the idea in mind that objects should not and cannot be removed from the information model while they still have children. This expectation opposes the tendency that an administrator should be able to perform configuration changes any time and therefore should be able to remove configuration objects even if they still parent other potentially runtime objects.

While for configuration objects the removal of the parent can be viewed as equivalent to the removal of all its children starting from the leaves, the operation becomes questionable if any of the child objects is a runtime object, which should only be removed by the owner service as appropriate for service users. One may argue that application processes may crash and considering the life-cycle of runtime objects – particularly persistent ones – they may pollute the information model by locking in potentially even a configuration object. So there is a need to get rid of such runtime objects. On the other hand experience shows that most outages are caused by operator errors and the information model is the main user interface of an SA Forum compliant system.

The specifications give few guidelines regarding the proper design of the information model tree. At the same time, as we have seen, this knowledge may be needed in some form at application development time as API calls are expected to provide the DN of the service entities they access. This may create the tendency of inserting objects at or close to the root of the information model as it is the safest place not to disturb and not to be disturbed by other operations. This flattens the model and provides little help to administrators in navigation.

The opposing tendency is driven by the containment relation, which maps well into the tree structure and usually also implies life-cycle dependency among represented service entities. We will see this in the case of the AMF information model.

The second major issue, which has no resolution, is related to IMM representation of the information model. As noted, some features of the information model defined in its UML representation are not mapped into IMM features defined as up to date. Some of these may therefore limit the extensibility of the information model for applications. The features of the model not reflected in the IMM representation include the administrative operations, some associations between object classes, and attribute relations. The main limitation they impose is that they need to be known by all parties using the model, that is, the application owning the given portion of the information model as well as the system administration and any management applications manipulating the model. The IMM cannot assist in discovering these features.

4.4   Conclusion

In this chapter we presented an overall view and the main concepts of the SA Forum information model. We will explore its different portions further as we introduce the SA Forum services, defining their own parts of the information model, all of which follow the concepts presented in this chapter.

The purpose of the information model is to provide an interface between the system management and the middleware through which the administrator can monitor the system status and configure it as required. Accordingly the model needs to suit both parties – the needs of system management as well as the configuration needs of the middleware. The third party typically not considered from this perspective is the applications running in the system which may act in a similar manner as administrators with respect to the services they use, that is, they may set up a particular environment for themselves, which is essentially a configuration that needs to be preserved in similar ways as the administrator provided configuration is preserved. Yet this environment setting needs to be distinguished from the administrator's realm as it is the application that maintains it and controls its life-cycle. One consequence is that there is no need for additional consistency checks and a lightweight API satisfies all the access needs.

We have introduced the concept of the physical resource, service entities, and managed objects as they reflect the different layers of abstractions and different perspectives. The services manipulate the physical resources, which are abstracted into logical service entities that provide the basis for the service definitions themselves. Service entities are mapped into management objects of the information model to expose these concepts toward the system management. These management objects provide the window into the service operation through which service behavior can be observed – and manipulated, if desired. Applications may extend the information model for their own needs following these principles.

To match the needs of configuration and observation the objects and their attributes are classified into two main categories: configuration and runtime.

Configuration objects and attributes are used to prescribe the expected behavior of the services and the system as a whole, while runtime objects and attributes are the reflections of the actual behavior, the status of the represented system resources.

To allow for the application side environment setup, the notion of persistent runtime objects and attributes has been introduced, which is a distinguishing feature of the SA Forum information model. It reflects the autonomy of such systems that they are able to run continuously without the need for administrative interventions, even in the presence of certain failures.

In this chapter we have presented the conventions defined for the SA Forum information model with respect to its organization, the naming of its objects as well as their classes and class attributes. We described the perspective from which multiplicities and defaults have been defined in the information model which helps their correct use, whether it is about interpreting the model defined for the SA Forum services or extending it for applications. This extensibility allows for similar management of the system as its applications, which facilitates the creation of a unified system view that should ease the task of system management.

The SA Forum information model is based on UML even though it does not adopt all its conventions and also incorporates other technologies. Nevertheless this definition is 'formal enough' for tools to be built to further aid, validate and automate system management functions.

5

Consistent and High Level Platform View

Maria Toeroe

Ericsson, Town of Mount Royal, Quebec, Canada

5.1 Introduction

The first prerequisite necessary for managing service availability (SA) is reliable information about all the resources available in the system that can be used to maintain the availability of the services offered by the system. The goal of the platform services is to provide the availability management with this information about the platform, that is, about the execution environments (EEs) – encompassing the SA Forum middleware itself – that can be used to run applications that deliver the services in question. In the process of providing this information the platform services also manage the resources at each level and even expose some of the control functionality to their users.

The SA Forum system architecture splits this task into three parts:

The Hardware Platform Interface (or HPI) [35] deals with monitoring and controlling the hardware resources themselves. The approach is that hardware components are given; these resources cannot be modified from within the system. Appropriately HPI has a discovery mechanism to find out exactly what hardware components are present, in what state they are and what management capabilities they have. It exposes these components and their capabilities in an implementation independent way so HPI users can monitor and control them as necessary for their purposes. HPI also provides timely updates on changes that happen to the system or to its components regardless of whether these changes are the result of internal events such as failures, programmatic management operations from users such as resetting a resource; or external (physical) interventions such as the insertion or removal of a blade in a server blade system. The HPI makes accessible the manageable capabilities of hardware resources such as fan speeds, sensor thresholds, but it also offers interfaces to control firmware upgrades and to initiate diagnostics.

The Platform Management service (PLM) [36] uses the information provided through the HPI. The main task of the PLM is to map the hardware discovered by the HPI into the PLM configuration which is part of the system information model. The configuration indicates the expected entities and their particular locations within the system. PLM then compares this information with the discovered

Service Availability: Principles and Practice, First Edition. Edited by Maria Toeroe and Francis Tam.
© 2012 John Wiley & Sons, Ltd. Published 2012 by John Wiley & Sons, Ltd.

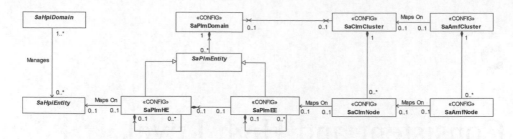

Figure 5.1 Mapping of entities of the different layers of the platform services [62].

hardware entities. If there is a match, it allows the entity to become part of the PLM domain and to boot the EE together with the SA Forum middleware as appropriate. The PLM information model also reflects the virtualization facilities such as hypervisors with their hosted virtual machines (VMs) potentially each providing an operating system (OS) instance as an EE for applications.

The PLM monitors and performs some basic life-cycle management of these entities primarily for the purpose of fault isolation. It also exposes to its users the up-to-date status information on the entities within its domain and some management capabilities on them.

Once the inventory of the hardware elements (HEs) and the EEs running on them has been established, the Cluster Membership service (CLM) [37] forms a cluster. Each PLM EE may host a CLM node and the cluster may include only the nodes configured for the membership.

Based on the information received from PLM about the PLM entities, the CLM is in charge to decide whether a configured node residing in a given EE is indeed healthy and the node is reachable so that distributed applications can use the node as part of the cluster. The CLM guarantees the formation and maintenance of a single cluster and up-to-date and reliable membership information. All Application Interface Specification (AIS) services with cluster-wide services rely on this information for their decisions on their resource handling to best support SA.

The platform services form a stack, one layer using the services of the other in a well-defined way for a well-defined purpose. However their interfaces are not exclusive to AIS services. Any application requiring up-to-date platform information may link the appropriate interface and obtain the information from any of these services directly. Figure 5.1 shows the mapping between the entities of the different layers of the Platform services.

In the rest of this chapter we take a closer look at each of these services and the way they work together to enable SA.

5.2 Hardware Platform Interface

5.2.1 Background

Relying on a feature rich hardware platform is essential for availability management. This feature richness includes the capability to detect hardware, monitor, and control its state. Since the availability management is a software solution it is desirable that these functions are offered through a programming interface.

The development of the HPI specification was the result of the growing trend of building hardware platforms modularly as a loosely coupled set of compute and I/O blades. These platforms required a centralized management and a common interface that could be used by the availability management.

Several vendors started the development of their own proprietary solutions. At the same time the PCI Industrial Computer Manufacturers Group (PICMG) initiated the development of the CompactPCI [65] and later the AdvancedTCA [66] then MircoTCA [67] specifications, all of which for management relied on the Intelligent Platform Management Interface (IPMI) [68].

The IPMI effort was led by Intel who also initiated the work on a complementary hardware platform management application programming interface (API) named the Universal Chassis Management Interface (UCMI). It addressed the gaps identified by the High Availability Forum, an industry group discussing the issues of open architecture based high-availability computer systems.

From its formation the SA Forum drew on the heritage of the High Availability Forum, and the UCMI initiative became the core part of the first release of SA Forum HPI specification.

While exposing all the richness of the underlying technology HPI defines a platform independent management API that the high-availability middleware and applications can use to monitor and control the platform hardware in a uniform way.

HPI is designed to be flexible and to provide management capability for any type of hardware and not only blade systems typically used for high-availability systems. Its key feature of discovering the hardware topology and the management capabilities first allows for the creation of management applications without advance knowledge of the targeted hardware.

Since its first release, the SA Forum has published several updates to the HPI specification. It was complemented with specifications standardizing the access of an HPI implementation via Simple Network Management Protocol (SNMP) [69] and also with the mapping between HPI and xTCA compliant platforms [70].

5.2.2 Overview of the Hardware Platform Interface

The HPI specification defines a set of functions that allows applications and the middleware to access the management capabilities of the underlying hardware platform in a platform independent and uniform way. The specification defines a model that guides the access to these management capabilities. The basic concepts of the model are: the HPI entity, the management instrument, the resource, the session, and the domain.

HPI entities are the hardware components in the system that compose the hardware platform ranging from fans to compute blades and processors. HPI entities reflect the physical organization of the platform and accordingly HPI entities are identified based on their location information tagged with their type.

HPI entities expose different management capabilities allowing one to determine and/or control their status such as reading their temperature or performing a power cycle. These different management capabilities are modeled as *management instruments*.

There are different types of management instruments: as simple as a sensor or a timer, or as complex as a firmware upgrade management instrument (FUMI) or diagnostics initiator management instrument (DIMI). The HPI specification classifies them and for each class defines an appropriate set of functions that manages the particular aspect of the hardware components they represent.

For example, the sensor management instruments provide readings for different measurements regardless what is being read. The sensor type returned as part of the reading determines the interpretation of the values read. Besides reading sensor values HPI allows one to define and adjust thresholds on them so that when violated HPI generates events – essentially alarms. In turn management applications can react to them potentially using other management instruments associated with the same HPI entity to resolve the situation. For example, if the temperature is too high the user may use an available control management instrument to turn on a fan.

It is important to see that the management instrument physically does not have to be part of the associated HPI entity. For example, if the HPI entity we would like to control is this fan in the chassis and the triggering event to turn it on or raise its speed is a rise in temperature above a given threshold, the temperature sensor generating this triggering event is not likely to be part of the fan itself. It might be located at the farthest place that this fan needs to cool.

The control and the sensor used together to control this fan would compose an *HPI resource*, that is, a set of management instruments which are used together to manage some HPI entity representing some platform hardware. HPI resources reflect the logical organization of the system for some management purposes.

The management application gains access to the different HPI resources by opening a *session* on an *HPI domain*. The domain groups together a set of resources and the services necessary to discover these resources as well as additional domains. Furthermore, it provides event and alarm management and event logging services.

Accordingly, an HPI user is expected to discover the management capabilities available on the HPI implementation it is connected to. A possible starting point for this is the 'default domain,' which is a concept supported by all HPI implementations. From this and using the domain services the HPI user can discover the different resources available in the domain and any additional domains it can further explore. Different HPI implementations may interpret the 'default domain' concept differently. For example, one may associate it with one particular domain within the platform, while another implementation may associate it with a different domain for each subsequent user request.

For each of the resources in the domain the HPI user can find out the following information: The entity the resource is associated with including its inventory data; the management capabilities the resource itself exposes for the entity (e.g., whether the entity is a field replaceable unit (FRU) that supports hot-swap); and any additional management instruments included for the resource (e.g., sensors, controls, FUMI, DIMI, etc.) in the domain.

Although it is typical, resources are not in an exclusive relationship with their associated entities. They may represent only a certain management aspect of that entity, while another resource may expose additional aspects. For example, if the earlier-mentioned fan is hot-swappable the hot-swap capability may be exposed through a different resource which reflects all the features impacted through the hot-swap.

Once the HPI user has discovered the current platform configuration it may subscribe to the events generated in the domain it has a session with. Among others these include notifications about changes in the domain configuration, that is, the HPI implementation indicates when resources are added to or removed from the domain. There is no need for periodic rediscovery of the platform.

The HPI specification does not cover any details on how the HPI implementation itself collects the information exposed through the API or how the hardware entities and the related model concepts are configured, for example, what features of an entity are exposed through a particular resource. All this is left implementation and therefore vendor specific.

The HPI specification also does not mandate any synchronization among different instances of HPI implementations. For example, if implementations of different vendors are combined into a system, they will coexist without any knowledge of each other – at least with respect to their standard capabilities. Any coordination between them is beyond the scope of the current HPI specification.

In the following section we elaborate further each of the mentioned HPI model concepts starting with the HPI domain.

5.2.3   The HPI Model

Figure 5.2 presents an overall view of the HPI model concepts that we elaborate on in this section.

5.2.3.1   HPI Domain and Session

As we have already seen, an HPI domain exposes a set of services common to the domain, which are collectively referred to as the *domain controller* as they cover administration aspects; and a collection of HPI resources accessible through the domain.

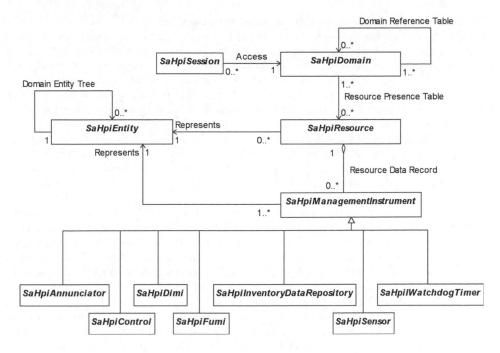

Figure 5.2   HPI model elements and their relations.

Although the wording 'collection of resources' could suggest that the HPI domain is a set of physical entities, it is more appropriate to think of the domain as an access control mechanism, or even filtering mechanism. The same resources may be accessible through several domains while others may be dedicated to a particular one.

An HPI user initiates the access to a domain by opening a session on the domain. The user will see only the resources of this domain and it can access only these resources within the given session.

The domain controller includes two tables: the resource presence table (RPT) and the domain reference table (DRT). The former allows the HPI user to discover the domain itself while the latter can be used to discover other domains in the system.

Domains may be related to each other depending on what can be discovered from these tables:

Peer domains are domains that allow access to a single set of resources and discover a single set of domains. That is,

- the RPT of each domain in the peer relationship lists exactly the same set of resources; and
- the DRT of each peer domain references all the other peer domains and the same set of nonpeer domains.

For example, assuming domains X, Y, Z, and W in an HPI implementation; domains X, Y, and Z are peers if they all list in their respective RPT the same set of resources {a, b, c, d} exclusively. In their DRT X has {Y, Z, W}, Y has {X, Z, W}, and Z has {X, Y, W}. W is a nonpeer domain that can be discovered from any of these peer domains.

Essentially peer domains provide redundancy of the domain controller functionality for a given set of resources.

There are subtleties, however, that HPI users need to be aware of:

Initially, the HPI implementation will populate the RPT of each of the peer domains X, Y, and Z with exactly the same information with respect to the set of resources {a, b, c, d}. The HPI users of each of the domains may manipulate this information, and they may manipulate it differently. For example, users may tag the resources differently in the different domains. It is like 'personalizing the browser' through which they see the domain. The HPI implementation does not synchronize such changes across peer domains; only users of the same domain will see the same information for a given resource, that is, all users of the domain use the same 'browser.'

Subsequently, as changes happen in the system – some resources may go away, others may appear, the configuration, the state of the controlled hardware may change – the HPI implementation will update the RPTs of all the domains and peer domains will continue to see the same set of resources. However there may be some differences in the timing and the order as if the browsers of the peers domains show the same scenery, but from different angles.

Domains may also be in a *related nonpeer domain* relation as we have shown W in our example. These are those domains that can be discovered by an HPI user who initially had only a single entry point to the set of related domains. Related nonpeer domains also have disjoint sets of resources, which guarantees that the same resource will not be discovered twice, through two different domains except if these are peer domains, in which case the entire domain does not need to be discovered the second time. Related nonpeer domains compose a loopless tree structure in which there is only a single path leading from one related domain to another.

One may interpret related nonpeer domains as different management aspects of the system. For example, one domain may collect all the resources providing temperature control capabilities for the system, while another would allow hot-swap management only. These related nonpeer domains may be complemented by a set of peer domains that expose the same set of sensors so that any of the managers using the nonpeer domains can observe how the system state changes as a result of the interactions.

Finally domains that do not fall into either of these categories are called *unrelated domains*. Unrelated domains have no restrictions on what resources may be accessed through them. Typically an HPI user needs to have advance knowledge of an unrelated domain as by definition it cannot be discovered through the discovery process and therefore their use is discouraged by the specification.

Note also that this means that domains accessing overlapping but not equal sets of resources cannot be related through the DRT. They need to be unrelated. So if we want our temperature and hot-swap control managers to access each their own single domain we can only use unrelated domains as they would need to share the access to the overlapping set of sensor resources while also accessing the appropriate control resources.

HPI users may open multiple simultaneous sessions to the same or different domains. Also a domain may be accessed through multiple sessions simultaneously. However HPI does not control concurrency. User may detect concurrent changes by checking the respective update counters maintained by an HPI implementation as well as by listening to the events it generates.

Besides the RPT and the DRT, the domain controller maintains a single event channel specific to the domain. Events published on the channel usually indicate some kind of change within the domain or within the HPI resources accessible through the domain. This event channel is a very simple interpretation of the publish/subscribe paradigm.

The publishers of the domain event channel are: the domain controller itself, the resources accessible through the domains, and HPI users with open sessions to the domain.

To receive the events published on the channel, an HPI user needs to subscribe with its session id. Once done so it will receive all events published in the domain. That is, no filtering capability is provided by the HPI event manager; hence all subscribers of the domain will receive all events published on the domain's event channel.

Logically the subscription results in the creation of an event queue dedicated to the particular subscriber. The HPI implementation places a copy of an event to each of these queues. The copy remains in the queue until the consumer HPI user retrieves it. To read the events a subscribing HPI user needs to call the get API function, which returns the first available event in the queue. This call can be invoked in blocking mode so that events are delivered as they occur.

If it is a resource that generates an event, the same event will be published in all domains through which the resource is accessible. If more than one such events is generated simultaneously in the system, the order of their publication may be different in each of the domains that the generating resources are visible from.

One of the subscribers to the domain event channel is the domain event log. As one may expect, it collects the events occurred in the domain, however, there is no requirement on logging all the events. Each HPI implementation may implement its own filtering criteria. The HPI provides no API to set or modify these criteria.

Finally the domain controller maintains a third table, the domain alarm table (DAT). Alarms indicate detected error conditions in the system.

When an error is detected within the domain or in a resource accessible through the domain an entry is added to the DAT. The entry remains in the table as long as the error condition persists unless it is a user generated alarm, in which case the HPI user can remove the entry. System generated alarms can only be cleared by the HPI implementation. Their entry is removed from the DAT when the HPI implementation can no longer detect the related error condition.

The HPI specification does not mandate any method for announcing the presence of an alarm in the domain. An HPI implementation may choose its own mechanism, for example, it may light an LED or use an annunciator management instrument, which then generates an event that users subscribing to the event channel may receive in real-time.

5.2.3.2 HPI Entity

The actual hardware components managed through a domain and its resources are reflected in the HPI model by HPI entities. These are the hardware platform components capable of providing some services like housing other entities or cooling them; and carrying out some jobs such as running an OS instance or storing information.

HPI entities do not have an identity outside of the system; they have only types such as being a power supply, a cooling device, a central processing unit (CPU) blade, or a chassis. When an entity is inserted into the system it becomes uniquely identifiable for the HPI implementation based on the location it has been inserted. Accordingly, HPI entities are identified by a construct called the *entity path*.

The entity path is the sequence of {entity type, (relative) entity location} pairs starting from the entity and toward the system root. An example of an entity path is the following: {{power supply, 2} {subrack, 3}{rack, 1}{root, 0}}. The location numbering starts with zero, so this entity path identifies the third power supply in the fourth subrack of the second rack within the system.

The structure of the entity path implies that the HPI entities are organized into a tree – the domain entity tree (DET). This tree structure serves like a map for the HPI implementation. It knows the way it is built and therefore all the locations on the map that may house entities of different kind. When the entity is inserted, the type and the location information create its identity specific for the system into which it was inserted.

This tree organization allows for some flexibility, that is, the HPI implementation does not need to know the entire tree structure from the beginning; the tree can be extended at real-time with sub-trees for entities that nest smaller entities. The numbering of peer nodes within the tree, that is, the locations on the map is implementation specific. The specification does not mandate such details as whether it is left-to-right or right-to-left, and so on.

The tree organization also implies that when a node goes away, all its children go away too: For example, when a carrier blade is removed from the system all its mezzanine cards go away with it.

The entity path is unique within the system and it reflects the physical location of an entity regardless how it is obtained through which domain. The HPI implementations must guarantee this. In other words, all resources managing an entity should be able to use the same entity path, even if they access it via different domains.

For maintaining SA, it is essential to know the status of the different HPI entities at any moment in time and to be able to manage them so they best meet the availability requirements. For this HPI entities may expose some management capabilities. The exposed management capabilities compose the HPI resources that we are going to discuss next.

5.2.3.3   HPI Resource

An HPI entity typically has a number of management capabilities: its power supply may be turned on or off, its temperature may be read. These capabilities are exposed as HPI resources.

The HPI resource concept is very specific as it only represents management capabilities. HPI resources are management resources only as opposed to the HPI entities themselves that may expose resources in the generic sense such as computing or storage capabilities. HPI however has no direct concern of these.

An entity may expose its management capabilities by one or more resources. Typically each HPI entity is associated with at least one HPI resource.

On the other hand, each resource is associated with a single HPI entity for which it exposes a basic set of management capabilities. These resource capabilities include:

- resource configuration management – allowing to save and restore the configuration parameters of the resource manager itself;
- load management – allowing to control the software loaded into the entity associated with the resource;
- power management – controls whether the entity associated with the resource is powered on, off, or power-cycled;
- reset management – which allows one to perform different types of resets (e.g., warm, cold, etc.) on the entity associated with the resource;
- resource event log management – provides the same functionality as at the domain level, but for the particular resource only;
- hot-swap management – indicates whether the associated entity is hot-swappable; and
- additional management capabilities accessible through management instruments that are available in association with the resource. These additional management capabilities may be associated with the HPI entity the resource is representing or they may be associated with other entities in the system.

The currently available resources, their associated entities, and their resource capabilities are listed in the already mentioned RPT of each domain through which these management capabilities are accessible. The HPI implementation updates this table dynamically as changes occur in the system. As indicated in Section 5.2.3.1, the same resource may be exposed through different domains.

Also, an entity may expose different management capabilities in different domains, which is reflected in the appropriate setting of the resource capabilities information in the RPT of each domain. For example, even if a hot-swappable entity can be managed through several domains, typically managing its hot-swap is dedicated to and therefore exposed only in a single domain. In the other domains through which it is visible the setting of the resources associated with the entity would show no hot-swap capability. Such exposure of resources would be typical for related nonpeer domains, each of which would collect platform resources for a particular management aspect.

Note also that the RPT lists the resources as a flat structure, while in reality the entities whose management capabilities they represent are typically nested and comprise the hierarchical structure of the DET.

As 'container' entities are plugged in or removed from the system resources associated with them and their children show up in the RPTs or disappear from them in seemingly unrelated groups. Only the analysis of the entity path of the entities associated with these groups of resources would reveal the relationship, the nesting of entities.

The specification recommends that HPI implementations add first to the RPT the resources associated with the 'container' entity and remove them last.

Additional management capabilities of a particular resource are listed as resource data records (RDRs). Each of these records describes a management instrument associated with the resource. In turn each management instrument is also associated with an HPI entity, which may or may not be the same HPI entity with which the resource has the association.

5.2.3.4 Management Instruments

The HPI specification defines different management instruments for the management of HPI entities. A management instrument is a generalization of some management capabilities typically available in hardware platforms. HPI defines seven different management instruments and their appropriate API that represent classes of management capabilities. These management instruments are:

- the control management instrument – allowing for setting the state of the associated entity;
- the sensor management instrument – providing a mechanism to query some conditions;
- the inventory data repository management instrument – returns information identifying the associated entity such as serial number, manufacturer, and product name;
- the watchdog management instrument – provides timer functionalities;
- the annunciator management instrument – replicates the functionality of the DAT, that is, announces error conditions existing in the associated entity. As opposed to the DAT the content of which is standardized, annunciators can be tailored to the HPI implementation and even by the HPI user;
- the diagnostics initiator management instrument – exposes diagnostics capabilities for the associated entity; and
- the firmware upgrade management instrument – allows the management of upgrades of the firmware of the associated entity.

In the RPT it is indicated which types of management instruments a resource contains and for each of them a set of RDRs describes the actual management instruments.

RDRs have some common fields that include the RDR type (implying the management instrument type) and the entity associated with the management instrument, and some fields specific for each management instrument.

5.2.4 HPI Capability Discovery

When an HPI user would like to manage the hardware platform it first opens a session toward a domain. The user may be aware of a specific domain id or – most often – it opens the session on the 'default domain.' The HPI implementation then decides on which domain the session should be opened. This may depend, for example, on the security parameter the user provides in its call or other, HPI implementation specific details.

Once the user has an open session with a domain it is ready to discover the domain. It can find out on which domain the session was opened, and what the domain current status is. This includes:

- the domain id, if it participates in a peer relation, and some informal text associated with the domain;
- the DRT with a list of other domains;
- the domain resource table with the list of resources accessible through this domain;
- the DET with all the entities manageable via the resources of the domain;
- the DAT with the list of currently outstanding alarms.

To find out the information in each of the tables and in the DET, the HPI user needs to iterate through the tables and traverse the DET.

Iterating through the DRT allows the user to discover all the resources it can use to manage their associated entities through the given domain. This as we have seen in Section 5.2.3.3 covers the basic management needs such as power and reset management.

Each of the resources may indicate additional management capabilities such as watchdog timers and sensors as presented in Section 5.2.3.4. To discover these, the user needs to iterate through the RDRs associated with each resource. Once the user has discovered the management instruments from the RDRs, it can access them directly.

To find out about the current conditions and to perform management actions the user can access each of the management instruments using the methods appropriate for the type of the management instrument. If only a single management instrument of the given type can be associated with a resource the user only need to indicate its session id and the resource id. Otherwise the user needs to indicate the management instrument's number or may need to iterate through the management instruments of the given type to find out how many of them are available.

Besides the management capabilities, the user may also be interested to find out about the actual HPI entities composing the platform that are manageable through the domain. Indirectly some of this information is already included in the tables we described so far as the resources and management instruments all reference the HPI entities they are associated with. Of course, the user would need to correlate the different entity paths to reconstruct the organization of the managed entities. In the early versions of the HPI specification this was the only way to find out about the entities. Now the HPI API includes functions that allow the discovery of the DET directly.

Finally the user may subscribe to receive events generated within the domain. These will announce the changes as they occur within the domain.

Using the information of the DRT, the HPI user may continue the exploration by opening sessions and discovering other domains in the same way until they have discovered all related domains with all the management capabilities available.

5.2.5 Error Handling and Administrative Operations

One may have realized by now that the entire HPI is about the administration and management of the components within the hardware platform. Much of its functionality is or can be used by higher level administrative operations to carry out an intended operation. For example, higher level restarts or resets may map into an appropriate mode of the reset of the HPI resource associated with some hardware entity. As a result HPI does not expose any additional standard administrative API.

The interesting and additional aspect that an HPI implementation needs to deal with is that an administrator may physically manipulate the platform, open or close latches, remove or plug-in boards. In fact, such a manual operation is the required administrative repair action for many hardware error conditions or may be part of some upgrade operations when some old HEs are removed and replaced with new ones.

The HPI specification does not address this issue specifically in the context of administrative operations, but more from the perspective of error handling. It distinguishes the handling of either situation depending on whether the entity in question is a FRU and can be expected to report hot-swap events or it is not an FRU.

For non-FRU entities manual extraction and insertion are viewed as sudden failure and recovery of the entity. Some systems may also be capable of distinguishing whether the entity is physically not present or just inaccessible. Accordingly, the events that an HPI implementation may report for such entities are 'resource failed,' 'resource inaccessible,' 'resource removed,' 'resource restored,' 'resource added,' and 'resource updated.'

For entities that can be replaced in a live system, that is, they are FRU; this fact is indicated as part of the resource capabilities. For such entities, or more precisely the resources representing them, the HPI implementation reports the hot-swap states. These are: 'not-present,' 'inactive,' 'insertion-pending,' 'active,' 'extraction pending.'

The main difference between a manual administrative replacement operation and a failure recovery is that after recovering from a failure an FRU may not report the sequence of state transitions expected during a properly performed hot-swap reported at a replacement. It may transition directly from the not-present state to any other state.

In the opposite direction both a failure and also a surprise extraction may transition the resource from any state to the not-present state.

In either case when a resource is restored or added, HPI users cannot assume that any of the earlier discovered information is still applicable to the resource even if the resource id is the same as of the resource that was part of the system before.

The failure/removal and the repair/insertion of entities of the system may result in the reconfiguration of resources and reported as such by the HPI implementation.

The specification does not mandate whether the failure/removal of a resource results in only reporting its state as 'not present' or also in the removal of the associated record from the RPT completely.

On the other hand, if the resource associated with a newly inserted/repaired entity has not been discovered yet then once the HPI implementation detects the resource HPI reports it as a newly added resource and inserts a record into the RPT of the relevant domain. The specification does not cover how an HPI implementation determines the domain or domains through which the new resource needs to be exposed.

An HPI implementation reports any updates to the RPT, for example, as a consequence of some firmware upgrade. Each time it also increments the update count associated with the domain, so any user can detect the fact of a change.

5.2.6   Open Issues and Conclusions

From the presented features of the HPI one may understand that the software wanting to manage the hardware platform can be written independently of which HPI implementation it needs to interact with. The HPI user code is portable between HPI implementation.

With the tendency toward specialization and modularity, more and more system integrators face not only the need to port a management application from one homogeneous platform to another, but increasingly see the requirement of integrating such an application with heterogeneous hardware systems in which the different hardware modules come from different vendors, each of which provides their own HPI implementation.

The problem is that in this situation the HPI user code would need to be linked with the HPI libraries of the different vendors, which obviously does not work and another solution is required.

To address this need the SA Forum Technical Workgroup launched the work on the definition of an extension to the existing HPI specification allowing the modular use of HPI implementations.

Meanwhile the developers of the OpenHPI [71] implementation of the SA Forum HPI specification resolved the issue by using the plug-in Application Binary Interface (ABI) in their HPI implementation.

OpenHPI is an industry collaboration toward an open source implementation of HPI.

The OpenHPI plug-in solution remains within the framework of the current HPI specification. It works as a dynamic library. The OpenHPI plug-in handler maps the ABI functions to the HPI API. In this mapping a plug-in may not implement the entire ABI, only a subset supported by the hardware or the interface. OpenHPI requires only the implementation of the 'open' and the 'get_event' functions for proper operation.

This OpenHPI solution questions whether further standardization efforts are necessary or the HPI specification should simply recommend the ABI-based plug-in solution.

As pointed out earlier, the HPI specification does not specify how domains are configured, that is, there are no rules to determine how many domains a system would have and the visibility of resources from these different domains, and so on. All this is implementation specific.

We have also seen that the domain definition is the only defined standard filtering mechanism available in HPI compliant systems.

Considering that an HPI implementation cannot be aware of the applications' needs in advance there are only two possibilities: Either the implementation exposes some kind of nonstandard configuration interface for the system integrator – and this is the route OpenHPI took – or it needs to expose all the information available about the system. In either case an HPI user who opens a session on a domain will get all the information available for that domain; it has no option of filtering it.

In the first case the problem is that each HPI implementation may come up with its own configuration solution that makes the integration difficult. In the later case the amount of information generated in the domain can be overwhelming, for example, if the piece of software needs to manage a piece of hardware in the system without the need of understanding of the entire context of that piece of hardware.

Addressing this gap may be a direction to continue the standardization efforts within the SA Forum.

We can summarize that the HPI provides a platform independent representation and management of hardware components. This encompasses portable fault, alarm, and hot-swap management policies and actions, all essential for availability management.

The HPI solution can utilize different form factors ranging from rack mounted servers to xTCA systems. To help the adoption in these different systems, the SA Forum complemented the basic specification with mapping specifications to xTCA [66] and to SNMP [53].

5.3   Platform Management Service

5.3.1   The Conception of PLM

HPI users need to discover at least one HPI domain but maybe many to find the management capabilities of the piece of hardware they are interested in and to be able to obtain its status and receive subsequent updates. These updates are not filtered by the HPI and the users need to sort out themselves the relevant items from the stream of events delivered by HPI about all entities manageable through the domain or even domains.

Moreover, users need to be aware of the dependencies they may have toward HPI entities and also between different entities and their management resources to figure out whether the HPI entities indeed can be used and relied on for providing their own services.

All this may be a daunting task depending on the relation of the information the user is interested and the size of the domain. It may also consume significant computing resources if the procedure of obtaining the same information needs to be replicated in a number of places within the cluster.

On the other hand, the CLM (which was at the bottom of the AIS 'stack' until 2008) forms the cluster and provides its users and other AIS services with the membership information. While the information obtainable via HPI is essential for the cluster formation, it is not enough.

Between CLM and HPI was lying a no man's land encompassing the OS and the quickly emerging virtualization layers about which no information could be obtained and no control exercised that would fit the AIS stack and satisfy the high availability requirements.

This disconnect was further widened by the difference in the approaches used by HPI and AIS.

The basic philosophy of HPI is that the hardware is a given: One needs to discover the entities present in the platform and available for service provisioning. The HPI specification does not cover any configuration aspects, not even the configuration of domains and their content, which are used to control the visibility of the entities for HPI users.

On the other hand, since AIS deals with software entities it is based on configuration, namely the system information model. In high-availability clusters it is tightly managed what software entities are allowed to come up, where and when. The information model covers these aspects for the AIS stack and each service processes its relevant part to obtain the required configuration information. They also use the model as an interface to expose any status information for the system administration.

To bridge the gap between these two worlds, the option of extending the CLM came up; however, it was quickly turned down by the amount of jobs it would need to perform. Such a solution would go against the modular design of the SA Forum architecture.

Instead the PLM was defined to connect HPI's dynamic discovery of hardware with the configuration driven software world of AIS. The definition of a new service was also an opportunity to address the rapidly emerging virtualization layers, and to provide some representation and control over them and the OS that was not covered by any of the existing at the time specifications.

The PLM appeared within the AIS quite late and at the time of writing only the first version was available.

5.3.2   Overview of the SA Forum Platform Management

The main tasks of the PLM can be described as follows:

- it connects the information collected from the HPI about the available hardware entities with the system information model used by the AIS services and their users;
- it complements the information about the hardware with the information about the virtualization layers and the OS;
- it exposes an API through which user processes including the availability management can track the status of selected PLM entities; and
- it provides an administrative interface for the system administration to verify and perform lower level operations in a simple platform independent way.

To achieve these tasks PLM defines an information model with two basic categories of entities and their types. HEs are used to represent HPI entities in the system information model, while EEs may depict elements of the virtualization facilities (e.g., a VM, a hypervisor), OS instances or combinations of them.

An important feature of the PLM information model is that it does not necessarily present the reality in all its details. It paints the platform picture in 'broad strokes,' reflecting the significant aspects and abstracting from the rest.

The selection of what is significant and what can be abstracted is based on the configuration information provided to PLM as part of the system information model. PLM takes this configuration and compares it with the information collected from the HPI implementation.

The configuration describes in terms of HEs the types of HPI entities and their expected locations in the hardware platform. PLM matches this up with the detected HPI entities. If there is a match, the HE defining the criteria is mapped to the HPI entity with the matching characteristics. For example, the configuration may indicate that in slot 5 there should be an I/O blade. If HPI detected and reported a CPU blade for slot 5, there is no match and PLM would not allow the CPU blade to become part of the PLM domain. It even issues an alarm that there is an unmapped HE. If the detected blade in slot 5 is indeed an I/O blade the configuration element specifying the I/O blade in slot 5 is mapped to the HPI entity of the blade.

On a HE, which has been mapped and which is configured to host an EE PLM monitors if the booting EE is indeed matches the one indicated in the configuration. If yes, PLM lets the element to continue the boot process and bring up the EE. If the EE does not match, PLM may reboot the HE with the appropriate EE or if that is not possible it will prevent the EE to come up using lower level control, for example, keeping the hosting HE in a reset state or even powering it down.

The PLM specification also defines the state model maintained by a PLM implementation for its entities. The PLM state definitions resemble, but are not identical to, the state definitions of the X.731 ITU-T recommendation (International Telecommunication Union) [72]. A mapping between the PLM and X.731 states can be established easily however.

A major role of PLM is that it evaluates the state interaction between dependent entities. The basic dependency is derived from the tree organization of the information model. In addition the configuration may specify other dependencies between PLM entities. The PLM correlates this dependency information with the state information of each of the entities and determines the readiness status for each entity reflecting whether the entity is available for service provisioning.

PLM calls back client processes when there is a change in the readiness status of entities of their interest. Within the AIS stack the CLM is the primary client of PLM and uses this information to evaluate the cluster membership.

To register their interest with PLM, client processes need to be aware of the PLM information model, its entities. A PLM user first creates an entity group from PLM entities it wants to track and then it asks PLM to provide a callback when the readiness status of any entity within the group changes or when it is about to change.

The PLM readiness status track API supports the multi-step tracking to ensure if necessary that

- an operation proceeds only if it causes no service outage; and also that
- changes to the system are introduced whenever possible in a graceful way, that is, the assignments of service provisioning entity of the system are switched out before, for example, the powering of this entity.

With these options PLM takes into consideration the responses received from its users to allow and disallow different administrative operations, and to stage their execution appropriately regardless of whether they were initiated through physical manipulation (e.g., opening the latch) or by issuing a command through the administrative API.

In the following sections we take a closer look first at the PLM information and state models and then we present the ways PLM users and administrators may use it in their different interactions with the service.

5.3.3 The PLM Information Model

The information model of the PLM shown in Figure 5.3 is part of the system information model [62] maintained by the SA Forum Information Model Management service (IMM) [38] and follows its conventions.

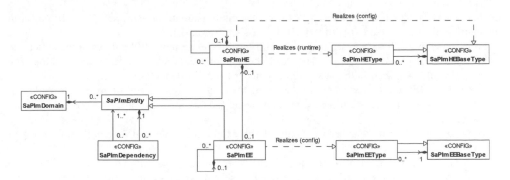

Figure 5.3 The PLM information model [62].

5.3.3.1 PLM Domain

The PLM information model is rooted in the PLM domain object. The PLM domain represents the scope of entities a given PLM implementation manages. The hardware expected to be present in this PLM domain is configured as sub-trees of HEs rooted in this domain object. We discuss HE in details in Section 5.3.3.2.

The system software running on these HEs is represented as EEs. They may compose the last few tiers of the HE sub-trees or may compose separate sub-trees of EEs rooted directly in the PLM domain object. EEs are presented in Section 5.3.3.3.

The tree organization of the represented entities describes their dependency in terms of containment or hosting, for example, a compute blade hosting an OS instance is the parent of the EE representing the OS. Additional dependencies can be defined through the dependency class described in Section 5.3.3.4.

The model also permits that the PLM domain contains only EEs or only HEs.

5.3.3.2 Hardware Elements, Their Types, and Status

The HE represents some hardware within the PLM domain which in turn is mapped into one or a group of HPI entities (Section 5.2.3.2).

The HE is a configuration object describing for PLM the hardware expected to be present in the system. PLM will try to map it into an HPI entity among those discovered through the HPI. The HE is a purely logical entity, which does not manifest in any way in the realms of PLM. The actual manifestation is part of the HPI realm, where the HPI entity maps it into a piece of hardware.

The purpose of this logical HE entity is to provide PLM with enough information to perform the mapping, thus verify the hardware configuration of the system. Once the mapping is established the HE object is used

- to reflect the runtime status of the mapped physical hardware; and
- as a reference to this hardware in the interactions with PLM clients and administrators.

Let's say we have a CPU blade plugged into the fourth slot of our Advanced Telecommunication Computing Architecture (ATCA) chassis; this is the physical hardware. The HPI entity representing it at the HPI level has the entity path {{CPU blade, 0}{slot, 4}{ATCA chassis, 1}{root, 0}}, which is used to identify our blade within the HPI realm. At the PLM level PLM may have mapped it successfully into the CPU3 HE object within the information model and by that PLM verified the correctness of the hardware configuration. As a result PLM sets the entity path attribute of the CPU3 HE object to the entity path {{CPU blade, 0}{slot, 4}{ATCA chassis, 1}{root, 0}}.

From the entity path we can see that at the HPI level there are more HPI entities in the system: at least the slot and the ATCA chassis. The PLM configuration does not necessarily reflect these. The CPU3 HE could be placed in the model directly under the PLM domain object.

The CPU3 HE object may represent the entire blade as a single entity or may have child HE objects representing the different hardware components such as memory, processor, or the firmware on the blade.

The choice is based on whether we would like to expose the states and the control of these HPI entities via PLM.

Each HE has two important attributes: the HE base type and the HE type, which represents a particular version within the HE base type. A HE base type specifies an HPI entity type and it is a collection of different implementation versions of this entity type, each of which is described as a separate HE type. More precisely, each HE type qualifies the HPI entity type of the HE base type with information found in the HPI inventory data record (IDR) management instrument, for example, indicating the vendor, the product line, the model, and so on.

Considering the example CPU3 above it may indicate that it needs to be a CPU blade by referencing a HE base type specifying this feature. PLM will accept and map any CPU blade in slot 4 as long as it matches the IDR characteristics of one of the versions belonging to the 'CPU blade' HE base type in the PLM model. Their characteristics may be very specific or just a few depending on the range of implementations that need to satisfy the relation.

Among the attributes of the HE the attribute showing the HE base type is configured, while the attribute holding the HE type of the HE is set by PLM as the result of the mapping. It tells the HE type matching the IDR information of the piece of hardware discovered by HPI at the location that has been mapped to this HE object. For more details on the mapping procedure itself see Section 5.3.5.1.

To reflect the status of the mapped hardware, PLM defines the presence, the operational, and the administrative states as runtime attributes for HEs. The PLM implementation sets the appropriate values for each of these states based on the information it receives from HPI about the mapped HPI entity and also information gathered at the PLM level.

HE Presence State
The presence state of a HE indicates whether the piece of hardware it represents is present in the system and whether it is active. Its values follow the HPI hot-swap state values; however this does not mean that the hardware actually needs to be hot-swappable. In fact it may not even support all the values defined for the state. The PLM implementation maps any information received from HPI to the specific values of the presence state based on their general interpretation. These values and interpretations are the following:

- **Not-present:** PLM is not aware of any piece of hardware that would match the configuration of the HE entity. For example, there is no blade inserted into the slot where a CPU blade is expected to appear, or the blade inserted is an I/O blade.
- **Inactive:** PLM has mapped successfully the HE to a piece of hardware in the system, but it provides no functionality. Effectively the hardware is isolated from the rest of the system. For example, HPI detected a CPU blade, which at the PLM level matches the HE configured for the slot; and HPI reports the inactive hot-swap state for the blade.

- **Activating:** The activation of the piece of hardware represented by the HE has been initiated, but it has not reached its full functionality yet. For example, HPI reports the insertion pending hot-swap state for the CPU blade after the latches have been closed.
- **Active:** The hardware represented by the HE has reached its full functionality. It can be used for service provisioning as intended. For example, HPI reports the active hot-swap state for the CPU blade.
- **Deactivating:** The deactivation or removal of the hardware represented by the HE has been initiated. For example, HPI reports the extraction pending hot-swap state for the CPU blade when someone opens its latches.

HE Operational State

The HE operational state reflects whether PLM is aware of any kind of error condition regarding the piece hardware the HE represents. PLM sets the operational state to *disabled* when it learns about an error condition in the piece of hardware represented by the HE. Otherwise the operational state is *enabled*.

PLM may find out about error conditions from several sources: From HPI by monitoring the relevant HPI resources and by analyzing the events reported by HPI; from PLM users who may submit error reports on PLM entities; and by implementation specific means.

Whichever way PLM detects an error, it sets the operational state to disabled and, to protect the rest of the system and prevent fault propagation, it isolates the related piece of hardware from the system. HEs are considered intrinsic fault zones.

To isolate, the PLM implementation performs the operations appropriate for the HE to put it to the inactive presence state. How it does so depends on the HPI resources available for the HPI entity representing the hardware at the HPI level. For example, for an FRU with manageable hot-swap states this would mean setting the inactive hot-swap state. For other HPI resource this could be equivalent to a power-off.

Sometimes PLM cannot initiate or verify the isolation as the hardware is inaccessible. PLM indicates this condition by setting the management-lost and the isolation-pending flags.

The HE operational state becomes enabled when the represented hardware is physically removed from the system, or if PLM receives a report that the error condition has been cleared.

HE Administrative State

The administrative state reflects to what extent the HE may offer its functionality within the system. It is controlled solely by administrative operations that PLM receives via the administrative API. Here we only provide the values that the administrative state may take. We will look at its control in Section 5.3.4:

- **Unlocked:** The HE is not directly blocked from providing services.
- **Locked:** The HE may be in the active presence state, but is being blocked from providing service. As a result all of its child entities are also blocked. For example, if the HE hosts an EE when the HE is locked PLM makes sure that the EE does not come up or if it is running PLM shuts it down. However if the HE is active one may run diagnostics on it as it is not related to service provisioning.
- **Locked-inactive:** The administrator forces the isolation of the HE and all its dependants from the system. It is put into the inactive presence state.
- **Shutting-down:** The HE may continue to serve existing users, but it may not accept new service requests. When the last existing user completes the administrative state of the HE changes to locked.

5.3.3.3 Execution Environments, Their Types, and Status

An EE may represent an instance of an OS, a VM hosting an OS, or a hypervisor hosting, and managing some VMs.

The common theme among these entities is that they are often referred as system software as each of them is a kind of container, which is capable of running some other software entities while managing system resources. They also expose some control interface through which the hosted software entities can be controlled individually or as a collective.

Since within the container itself, the different software shares the different resources fault propagation may occur even between otherwise independent pieces. Consequently the representation of the different containers, the EE is considered as a barrier to fault propagation. It is an inherent fault zone and therefore it is used for fault isolation.

In the PLM information model a HE representing some HPI entity may host at most one EE and vice versa an EE is hosted on exactly one HE.[1] This single EE may be an OS instance or a hypervisor – also referred as virtual machine monitor (VMM).

This single EE may host additional VM and OS instances all sharing the computing facilities of the HPI entity and consequently all impacted by a failure of this HPI entity. Obviously any software running in these EEs will also be impacted, therefore for availability management it is essential to be aware of this lower level organization of the system when, for example, one decides about the distribution of redundant entities among EEs.

Being a fault zone and reflecting the organization of fault zones are the features that the EE logical entity introduces to the system information model. It also exposes a simple and uniform management interface for these different software entities that PLM users can rely on without the need to be aware of the specificities of each of these system software entities.

The PLM implementation is responsible for mapping the administrative control operations and the state information between the particular software instance represented by the EE entity and the PLM administrator. Doing so PLM may interact with the software instance using a Portable Operating System Interface for Unix (POSIX) [73] like interface in case of an OS instance, or use some API such as libvirt [74] for managing the virtualization facilities.

In either case, PLM does control the life-cycle of the EE software entities. So we can say that there is an actual entity in the system, which corresponds to the configuration object in the information model and which is managed by PLM. PLM also controls the type of the EE instance starting up, which means that if the booting software does not correspond to the expected EE, PLM is at least able to prevent it from the completion of booting, but typically it is able even to initiate the boot of the correct software.

To determine whether an EE is the correct one, the PLM information model defines EE base types and EE types. The base type specifies at least the vendor of the EE, but it may also indicate the product and the release. The EE type identifies a particular version within a base type. It also provides some instantiation and termination parameters for the type applicable to all EE instances of the type. For example, the EE base type may say that it is a particular vendor's Enterprise Linux line. It may have several EE types depending on the various versions used in the system, and for each it would indicate the appropriate instantiation and termination timers.

Looking at the EE object class used to configure each EE instance in the system; it references the EE type as a configuration attribute based on which PLM can determine if the booting instance is the expected one.

As we have seen in Section 5.3.3.2, this approach differs from the approach used with HEs which only configures the HE base type. This difference reflects the fact that PLM controls the

---

[1] In reality it may be the case that more then one hardware element is required for the execution environment. The PLM specification calls for selecting the most important HE as the hosting entity. See also discussion on dependencies in section 'PLM Entity Dependencies.' on page 105.

version of the EE booting as opposed to just detecting the version of the HE plugged into the system.

Each EE object also provides the state information for the EE instance it represents. The PLM specification defines the same set of states for the EE as for the HE. That is, it defines the presence, the operational, and the administrative states. The interpretation of these states is slightly different though than it is in the case of HEs, reflecting the EE's software nature.

EE Presence State

The presence state of an EE reflects its life-cycle. The EE presence state may take the following values:

- **Uninstantiated:** The system software represented by the EE object is not running, PLM has not started or detected its instantiation.
- **Instantiating:** The instantiation of the system software has started either by PLM initiating it or automatically after the hosting hardware reached the active presence state.
- **Instantiated:** The system software represented by the EE object has completed its startup procedure and reached the state where it is capable of providing the desired operating environment.
- **Instantiation-failed:** The system software did not reach the fully operational state within the applicable time limit. It failed to instantiate.
- **Terminating:** The EE presence state is set to terminating as soon as the procedure to stop an already instantiated EE has been initiated. The termination typically starts first with the graceful termination of the software entities hosted by the EE followed by the termination of the EE itself.
- **Termination-failed:** PLM sets the presence state of an EE to termination-failed if after the applicable timeout it still could not verify the successful stopping of the EE instance.

As opposed to the HE presence state which follows the HPI hot-swap state definition the EE presence state definition is aligned with the presence state of Availability Management Framework (AMF) [48] entities. The main difference compared to AMF is that there is no restarting state for EEs. In AMF the restarting state reflects that any assigned service function logically remains with the entity during its restart. A lack of this state at the PLM level reflects that there is no such expectation for PLM entities mainly because AMF and the other AIS services manage themselves the life-cycle of their entities and perform service recoveries as appropriate. A PLM level service recovery could contradict to these higher level actions.

EE Operational State

The EE operational state is very similar to the HE operational state. It reflects whether the PLM implementation is aware of any type of error condition of the represented EE instance.

PLM sets the operational state to disabled

- if the EE presence state becomes instantiation-failed or termination-failed;
- if it detects an error condition in the EE in some implementation specific way; or
- if it receives an error report for the EE.

Whenever the EE enters the disabled state, the PLM implementation is responsible of isolating it from the rest of the system to protect the system and prevent any fault propagation. This means the *abrupt* termination of the EE instance, which PLM carries out via the entity hosting the faulty EE. It depends on the management capabilities of this host entity how PLM achieves this task.

For example, if the EE resides directly on a HE, PLM may power off or reset the hosting HE. If the EE is a VM hosted by a hypervisor, the hypervisor typically offers the control to terminate the particular VM and the PLM implementation would use that.

As in the case of HEs, when PLM cannot initiate or verify the isolation due to the relevant entity being inaccessible it indicates the condition by setting the management-lost and the isolation-pending flags.

PLM re-enables the EE operational state if it receives a report that the EE has been repaired by the administrator or the error condition was cleared.

EE Administrative State
As for HEs, the EE administrative state reflects whether the EE may offer its functionality for its users. The EE administrative state is controlled via administrative operations, which we will discuss later in Section 5.3.5.2.

The EE administrative state has the following values:

- **Unlocked:** The EE is not prevented from providing service.
- **Locked:** The EE may be instantiated, but it is prohibited from providing service. For example, for an EE representing an instance of the Linux OS the locked state may mean that it runs at a lower init level.
- **Locked-instantiation:** The EE must not be instantiated. For example, PLM may use the hypervisor's control capability to block the instantiation of a VM, or for an OS hosted directly on an HE assert the reset state of this hosting HE.
- **Shutting-down:** The EE may serve existing user, but it may not accept new service requests. Once all existing users have been served the EE enters the locked state.

5.3.3.4 Entity State Interaction

As PLM entities change their states they impact each other, allowing or preventing the other to provide services. Whether an entity can provide any service is reflected in its readiness status. The path along which the entities impact each other is determined by the dependencies among the entities.

In this section we present how this essential for availability management information is defined and maintained by PLM. Subsequently in Section 5.3.4, we present how PLM users can obtain this single most important information.

Readiness Status
The readiness status reflects to what extent an HE or an EE is available for service. It is composed of the readiness state and the readiness flags.

The **readiness state** summarizes the presence, the operational, and the administrative states of a PLM entity itself and the readiness state of the entities it depends on. It has three values:

- **In-service:** When the readiness state is in-service it indicates that
  - the PLM entity is
    - o healthy – its operational state is enabled;
    - o capable – its presence state is active or deactivating in case of an HE, and instantiated or terminating for an EE; and
    - o permitted of providing services – its administrative state is unlocked. And
  - all the entities the PLM entity depends on are also in the in-service readiness state.
- **Out-of-service:** A PLM entity is out-of-service when it is
  - not capable – it is disabled or it is in a presence state other than
    - o active or deactivating in case of an HE;
    - o instantiated or terminating in case of EEs; or

- administratively prevented from supporting service provisioning – it is locked, locked-inactive, or locked-instantiation; or
- missing a dependency such as its container entity being out-of-service, for example, due to being locked.

- **Stopping:** A PLM entity enters the stopping readiness state when the administrator moves it or an entity it depends on to the shutting-down administrative state.

The **readiness flags** further refine the readiness state. There are six flags qualifying the information provided by the readiness state:

- **Management-lost flag:** It indicates that the PLM implementation has no management access to the PLM entity; any state information for the entity needs to be taken with a grain of salt as PLM cannot verify the information and the state values which represent the last known values may be old data.
  The following two flags can only be used in conjunction with the management-lost flag:
  - **Admin-operation-pending:** The PLM could not perform or verify the execution of an administrative operation due to the loss of management access.
  - **Isolate-pending:** Because of the loss of management access, the PLM could not isolate or verify the isolation of the PLM entity.
- **Dependency:** By setting this flag PLM indicates that the PLM entity is not in-service due to a state change of an entity that the PLM entity depends on, rather than due to a state change in the PLM entity itself.
- **Imminent-failure:** PLM uses this flag when it receives or detects some indications that a currently enabled PLM entity may become disabled any moment (e.g., a sensor may have signaled the stepping over of some threshold). However at the moment there is no state change yet. Effectively it is a warning recommending users to abandon the use of the entity for which it is set to prevent sudden service impacts.
- **Dependency-imminent-failure:** Same as imminent-failure but rather than by its own fault the PLM entity is expected to become unavailable due to a dependency.

PLM Entity Dependencies
PLM entities are organized in sub-trees under the PLM domain object. This tree organization is interpreted as a dependency resulting from containment and it also implies a top-down propagation of the state change implications.

Let us assume a chassis having slots which house blades that are carriers of mezzanine cards, each of which may host an OS instance. If a mezzanine card fails it can no longer run the OS it hosts; if the carrier blade breaks none of the mezzanine cards can run their EEs or provide their other functionalities, and so on. The presence and operational states of these entities interact in this manner according to the specification, which reflects reality.

Depending on the hardware architecture, the representation of hardware dependencies may not be this straightforward. It is also influenced by the choice of which HPI entities one wants to reflect in the PLM model.

For example, the CPU, the memory, the hard drive, the I/O card all may be necessary to run an OS instance, but they may be different pieces of hardware, which means that they show up as different HPI entities and accordingly they may be represented in PLM as different HEs. In the tree representation these HEs could be children of an HE representing the board. The question is where in the tree should be the EE object reflecting the OS requiring all these elements.

To properly reflect this relation besides the tree organization the PLM information model also includes the configuration object class that indicates dependencies among PLM entities. Using objects of this class one could define that although it is the HE mapped to the CPU HPI entity that hosts the OS, this EE also depends on the HEs representing the other devices (i.e., memory, hard drive, and I/O card).

Hardware platforms built for high-availability systems typically encompass some – a lot of – redundancy. Among others, this is the reason why they are often built as a chassis capable of housing many identical blades. These blades then are connected through redundant communication paths including redundant switches and powered through redundant power supplies.

For the correct operation of the system, however, at any moment in time not all of these redundant devices are needed. It is enough that at least one of the duplicated devices is available. The dependency class allows the reflection of this constraint.

The PLM dependency object defines the way a PLM entity (HE or EE) depends on a set of other PLM entities. It is used for each set of redundant entities and it specifies how many entities in the set need to be available for service to satisfy the dependency.

'Available for service' here means that the readiness state is not out-of-service. In other words, at least the indicated number of entities of the set needs to be either in the in-service or in the stopping readiness state.

For example, for the chassis with dual power supply we could specify that the chassis has a dependency on two power supplies and at least one of them must not be out-of-service.

The dependency – whether it is implied through containment or defined through dependency objects – impacts the readiness state of the dependent entity.

In case of containment all ancestor entities need to be in the in-service state for the PLM entity to be able to stay in the in-service state. If any of them moves to the stopping or the out-of-service readiness state, the dependent entity moves with them. In other words, if an entity moves to the stopping or out-of-service readiness state, all its in-service descendant entities move with it immediately.

In case of a dependency relation configured using the dependency object class, if an entity requires m out of the set of n entities, then for the entity to be in-service at least m of these n entities need to be in-service. If at least one of these m entities moves to the stopping readiness state the dependent entity also moves to the stopping readiness state.

In conjunction with the readiness state the dependency readiness flag indicates that the entity is stopping or out-of-service due to a dependency (regardless of whether it is specified through the tree hierarchy or the dependency objects) and not as a result of a change of its own state values.

The dependency-imminent-failure flag is used to indicate that an imminent failure may occur due to a dependency; however, this does not change the readiness state of the entity for which it is reported. It is a warning. An HPI sensor stepping over a sensor threshold may set the imminent-failure flag for the related HE. PLM propagates this to all its dependent entities by setting their dependency-imminent-failure flags. For the propagation PLM takes into account both the containment and the dependency objects.

It may be interesting to look at how some virtualization features may be represented using dependencies. As we mentioned earlier, if an EE is hosted on a HE it can be configured as the child of this HE. If it is a hypervisor it could be the parent of a set of dependent child-EEs. We can represent each VM and its OS as such a child-EE of the hypervisor EE. These will be the leaf objects in the PLM information model.

Such a representation, however, fixes the relationship of these child-EEs with the HEs and the hosting hypervisor due to the naming convention used in the information model discussed in Section 4.3.1.3. This conflicts with one of the desirable feature of virtualized environments, namely that VMs may migrate.

Instead, one may configure such migrating VMs using dependencies. In this case the migration enabled VMs would be configured as direct descendants of the PLM domain, each of which depends on a set of hypervisors which are configured as leaves of the HE sub-trees. As long as at least one of the hypervisors is in-service and ready to host the VM, the VM can be migrated to it and can be in-service as well. Note however that at the moment PLM is not required by the specification to manage this migration nor it is specified how the dependency object reflects which hypervisor hosts the VM at any given instance of time.

Dependencies just like the PLM domain are not typed objects in the PLM information model.

5.3.4   Tracking of PLM Entities

For PLM users the most important aspect of platform management is that through its track interface it can provide them with up-to-date information about the readiness status of the different PLM entities within the system. This includes readiness state changes, readiness flag changes, but also allows the users to

- verify planned administrative operations to ensure that they do not cause service interruption;
- gracefully execute administrative operations; and
- detect domain reconfigurations, such as adding or removing PLM entities.

The PLM track API supports single time inquiries as well as continuous monitoring.

To receive such indications first of all a PLM users needs to identify the set of PLM entities that it would like to track. The PLM does not specify any default set of entities like we have seen in HPI, that is, all entities visible in a domain.

The PLM user may create this *entity group* by either listing the distinguished names of all included PLM entities explicitly, or naming the root entities for the sub-trees within which optionally all, only HEs, or only EEs entities are tracked.

Tracking a sub-tree allows the detection of PLM entities added to the sub-tree, while tracking of status changes and removals is possible regardless of whether the entities are listed explicitly or implied by a sub-tree.

A PLM user may choose to receive the readiness status information for the entire entity group in each callback or only for those entities that have a change in their status.

The tracking may involve three steps: the validation, the start, and the completed.

They enable the validation and graceful execution of different controlled operations as well as the tracking of spontaneous changes as they occur in the system. A PLM user may subscribe to track all three steps, the last two or only the completion step.

The user requests the different options and the entity group to be tracked when it initiates the tracking.

As changes occur or being proposed in the system, a PLM implementation continuously evaluates their impact on the PLM entities and sets their different states and status flags as appropriate. Whenever there is a potential impact on the readiness status of some entities being tracked, PLM informs those subscribers that have registered their interest for the impacted entity.

If the change is a proposal – such as a try-lock discussed in details in section 'PLM Lock and Unlock' on page 114 – PLM considers the validate step first.

If the change is a request to carry out a controlled operation gracefully – such as a shutdown – then PLM moves directly to the start step.

If the change is abrupt such as a failure, PLM can only inform the interested parties for which it uses the completed step.

In any case first, PLM calculates the list of PLM entities that have changed or would change their readiness status. This is the set of impacted entities.

5.3.4.1   Validate Step

The goal of the validate step is to negotiate whether the proposed change is acceptable for all the parties that registered their interest for such negotiation. The PLM specification defines the acceptability criterion as whether the proposed change would cause any service outage.

The validate step implies the following actions:

- PLM evaluates if there is any PLM user process tracking any of the entities the readiness status of which is impacted by the proposed change. If there is a PLM user that requested a callback in the validate step for any of the entities, PLM shall inform this user about the proposed change.
- PLM calls back each user who tracks an entity group that contains at least one impacted entity. For each of the entities, PLM indicates its current readiness status and the status it will have after the change is accepted.
- The tracking user has the choice either to accept the proposed change or to reject it depending on whether it expects any service outage as a consequence. In either case it needs to respond to PLM indicating its choice. Accepting the operation also means that the user has made all the necessary preparations to avoid service outage if the change is carried out. For example, the user may make some resource reservations.

  Note that these preparations cannot still guarantee that there will be no outage during or after the change as failures may happen in the system any time. However it should ensure that if a second operation is validated before this change takes place; the validation will take into account the combined effect of the two operations.
- When all users that received a callback from PLM for the validate step have accepted the proposed change, PLM proceeds with the start step to deploy it. If any user rejects the proposed change, the operation is *aborted*:
- PLM calls back the same set of users as for the validate step to inform them that the proposed change has been cancelled and they may release any resource that they reserved in response to the validation request.

  PLM also communicates the rejection to the initiator of the change (e.g., HPI, the administrator).

Once the proposed change has been validated successfully the change cannot be canceled any more and PLM proceeds with its deployment through the start step.

5.3.4.2   Start Step

The start step guarantees the graceful deployment of the changes by executing it in a top-down manner. This is achieved by first initiating the change toward PLM users, that is, toward the upper layers, and then waiting until these users indicate that they have completed all the necessary actions. Only after this confirmation the operation is deployed at the PLM level.

The start step involves the following actions:

- PLM evaluates the set of PLM user processes tracking the impacted entities for the *start* step. That is, those users that requested callback for the validate step or for the start step.
- PLM calls back the identified set of users to requests them to carry out any action required to avoid service impact from the initiated change. To be able to evaluate the impact at their level, PLM provides them with the current and the new readiness status (i.e., when the operation is completed) of the impacted entities.

  For the PLM (or higher level) users (e.g., AMF) this means that they may need to re-arrange the provisioning of their service so that they avoid using the PLM entities going out-of-service. Any reservation made in the validate step would be used at this time.

- Once the evacuation is completed, users respond to PLM indicating that they are prepared for the change and PLM can apply it. That is, PLM can perform all the necessary actions related to the initiated readiness status change without service impact.
- When all users have responded toward whom the change was initiated, PLM deploys the changes at the PLM level and sets the different states and flags as appropriate for the change.
  The operation is complete and PLM reports this fact to the initiator.

With this, however, the tracking process is not over yet as there might be users who requested track callbacks only for completed operations.

5.3.4.3 Completed Step

The completed step is the bottom-up signaling mechanism of PLM to inform its users about readiness status changes in the system.
PLM notifies its users only in the completed step whenever

- entities become available for service;
- they fail unexpectedly;
- the administrator forces an operation (see Section 5.3.5.2 for more details); and also when
- a user tracks only completed changes.

With the completed step PLM also confirms the committed changes toward those users who track the validation or start steps.
The completed step involves only a single action: PLM calls back all its users that track any of the impacted entities regardless the tracking step option. With this callback PLM informs them that some entities they are tracking have changed their readiness status and it provides these changes in the callback.
PLM does not expect any answer to the completed step.

5.3.4.4 Tracking Side Notes

The most important thing to understand about the tracking API is that the tracking processes are not necessarily the same as the entities impacted by the readiness status changes at the PLM level. In fact in the AIS architecture, the tracking entity is the CLM and not the 'cluster node' itself, which is CLM entity impacted by the readiness state change at the PLM level.
This shows that the tracking API is geared toward management entities in the system and it provides means for these entities to obtain the information necessary to avoid service impact in a timely manner.
It is the responsibility of these management entities to find out the correct list of entities they need to track in order to be able to maintain SA and to start the tracking process with PLM.
To track the correct PLM entities these management entities need to understand the relation between the entities they are in change of and the PLM entities. The PLM has no idea about this relation or any other upward relation. It only handles relations between PLM entities and toward HPI entities. Doing so, however, PLM releases the higher level management entities from processing this low level information in all its details.
It is also the PLM users' responsibility to interpret the changes, that is, their consequences. For example, considering the locked administrative state, which is defined as the entity out-of-service but instantiated or active, it is implementation dependent how or whether PLM can achieve it at all that such an OS instance is not used by any nonsystem process to fulfill the 'out-of-service' status.
Similarly for a locked HE not hosting an EE what it means that it is not allowed to provide services while being fully functional otherwise is not defined.

This means that while PLM changes and signals the readiness status of the impacted PLM entity, PLM may not guarantee that all the dependent entities indeed will experience the change immediately. It is a management entity tracking to the readiness status that can enforce the proper reaction of higher level entities.

Finally, it is also interesting to note that neither the validation nor the start steps are time limited. The PLM specification does not define timers for them, so they may potentially take a very long time.

The reasoning behind this is that these steps are used in cases when there is a way to escalate the operation. As we will see in the following sections, if the try-lock takes too much time, the administrator may escalate it to a normal or a forced lock; or if after opening the latches the signal indicating the completion of the evacuation does not come in time, the board can be taken out unexpectedly for the system, which will perceive this as a failure.

5.3.5 Administrative and Management Aspects

There are two aspects of the administrative control of the PLM that require discussion:

- **Configuration:** As mentioned earlier, the expected configuration is described in terms of HEs and EEs and their types. PLM uses this information to match the HPI entities detected in the system with the HEs and to control the EEs booting on them.
- **Administrative operations:** The administrator may issue different administrative commands on PLM entities or manipulate the platform physically. These actions control the administrative and other states of the targeted entities. The key feature of the PLM administrative API is that it offers the *try* option which succeeds only if the operation causes no service outage.

In this section we look at each of these aspects.

5.3.5.1 Mapping HPI Entities to PLM Entities

In the course of our discussion of the HPI we have seen that it detects the hardware composing the system and presents this information to each of the domains defined in the system as a DET and a RPT (see Section 5.2.4). The first lists all the HPI entities visible in the domain, while the second lists the management resources available on these entities including the IDRs. In addition the DRT lists additional domains in the system.

Conversely, the PLM information model includes three relevant sets of management objects: (i) the objects representing the expected in the configuration HEs; (ii) these reference the objects representing HE base types that indicate the expected HPI entity types; and (iii) for each HE base type there is a set of HE type objects, each of which refines further characteristics of the different implementations of this HPI entity type.

These management objects of the PLM information model describe the expected configuration of the hardware.

The task of a PLM implementation is to match up the expected configuration with the actual one, that is, the HEs of the information model with the discovered HPI entities.

The PLM specification outlines the matching procedure only at a high level. It implies approximately the following procedure:

PLM opens a session on the default HPI domain to gain access to the DET, RPT, and IDRs. It also reads the PLM information model:

1. For each HE in the PLM information model starting at the PLM domain, PLM sets the *parent entity path* to the root.

2. PLM reads the relative entity path attribute of the HE, which together with the parent entity path defines the location where the matching HPI entity is expected to be. The entity path may be an exact location or may indicate alternatives, for example, a range of locations.
3. PLM reads also the HE base type of the HE, which indicates the HPI entity type.
4. From the DET PLM finds the HPI entities that may match the given location and entity type.
5. PLM compares the IDRs of these potentially matching HPI entities with the different HE types of the HE base type the HE references. An HE type contains an attribute that indicates the IDR values that characterize the entities of the give HE type.
6. There is a match if the IDR of the HPI entity matches all the IDR values required by the HE type. Note that the HPI entity may have more values in its IDR, or it may not have an IDR management instrument at all. In this latter case the HE type should be empty too for a match.
7. Once PLM finds a match it sets the runtime entity path attribute of the HE and the *parent entity path* variable of the matching procedure to the entity path of the matched HPI entity. The matching procedure continues with step #2 for the child entities of this matched HE unless this was a leaf HE or its children are EEs. In this case PLM proceeds with step #1 to process the next HE sub-tree. If there is no more HE in the PLM domain the procedure is completed.

For the matched HEs PLM sets the state information based on the readings of the RPT for the matching HPI entity. In particular PLM changes the HE presence state from 'not-present' to the one matching the detected status.

If after investigating all HPI entities in DET there are still unmatched HEs and there are also other nonpeer HPI domains present in the DRT, the PLM implementation moves on to investigate those domains if there are more entities visible from them that could match the still unmatched HEs.

It is possible that PLM cannot match all the configured HEs. They will remain in the PLM information model with missing runtime entity path attribute until HPI detects a new entity that would satisfy their matching criteria. PLM subscribes to the HPI domain events to get informed about such events.

It is also possible that PLM discovers that the HPI entity at a particular location is different from the expected one. PLM will generate an alarm for such a case.

As we noted, the matching procedure described in the first release of the PLM specification only at a high level and therefore it can be implemented in different ways. This may result in a situation in which taking the same HPI information and the same PLM configuration, two PLM implementations will match up the entities differently and may find different sets of unmapped HEs and/or unmapped HPI entities.

Once the HPI entities are mapped to the HEs the PLM implementation figures out from the RPT of the entities domain what management capabilities these entities have and how these relate to the control operations PLM needs to perform.

We have already mentioned that the most important operation is to be able to isolate the HPI entity from the rest of the system whenever PLM or anyone else in the system detects an error on it. This of course also requires that PLM identifies how to detect errors on the different HPI entities.

At the moment of writing the PLM specification does not describe how this functional mapping is performed.

5.3.5.2 PLM Administrative Operations

Due to their different nature, the PLM defines slightly different sets of administrative operations for HEs and EEs. Table 5.1 gives a summary of the PLM administrative operations.

Except for the operations marked as physical, an administrator issues these administrative operations through the IMM object management (OM) interface [38] on the objects representing in the PLM information model the targeted PLM entities.

Table 5.1 PLM administrative operations

Hardware element	Execution environment	Effect
Physical removal of hardware	n/a	Change the HE presence state to not-present and uninstantiated for its EEs
Physical insertion of hardware	n/a	Attempt to match the HPI entity and change the states of the mapped HE as appropriate
Unlock		Change the administrative state to unlocked
Lock		Change the administrative state to locked
Shutdown		Change the administrative state to shutting-down
Deactivate	Lock-instantiation	Change the administrative state to locked-inactive/locked-instantiation
Activate	Unlock-instantiation	Change the administrative state to locked
Reset	Restart	Cycle the presence state
Repaired		Change the operational state to enabled
Removed		Unbind the management object from the entity

Physical Insertion or Removal of Hardware

In Table 5.1 we have taken into account as PLM administrative operations the physical insertion and removal of hardware entities. The world of hardware and software meets at the PLM level also in the sense that we can talk about 'hard' and 'soft' administrative operations.

When carrying out a 'soft' operation, the administrator interacts with the PLM implementation via the administrative API. This gives PLM the possibility of either accepting or refusing the operation and also of enforcing a particular way of its execution.

On the other hand, when someone decides to physically manipulate the hardware, that is, to perform a 'hard' operation, the PLM and HPI implementations have little chance to control the actions. They may need to resort to detecting the physical manipulations. These may or may not follow the intended execution order. If they do not, then the 'offending' action is mapped into a failure condition and handled accordingly as we already noted at the discussion of HPI in Section 5.2.5.

In any case, the PLM implementation finds out about any physical manipulation of the hardware from HPI, which detects and reports them to its subscribers as events.

When a new hardware is added to the HPI domain HPI generates an event and the detected HPI entity with its management capabilities is added to the DET and its resources to the RPT. In the subscriber PLM implementation, this event triggers the HPI entity matching procedure as described in Section 5.3.5.1.

The mapping can only be successful if the PLM configuration contains a HE matching the properties of the inserted HPI entity. The mapping may fail for different reasons, such as the entity is not represented in the PLM model or the previous mapping was not deterministic and now it conflicts with the hardware configuration. The key point is that PLM is not required to remap HEs. In fact to avoid any service impact, it must not remap them while they may be providing service. Remapping is only allowed for entities in the out-of-service readiness state.

When PLM cannot map an HPI entity it may generate an alarm. In the alarm PLM is not required to indicate the reason for the mapping failure or whether remapping may clear the alarm. Moreover, the first version of the specification also does not specify any administrative operation to initiate a remapping.

In case of successful mapping, PLM sets the different states and flags for the mapped PLM entities based on the reads it collects from resources managing the HPI entity. PLM calls back any of its users tracking any of the PLM entities that change readiness state as a result of the insertion. The callback is issued in the completed step only (see Section 5.3.4.3).

When a piece of hardware is being removed from an HPI domain, depending on the type of the HPI entity, HPI may generate one or several events. As we have noted in Section 5.2.5, if the entity is not a hot-swappable FRU, an HPI implementation may interpret the removal as a failure of the HPI entity and signal it accordingly.

If the HPI entity is hot-swappable, HPI signals a series of hot-swap state changes. These changes may or may not be manageable and in addition the person performing the operation may or may not take into consideration the signals reflecting the state control.

PLM processes the signaled HPI events and if the entity being removed maps to a HE or impacts PLM entities it changes their state information according to the signaled event and generates the appropriate track callbacks.

In particular when HPI signals an extraction pending event for a hot-swappable entity (e.g., as a result of opening the latches of a board), which is mapped into a HE currently available for service, PLM attempts to cancel any HPI deactivation policy to gain control of the extraction process and if successful PLM tries to drive the extraction through the three-step tracking procedure as described in Section 5.3.4:

It initiates the validate and starts steps toward its users to evacuate all the services from the impacted HE. When this is completed, PLM also terminates any EE requiring the HE. At completion finally PLM sets the HPI resource hot-swap state to inactive thus allowing HPI to complete the operation at its level.

However, any time during this process the person performing the extraction may ignore PLM's attempt to control the process and may proceed with the extraction. This takes the HPI resource representing the HPI entity's hot-swap state to not-present which is reflected at the PLM level by setting the same value for the presence state of the mapped PLM HE and with that the readiness state goes out-of-service for all impacted entities. This results immediately in a track callback in the completed step notifying about the change, which cancels out any previously initiated validate and start steps for the entities.

If the HPI entity does not have the hot-swap capability or it is unmanaged, PLM only signals the completed step to its users.

Note that if the PLM entity is out-of-service at the insertion or removal of the mapped HPI entities then no readiness status change occurs and therefore no track callback is needed.

HE Deactivate and Activate

The 'soft' version of the insertion and removal of HEs are the HE activate and deactivate commands respectively, which control the administrative and if applicable the presence state of the HEs.

The biggest difference however is that while physical manipulation bypasses the administrative state of the PLM entities, the HE deactivate and activate commands do not. As a result these operations can be issued even on HEs whose presence state is not-present, in which case only their administrative state is changed. In addition PLM may set the management-lost and the admin-operation-pending flags for the HE indicating that it could not execute some part of the operation as in reality PLM may not always be able to tell apart if the entity is not there or it is inaccessible.

When the entity is inserted to the system, its state is controlled by the setting of the administrative state of the HE it is mapped to. That is, if it is locked-inactive the presence state can only change from not-present to inactive and PLM will maintain it there as long as the administrative state remains so. The presence state may change to activating only after the administrative state has been changed.

The activate command changes the administrative state of the HE from locked-inactive to locked. The operation is invalid for HEs in any other administrative state.

Moving to the locked administrative state allows the mapped HPI entity to start activating and eventually become active. However, it still may not offer any services, so its readiness state remains out-of-service, which means, for example, that no EE is allowed to boot. It also means that no change is signaled on the tracking API.

The counterpart of the activate command is the deactivate, which is only valid on a HE, which is in the locked administrative state and therefore it should not provide any type of service. Accordingly, the operation has no service impact. The deactivate command changes the HE's administrative state to locked-inactive and its presence state to deactivating followed by inactive.

Due to the required for the operation locked state, the entity is already out-of-service and its deactivation does not cause any change in the readiness status that would initiate any track callback.

PLM Lock-Instantiation and Unlock-Instantiation

The deactivate and activate commands apply exclusively to HEs. Their equivalents for EEs are the lock-instantiation and unlock-instantiation commands.

The same way as for the HE deactivate and HE activate, the EE lock-instantiation operation is only valid for EEs in the locked administrative state, and the unlock-instantiation only for EEs in the locked-instantiation state. As a consequence the state changes caused by these operations do not cause readiness status change and do not need to be signaled via the track API. Only the presence state may be impacted.

When the administrator initiates the lock-instantiation of an instantiated EE PLM terminates the EE and its presence state transitions to the uninstantiated state in addition to the administrative state changing to locked-instantiation.

When the administrator issues the unlock-instantiation command, PLM allows the instantiation of the EE in question; however, since it still remains in the locked administrative state it is still not allowed to provide services. For example, if this is a hypervisor, its child EEs remain uninstantiated.

The same way as the HE deactivate and activate commands do not require the actual presence of the mapped entity, the lock-instantiation, and unlock-instantiation also can be issued regardless of the EE presence state. In these cases PLM will set the management-lost and the admin-operation-pending flags for the EEs as well.

PLM Lock and Unlock

The lock and unlock operations apply to both HEs and EEs. They control whether a PLM entity may offer its services or not, which is reflected by the administrative state of the entity. Due to this state change the readiness state may also change which is most important for SA. That is, the administrator needs to be aware of its action, whether the entity goes out-of-service and potentially impacts others too.

From the perspective of the SA Forum services of the system that need to be available are managed by the AMF [48] as described in Chapter 6. AMF reflects in its information model the distribution of assignments among the AMF provider entities as a dynamically changing mapping, which follows the changes that occur in the system.

The relations between the AMF service provider entities and the PLM entities are also not obvious even though they are not changing dynamically. The reason is that they are represented in separate sub-trees of the system information model and they are mapped through the CLM entities composing a third sub-tree as shown in Figure 5.1 and described in Section 5.4.3. As a result figuring out the impact of a PLM entity on the services managed by AMF is not straightforward.

In addition, the service outage we are interested in is not in the current state of the system, but the way it changes due to a lock operation, that is, once the target entity and entities requiring it become locked and stop servicing. All of this depends on the actions AMF and the other services will make as a result of the lock.

This makes it hard if not impossible for the administrator to foresee if there is a service impact.

As a solution PLM offers different flavors of the lock operation:

- Try lock, which evaluates whether the operation causes service outage and if it does, then the operation is aborted.
- (Normal) lock, which gracefully carries out the operation to its end. That is, PLM informs its clients about the operations and waits until the services are evacuated from the impacted entities at higher levels and completes the lock only after it receives a confirmation from its clients.
- Forced lock, which carries out the lock without informing anyone in advance. PLM provides only postmortem notification.

PLM uses the appropriate options of the three-step track to communicate to the interested processes the changes occurred or about to happen.

For the try-lock operation PLM uses the track callback starting with the validate step as describe in Section 5.3.4.1.

If any user rejects the lock operation, it is *aborted* by PLM also using a track callback. PLM also informs the administrator about the rejection of the try lock. Once the validate step has been accepted for a try-lock, PLM carries out the operation continuing with the start step as if it was a normal lock operation.

For the normal lock PLM invokes the track callback in the start step requesting its users to carry out at their level any preparation necessary to prevent service impact caused by the targeted PLM entity going out of service. It is up to each PLM user to interpret what this means. We will see that, for example, the CLM simply passes on this information to its users, such as the AMF, which needs to move the service assignments away from the AMF entities going out-of-service as a result of the PLM entities going out of service. Ideally if this step has followed the validation step, AMF is able to evacuate all the services.

Once the evacuation is completed, the PLM users are expected to confirm that PLM can proceed with the lock of the PLM entities. Thus, PLM completes the administrative operation at its level and locks the target entity's by setting its administrative state to locked. With this, the entity's readiness state and the readiness state of all entities requiring it go out-of service. As a finishing touch PLM informs all its users tracking any of the impacted entities about the change by a final callback in the completed step.

As a result of a force administrative lock, PLM immediately executes all the actions required at its level to lock the target entity and by that PLM takes it out of service together with all impacted entities. Only after this, PLM informs its users tracking any of the impacted entities in the completed step that the target entity has been locked. Obviously in this case no service impact is taken into consideration at the execution of the operation.

Even in the earlier cases the service impact can only be considered to the extent that the PLM track API is used by the availability management.

The lock operation may be issued on PLM entities which are in the unlocked or the shutting down administrative state. It may also be repeated on the same entity as an escalation of the operation to a stronger option. That is, a try lock may be escalated to a normal or forced lock, and a normal lock to a forced. One may consider this escalation if the completion of the weaker option takes too much time since these operations are not time limited within the PLM.

If the entity is already out-of-service at the time of the lock operation – this may happen if it is disabled or missing a dependency – then none of the track steps is needed at all. The operation completes right away with PLM setting the locked administrative state for the entity.

The unlock operation moves the administrative state of a PLM entity to the unlocked state. This may or may not change the readiness state of the entity depending on its dependencies and operational state. If there is a missing dependency still or the entity itself is disabled the readiness state remains out-of-service.

Otherwise it becomes in-service, which results in a callback to all users tracking the target entity and any other PLM entity which becomes in-service together with it.

As a result of becoming in-service dependent entities may also change their states, for example, a HE hosting an EE now can start up this EE provided the EE itself is not locked for instantiation. If the unlocked entity is a hypervisor, it is also allowed to start up its child EEs that are allowed to instantiate.

PLM Shutdown

The interpretation of the administrative shutdown operation is that the target entity may continue to provide its services to existing users, but it has to reject new user requests. This is reflected with the shutting down administrative state.

When the service is completed for the existing users the shutdown operation is completed and the entity moves to the locked administrative state. The operation is valid only for entities in the unlocked administrative state.

As we have seen already the notion of the 'service' is rather vague for PLM entities, so distinguishing their existing and new users becomes virtually impossible – at least for the PLM itself. For example, for an OS the shutdown state could mean that no new processes can be started, which may be hard for PLM to enforce as the underlying OS does not typically support such a notion.

As a result the PLM shutdown operation is seemingly identical to the 'normal lock' operation except for the administrative state value.

If the entity was in-service at the moment the shutdown is issued, its readiness state changes to stopping, which means that PLM needs to inform users tracking the entity about the state change. As for the 'normal lock' operation PLM indicates to its users in the start step that a shutdown has been initiated. It proceeds with locking the PLM entities only when all called back users confirmed that they have completed the shutting down operation.

The key difference is not in what PLM does, but what it requests from its users. In the case of the lock operation the users are expected to evacuate the services from the impacted entities right away, while in the case of the shutdown they are expected to perform a shutdown operation by rejecting new users at their level. This does not require evacuation. They continue to serve existing service users. Thus, the time to perform a shutdown operation may be significantly longer than that of the lock operation.

As mentioned earlier, the PLM specification does not have a time limit for the start step, neither the lock nor the shutdown are restricted. The specification also does not mandate any particular behavior for the case when a PLM user cannot evacuate all the services due to lack of resources. It is left up to the user what strategy it defines for such a case. It may respond to PLM as soon as it evacuated all the services that were possible to evacuate potentially resulting in a service outage due to the completion of the operation at the PLM level. Or it may wait, and wait, and wait if the evacuation eventually becomes possible delaying by that the completion of the PLM lock.

The same applies to the shutdown, except that in this case no evacuation is attempted at all and the PLM user is expected to respond only when all the services 'die off naturally' on the impacted entities.

In either case as we have indicated, the administrator may need to deal with the situation that the operation takes too much time. The shutdown operation can be escalated to any option of the lock, which itself may be escalated to a forced lock that applies the operation right away.

On an entity in the shutting down administrative state the administrator may issue the unlock operation to cancel the effect of the shutdown. This returns the entity to the unlocked administrative and in-service readiness state. As always the readiness state change is signaled by PLM via the tracking API. In this case again only the completed step is used.

PLM Restart and Reset

The reset and restart administrative commands cycle the presence state of the targeted entity. Reset is applicable only to HEs and restart is applicable only to EEs. The goal of these operations is typically a repair of some kind.

PLM maps the reset operation to the HPI function setting the reset state of the resource controlling the reset of the HPI entity to which the HE was mapped. The specification does not mandate and the administrator has no control over the type of the reset (e.g., warm, cold, etc.) PLM performs. The mapping depends on the PLM implementation.

The HE reset brings the HPI entity to a known state and therefore it may be used to repair the entity. The reset may however be performed to repair entities hosted on the HE.

An administrator has the choice to restart an EE in a normal fashion or abruptly.

The *normal restart* means the termination and re-instantiation of the EE itself through its own means. That is, the normal restart is mapped to the appropriate operation of the *target* EE (e.g., to the 'reboot' command in the case of Linux). Note that this typically means an abrupt termination for the entities hosted by this target entity; hence the restart operation isolates them from the rest of the system and may even repair them. However, the EE itself is still trusted to be able to execute the restart correctly.

The *abrupt restart*, on the other hand, is mapped by PLM to the appropriate operation of the entity *hosting the target* entity. If the EE resides on an HE directly then the abrupt restart is mapped to the mentioned above HPI reset. If the EE is a child of another EE such as a hypervisor, then the abrupt restart is mapped to this hypervisor's appropriate operation to restart only this given child EE.

As a result the abrupt restart isolates and it may also repair the target entity itself. This operation handles the case when the target EE is assumed to be faulty and cannot be trusted any more.

Whether it is an HE reset or an EE restart, the operation impacts the dependent entities. All child entities may go through their own presence state cycle and the same may happen to entities defining a dependency on the target entity or its children.

The most important impact of the operation is that after successful completion the operational state of disabled entities becomes enabled again.

As a result of the presence state cycle, the readiness state of the target and all dependent entities may also go through a cycle: It changes to out-of-service first and then to in-service if the operation is successful. This means that PLM calls back all the processes that track any of the impacted entities, but all readiness state changes are made only in the completed step regardless of whether the operation is normal or abrupt.

PLM Repaired and Removed

The common feature of the repaired and removed operations is that they inform PLM about some change about which PLM would have no information otherwise.

The repaired operation is applicable only to PLM entities that have their operational state disabled. By applying the operation to such an entity the administrator informs PLM that the entity has been repaired and PLM should try to take it back into service. To ensure correct initialization the entity should be handed over in the inactive/uninstantiated presence state.

If PLM does not detect the error condition any more and it can activate or instantiate the entity then PLM clears all of the readiness flags and sets the operational state to enabled.

If the circumstances evaluate so it also sets the readiness state to in-service. In turn, entities dependent on the repaired entity may also become in-service.

For all the entities changing their readiness status PLM calls back the users tracking them in the completed step.

With the removed operation the administrator indicates that while PLM had no management access to some entity it has been physically removed from the system, while the management object representing it remains in the system. Accordingly, it only applies to entities for which the 'management-lost' readiness flag is set. As discussed in section 'Readiness Status' on page 104 when PLM loses management of an entity it only sets the management-lost flag for it and leaves all the states including the readiness state with their last known value. So for a board that was unreachable before its removal the system may still report that it might be still in-service, as are its dependent entities. The 'might' is implied by the setting of the management-lost flag.

Thus, this action corrects the state information in the PLM information model that PLM cannot detect by itself since it has no access to the entity. (Recall that normally PLM detects the removal of HPI entities via HPI and of EEs by implementation specific ways.)

As a result of the operation PLM sets the presence state of the impacted HEs to not-present and for EEs to uninstantiated. It also corrects the states of the dependent entities. For entities that change their readiness state as a result of these actions, PLM also informs about the change the users tracking them through a callback in the completed step.

5.3.6   Service Interaction

5.3.6.1   PLM Notifications

The PLM generates a set of alarms and notifications for its entities using the SA Forum Notification service (NTF) [39].

Optionally, PLM may also convert HPI events into a format appropriate for NTF, so they are delivered by the same channel as the notifications of all of the AIS services. This is done in addition to the applicable PLM notifications. That is, an HPI event which triggers some changes at the PLM level is reported by PLM on behalf of HPI as well as on its own, reporting the PLM level change.

As opposed to the HPI event subscription, which delivers all events of the domain, the HPI events converted by PLM into NTF notifications can be filtered based on the event type and some other attributes because these attributes of the HPI event are converted into NTF fields. The rest of the content is provided as raw data in one of the three standard formats or in implementation-specific way. The standard formats are:

- as binary with the most significant bit first;
- as binary with the least significant bit first; or
- in the external representation format defined in [75].

For its own entities the PLM generates the following alarms:

- HE Alarm;
- EE Alarm;
- HE Security Alarm;
- EE Security Alarm;
- Unmapped Hardware Entity Alarm.

Except for the unmapped hardware entity alarm, the PLM specification leaves it implementation specific the exact conditions that trigger the alarms. An alarm is cleared by PLM when it cannot detect the alarm condition any more, for example, because a repair cleared it or the entity was removed from the system.

PLM generates the unmapped hardware entity alarm when it detects an HPI entity, which it cannot map to any HE when applying the matching procedure described in Section 5.3.5.1. There we mentioned the case that the alarm is generated when the HPI entity at a location does not match the HE expected at that location.

PLM also generates this alarm when it discovers HPI entities that cannot be mapped to any HE. The exact criteria of how a PLM implementation determines this fact is not straightforward since the relation of HPI entities and PLM HEs is not 1 : 1, but n : 1. That is, several HPI entities may be mapped to a single HE. For example, the PLM model may only represent the carrier blade, but not the individual mezzanine cards it carries. While HPI has an entity path identifying each of the later ones, they are not considered unmapped by PLM. Instead, all of them are mapped to the carrier blade.

5.3.6.2 Monitoring and Controlling HPI Entities

With respect to the platform hardware, the PLM specification was written with HPI in mind. Nevertheless PLM does not mandate the use of HPI. Any other solution may provide PLM with similar information. In this section we look at the case when PLM uses HPI.

As described in Section 5.3.5.1 when PLM uses HPI, it finds out about the different HPI entities composing the hardware platform using the HPI discovery procedure and maps them to the HEs in its own information model. Since the PLM specification does not describe exactly how PLM implementations should use HPI to accomplish their task, different PLM implementations may have different strategies for the discovery. It is important, however, that a PLM implementation detects all the HPI entities represented as HEs in the PLM configuration and also all the management features offered by this underlying hardware.

An HPI implementation may expose the resources for the same HPI entity in several domains and not necessarily the same way. As mentioned in Section 5.2.3.3, for example, the hot-swap capability of an entity is typically indicated in only one of the domains through which it should be controlled. Other resources in other domains related to the same HPI entity will not indicate it as hot-swappable.

This means that a PLM implementation needs to discover all these domains to become aware of all the available resources for each HPI entity, that is, not just to map them properly to the HEs to the HPI entities, but also to discover correctly their states and the available operations.

To remain up to date with the status of the HEs, a PLM implementation also needs to subscribe to the HPI events of these different domains. Through these subscriptions PLM can learn in a timely manner about the changes occurred in the different domains. These may be state changes of the resources, but also addition or removal of entities.

While the addition and removal events are relatively straightforward to map to PLM, the mapping of other HPI events is not necessarily so.

For example, an HPI resource may report that the temperature measured by a sensor exceeded some threshold, but whether this means a failure and the entity needs to be isolated or it is a warning that indicates an imminent failure, neither of the specifications (i.e., PLM or HPI) defines.

Moreover, since each entity may be exposed through several resources of several domains, it is also possible that the same event needs to be interpreted differently depending on the context of other events reported by other resources in those other domains.

Since no standard guidelines exist, each PLM implementation may have its own slightly (or not so slightly) different mapping of HPI events to the status of PLM entities. They may also react differently to some of these events.

Even for PLM operations (such as the administrative operations) the specification does not elaborate in detail how they map to HPI control operations. It only describes the intended result and the PLM implementation is responsible for the 'correct' mapping to the HPI entity to the target HE and its detected management capabilities.

In any case, the complexity of mapping of entities, events, and operations between PLM and HPI is one of the main reasons the PLM was defined and became part of the AIS. It guarantees that at least within a single PLM domain these aspects are handled uniformly and it also frees higher level entities from this burden.

5.3.6.3 PLM Interaction with IMM

As we mentioned already in Section 5.3.3, the PLM obtains the PLM information model from the IMM [38]. IMM also provides the administrative interface for all AIS services including PLM.

All the 'soft' administrative operations discussed in Section 5.3.5.2 are issued using the OM interface of IMM (IMM OM-API) and target a particular object representing a PLM entity. In turn IMM invokes the PLM implementation to pass on the request of the administrator. When the operation completes PLM returns the result to IMM, which forwards it to the administrator.

The challenging part in this interaction is PLM acquiring its information model as according to the specification, the IMM provides its services only within the cluster, but the PLM is below the CLM in the architecture; thus, it does not require the presence of the cluster at all.

Therefore we need to make the assumption that there will be a cluster forming on top of the PLM domain. However for the bootstrapping of this cluster the PLM specification does not make any suggestions. We will look at the issue in a bit more details in Section 5.4.3. Here we will assume that the PLM is able to obtain its information model in some implementation specific way at the system startup.

Once the cluster is up and the IMM is available, the PLM implementation registers with it as the implementer of the PLM information model objects and uses this configuration information to bring the platform to the state best matching it. The result is reflected in the runtime attributes of the objects that represent an entity in the system so their status can be monitored by administrators.

Whenever an administrator manipulates the PLM configuration objects, the PLM implementation receives callbacks from IMM to deploy these changes in the system.

5.3.7 Open Issues and Conclusions

At the time of writing only the first, the A.01 version of the PLM has been published yet. This release could not and did not address many of the issues that are in the intended scope of the service. In the course of our presentation of the service we have already pointed out a number of the areas where the current specification is sketchy or incomplete.

One such area of incompleteness is the PLM information model, and in particular the object class describing the EE and its types. They provide currently the basic administrative operations and some state information, but very little configuration information that one may associate with OSs or VMs.

In the future the concept of EE may be refined to potentially address the following:

- **Virtualization:** It might be advantageous to distinguish the hypervisor, the VM, and the OS instance. For example, one could then associate migration only with the VMs and not necessarily with all three. At the same time, however, there are OSs that encompass the functionality of one or the other, or both.
- **Alternative EEs:** In real systems particularly for diagnostic purposes, one would like to be able to perform diagnostics without changing the configuration, to have an alternative configuration readily available for this task. The current model provides no solution for this.
- **Boot image:** Currently the EE class has no information that PLM can use to figure out where to find the boot image for a particular EE instance yet it is expected to control its life-cycle and potentially even switch to the correct EE if the one booting is not correct. There is no information about whether there is only one image or if there are alternatives. For example, if PLM fails to boot the OS from one location whether there is another source to boot it from. Whether the types of these boots are the same or different, for example, network boot versus boot from local disk. All this is left up to the implementation.
- **File system, backup, and restore:** With respect to the file system available in the SA Forum cluster there is no standard solution. It is not even clear whether this should be part of PLM or a

separate service. Its close relation to the boot image and the need of high-availability systems to be able to fall back to a working image after system failures suggest that PLM could be the right places to address these issues too.

With respect to the HEs it is also not clear whether the current model contains enough information for PLM and for system management with respect to the hardware.

The primary goal of the PLM information model is to allow the availability management to evaluate the distribution of logical entities on the physical hardware to minimize the impact of hardware failures. The model also provides a more homogeneous AIS view. It brings the hardware entities into the picture which had to be handled differently before. This means that some functionality defined for PLM seemingly overlaps with the functionality provided by HPI. Yet, PLM provides a higher level of abstraction. It generalizes the many functions that HPI provides into a few essential ones. The level of this abstraction may require refinement in the future.

For example, in HPI there is the FUMI which reflects and controls the firmware version loaded in the hardware. It is a question to what extent this needs to be exposed through PLM in addition to HPI, and whether the firmware needs to be visible at all through PLM when it comes to the management of its upgrades, for example, by the Software Management Framework [49].

There are a lot of open issues, but at this point of time the Technical Workgroup of the SA Forum has decided to take a step back from the specification work and see what requirements will emerge in real implementations of PLM, in their real life deployment. The first release of the specification has set the direction and it is satisfactory to start the implementation work. A PLM implementation will face many of these and even other issues; it needs to provide some solution to them. Armed with the practical experience the SA Forum intends to revisit the specification and refine it to address the issues that indeed require standard resolutions. After all it is a question of why or to what extent would one want to make the platform management service itself platform independent?

As we mentioned at the beginning of the discussion of PLM there was some debates whether the functionality intended justified the creation of a new service or it should have become part of the CLM. The definition of the first release showed that indeed the separate service is justified and we believe that these sections showed the correctness of this choice to the reader as well.

This hesitation, however, somewhat determined the status of the PLM specification within the SA Forum architecture:

The SA Forum architecture does not define the PLM as mandatory. It may or may not be part of a middleware implementation. Moreover the PLM information model is defined so that the hardware and software parts can be used by themselves. They do not require their counterpart.

Also while the specification of the HE is based on the concept and functionality of the HPI, its use is also optional and a PLM implementation may collaborate with any other hardware management interface.

Regarding the EEs there is hardly any reference to any standard interface, although the OS is generally considered to be POSIX [73] compliant.

With respect to the virtualization facilities there were no standard interfaces to rely on. At the time of writing there is already work going on in the Distributed Management Task Force to define some standard virtualization management interfaces [76]. Future releases of the PLM specification may take advantage of their results.

5.4   Cluster Membership Service

5.4.1   Background

The information about the platform that applications and middleware services really would like to know is the set of 'nodes' they can use reliably in the provisioning of their services. They need to know the EEs that are healthy, well connected, and provide the computing resources they need; the ones that compose a set presentable to the users as a single logical entity – a cluster.

Most AIS services including the AMF need this information; and applications using AIS services may also require it. Relying directly on the PLM, all of them would need to implement a similar logic that generates this information. At least now that PLM has actually become part of the SA Forum architecture. In reality the CLM [37] was defined first. It was one of the first services defined indicating its essential role and the importance of the information it provides.

The problem of cluster or more generally group membership has been around ever since distributed systems have been used for distributing the tasks they performed. It gained even more significance with the use of such systems to provide fault tolerance.

Chandra *et al*. [77] provided the following definition: 'A group membership protocol manages the formation and maintenance of a set of processes called a group. A process group is a collection of programs that adapt transparently to failures and recoveries by communicating amongst themselves and exchanging information.'

Fischer *et al*. [78] however showed by 1985 that such a consensus could not and cannot be achieved in asynchronous systems containing at least one faulty process.

Luckily the problem of membership is not the same as consensus since to make a decision we do not need to obtain the agreement of the faulty process. That is, the faulty process can be expelled from the group. Nevertheless in [77] it is proven that even with this relaxed model, the problem remains unsolvable for asynchronous primary-partition systems – those that try to maintain a single agreed view of the current membership – as it is impossible to decide whether a process is infinitely slow compared to others or faulty.

The authors of [77] came to the conclusion, however, that any technique that proved to make the consensus problem solvable was applicable also to the membership problem with the same result. These techniques weaken enough the problem specification to make it solvable. They are: randomization, well-defined probability assumption on the behavior of the environment, and use of failure detectors with well-defined properties.

In the context of the SA Forum architecture one may look at the PLM – or any other even proprietary technique – as such a failure detector and therefore making the problem solvable. This still means that CLM is left with the task of the formation and maintenance of the membership in this case from a set of EEs – or in CLM terms *nodes;* and the group in CLM term is called the *cluster*.

CLM also needs to provide a distribution mechanism to update with this information all the interested parties – the AIS services and the potential user applications – in such a way that they can synchronize on the content.

5.4.2   Overview of the Cluster Membership Service

The SA Forum Application Service Interface identifies the CLM as the authority on the cluster membership and all services and applications requiring this information should depend on the CLM. It is in charge of forming the membership from the nodes configured for the cluster and of providing its users with up-to-date information about the membership.

From CLM perspective a *node* is a logical entity that represents all the software executing on an instance of an OS that what in PLM terms is an EE.

The group of nodes intended to collaborate in a location transparent manner is called the *cluster*. Nodes are configured for cluster membership, but at any given moment it is possible that not all *configured nodes* are part of the membership. A node which is part of the membership is referred to as a *member node*.

To become a member node, a node needs to satisfy a number of criteria. Namely,

1. The EE of the node needs to be *sufficient for the execution* of application software.
   On practice this means that in addition to the OS some daemons of the middleware need to be up and running. It is implementation and configuration dependent exactly what daemons need to be available to make, for example, the AIS services available for user applications.
2. The node needs to be *reachable* within the cluster.
   The reachability needs to be such that it allows for the location transparent collaboration of middleware services and applications. This means that the users of these applications and services should not be aware from where and from which node they are being served within the cluster; they should obtain the same service with the same result.
3. The node needs to be able to provide the Cluster Membership *API*.
   Among the AIS services at least the CLM needs to be available for application software executing on a node. As mentioned earlier, other services also might need to be provided depending on the middleware implementation and the system configuration.
4. The node needs to have the *permission from the administrator* to participate in the membership.
   The CLM node itself and all PLM entities it relies on need to have the permission of the administrator to provide services. In sections 'HE Administrative State' and 'EE Administrative State' we have discussed the administrative state of the PLM entities. In Section 5.4.5 we will take a look at the administrative control of the CLM.

If any of these criteria is not satisfied the CLM does not allow the node to join the membership, or if the node is already a member, CLM expels the node from the membership.

The CLM specification does not describe how a CLM implementation verifies for each node that it meets the above list of criteria. However requirements #1 and #4 imply that the readiness state of the PLM EE the node resides on is in-service or stopping and CLM can obtain this using the PLM track API as described in Section 5.3.4. The method used to verify the rest of the criteria is left to each of the implementations.

It is also left to the CLM implementation how it forms the membership, that is, what protocol it uses to establish a single view of the membership.

A *view* is the list of nodes that are part of the membership and an associated view number, which is an integer. Any time a node joins or leaves – by whatever reason – the cluster CLM transitions the membership to a new view. Most importantly this new view is associated with a number greater then any previous view. This increment ensures that any receiver can easily identify the latest view among a series it has received. The view numbers of subsequent views, however, may or may not be consecutive numbers.

In the membership nodes are identified by their name and their node id. The two have different life spans.

The node gets a unique node id when it joins the membership and keeps this id for the time it is part of the membership. When the node leaves the membership its id may be recycled by CLM.

The node name, on the other hand, is independent of membership; it is unique in the cluster and remains constant even after a particular node has left the membership and then rejoined it again.

The CLM exposes the different views to its users through a tracking API similar to that discussed at the PLM in Section 5.3.4.

The main simplification, however, is that either a single predefined 'entity group' or a single entity are traceable via the CLM API. This single entity group is the cluster. Alternatively, users may also track the node on which they reside as an individual entity.

This means that the users do not need to be aware of the cluster configuration. Beyond these two options they do not need to define which entities they want to track. They can learn about nodes as they become part of the cluster from the views they receive from the service. That is, no prior knowledge of configured nodes is required at all.

As one can see from Section 5.4.1 it is a challenging task to establish a single view of the cluster membership. The challenge lies not only in timing issues, but in the fact that in distributed systems it is always difficult to distinguish whether the remote entity itself is down or it is actually up, but it is unreachable due to some communication failure or just very slow.

In high-availability circles the dreaded condition is when two disjoint subsets of cluster nodes are up, but cannot communicate with each other and therefore form independent memberships; this is called the 'split-brain' scenario. It may not be obvious, but it goes against SA.

The problem is that in this case we have two subclusters visible from the outside world each as the whole. For the users they are indistinguishable as both have the same identity. Unwittingly they present themselves as 'the cluster,' but obviously users access one or the other part, also knowing nothing about the existence of the second part. Since the two parts are unaware of each other changes made in one part are unknown to the other and over time they may show up for a user as discontinuity.

Imagine that the cluster is used to maintain your bank account and you deposit $1000 accessing let's say part A while the cluster happens to be split into parts A and B. Next moment when you want to transfer some of this amount to somewhere else you happen to access the other part B, which then informs you that your money has gone and that you have overdrawn your account. You freak out and with good reason.

Consolidating after such a scenario is rather difficult and may turn out to be impossible without any loss of information.

In the CLM specification there is one requirement related to the issue. It says that when a node leaves the cluster CLM needs to reboot the node to guarantee that no services are left running on the leaving node.[2] However, it does not specify how to determine which one is the leaving node and which one is remaining in a – let's say – two-node cluster; or under what circumstances one node is a cluster.

A CLM implementation needs to resolve these issues so it protects against all such scenarios and guarantees that at most one cluster entity is exposed to the external world at any given moment.

The CLM specification also does not address timing issues. For example, the time frame within which CLM users can or need to learn about new views is neither specified with respect to the actual membership transition, nor with respect to different CLM users between each other. The only requirement is that if two users receive two views from the CLM referring to the same view number then the list of member nodes must also be the same. The view numbers provide the synchronization point when parties compare their views of the membership.

Obviously a membership view is useful if the users receive this information in a timely manner. There is little use in knowing the list of member nodes once the actual membership has changed. In certain circumstances this timing issue may become critical.

Most of the AIS services should only be provided on nodes that are members and, as we indicated at the beginning of this section, CLM is the authority on determining the current membership.

Hence the AIS services are required to use the CLM to obtain the current membership information and provide or withdraw their services accordingly. These changes in the membership also trigger recovery actions such as service fail-over at the level of the AMF. The faster AMF learns that a node left the cluster the earlier it can react and recover the services provided by that node. So the timely signaling of membership transitions is a determining factor in failure recovery.

[2] Note that this behavior can be overridden by the configuration as for some software the startup time may be unforgivingly long and therefore the reboot needs to be avoided. However in such cases the particular solution has to provide the equivalent guarantees.

It is again left to the CLM implementation to make sure that it comes up with the new view in a short enough time that does not hamper SA and that the information is provided across the membership within a short time.

We have to clarify that as dire as it might sound we do not criticize the Cluster Membership specification itself. It is an API specification and users do not need to be aware of any of these issues; hence they do not need to be part of the specification.

It is the CLM implementation that needs to deal with these issues and provide an appropriate solution. As a result, however, depending on how successful different implementations are, for example, in the convergence of the membership view, they may provide quite different user experience and characteristics of SA.

5.4.3   CLM Configuration: The Bootstrap Trap

The information model of the CLM describes the configuration of the cluster. Accordingly it has only two configuration object classes: one to represent the cluster and another to represent the configured cluster nodes.

At the root of the system information model a single cluster object represents the CLM cluster and its child objects represent the CLM nodes configured in this cluster.

The cluster object is very simple. It has a single configuration attribute: the cluster name.

The CLM uses the information of the objects representing the configured nodes to identify which nodes of the system can participate in the membership. If applicable, two essential pieces of information are provided here: The PLM EE on which the node should reside and the address to access the node.

As in other services the CLM node configuration objects also provide the attributes that indicate the status of each node: First of all, whether it is in the membership or not.

If the node is a member then it has a node id and a current address through which it is accessible and which may be different than the configured address.

Additional attributes indicate the view number at which the node has joined the cluster and when it has booted. These pieces of information become decisive whenever a node wants to rejoin the cluster to determine whether it can do so or it should be rebooted first.

Even though the information model defines these different runtime attributes all as mandatory they need to be handled differently. Some of them have a meaning only when the node is a member. For example, the node id has such a life-cycle. Others, like the boot time, contain information significant after the node left the cluster. By keeping the boot time in the model, a CLM implementation can easily decide whether to admit the node to rejoin the cluster based on how its current boot time compares to the one kept in the model.

Note the chicken-and-egg issue regarding the CLM information model. It is about this requirement that AIS services should provide their service only on nodes that are members of the cluster. This requirement is applicable to the IMM as well, which appropriately holds the configuration information necessary to form the cluster. Obviously, the same applies to PLM, which also needs to learn its configuration from IMM before it can map the HPI entities to HEs that in turn host EEs hosting CLM nodes required to form the cluster before IMM can provide the service to obtain the necessary configuration information.

There are several generations of poultry involved here!

Again, the solution to this bootstrap is left to the Cluster Membership implementation and it will be different from one environment to the other, from one implementation solution to another.

Because of this variability the consensus within the SA Forum Technical Workgroup was not to put any requirements on the bootstrap process. In fact there is a consensus that the requirement on AIS services that they should provide their services on member nodes only is applicable only after cluster formation, but not during the cluster bootstrap.

5.4.4 Are You a Member?

Among the AIS services providing cluster-wide service, the CLM is an exception as it provides some of its services on configured nodes that are not members of the cluster yet. This allows CLM users to find out when the node becomes a member. This is useful as up to this moment they cannot initialize any of the services that should be provided only on member nodes. They are unavailable. To receive this information a user needs to start tracking the cluster membership.

CLM provides the following information on nodes: node name and id, network address, the hosting EE and its boot time, the membership status, and the view number when the node joined the cluster.

As mentioned earlier there are two options of tracking with respect to the scope of entities. A user may choose to track the local node, that is, the one it resides on or the entire cluster. In neither case it needs to know the actual name or identifier of these entities to start the tracking. The choice is controlled by the track local flag. If it is set only the local node is tracked; if it is not set, the entire cluster is tracked.

The other tracking options are similar to those available in the PLM track API discussed in Section 5.3.4. That is, one may receive information about the changes only or on each change the complete information for the selected scope.

As in case of the PLM track API, CLM also provides the three-step tracking option primarily to be able to pass on these steps to its users when they are initiated at the PLM level. We will not go through the details here again as they are essentially the same as discussed in Section 5.3.4.

The main difference is that CLM maps the PLM entities and the reported impact to CLM level cluster nodes and effects on them. In its callbacks CLM reports the affected nodes with the anticipated (for validate, start, and abort callbacks) or actual (for the completed step) CLM level impact.

In addition to passing on PLM reported events, CLM uses the track API to report also cluster membership changes independent of PLM. As we have seen, the CLM has its own responsibilities in evaluating and maintaining the membership. Whenever it detects or initiates a change, for example, by expelling a node that does not satisfy the membership conditions any more, CLM reports these changes to its users through the track API; typically in the completed step.

However, although the CLM administrative operation expels the node from membership, CLM uses the start step similar to the PLM normal lock option; therefore it is strongly recommended for applications and services responsible for SA to use the CLM track API at least with the start option. For more on the administrative operations please see Section 5.4.5.2.

In the same way as for PLM, CLM expects an answer from its users after callbacks in the validate and starts steps. In the validate step the response indicates whether the proposed change has service impact and therefore it is rejected by the user, or no service impact is expected and the user accepts the change. Again users accepting the change should be prepared for it when it is carried out subsequently in the start step. If the change is rejected by at least one user CLM aborts the proposal.

The response to the start step indicates whether the users completed the evacuation from the impacted nodes. If the entire membership is tracked, the user receiving the callback needs to evaluate itself whether the evacuation applies to it, while tracking the local node implies this, as the callback always contains the information with respect to the node where the user receiving the callback resides.

As opposed to PLM, the CLM uses some time supervision for its callbacks, and in the callback it indicates the time within which it expects the response. When this attribute is set to zero, it means that no response is required such as for the completed step. When the attribute is set to 'unknown time,' CLM will wait indefinitely for the answer the same way PLM does.

An important nuance is that within a single initialization of the CLM a user can have only a single tracking session through which it may either track the entire membership or the local node. Subsequent invocations to the track API only modify the options of this already initiated tracking by logical ORing (i.e., applying the logical OR operator to the current and the new flags). Subsequent invocations do not result in additional tracking sessions.

However, there is no limit to how many times a user initializes the CLM service. Through multiple initializations of the service the same user can start multiple tracking sessions each with different options, for example. Doing so however requires the correlation of callbacks as the same change may be reported by CLM in multiple callbacks. This should be possible as each callback includes the root cause entity.

In addition to the track API, which 'pushes' the information to the users at the time of the changes, the CLM provides an API by which users can pull the status of individual member nodes. This API is only offered on nodes that are members and applicable only to member nodes since the inquiry is based on the node id, which is valid only during membership.

5.4.5 Administrative and Management Aspects

The CLM defines three administrative operations; all three are applicable to cluster nodes only and manipulate its administrative state. These operations are the unlock, the lock, and the shutdown. We start our discussion with the review of the administrative state, which is the only state defined by CLM.

5.4.5.1 CLM Node Administrative State

In the CLM only cluster nodes have an administrative state, which indicates whether the node is eligible for membership. An administrator manipulates the administrative state through the administrative operations. The current value of the administrative state of each node is presented in the CLM information model as a runtime attribute.

The cluster node administrative state has three values:

- **Locked:** The cluster node is ineligible for cluster membership.
- **Unlocked:** The cluster node has not been administratively prohibited from joining the cluster. Whether the node will join the membership depends on whether it satisfies the membership criteria presented in Section 5.4.2.
- **Shutting-down:** The administrator requested that the cluster node gracefully transitions to the locked state. If the node is not a member node, it makes the transition immediately. If it is a member node, CLM user processes residing on the node are expected to follow ITU X.731 [72] shutting down semantics: 'Use of the resource is administratively permitted to existing instances of use only. While the system remains in the shutting down state the manager may at any time cause the managed object to revert to the unlocked state.'

To better understand the mechanism enforcing these states let's see the administrative operations.

5.4.5.2 CLM Node Lock and Shutdown

The lock and shutdown operations are very close relatives as the goal of both is to put the CLM node into the locked administrative state. Now let's see how this is achieved because this gives an interesting insight to the operations.

In both cases CLM uses the track API to inform its users about the issued operation. CLM invokes the callback in the start step to request its users to perform the actions necessary for the execution of the initiated administrative operation, which is indicated as the cluster change for the target node.

Then CLM waits for the responses from the users in case of the shutdown operation indefinitely, in case of the lock operation at most for a configurable time period. Whenever CLM receives all the responses or if the timer expires, CLM proceeds with the administrative operation: That is, it locks the node and reboots it to guarantee that no process using AIS services remains on the node. CLM informs in the completed step its remaining users tracking the membership that the node has left the cluster.

Table 5.2 Comparison of the lock and shutdown administrative operations

Lock	Shutdown
Applicable to administrative state	
The lock operation is valid on nodes in the *unlocked* and in the *shutting down* state. Hence it may be used as an escalation of the shutdown operation.	The shutdown operation is only valid on nodes in the *unlocked* administrative state
Administrative state	
For the lock operation the administrative state remains unlocked until all responses are received or the timer expires, at which point it is changed to *locked*.	For the shutdown operation the administrative state is changed right away to *shutting down* and it remains so until CLM receives all the responses when it changes to locked; or a new administrative operation overrides the shutdown operation.
Time supervision	
The lock operation is *time limited*. CLM waits at most a configured period of time to receive all responses. At the expiration of this period CLM proceeds with the operation regardless.	The shutdown operation is *unlimited* in time and CLM will wait as long as it takes to receive all the responses before it proceeds with locking the node.

In case of the administrative operations, the reboot on exit is applicable regardless of the configuration attribute disabling node reboot.[3] The rational of this is that the intention behind the option of disabling reboot is not to hamper applications that have a long startup up procedure when the node 'accidentally' drops out of membership, for example, due to intermittent connection. Since in case of the administrative operation the dropping out is not accidental, all the rules of leaving cluster membership should apply.

There are a number of differences between the lock and shutdown operations. We summarized them in Table 5.2.

The consequence of using the track API to perform these administrative operations is that the presence of tracking users determines whether the operation is applied right away or it goes through the start step of the tracking. On a member node if there is no user to call back, for both operations CLM expels the node from the membership immediately at the initiation of the operation. This means that services and application responsible for SA can avoid service impact by becoming CLM users and using the track API.

When there are tracking users, it is left up to the user how they react to each of the operations. The fact that the lock operation is time limited suggests that users should evacuate their services from the node as soon as possible. In case of the shutdown operation there is no such urgency at least from the perspective of the CLM. The reason for not having time limit for the operation is that the same way as for PLM, the administrator can always escalate the shutdown operation to a lock, which then guarantees that the operation will succeed within a time limit.

5.4.5.3 CLM Node Unlock

The node unlock administrative operation reverses both the lock and the shutdown operations in terms of the administrative state. It moves the node to the unlocked administrative state.

---

[3] Disabling node reboot may prevent a CLM implementation from protecting against the split-brain and other scenarios, which need to be considered before setting this option for any CLM node.

In case the node is still in the shutting down administrative state, it has not left the cluster yet and therefore the only action CLM needs to take is to inform tracking users about the state change. It does so in the completed step.

An already locked node has already left the membership. Once the node is unlocked the CLM evaluates whether it satisfies the membership criteria as described in Section 5.4.2. If it does, the node may rejoin the cluster and when it succeeds CLM announces it to its tracking users in the completed step.

5.4.6 Service Interaction

All AIS services providing cluster-wide service are required to use the CLM to determine whether a node is a member or not and adjust their services accordingly. This means that from the perspective of CLM all these services are user processed. CLM is not aware of their identity. It is the responsibility of the AIS service implementation to use the CLM therefore we do not discuss this interaction any further.

The CLM is a user to three AIS services: The NTF, the IMM, and the PLM. Here we take a look at these. In addition to these services a CLM implementation may also use the Log service (LOG) and other AIS services, however the specification requires only the mentioned three.

5.4.6.1 CLM Interaction with NTF

The CLM does not generate any alarms, but it generates notifications on nodes entering or exiting the membership, on node reconfiguration and on administrative state changes.

All these notifications follow the generic structure of the state change notifications defined in the NTF specification. They allow one to follow the membership changes without the use of the track API. The main difference is of course that there is no possibility to influence these changes and that notifications may be delayed or suppressed compared to the track callbacks.

Since the NTF is also a cluster-wide service the bootstrapping issue discussed in Section 5.4.3 is very much applicable. It also means that the monitoring of cluster membership via notifications is available only on the member nodes as opposed to the track API.

5.4.6.2 CLM Interaction with IMM

We have already pointed out in Section 5.4.3 the bootstrap trap with respect to the CLM configuration. The CLM is also expected to obtain its configuration from the IMM, which offers its services only on member nodes – it is a Catch-22 [79].

Assuming that a CLM implementation resolves this catch in some implementation specific way and completes the bootstrap it registers with IMM as the implementer for the CLM node and the CLM cluster configuration objects which represent respectively the cluster node entities and their collection, the cluster. The CLM implementation maintains these logical entities according to the configuration requirements of the representing objects. It also updates the configuration objects' runtime attributes as necessary, so administrators (and management applications) can monitor the cluster status via IMM.

The IMM also mediates any administrative operation and configuration change between an administrator and the CLM implementation. Whenever the administrator initiates an action IMM calls back the CLM implementation to deploy the action in the system. CLM reports back the changes to IMM, which then delivers the results to the administrator.

5.4.6.3 CLM Interaction with PLM

The trickiest of all is the CLM interactions with the PLM. Obviously CLM needs to be a user of the PLM track API. The question is what entity group it needs to track. The specification does not specify this and lets implementations choose their own solution.

To see the nuances that an implementation needs to consider let us assume a configuration of four cluster nodes residing on four PLM EEs. Each of these EEs are an OS instance running in a VM EE that are hosted by two hypervisors residing on two physical nodes represented as HEs. This means that if any of the HEs are turned off, two of the CLM nodes leave the membership. For a CLM user (such as AMF), providing highly available services, it is essential to know this consequence.

PLM reports the track callback based on the entity groups. This means that if the two nodes residing on the same HE form different entity groups, PLM will provide two callbacks one for each entity group, and CLM may map them into two independent membership changes, so in turn the CLM user may start to evacuate its services from one leaving node to the other.

Although with all the parameters provided in the track callbacks it is possible to correlate subsequent callbacks with each other and identify the scope of impact, this can only occur at the moment when all callbacks triggered by a single event have been received. Unfortunately a receiver cannot know if there is a subsequent related callback still to be received. As a result it should react at the first callback, which means that it may initiate the wrong reaction.

If CLM creates and tracks an entity group containing all the EEs hosting cluster nodes, it receives in each callback the complete picture for the cluster, that is, all the EEs impacted by the event being reported.

It is also essential that CLM itself does not split up the information received in a single callback into multiple callbacks to the same user, that is, it should report all leaving nodes in a single callback and not generate a callback for each node.

The reason the CLM tracks PLM entities obviously is that the Platform Membership service provides essential information about the health of these entities, which is a key factor in the evaluation of the eligibility of each node for membership. However, the specification does not mandate that this is the only determining factor. As we have seen, PLM may report that it lost management access to an entity, which does not necessarily mean that entity is faulty and cannot be used for service provisioning. This means that CLM may need to resort to its own resources to figure out whether the node using the entity remains eligible for membership.

PLM may also represent in its model entities that do not map into cluster nodes, but which represent resources enabling the cluster formation. For example, the switch or the network interface cards may be reflected in the PLM model and a CLM implementation may track these PLM entities in addition to those hosting CLM nodes to evaluate the health of member nodes and the cluster as a whole and to use this information for the cluster formation and maintenance.

5.4.7 Open Issues

The CLM is among the services defined first within the AIS. In addition as Section 5.4.1 pointed out, the problem has been studied for decades. As a result the CLM specification can be considered mature at this time.

There are only a few open issue not addressed in the specification available at the time of writing. The first one is the dynamic configuration of the cluster and cluster nodes. Considering the relatively few configuration attributes this is quite straightforward and the next release should remedy it.

The case is similar for the next issue: There is a need to have more than one address represented in the information model for each cluster node.

A more complicated issue is the relation of the PLM domain and the CLM cluster. The current AIS specifications limit this to 1 : 1. However one can easily imagine cases when a single cluster is spread over multiple physical locations that are managed at the PLM level as different domains; or the opposite, when a single PLM domain hosts multiple CLM clusters; and any configuration in between.

These configurations are increasingly interesting for real deployments. They do not necessarily contradict to anything defined by today's CLM. The reason they are not covered by the specification is merely that these cases were not investigated thoroughly as one would like to do when it comes to guarantees about SA.

5.4.8 Recommendation

Since the CLM is the authority on the cluster membership, all applications and services that need this information (because e.g., they provide cluster-wide services) should use the CLM.

In addition applications and services requiring any piece of information returned by CLM such as a node's network address or hosting EE may prefer to use the CLM API instead of finding out the information via the IMM OM-API.

5.5 Conclusion

This chapter presented the services providing information about the platform at different levels and granularities. We presented the way this information is collected and presented to the users. We started out with the details on the discovery of the hardware platform through the HPI, the information it presents to its users, among them to the PLM.

PLM uses the received information to match it with the expected hardware configuration and complements it with the information available on the EEs providing the runtime platform for software entities. PLM exposes the results in the information model available as part of the system information model and also through the PLM track API.

The CLM forms the cluster among the configured nodes residing on EEs reported by PLM as healthy and ready for service provisioning. CLM checks and monitors additional conditions to continuously verify each nodes eligibility for the cluster membership. It provides the membership information to its users who should rely on this information in all decisions related to membership, such as whether they can use any given node to host their own entities or provide their services to users residing on a particular node.

As the information percolates up through these layers it becomes more and more abstract focusing on the key objective: To provide the platform users with reliable information on the nodes available for service provisioning at any moment in time. This is the single most important piece of information most applications and services need to achieve SA and it is provided by CLM.

Users requiring more details and tighter control over particular platform entities can obtain this in the appropriate details yet in a standard platform independent way using any of the lower level platform services, that is, PLM or HPI. Hence this layered platform solution provided by the SA Forum specification can satisfy the needs of a wide variety of applications whether they are geared toward SA or not.

6

Model Based Availability Management: The Availability Management Framework

Maria Toeroe
Ericsson, Town of Mount Royal, Quebec, Canada

6.1   Introduction

This chapter introduces the Availability Management Framework (AMF) [48], which is the key enabler for achieving service availability (SA) with SA Forum systems.

The chapter focuses on the main aspects that are essential for understanding AMF and it is based on the general availability concepts introduced in Chapter 1 of the book.

In particular, this chapter introduces first the perspective that application developers will face, the concepts of components and component service instances (CSIs), their interpretation and how they are exposed through the AMF application programing interface (API).

This is followed by the view of the AMF on these same concepts and their organization into a hierarchy of logical entities composing the AMF information model which forms the basis for the availability management performed by the AMF. This is also the view that site designers, site administrators need to understand as the information model is the interface through which AMF is instructed what it needs to manage, that is, the AMF entities and their features composing the system and the different policies applicable during the management such as the applicable redundancy, the error detection mechanisms, and the error recovery policies.

The information model also provides status information on the health and readiness of the different entities to participate in the service provisioning. The AMF itself uses this information to distribute the role assignments for this service provisioning among the different entities. This in turn is also reflected as state information that reflects the role the different entities take in this task and also as the status of provisioning for the services.

The different redundancy schemas that can be used in conjunction with the AMF demonstrated on simple examples of single-failure scenarios to provide an easier grasp on them as they are one of the

Service Availability: Principles and Practice, First Edition. Edited by Maria Toeroe and Francis Tam.
© 2012 John Wiley & Sons, Ltd. Published 2012 by John Wiley & Sons, Ltd.

key concepts of AMF. They also demonstrate the wide range of availability needs that can be covered by the AMF.

A significant part of the chapter discusses the administrative operations applicable to different AMF entities and their impact in the system. Finally the interactions between the AMF and other Application Interface Specification (AIS) services are summarized.

Considering the size of the specification this chapter cannot embark on an in depth presentation and analysis of AMF. So the main goal we set out was to try to convey to the reader the logic behind the solutions provided by the specification. To teach the way of thinking about availability management as we see this is an easier way to grasp this complex subject.

Probably the most ingenious steps in defining a standard for availability management was the abstraction from the services as perceived by 'end-users.'

An end-user is interested in the functionality and performance of the service, while the AMF abstracts from these aspects and defines services as units of workload that can be assigned to some provider entities in order to provide or protect the end-user perceived functionality. The AMF service instance (SI) concept is the control mechanism used by AMF similar to the way a faucet knob is used to control the water flow – the SI being the knob and the water the end-user perceived service.

In addition these SIs are also logically separated from the entities that are providing them, allowing AMF to move around the SI assignments as an independent entity for which then the continuity is based on the existence of the appropriate assignment.

The logical organization of the entities and their state models, the policies and schemas defined in the specification all serve this single purpose: to maintain the appearance of this service continuity; and as long as a user can initiate successfully a service request this attempt is successful and SA is provided.

After a short overview that links the AMF concepts to the basic concepts of fault tolerant systems discussed earlier in this book we will embark on our adventure of familiarizing the AMF.

6.2   Background

Chapter 1 has introduced the difference between fault, error, and failure. One may realize that from the perspective of SA, availability is only affected when an error manifests externally as a failure. This means that there is a window of opportunity for availability management to detect an internally manifested error and take some corrective measures to prevent the externally manifested failure to happen or at least to lessen its impact. This window of opportunity lasts from the moment a fault is activated until its external manifestation; and within this period the earlier one is able to take the corrective actions the lesser is the service impact, it may even be avoided completely.

6.2.1   Error Detection and Repair

The prerequisite of early intervention is the early detection of errors. This means that the AMF must provide tools appropriate for error detection. As we will see in the subsequent sections the specification indeed includes a range of tools that follow today's best practice in error detection. The different options suit different circumstances. The AMF uses some of them in all cases, while others are configured and activated on request. These different techniques can be tailored for different AMF managed applications depending on their capabilities and needs in the particular circumstances of a deployment.

Once it has detected the error, the AMF takes actions to correct the situation. These actions have three distinct goals:

• Most importantly they provide fault isolation, so that the fault cannot propagate any further.
• Secondly they repair or replace the faulty element to return the system to the healthy state.
• Finally, if the faulty element was providing some services the actions try to restore those services.

A particular corrective action may achieve more than one of these goals. For example, a restart of some faulty software isolates the fault as it removes the faulty element from the system. At the same time it also repairs the element as at the completion of the restart a new healthy element becomes available.

Note that healthy in this case is relative as restarting the software does not eliminate any bugs from it. As earlier chapters indicated, it is essential that the software is sufficiently tested before deployment in such systems and it fails only under somewhat peculiar circumstances. The availability management cannot eliminate software bugs and other faults in the 'workmanship' permanently.

The AMF relies on the termination of a managed entity for fault isolation and after successful restart it considers the entity healthy again. This implies that the AMF is in charge of the life-cycle of the entities it manages.

6.2.2   Fault Zones and Error Escalation

The problem is, however, that faults may or may not manifest in the faulty entity. For example, a memory leak caused by one process may cause the malfunction or failure of another independent process and the error is detected in this innocent second process.

In such cases obviously the termination and restart of the second malfunctioning process neither isolates the fault nor repairs the first process at the root of the problem. It only isolates any derivate fault propagated to the second process in which the fault manifested and repairs the second process, more precisely the logical entity represented by the second process (as after the restart we cannot talk about the same process instance any more).

To deal with this issue of fault propagation, the AMF defines an incremental set of fault zones that encapsulate more and more entities based on the expected fault propagation. This allows the definition of escalation policies to a wider fault zone if repeated errors are detected within some probation in this encapsulating fault zone. Accordingly, a successful restart becomes a successful repair only after this probation period.

Using our example of the memory leaking process, this means that since the first process capable of propagating its fault to the second they are put in the same wider fault zone, which also becomes under suspicion with the detected error. The successful restart of the malfunctioning second process does not complete the fault isolation and repair for this wider fault zone. Instead, it starts a probation timer for it. While this timer is running, AMF considers all errors detected within this fault zone related.

This means that if either of the processes fails during the probation period, the error is escalated to the entire fault zone and AMF will terminate and restart both of the processes. This allows us to deal with the root cause of the problem, with the faulty process regardless whether any error was detected in it or again it impacts its neighboring process.

The AMF information model reflects these incremental fault zones as we will see in Section 6.3.3 and it forms the basis for the implementation of the error escalation policies.

Obviously while AMF is terminating and restarting a process or the encapsulating fault zone none of the enclosed entities can provide service, which in turn adversely impacts our main goal of providing highly available services.

It is often desirable to have some replacement entities readily available in the system, which then can take over the service provisioning as soon as the faulty entities have been isolated from the system. Hence as discussed in Chapter 1, redundancy is the standard solution for this problem and it has been widely used in systems that require any level of availability. The AMF specification follows the same general solution, but in a somewhat special way that may be hard to grasp on first read.

6.2.3   Separation of Services from Serving Entities

The distinctive feature of the SA Forum's AMF is that it logically separates the entities providing the service from the services they provide.

Let's consider an application implementing an alarm clock. As one would expect when the alarm clock is started it is set for the time it should raise the alarm. From this moment on the process will compare the current time with the time for which it is set and decide whether it is time to raise the alarm. If the process fails by any reason before the set time and if we are able to replace it with a new process for which the same time is set, we will be able to raise the alarm at the requested time and fulfill the requested service.

Thus, the failure of the process providing the service does not mean the failure of the service itself and another process receiving the same service assignment can complete the task. This justifies the separation of the service from the entity providing the service.

Furthermore, the same alarm clock program can be started multiple times with different time settings which will result in multiple processes, each providing its own alarm service, each of which could be considered as a different SI. Even if we start two processes with the same time setting, they will each raise the alarm separately and distinguishably, so there are still two different instances of the service. On the other hand giving the same time setting to two consecutive processes – as we discussed above – creates the impression of service continuity for the service user as the alarm will be raised only once.

From these considerations, the AMF distinguishes service entities and service provider entities. Service provider entities are organized in a redundant way to provide and protect the nonredundant service entities.

Note that failure may occur in both service provider and service entities; however, the failure of a service provider entity is considered as an internal manifestation of a fault, that is, an error as long as there is another service provider entity that can take over the service provisioning from the failed provider. Only if there is no available service provider entity that can take over the service provisioning can we consider the failure as a service impacting external failure.

The action of taking over the service provisioning from a failed entity is referred to as failing over the service from the failed entity to the new provider entity, or fail-over for short.

6.2.4   Service Provisioning Roles

At this point we need to look at the service from a user's perspective. Considering the alarm clock implementation, it is easy to see that no matter how many times we need to replace the alarm clock process by failing over the service to the next available process, we can do so successfully using the same initial data – the time at which the process has to raise the alarm. As long as there are no two processes running at the same time with this same time assignment, the service will be provided exactly once regardless whether the availability manager is aware of the actual meaning of the service and its status as seen by the service user.

Now let's consider instead of this alarm service another application which implements the stopwatch service. Again when a process is started it starts to measure the time, however, to be able to replace a failed process without service impact the replacing process needs to know the elapsed time the failed process measured and be able to continue the counting as increment of this initial value. Without this even if we are able to replace the process providing the service, the delivered to service user result will not be flawless.

One of the solutions is that we assign two processes to the task from the beginning, but with different roles. One of the processes plays a primary role and actively delivers the service, while the second stands by and regularly synchronizes the measurement of the elapsed time with the first one in case it needs to take over the service provisioning. In this setup the user of the stopwatch service interacts only with the active process. The user is not aware of the existence of the second process. If the user suspends or stops the timer, it is the active process that informs its peer standby process about these events.

Notice that in this solution it is enough if the processes know their responsibilities depending on the role they have been assigned. The AMF does not need to be aware of any of the user aspects of the service provided by the application. It only needs to coordinate the roles the different processes play in the provisioning of the service, monitor their health, and if any of the processes fail reassign the roles and repair the failed entity by restarting it.

Obviously the application now needs to be prepared to handle these different roles. This raises several questions ranging from the benefits of using the AMF to the appropriate synchronization that needs to be implemented between these peer processes.

6.2.5 Delicacies of Service State Replication

If there was no availability manager to use, the application would need to include a part that monitors the health of its different parts. While this is relatively simple when only two redundant entities provision a single service in an active-standby setup, it becomes more complicated when other considerations are taken into account. For example, the active-standby setup always means 50% utilization as one of the entities is there just in case it needs to take over; until this, it does not provide any service.

When we have relatively reliable entities, this may be a huge waste of resources and we may want to have one entity acting as standby for many others. In other cases when, for example, the recovery is long and the service needs to be very reliable, one standby may not be enough and we may want to have multiple standbys.

From the service provisioning aspect all these scenarios mean the same:

- acting as an active and provide the service; or
- acting as a standby and protecting the service whatever that means for a particular service; and
- switching between the roles as necessary.

It is the coordination of these roles which is different for the different scenarios; however this coordination does not require any knowledge of any of the user aspect of service. This is the reason why the AMF was defined and its specification incorporates a wide selection of redundancy models that application designers or even site designers may use. Note that since the application relying on AMF's management only needs to implement a well defined set of states and state transition which apply to most of the redundancy models offered by the AMF, the actual selection of a particular redundancy model can be postponed till the moment of deployment and therefore tailored to the concrete deployment circumstances.

The question is whether it would be possible to eliminate even this need for the application's awareness about the redundancy.

As we have seen in the case of the alarm clock service, since the service did not require any state information additional to the initial data, there was no need even to assign a role to the process protecting the service. However, in the case of stopwatch service there is hardly any way around for providing service continuity than synchronizing the standby with the active process, so it is aware of the current status of the provided service.

We can see that there will be always a class of applications that have some state information which needs to be transferred to the standby in order to provide the impression of continuity for the service user.

This means that to release the application from the awareness of this state synchronization requires that an external entity provides some mechanism for the state replication. The problem is, however, that as soon as state replication is done outside of the application's control the 'replicator' cannot judge the relevance of the information, so it needs to be a complete state replication, otherwise information important for the application may be lost. Unfortunately this opens up the fault zone: The more state information is copied from a faulty entity to a healthy one, the more likely it is that the information encapsulating or manifesting the fault itself will be copied as well, which in turn corrupts the standby entity.

There is a delicate balance between state synchronization and fault isolation.

The AMF, however, does not deal with this aspect of state synchronization between active and standby entities. It was defined so that it only coordinates the roles between the redundant entities protecting the service and it is left to the application to decide what method it will use for the state synchronization. It may use databases or communication mechanisms for this purpose; however, general purpose solution may require adjustments to the clustered environment (e.g., to achieve location transparency).

There are a number of AIS services that address these issues. The service targeting exactly this need is the Checkpoint Service [42]. It allows its users to define cluster-wide data entities called checkpoints that are maintained and replicated by the service and that an application can use to store and replicate state information so that it becomes accessible across the cluster. Services providing location transparent communication mechanisms within the cluster can also be used to exchange state information between entities. We will review these different services in Chapter 7.

In this chapter we continue with the overview of the AMF as defined by the SA Forum, introduce its terminology and elaborate more on the solutions it offers to the different aspects of availability management we have touched upon in this section.

6.3   The Availability Management Framework

6.3.1   Overview of the SA Forum Solution

As we have seen previously the tasks associated with availability management can be generalized and abstracted from the functionality of the applications themselves. On the one hand, this makes the application development process shorter, simpler, and focused on the intended application function-ality. On the other hand, this requires an appropriate well defined availability management solution that application developers can rely on.

Such a generic availability solution is offered by the SA Forum AMF specification [48]. It defines an API that application processes should use to interact with an AMF implementation and an information model, which describes for the AMF implementation the entities it needs to manage and their high-availability (HA) features. At runtime it also uses the information model to reflect the entities runtime status. In addition for operational and maintenance purposes, the specification defines the management interface in terms of administrative operations AMF accepts and notifications AMF generates.

Within an SA Forum middleware, the implementation of the AMF specification is the software entity responsible for managing the availability of the services offered by applications running on the SA Forum middleware implementation. From this perspective, the AMF interfaces are the only way the middleware can impose availability requirements toward the applications it supports; there-fore only the services provided by applications integrated with AMF are considered to be highly available services.

To understand how the SA is achieved by AMF we need to look deeper into the functionality of the AMF. However, it is very important to understand that an application developer does not need to deal with or even understand all these details to be able to accomplish his or her task of writing an application to be managed by AMF to provide highly available services. Much of the AMF functionality is hidden from the application, which only needs to implement appropriately the API and the states controlled via the API. Similarly, most parts of the AMF information model are also irrelevant for application developers.

The information model is important for site designers or integrators as it is the main interface toward AMF from their perspective. It is the means by which they describe for an AMF implementation what entities compose the system that it needs to manage and what policies apply to those entities.

Finally site administrators need to understand the information model as AMF exposes the state of the different entities through the objects in the model and their attributes. Administrators can also exercise administrative operations on the subset of entities for which such operation have been defined. In addition, they also need to have a general understanding of the consequences different events and actions may cause in the system and the notifications and alarms AMF generates that may require their attention. This chapter provides an insight into these different perspectives.

Let us now dive into the depth of availability management as defined by the SA Forum AMF specification.

6.3.2 Components and Component Service Instances

The only entities visible through the AMF API are the component and the CSI. They represent the two sides distinguished by the AMF as discussed earlier: the service provider side and the provided service side. Since they are visible through the API, application developers need to understand these concepts. In fact these are the two concepts that AMF managed applications need to implement.

So what are they: the component and the CSI?

6.3.2.1 The Component

The component is the smallest service provider entity recognized by the AMF. It can be any kind of resource that can be controlled directly or indirectly through the AMF API or using CLI (command line interface) commands. Components can be software resources such as operating system processes or Java beans or hardware resources such as a network interface card as long as there is a way for AMF to control them by one of the above methods or their combination.

We can distinguish different component categories depending on the method AMF controls them. We will look at them in due course (see Section 6.3.2.5). For the time being we will focus on the simplest component that implements the AMF API, which is referred as a regular SA-aware component.

A component may encompass a single or multiple processes, or in other cases a single process may encapsulate a number of components. In general, however, we can say that there is always one process which is linked to the AMF library and implements the AMF API. It also registers the component with the AMF.

This process is responsible for the interaction between AMF and the component the process represents. Note that this allows the process to represent more than one component, therefore the API ensures that it is always clear which component is the subject of any interaction through the API.

The decisive criterion on what constitute a component is the particularity that it is also the smallest fault zone within the system. If any part of a component fails AMF fails the entire component and applies fault isolation, recovery and repair actions to the entire component. This means that the component boundaries need to be determined such that they provide adequate fault isolation, for example, fault propagation through communication is kept to the minimum. At the same time it is desirable to keep the disruption caused by these recovery and repair actions also to the minimum, that is, to keep the component small enough so it can be repaired quickly, particularly if such repair is needed relatively often. A car analogy would be the spark plug; we want to be able to replace the spark plug whenever it becomes necessary and not define the whole engine as a single undividable component.

6.3.2.2   The Component Service Instance

The reason of having the components in the system is of course that they can provide some services we are interested in. The AMF is not aware of the service itself a component can provide, it is only aware of the control mechanism that allows the management of the service provisioning. This is what the CSI is defined for. It represents a unit of service workload that a component is capable of providing or protecting.

For example, a component may be able to provide some service via the internet and therefore it expects an IP (internet protocol) address and a port number as input parameters to start the service provisioning. By assigning different combinations of these parameters different instances of this internet service will be created. They represent different workloads as users knowing a particular combination will be able to access and generate workload only toward that one component serving that particular combination.

Hence, one may also interpret the CSI as a set of attributes that configure a component for a service it is capable of providing. By providing different configurations, that is, different sets of input parameters, different CSIs can be provisioned by the same component.

Some components may even be capable of accepting these different configurations simultaneously and therefore provide multiple CSIs at the same time. Others may not have such flexibility. Some components may be able to provide more than one type of service, each of which would be configured differently.

The AMF abstraction from all this is the CSI. It is a named set of attributes that AMF passes to the component through the API to initiate the service functionality the CSI represents, that is, to assign the CSI to the component.

AMF does not interpret any of these CSI attributes; they are opaque to AMF. The same way the actual service functionality they initiate is also completely transparent for AMF.

The component receiving the assignment from AMF needs to be able to understand from the attributes themselves what service functionality it needs to provide and its exact configuration.

AMF uses the CSI name passed with the attribute set to identify this combination of service configuration. The CSI name, however, is not known at the time the software implementing the component is developed. If another component receives the same named combination of attributes, it should provide the exact same service functionality indistinguishably to the service users.

Hence we can perceive the CSI as a logical entity representing some service provisioning.

It is interesting to note – and this further explains the difference between the user aspect and the availability management aspect of the service – that from the perspective of the service user a completely different software may be able to provide exactly the same service (let's say this internet service mentioned), but a component running this software may require a different set of attributes (e.g., an additional parameter is needed that the service to be provided is ftp – as it can also provide ssh). Since the CSIs need to be different for each of these components, AMF will see these as CSI of different types. One composed of two the other of three attributes regardless that the user will receive the same ftp service.

It is also true that the same set of CSI attributes when they are assigned to different components may result in different services as perceived by the user. Many internet services require the address and the port as input parameters. AMF, however, will perceive these as different CSIs if these attribute sets are associated with different names regardless of whether the attribute values are the same or not. If they could have the same name then AMF would handle them as the same CSI and therefore the same associated service functionality. This however is not permitted.

6.3.2.3   Component High Availability State

To achieve HA we need to introduce redundancy on the service provider side, so in addition to service provisioning we also protect that service. Translated to the AMF concepts learnt so far this means that we need multiple components assigned to the same CSI at least one of which is actively providing the service while at least one other stands by in a state that it can take over the service provisioning in case of the failure of the first one. In the assignment the roles of these two components need to be distinguished. The AMF accomplishes this through defining different HA states in which a CSI can be assigned to a component.

As a minimum we need to distinguish the active assignment from the standby assignment; that is, when a component receives an active assignment AMF it must start to provide the service functionality according to the configuration parameters indicated in the CSI assignment. When a component receives the standby assignment for the same CSI, it must assume a state that allows a timely takeover of the service provisioning from the active component should it become necessary. It should also initiate any functionality that keeps this state up to date. This typically requires state synchronization with the component having the active HA state assignment for the same CSI; therefore when AMF gives a standby assignment to a component, it always indicates the name of the component having the active assignment for the same CSI. Based on this information, the component must know how to obtain the necessary information and maintain this state.

The AMF does not provide further means for state synchronization and other methods; for example, other AIS services such as the Checkpoint Service can be used to achieve this goal.

Having active and standby assignments for the same CSI is sufficient to handle failure cases. If the component having the active assignment fails AMF can immediately request the component having the standby assignment to assume the service provisioning starting from the latest service state known by the standby. At this point since the active component is known to be faulty, any interaction between the active and the standby components including state synchronization is undesirable. Note that this may mean the standby lagging behind depending on the frequency and used state synchronization mechanism between the active and the standby components. This may even mean a service interruption from the perspective of a particular service user, for example, a call or a connection may be dropped as a result. All this depends on how up-to-date the standby is at the moment it receives the active assignment.

The term for this procedure is CSI fail-over.

There are cases when we would like our components to change roles. For example, we may have received an update of our software and we would like to upgrade our components. We could do this by upgrading one component at a time, which means that we would like to be able to transfer the assignment between the components so that this results in the absolute minimum service disruption. To achieve this we would like to force a state synchronization between the components assigned active and standby for the CSI at the time of the role transfer.

For this purpose AMF defines the quiesced HA state.

It is always the component having the active assignment which is requested to assume the quiesced state by suspending the provided service functionality and holding the service state for the component having the standby assignment. As soon as the component confirms that it has assumed the quiesced state for the CSI, AMF assigns the active state to the component currently holding the standby assignment. In the assignment AMF also informs the component taking over the active role about the availability of the now quiesced former active component for the CSI. As part of the transfer the component assuming the active role should obtain the current service state from the quiesced component (directly or indirectly) and resume the service provisioning. When this is completed and the component confirms the takeover to AMF, AMF can change the HA state for the CSI of the other component from the quiesced to standby (or if needed the assignment can be completely removed without service impact).

This procedure of exchanging roles between components is called component service instance switch-over.

Note that even though we used the expression 'quiesced component' it needs to be interpreted as 'the component assigned the quiesced HA state for the CSI' as for other CSIs the component may maintain other HA state assignments.

The AMF defines different component capability models depending on the number and the variation of HA states a component is able to maintain simultaneously. If a component is capable of providing different types of services, each of them can be characterized by a different component capability model.

The component capability model defines the number of active assignments and the number of standby assignments that the component is able to maintain simultaneously for different CSIs, and whether it is capable of handling active and standby assignments simultaneously. The component capability model is one of the most important characteristics the developer needs to specify about the software implementing the component. The list of possible component capability models is given in section 'SU HA state' on page 163 where we discuss this feature in a wider context.

A different type of service suspension is necessary when we would like to gracefully terminate the provision of some service. Graceful termination means that users that have already obtained the service and are being served currently should be able to complete their service request. However in order to terminate the service we do not want to allow new users to be able to obtain the service by initiating new requests.

Let's assume an http service that we configure by giving the server's IP address and port number. From AMF's perspective this is a single workload as AMF uses a single CSI assignment to assign it to a component; but from the service user's perspective this workload is perceived as hundreds and thousands of service initiations each started by an http request and completed when the page is completely served to the user. In the active state the component would accept each request addressed to the IP address and port assigned in the CSI assignment and serve it till completion. These http requests arrive to this active component continuously. To gracefully terminate this service we want to let already initiated requests complete, but prevent the component accepting any new request. The AMF informs the component active for the CSI about this behavior change by changing the HA state assignment for the CSI to quiescing. This assignment also requests the component to inform AMF when it has completed the service of the last ongoing request. When the component completes the quiescing it is assumed that it quietly (i.e., without AMF instructing it) transitions to the quiesced state. In the quiesced state the CSI assignment can be completely removed or switched over as we have seen.

The quiescing state is a variant of the active state therefore it is protected by a standby assignment the same way the active assignment is protected. In cases when at most one component may have the active assignment this restriction applies to the quiescing state as well, that is, only one component may have the active or the quiescing state for a CSI at any given time.

To summarize the difference between the quiescing and quiesced states: quiescing counts for an active HA state assignment, quiesced does not; in quiescing the service is provided by the component for existing users, in quiesced the service is suspended for all users; as a consequence a quiescing assignment is typically protected by a standby assignment, while quiesced is the state protecting the service state during a state transfer.

6.3.2.4 Error Handling and Component Life Cycle

After the service provided by the component has been gracefully terminated, we may want to remove also the component itself from the system. AMF terminates the component using the API, namely using the terminate callback. The component receiving the callback should wrap up its actions as necessary for the service functionality (e.g., save any necessary data, close connections, etc.), terminate all of its constituents (e.g., child processes) and exit while confirming the completion of the operation to AMF.

Since the AMF defines the component as the smallest fault zone it makes the assumption that with the termination of the component the error that was detected in the component becomes isolated and removed from the system. This assumption contradicts the termination described above as it preserves some state and also it is performed via the API, which cannot be trusted any more.

For the termination of a component on which an error has been detected AMF uses tools available at the level below the component, for example, at the operating system level, which still can be trusted and which prevents any state preservation. It expects for a component executing on the operating system a so-called clean-up CLC-CLI (component life cycle command line interface) command, which when issued must abruptly remove all the constituents of the component from the system. It needs to prevent any fault propagation: The execution of the cleanup command must be as quick as possible to prevent any interaction with other components and must not preserve any of the service state information of the faulty component.

As we mentioned earlier, the service state synchronization between the active and the standby components counteracts to the intention of the component being a fault zone. This and any data exchange carries the risk of propagating the fault; they open up the fault zone. One has to take special care when determining the information exchanged even while the component is considered healthy. The exchanged data should be the necessary and sufficient amount as the lack of error does not indicate the absence of faults, it means only that a fault has not manifested yet so it could be detected.

As soon as an error has been detected any data exchange must be prevented immediately. Therefore AMF performs right away the cleanup procedure associated with the component – typically at the operating system level.

Once the component has been cleaned up to restore the system health the AMF will try to restart the component and if it is successful the component becomes available again.

All this means that besides managing the HA state assignments, AMF also controls the life-cycle of the components it manages. It initiates the component instantiation normally using the designated instantiate CLC-CLI command, which may, for example, start a process as the manifestation of the component.

Once all the constituents of the component have initialized and the component is in the state that it can take CSI assignments, one of its processes registers with AMF using the AMF API. From this moment on the component is considered available and healthy. That is, in general the successful re-instantiation of a faulty component is perceived as also a repair action. AMF may or may not assign a CSI to the repaired component.

Note that the AMF is not aware of the component and process boundaries. The registration creates an association between AMF and the registered process (represented by a handle) and AMF uses this association to identify or to communicate with a given component.

Since usually it takes some time to clean up a faulty component and instantiate it again, AMF fails over to their appropriate standby component(s) the CSIs that were assigned to the faulty component at the moment the error was detected. The successful assignment of the HA active state of the CSI to the component with the standby assignment recovers the service, so the CSI fail-over is a recovery action.

Failing over the assignment also takes some time. If we compare the time needed for these two operations the component restart and the CSI fail-over we may see that for some components actually the re-instantiation may take less time than the fail-over.

The CSI fail-over can only be executed after successful cleanup, otherwise it cannot be guaranteed that the faulty component indeed stopped providing the assigned CSI. Hence the cleanup is part of the recovery time regardless of the recovery method.

Restart may take less time in cases when failing over the assignment would mean significant data transfer to the standby component's location, which does not need to be performed if the component is restarted locally. For such cases the AMF includes the concept of restartable components.

If such a restartable component fails, the AMF first proceeds as usual with the cleanup of the component. But, when the cleanup successfully completes rather than moving the active assignments to the appropriate standby components, they are logically kept with the failed component:

AMF attempts to instantiate the failed component to repair it; and if it is successful AMF reassigns to the component the same CSIs assignments that it had at the moment the error was detected.

Consequently such a restart operation, which is executed as a cleanup followed by an instantiation, becomes not only the fault isolation and repair actions, but also the service recovery action.

The AMF performs similar fault isolation and repair actions on components that have no active assignment. In these cases there is no need for service recovery; however to replace a failed standby AMF may assign the role to another suitable candidate.

Components that have an HA state assignment on behalf of a particular CSI are collectively referred to as the *protection group* of this CSI.

Any process in the system may receive information from the AMF about the status of a protection group using the protection group track interface and referencing the name of the CSI in question. Depending on the track option AMF will provide the list of components and their HA state assignments that participate in providing and protecting the CSI. It will also report subsequent changes to this status information such as state re-assignments, addition, and removal of components to the protection group.

6.3.2.5 Component Categories

Component categories ultimately reflect the way the AMF is able to control the component life-cycle. Table 6.1 summarizes the component categories that have been defined for the AMF. The category information is part of the information model AMF requires for proper handling of components.

In the table we indicated in parenthesis implied categories deduced from other features. Italic indicates the main component categories we distinguish for availability management.

Except for the proxied components, components of all categories run within the cluster and referred as local components. Proxied components may run within the cluster locally or outside of the cluster externally.

As long as the component runs within the cluster (i.e., it is local), AMF can use system tools to instantiate, terminate, and cleanup the component. Essentially AMF can control any process by controlling its life-cycle through these CLC-CLI commands provided that the process when it is started it starts to provide its service immediately. Therefore such a process can be integrated with the SA Forum middleware even for availability management as an AMF managed *non-SA-aware-nonproxied* component.

The assumption is that such components do not implement the AMF API. They are also not aware of the HA states, therefore they cannot maintain an idle state after instantiation, which means that the AMF instantiates such a component only at the moment it needs to perform the active role and AMF terminates it as soon as the assignment needs to be withdrawn. The instantiation is considered to be a simultaneous CSI assignment operation; and similarly the termination is considered to be a simultaneous CSI removal operation. Consequently always a single CSI summarizes and represents all the services such a non-SA-aware-nonproxied component provides.

Table 6.1 Component categories

	SA-aware (pre-instantiable)			(Non-SA-aware)		
Contained	*(Regular SA-aware)*		*(Non-proxied-non-SA-aware,* non-pre-instantiable)*	Proxied		
	Container	*Proxy*		*(Pre-instantiable)*	*Non-pre-instantiable*
	(Local)			Local	Local	
					(External)	

Note that the termination even though it is performed as a CLC-CLI command is a graceful operation and the component may perform any action required for the orderly termination of the service it provides.

We have mentioned already the category of components that implements the AMF API and therefore referred to as *SA-aware*. This also implies that they are aware of the HA states and after instantiation they are capable of staying idle without providing any service until AMF decides to assign to them a CSI assignment.

Depending on whether AMF can control directly their life-cycle, SA-aware components can be still of two kinds. When the component executes on the operating system accessible for AMF and therefore AMF can use the instantiate and cleanup CLC-CLI commands in addition to the terminate API callback, AMF is in complete control of the component's life-cycle. *Regular SA-aware*, container and proxy components fall into this group.

AMF cannot directly control the life-cycle of components that require an environment different from the operating system accessible for AMF. In this case, this different environment is considered as a container of such components. For this reason, the components executing in this special environment are called *contained* components and the environment itself is also a component of the *container* category. For AMF to manage the life-cycle of the contained components, the container in which they reside needs to act as a mediator. The container needs to implement a part of the API, which allows the instantiation and cleanup of contained components. In other words, the service that a container component provides is the functionality of mediation between AMF and its contained components and it is also represented as a CSI – the *container CSI*. Accordingly when AMF decides to use a container for some contained components, first it assigns the CSI associated with the mediation of those contained components to the container component. Only after this it will proceed with the instantiation of the contained components.

The container's life-cycle is also linked with the life-cycle of the contained components: The contained components cannot and must not exist without their associated container component. For example, if the java virtual machine (JVM) process is killed all the Java beans running in it cease to exist at the same time.

This means that if the container component is being terminated, all its contained components also need to be terminated. AMF orchestrates the termination via API callbacks as all these components are SA-aware. On the other hand, if the container is cleaned up, that implies that all its contained components are also abruptly terminated. The abrupt termination of contained components needs to be guaranteed by the cleanup CLC-CLI command of the container component as AMF has no other means to cleanup contained components when their container is faulty and cannot be relied on.

The last group of SA-aware components is the proxy. *Proxy* components perform a similar mediation on behalf of AMF as container components, but toward proxied components. In this case the mediated functionality goes beyond life-cycle control since *proxied* components are non-SA-aware components. Another difference is that the life-cycles of the proxy component and its proxied components are not linked. The termination of the proxy does not imply the termination of its proxied components, as a result for local proxied components the cleanup CLC-CLI command still applies and AMF may resort to it.

The AMF has no direct access to external proxied components or their environment. In fact AMF is not even aware of the location of such a component as it is outside of the cluster, the scope recognized by AMF. Such components can only be managed via a proxy.

Proxied components are also classified whether they are able to handle the HA states even though they are not implementing the AMF API. Components that are able to be idle, that is, provide no service while still be instantiated are pre-instantiable components just like all SA-aware components. For them AMF performs the exact same HA state control as for SA-aware components all of which is mediated by the current proxy.

This means that AMF requests the proxy component to instantiate such a proxied component regardless if it needs to provide services or not and when the instantiation completes the proxy component registers with AMF on behalf of the proxied component. Later when the proxied component needs to provide a service, AMF calls back the proxy component mediating for the component with a CSI assignment for the associated proxied component. The proxy needs to interpret this CSI assignment, initiate with the proxied component the requested functionality and inform AMF about the result. The HA state changes for the CSIs assigned to the proxied component and the component termination are similarly mediated.

Proxied components that are not aware of HA states, and cannot stay idle because they start to provide their services as soon as they are instantiated are *non-pre-instantiable proxied* components. This means that the AMF handles them similarly to non-SA-aware-nonproxied components except that it uses the proxy's services to mediate the life-cycle control.

Just as in the case of the container component the proxy functionality is the service a proxy component provides and it is controlled through a CSI – the proxy CSI.

A peculiarity of the proxy-proxied relationship comes from the fact that their life-cycle is independent. That is, the failure of the proxy has no direct implications on the state of its proxied component. It may or may not fail together with the proxy, it may even continue to provide its service. Without a proxy AMF has no information on its state. If we can make any assumption, it should be that the proxied component remains operating according to the last successfully mediated instructions.

So for the availability management of the proxied component it is essential that the proxy CSI associated with the proxied component is assigned and provided as the active proxy is the only means to maintain contact with and control the proxied component. When the current proxy component fails AMF will attempt to recover the CSI as appropriate, but during this time it has no control over the proxied component. When the newly assigned proxy re-establishes contact with a running proxied component it registers on behalf of the proxied component. Otherwise it reports any error it encounters.

Figure 6.1 summarizes the different component categories and the interfaces AMF uses to communicate with each.

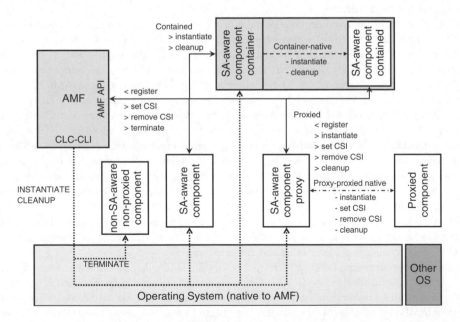

Figure 6.1 Component categories and related interfaces.

6.3.2.6 Error Detection

Reading thus far one should realize that the most important triggering event for the different actions carried out by the AMF is the detection of errors. So the question rises: how does AMF determine the health of the components under its control? How does it detect errors?

The obvious way of detecting errors is through the interaction with the component using the AMF API.

The AMF initiates most of the control operations by callbacks to the process which has registered itself on behalf of the component targeted by the operation. Whenever there is an error in this interaction, since the same process may register multiple components (e.g., a proxy) AMF correlates the component category information and the detected error to determine which one is the faulty component.

A component may respond to a callback with an error, or a faulty component may not reply at all. Both of these cases indicate an error for AMF and it should determine which component is the faulty one to be able to take the correct measures.

If the operation is not mediated – does not go through a proxy or a container – then no response reflects the failure of the component targeted by the operation.

If the operation is mediated (e.g., AMF requests the proxy to put its proxied component into active HA state for a CSI) then no response means the failure of the mediator component and not the targeted component (i.e., the proxy in the above case and not the proxied component).

If the AMF receives an error in response to the initiated operation then it interprets this as the component targeted by the operation is at fault.

A different type of interaction occurs when a component initiates an operation through the API. These operations must be in a particular order and appropriate for the state of the component. If an operation is inappropriate (for example, a component responds without being asked) the AMF deems the initiating component faulty and engages in the required recovery and repair actions.

Control operations occur in the system relatively rare and as we have seen they are typically already in reaction to some problem that has been detected. To continuously monitor the health of components, the AMF needs other solutions.

The first one is a special API dedicated for this purpose – the healthcheck.

Any process (and not only the registered process) may request AMF to start a healthcheck which means that the process will need to confirm its health at a regular interval either in response to AMF's inquiry or automatically by itself. If AMF does not receive such a confirmation at the expected time, it declares the component whose name was given in the healthcheck start request faulty and applies the recommended recovery action also indicated in the start request.

Note that the component name needs to be given at healthcheck start as AMF only knows about the association of the component with the registered process. Other processes of the component need to indicate their affiliation when they request the different operations.

The second option of health monitoring is the passive monitoring.

For passive monitoring the AMF uses operating system tools appropriate for the particular AMF implementation. It depends on the operating system what errors it can report; for AMF purposes it should at least be able to report process death.

A process may start and stop passive monitoring using the appropriate API functions. At start it needs to provide a process id and the level of descendants which initiates the monitoring of all processes satisfying the criteria. Again, since AMF is not aware of the process component associations, the component name and the recommended recovery action is given at the initiation of the passive monitoring.

The last health monitoring option is called external active monitoring (EAM). AMF initiates the EAM on all components for which the appropriate CLC-CLI commands are available. After the successful instantiation of such a component, AMF executes the am_start CLC-CLI command, which starts an external active monitor. It is completely application specific by what means this monitor determines the health or failure of the component. It may just be heart-beating the component, but it may also test the service functionality by generating test requests. The important part for the AMF is that the active monitor uses yet another AMF API dedicated to report errors.

The error report API can be used by any process in the system. The process does not need to be an active monitor or part of a component even. As a result the report needs to indicate the component on which the error is being reported and also the recommended recovery action. AMF uses this information to determine the appropriate reaction and engages it.

As the error reporting facility shows error detection is a 'collective effort' in the AMF managed system. Since the AMF is not aware of the different aspects of applications and their services, it has limited capabilities of detecting errors. Its main responsibility is to determine and carry out in a coordinated manner the most appropriate actions necessary to recover the services provided by the system it manages and if possible to repair the failed components.

This means that application developers need to consider the different error detection options and use for their components the most appropriate mechanisms so the application collaborates with the AMF in this respect too.

6.3.3 The AMF Information Model

The most important remaining questions are how the AMF would know what components and CSIs it needs to manage, what are their characteristics and what policies it should apply in their management.

The answer not surprisingly is that AMF requires a configuration, which contains all this information.

The AMF specification defines the UML (unified modeling language) classes used to describe the AMF information model. The UML class diagram is shown in Figure 6.2. It is part of the SA Forum information model [62].

This information model contains this configuration information in the form of configuration objects of the classes defined by the specification. As a result, the information model is the most important tool in the hands of site integrators and site designers, who decide how the system needs to be put together, from what components it should be composed of and according to what policies they need to be managed in order to provide the appropriate CSIs.

The AMF information model also contains information about the current status of the system, such as which components have been instantiated and what their assignments are at any moment in time. Therefore it is also an essential interface for site or system administrators who need to be able to obtain any relevant information about the status of different entities in the system and also to perform administrative operations whenever it is necessary.

Considering the potential size of systems managed by the AMF one quickly realizes that the organization of all this information is vital for easy comprehension, overview, and administration. As we will see these were the guiding principles when defining the AMF information model.

6.3.3.1 Component Type and Component Service Type

In fact components participating in the same protection group typically run the same software, which means that most of their configuration information (such as the CLC-CLI commands to instantiate and to cleanup, the different associated timers, the component categories, and so on) is the same for them.

In the recognition of this and to simplify the configuration of these components the concept of component type was introduced to the AMF information model. Since not only components of a protection group, but any component running the same software may have these same features, the concept is not limited to protection groups. Instead, a component type can be defined for any set

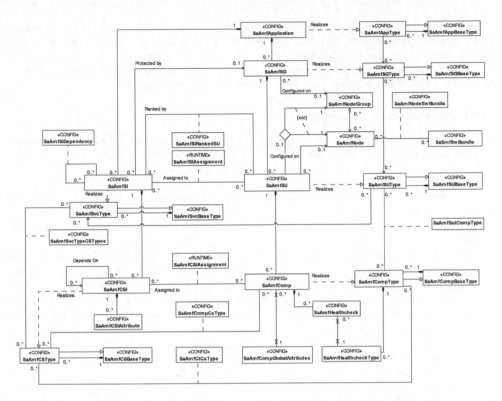

Figure 6.2 The AMF information model: AMF instances and types [62].

of components running the same software and requiring the same configuration information. This way the component type provides a single point of control for this common part of the configuration information.

Since the software associated with a component type typically has many versions, one may collect all these versions under a single umbrella, therefore the concept of a component base type became part of the AMF information model. Note that the AMF specification does not define what a version is, whether it reflects a different version of the software or a different configuration of the same software. It is left to the site designers to decide what the most appropriate use for these concepts for a given site.

For AMF to know how many and what components need to be instantiated, at least the component names need to be provided on an individual basis. Therefore for each of them there is a component object in the AMF information model describing at least the name of the component and the component type it instantiates. Other configuration attributes are associated with the attributes of the component type. If such an attribute is left empty that means that AMF applies by default the value given in the associated attribute of the component type object. Hence if the value is changed in the component type object, it applies to all components referencing this type and having their associated attribute empty. Otherwise when such an attribute is set for a particular component, AMF uses this value. Thus, setting such an attribute exempts the component from the control through the type with respect to this particular attribute.

These individual component representations are also convenient places to reflect the runtime status information of the component running in the system.

By similar analogy, the concept of component service base type and component service type were introduced for CSIs.

The component service type defines the set of the attributes – in terms of their names – that needs to be configured for each of the CSIs that instantiate this component service type. Each of these CSIs are configured by an object in the AMF information model which indicates: the name of the CSI which AMF uses to make the assignments, the component service type, and the attribute values for each of the CSI attribute names defined in this component service type.

6.3.3.2 Runtime Status of Components

To represent the status of components in the system a set of states were introduced to the information model as runtime attributes. The AMF API does not expose any of these states.

Component Presence State

To represent whether a component has been instantiated and is running in the system the presence state was introduced.

At system start no component runs in the system, all of them are in the *uninstantiated* presence state.

When AMF issues the instantiate CLC-CLI command (or the appropriate API callback for mediated components) the actions necessary to instantiate the component take place, and depending on the type of the application, it may take some time before the component becomes ready to take assignments. During this time the component is in the *instantiating* presence state.

If the component successfully registers with the AMF, the instantiation is successful and the component's state changes to *instantiated*. It means that the component has completed all necessary initialization procedures and it is ready to take CSI assignments.

Non-pre-instantiable components do not register, for them the successful completion of the instantiate CLC-CLI command indicates the completion of the instantiation.

When AMF decides that it does not need a component any more, it terminates the component to potentially free up the resources used by the component. It initiates such a termination via the API or the terminate CLC-CLI command. Alternatively, AMF may abruptly terminate faulty components using the discussed earlier cleanup CLC-CLI or its API equivalent. Regardless which way the termination is executed, the component's presence state becomes *terminating* until AMF can conclude that the termination or cleanup procedure was successful and therefore the component again is in the uninstantiated state.

The component is only fully functional in the instantiated state and it is not aware of any of these state values. The presence state is used by the AMF in evaluating the system state and within that the status of the different entities and also it is an important piece of information for system administrators.

The remaining three presence state values are all related to errors.

We have mentioned in Section 6.3.2.3 that AMF may decide to keep the CSI assignment with the component while it is being restarted in order to reassign it to the successfully restarted component. If this is the case then rather than driving the state machine through the terminating-uninstantiated-instantiating state transition sequence the actions are summarized in a single state value – *restarting*. This single value representation does not reflect the sequence of actions carried out by AMF to restart the component. AMF still terminates or cleans up the component – as appropriate – and then instantiates it.

The failure to instantiate a component is reflected by setting the component's presence state to *instantiation-failed* and the failure to terminate results in the *termination-failed* states.

The AMF may make a number of attempts to instantiate a component. The *instantiation-failed* state is set, when AMF exhausted all its attempts to repair the component by restart and therefore has given up on the repair.

AMF has only few tools to accomplish the termination and the *termination-failed* state is more critical. If the graceful termination does not succeed or it is not appropriate, the only option left is the cleanup CLC-CLI command. If this fails AMF has no other tools to handle the case at the component level.

The reason the *termination-failed* state is critical is that if the component was servicing any CSI when the termination attempt was initiated and this attempt has failed there is no guarantee that the component has stopped the service provisioning. This means that such a CSI becomes 'locked-in' with this failed component until the termination and therefore the withdrawal of the CSI can be guaranteed.

Component Operational State

The actual fact whether the AMF is aware of any error condition with respect to a component is reflected by yet another state, the operational state. The operational state may have the values *enabled* or *disabled*.

The enabled operational state means that AMF is not aware of any error condition related to the component. That is, none of the error detection mechanisms discussed in Section 6.3.2.6 has been triggered or if any of them was triggered, the restart recovery action is still in progress. In any other case, that is, if the recovery action is different from the component restart action then triggering the error detection mechanisms also results in disabling the component's operational state.

The operational state also becomes disabled if the component reaches the instantiation-failed or the termination-failed states.

Component Readiness State

If a component is healthy and therefore its operational state is enabled; it has been successfully instantiated, that is, the component has reported to the AMF that it is ready for assignments; all this does not mean that AMF can select this component for an assignment. It needs to take into consideration the environment of the component.

When both the component and its environment are ready for an assignment, the component is in the *in-service* readiness state. When either the component or its environment cannot take assignments, the component is in the *out-of-service* readiness state. And finally its readiness state is *stopping* if the component's environment – and therefore the component itself – is being shut down.

As we see the readiness state is a composite state. Its exact composition depends on the component category; however, for all components it includes the component's operational state and its environment's readiness state, which we will discuss later in section 'SU Readiness State' on page 162. In addition, for pre-instantiable components the readiness state also includes the presence state since these components register with AMF (directly or indirectly through a proxy component) to report their readiness to take assignments.

In other words, a non-pre-instantiable component is ready for service as long as it is healthy and enabled; otherwise its readiness state is out-of-service. A pre-instantiable component needs to be enabled and instantiated as well, otherwise it is out-of-service. To be in-service, however, components of both categories need to be in an environment, which is in-service. Both component categories change to stopping or out-of-service as soon as the environment's readiness state changes to stopping or out-of-service respectively.

Note that some AMF implementations may allow components to report their HA readiness state with respect to particular CSI assignments. By setting its HA readiness state to a particular value for a CSI indicates what HA state assignments the component is capable of handling for that particular CSI in its current state. It may refuse HA state assignments not aligned with this request without AMF evaluating it as faulty. It also does not impact the component's overall readiness state.

Next we investigate what we mean by the environment of a component.

6.3.3.3 Compound AMF Entities and Their Types

A component typically participates in two types of collaborations: We have already mentioned the first one, the collaboration of components to protect a CSI forming a protection group. The second type of collaboration combines the service functionality each component provides to a functionality more desirable for their end-user. In this relationship components form the so called service units (SUs) and the combination of their CSIs create the SIs which we will explore next.

The Service Unit

The component boundary defines a fault zone, which does not necessarily coincide with the boundary of the desired functionality. It merely says what can be isolated and repaired on its own. It is a little bit like with cars, one does not want to replace the car's engine because a part, the spark plug, for example, failed in it. So the engine is composed from pieces repairable and replaceable on their own, but the functionality we are after is the engine's functionality of powering the car. Similarly we put together components so from their combined partial functionality we can make up the desired service functionality boundary.

Such a group of components is called a service unit. It is a logical entity known to the AMF – components are not aware of it – and visible only in the AMF information model. There is no application code associated only with the SU, which is not part of any of its component's code. It is a set of components. The intention is to be able to develop components independently and combine their functionality later as needed and by that realizing the commercial-off-the-shelf (COTS) paradigm.

Since the SU is defined at the functionality boundary it is expected that within this boundary fault propagation can still occur relatively often due to the tight collaboration of its components. Therefore the SU forms the next fault zone within the AMF information model. The collaboration may be so tight that the AMF even provides a configuration option to escalate a component failure right away to the failure of its SU.

This implied tight collaboration is further reflected by the requirement that all components of a SU need to be collocated on the same node and in the same execution environment. Any type of collaboration requires communication, and it is usually easier and more efficient to communicate within an execution environment than between them. This however also means that faults can also propagate easier within the execution environment than between them.

This collocation requirement means that, for example, local and external components cannot compose a single SU. It also means that contained components that form a SU must be contained in the same container component and all components of such a SU must be contained components. Even the container component needs to be part of a different SU as it executes in a different environment.

To further isolate fault zones, SUs do not share components. Each component belongs to one and only one SU.

The Service Instance

The combination of the CSIs provided by the components of a SU, that is, the workload at the boundary of the service functionality is called the service instance.

The question is what is really the semantics behind this term 'service functionality boundary'?

It is not something easy to grasp or describe as the AMF – as we discussed earlier – is not aware of the user perceived service at all. As a result this functionality boundary can only be characterized by describing how AMF handles SIs.

As in case of SUs, there is nothing tangible associated with the SI beyond the CSIs it comprises from. When each of the CSIs in the SI is assigned to a component (as described in Section 6.3.2.3) in a SU it is said that the SI is assigned to that SU. AMF assigns all the CSIs of an SI to the components of the same SU in the same HA state. It selects the component for each CSI based on the component's capability model for the type of the CSI (i.e., component service type).

If a component is able to take more than one CSI assignment, AMF may assign more than one to it, but if there are more than one components available and capable of providing that CSI, AMF may distribute the assignments among them. That is, for CSIs being in the same SI does not guarantee their co-assignment to the same component. It only guarantees the assignment to the same SU, which may mean different components.

An SU may be able to provide more than one SI and AMF may use this additional capacity as needed. This also means that some components may remain unassigned in the SU when there is no need for their services, others may have assignments for CSIs of different SI.

Just like SUs are exclusive sets of components, SIs also cannot share CSIs even though this might be a very tempting idea when the functionality has a common portion. Instead, one should define CSIs of the same type in the different SIs, which may be assigned then to the same component when both SIs are served by the same SU.

The AMF moves around the assignments of SIs more or less independently therefore each SI needs to have the CSI definition independently to activate each of the required service functionality of the components.

To make these concepts more tangible let's assume this web-interface through which some database can be accessed. Typically the database software and the web-server software are developed by different software vendors and therefore we can assume that each of them integrated their solution with the AMF independently.

Now we would like to integrate them so they collaborate to provide our desired web-interface. To achieve this collaboration we configure together into this single SU:

- the web-server software with the page content for the interface as one component; and
- the database component with the data content as a second component.

We already know that none of the SA-aware components is allowed to provide any service unless they have an assignment with the active HA state for a CSI. This means that to activate them we need to specify for each of these components these CSIs, which in turn will compose the SI representing this web-interface service. Let's say the web-server requires the IP address at which it should listen to the enquiries and a file name the entry point for the interface page content; and the database requires the location of the data and the address it should listen to enquiries.

When we want to provide this web-interface the representing SI will be assigned by AMF to our SU, which effectively means that AMF will callback each of these components with their appropriate CSI attribute settings and in the HA active state. From that moment on, anyone requesting the page at the address indicated to the web-server component in the CSI assignment will receive the view of the interface through which a database enquiry can be performed. When an enquiry is submitted through this page, the web-server component will generate an enquiry toward the database component at the address provided to the database component in its CSI and the component returns the enquiry result from the data it finds at the location it received as the second attribute.

Obviously the page and the data content do not come from the vendor and we need to provide them. The address of the database enquiry interface toward which the web-server generates the request is part of the interface page.

All is well as long as both components are healthy, but what happens if our database component fails?

The Service Group

At the component level components form protection groups based on the CSIs they protect. These components belong to different SUs. This is because the SU itself is a fault zone and may fail. So if the protection group was within the same SU AMF would not be able to fail-over the CSI.

When AMF fails over a CSI of a SI to another component in another SU, it also needs to fail- or switch-over all the other CSIs of this SI, so that all the CSIs are assigned to the same SU.

SUs that work together to protect SIs form a service group (or SG). To provide any protection we usually need at least two SUs in a SG. It was the traditional approach to redundancy that we would have one active and one standby element in the system and only the active element provides the functionality required. This simple strategy is still widely used, but from an economical and efficiency perspective it obviously doubles the cost of everything as it works only at 50% utilization.

It turns out that we can define other redundancy schemas that are more efficient and depending in the reliability of the different components may still satisfy our required level of SA. The AMF specification defines five different redundancy models according to which SGs may operate and protect their SIs. These models are:

- the 2N redundancy model;
- the N+M redundancy model;
- the N-way redundancy model;
- the N-way-active redundancy model; and
- the no-redundancy redundancy model.

We will take a closer look at each of these models in Section 6.3.4, but before doing so first we describe the common features of SGs that apply to all redundancy models.

A SG may protect more than one SI and these SIs are assigned to the SG at configuration time. It is a long-term association, which cannot be changed without impacting the availability of the service.

Also at configuration time the redundancy model and all the appropriate values for the model attributes are given, and at runtime the AMF implementation decides based on these parameters:

How many and which SUs of the SG need to be instantiated. It is possible that not all SUs configured for the SG are instantiated right away. Instead only a sufficient number of them are started initially and AMF instantiates others only when this becomes necessary, when the number of available instantiated ones drops below the level required by the configuration attributes. This built into AMF feature often referred in other systems as dynamic reconfiguration. In case of AMF the configuration implies a certain level of dynamism and its extent depends on some configuration attributes as we will see.

After some waiting time or when the required number of SUs becomes available, AMF decides about the most appropriate distribution of assignments for the different SIs protected by the SG. This includes both active and standby assignments for each SI as required by the configuration of the SG and the SI.

AMF makes the assignments for each of the SIs a SG protects to the selected SUs. Doing so components of one SU get the assignments for one HA state (e.g., active) assignment while for the other HA state (e.g., standby) assignment the assignments are given to components of another SU.

Whenever an event occurs that requires a change in the assignments (e.g., an SU needs to be failed over), AMF re-evaluates the situation and first of all isolates the error and tries to recover the SIs for which the assignment is lost by reassigning them to other SUs. During this the main concern is to recover the active assignments for them by turning if applicable the standby assignments into active and then if it is possible also provide the standby assignments as required. This may require the instantiation of SUs that were not instantiated so far. AMF will typically also engage in the repair of the components that failed as we have seen in Section 6.3.2.4.

When AMF evaluates the situation it applies the policies implied by the redundancy model and if applicable the recovery recommendation received with the error report or the default recovery configured for the component on which the error has been detected.

If AMF reacts to a failure, the most important decision is the identification of the fault zone at which the recovery and the repair needs to be executed. The smallest recovery zone is the component on which the error is detected. However as we noted sometimes this level is inadequate and a different recovery recommendation may be configured for the component itself or the SU is configured as the first fault zone. Other times the detected error may indicate for the entity monitoring the component that more than the component was impacted and it may suggest to AMF a different recovery action in its error report. Finally AMF also correlates errors occurred within the same SU and within the same SG over time. If several of them happen within a configured period of time, it deems the failures related to each other and escalates the recovery to the next fault zone.

Any SU within the SG must be able to take the assignment for any SI associated with the SG. However, the number of SIs that may be assigned to a SU simultaneously may be limited. For example, if a SG is composed of three SUs {SU1, SU2, SU3} and it is assigned to protect the service instances {SIA, SIB}, then AMF should be able to assign any of these SIs to any of the SUs. However, each of the SUs may be able to take only one such assignment either SIA or SIB, but not both at the same time.

The remaining AMF entities have one primary role: they define even higher level fault zones encapsulating incrementally more entities.

The AMF Node

According to the specification an AMF node is the collection of all AMF entities on a cluster (Cluster Membership service (CLM)) node.

While this definition seems to be straightforward it is not as the AMF node does not require the presence of a CLM node. It may be configured independently. In addition the actual list of entities in this collection may not be known until runtime. So here we would propose a slightly different definition:

An AMF node is a logical container entity for AMF components and their SUs which is mapped to a CLM node for deployment. This mapping is 1 : 1.

Since a CLM node is mapped to an execution environment (see Figure 6.3) such as an operating system instance, all the AMF entities collected by an AMF node execute in this same execution environment and AMF uses the command line interface of this OS instance to control the life-cycle of the components associated with the AMF node.

As discussed earlier components that execute in the same execution environment, for example, that run on the same instance of an operating system cannot be completely isolated from one another as they need to use common resources, the same memory, the same set of communication interfaces, and so on. Therefore the AMF node is the next higher fault zone (after the component and the SU).

To demonstrate it let's reconsider our example with the process, which leaks memory. Let's assume it is part of a component. The memory-leak may impact the execution of all processes running in

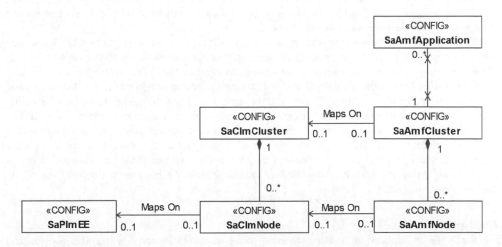

Figure 6.3 Mapping of the AMF node to the CLM node and to PLM execution environment (EE) [62].

this OS instance. Moreover, the shortage of memory may cause the failure of a different process in a different component before the leaky one; hence the restart of this failed component will be no remedy for the situation.

Not knowing the root cause of the problem or the justified memory use of each component it is impossible for the AMF to identify which component is the real offender, hence it will need to deploy the recovery escalation policies and widen the fault zones until it eventually captures the offending component. This is guaranteed in our case only at the AMF node level as this is the level where the common resources are shared. In particular, the same way AMF correlates failure for the SU and for the SG, it also correlates failures within the same node. When the number of recoveries within the same AMF node reaches a threshold the entire node is recovered.

While typically the AMF handles only AMF entities, when the recovery and the repair reaches the AMF node level AMF may attempt to repair the execution environment by restarting via the Platform Management service (PLM) [36]. This operation is referred as node fail-fast and depending on the configuration may range from operating system reboot to physical reset of a hardware element impacting therefore also the mapped CLM and PLM entities.

Indeed whenever a component (and therefore its SU) reaches the instantiation-failed or termination-failed states as discussed in section 'Component Presence State' on page 150 the only repair AMF can attempt (if configuration permits it) is the node fail-fast, which abruptly terminates all components running on the node.

By contrast, a CLM node may get disconnected or a PLM entity such as a physical node may fail independently from what happens to the AMF controlled entities. Such a failure impacts all the AMF entities on the AMF node mapped to these PLM and CLM entities and the AMF needs to be able to handle such a situation and maintain SA.

Therefore AMF when making the different HA state assignments for the same SI, it gives preference to SUs that are located on different AMF nodes (and if possible even on different physical nodes). In this respect AMF's responsibility may be limited by the configuration it is given to manage.

Namely, SUs may or may not be allocated to particular nodes in the configuration. If an SU is assigned to a particular AMF node by the configuration, AMF must instantiate this SU on that node only and if all SUs happen to be allocated to the same physical node, AMF has no way to distribute the assignments so that it would protect against the failure of this physical node.

If there is no such configured allocation, AMF chooses the AMF node from the list of eligible nodes – called node group – at the moment it instantiates the SU for the first time. As long as there is a node available in the node group without any SU belonging to the given SG, AMF will not select a node which already hosts such a SU. The node group is also configured and it is a list of nodes that are equivalent for the purpose of hosting some SUs.

With all these considerations the AMF node is still a fault zone, which in most cases can be isolated without any service impact. This prospect changes with the last two fault zones that we will look at.

Before doing so as a side note, it is interesting to compare the AMF node with the container component since both of them are associated with a single execution environment yet these execution environments are represented by different AMF entities. The key distinguisher is the AMF's ability to access the execution environment. As we have seen in case of container components AMF can use the AMF API to communicate with the container component itself, but it has no direct access to the environment this container provides for the other components therefore it cannot execute the life-cycle control operations directly. In case of the AMF node, AMF has direct access to the execution environment; in fact the SA Forum middleware including the AMF implementation may run within this same environment.

The Application

For the AMF an application is a logical entity composed of SGs and SIs the SGs provide and protect.

The AMF application logical entity reflects mostly the boundaries of a software implementation delivered by a particular software vendor, that is, the generic interpretation of the word 'application.'

As a result applications are considered relatively or completely independent from each other, which makes them good fault zones that can be isolated without serious impact of the rest of the system. However, since the application includes both the SIs and the SGs providing and protecting them, isolating this fault zone impacts at least all those SIs included in the application entity.

The questions are how the application fault zone is used and when it is appropriate to use.

As in other cases the isolation mean termination and restart of the entity; however, since the only tangible entities in our system are the components, the restart of an application is executed as the abrupt termination and restart of all the components composing the application. That is, all the components which are part of any SU whose SG belongs to the application. While executing this restart no service recovery is performed for the associated SIs. In fact, the goal is to terminate and restart the service functionality by terminating and restarting the SIs themselves so to speak. This obviously creates an outage for these SIs and the functionality they represent. If there are other applications depending on the functionality of the faulty one, they may suffer outage too. We discuss dependencies in detail in Section 6.3.3.5.

Such a drastic measure is appropriate only when the fault has manifested not only as an internal error (such as a failure of a redundant entity like a component or a SU), but it is visible already externally as a failure of the provided service functionality, which in turn is associated with the SIs of the AMF application. This also means that the error is likely to be part of the state information of the service functionality – the information exchanged between the active and standby entities for synchronization. Therefore its preservation – which is normally the goal – should be prevented in this case.

To guarantee that no state information is preserved by the redundant entities of the AMF managed application, the restart is carried out by first terminating all the redundant entities (SUs and their components) completely before starting to instantiate any of them again.

The AMF Cluster

The AMF cluster is the complete collection of AMF nodes and indirectly it encompasses all the entities managed by the AMF.

As each of the AMF nodes is mapped onto a CLM node the AMF cluster is mapped to the CLM cluster, which encompasses these CLM nodes hosting AMF nodes. However, it is not required that each CLM node of the CLM cluster hosts an AMF node, that is, the AMF cluster may be mapped only to a subset of the CLM nodes. The important point is that all the cluster nodes to which the nodes of an AMF cluster are mapped must belong to the same CLM cluster.

The reason for this is that the CLM is the authority to decide whether a CLM node is part of the membership. When a CLM node becomes part of the membership, it implies that it is healthy for running applications and it can communicate with the rest of the cluster.

When a CLM node joins the CLM cluster, the AMF node mapped to it becomes automatically available for AMF entities and AMF may start to instantiate them as necessary.

If a CLM node leaves the cluster membership the AMF is responsible for terminating all the components on the node.

From the AMF perspective the AMF cluster is the ultimate fault zone as it includes all the entities managed by an AMF implementation. The recovery action performed at this level is the cluster reset which goes beyond AMF entities. It is equivalent to the node fail-fast operation (discussed in section 'The AMF Node' on page 155) applied to all nodes of the AMF cluster simultaneously. Typically this means the reboot of the operating system of the nodes, but the exact meaning depends on the PLM configuration for the execution environment of the node.

The same way as we have seen for the application, the intention is to preserve no state from the AMF cluster requiring the reset. This means that no node may start booting before all of them have been terminated.

The Types of Compound AMF Entities

For components and CSIs the component type, the component base type, the component service type, and the component service base type were introduced to collect the sets of entities that have common characteristics and to simplify their configuration. The same approach applies to the compound AMF entities with the exception of the AMF node and the AMF cluster. Thus, the AMF information model includes the concept of the service type, the SU type, the SG type, and application type and their base types. The base types have no additional semantics than collecting the set of entity types that can be considered as versions of the same software concept.

The versioned entity types have the following interpretation:

- The *service type* defines the common characteristics of a set of SIs and, in particular, the set of component service types instances of which can compose these SIs. It defines for each of these component service types the maximum number of instances that a SI may contain. That is, a SI of a particular service type may include zero or more – up to this specified maximum number of CSIs for a particular component service type.

  This characterization of the service type facilitates the comparison of the capacity required from a SU to take an assignment for a SI of the type. Since no more than the maximum number of CSIs may be configured in an instance of the service type this defines how much capacity in terms of CSIs the components of the protecting SUs should provide.

- The counterpart of the service type for the entities of the provider side is the *SU type*. Its most important attribute is the list of service types the SUs of the type can provide and protect. To allow the capacity comparison it also defines the component types that should compose the SUs of the type and for each of the component types the minimum and the maximum number of components that can be included. The minimum number must be at least one so that there is at least one component in each SU that can provide each of the component service types of the SI the type of which is listed as supported service type.

- Two of the attributes defined in the SG *type* are important to mention.

  The first one is the redundancy model. The SG type specifies the redundancy model for all of its instances. For each of the SGs belonging to the type only the configuration attributes of the particular redundancy model are given which tailor these SGs for the needs of the deployment site. Future will show whether the control of these attributes from the type is a simplification.

The second attribute is the list of valid SU types instances of which may compose the SGs of the type. Since any SU of a SG must be able to take the assignment for any of the SIs assigned to SG it is typical that all SUs of the SG are of the same type. In addition the same SU type means same component types, which boils down to the same set of software implementations, which obviously means less software cost.

However there are situations when a SG may be composed of SUs of different SU types. The most obvious one is during upgrade. Since the units of redundancy are the SUs the way to upgrade the SG without service impact is one SU at a time. Thus after the upgrade of the first SU it will typically belong to a new SU type and the SG will be heterogeneous from the moment the upgrade starts till the moment it completes.

Another reason may be that the AMF cluster is heterogeneous, for example, composed of nodes running two different operating systems. There might be some functionality that needs to be available across the OS boundary. The most obvious example of this is a middleware service. Since the components for the two different operating systems will be different implementations of this same service, they will belong to different component types and therefore different SU types. This is because for the SU type no alternative component types can be indicated. All component types listed in the SU type must be instantiated at least by one component in each SU of the type. Note however that even in case of a heterogeneous SG, all the SUs must be able to take any of the SIs, that is, components of either side should be able to take the CSI assignments for any of the SIs. However the distribution of the CSIs can be quite different, that is, the relation of components does not need to be 1 : 1.

- Besides its version the application type defines a single attribute for its application instances, which is the SG types the applications may be composed of. This list is not qualified further, that is, a given application instance may include SGs of all of the listed types or maybe a subset. However they may not include an SG whose type is not listed in the application type.

  As we have seen the application AMF entity includes the SGs and also the SIs, so one may wonder why there is no service type list provided for the application type. The reason for this is that the SU types list service types their instances are capable of providing. As a result the list of valid service types is deducted from the inclusion of SG and SU types.

As we mentioned there are no types provided for AMF nodes and the AMF cluster. In case of the AMF nodes the node group plays a somewhat similar role to a type as it characterizes the equivalency of nodes from the perspective of hosting some SUs. It is not abstracted as a type because a node may belong to more than one node group, while in case of the types each entity belongs to one and only one of them.

Just like in case of the SG more experience may show that it is practical to introduce the type and base type for nodes as well. In case of the cluster since there is only one cluster in the system managed by a single AMF implementation there is no need for types. It may have a use when there are more than one AMF instances at a particular deployment site or the information model covers more than one site. These, however, are currently beyond the scope of the AMF specification.

The same way as in case of component types and component service types, the entities of these compound AMF entity types are configured individually by an appropriate configuration object. This configuration object at least provides two attributes: the entity's name and the entity type it belongs. Additional attributes may provide further configuration information some of which may override the configuration value provided in the type if they are provided.

These configuration objects also include runtime attributes that characterize the runtime status of the entity they represent. Next we are going to review the state information provided through the information model for the AMF compound entities.

6.3.3.4   Runtime Status of Compound AMF Entities

The AMF maintains the status information for the different compound AMF entities as a set of states. All these states are visible in the AMF information model as runtime attributes. They are not exposed through the AMF API.

This section does not cover the administrative state, which is discussed in Section 6.3.5 together with the administrative operation controlling it.

SU Presence State

The SU presence state reflects the life-cycle of the whole SU. Since the SU is a composition of components each of which has a presence state, the SU presence state is defined as the composition of these component presence states. It has the same values as the component presence state.

Since an SU may combine pre-instantiable and non-pre-instantiable components and their time of instantiation is different, the specification declares a SU pre-instantiable if it contains at least one pre-instantiable component, which in turn drives the presence state of such an SU. A SU is non-pre-instantiable if it consists of non-pre-instantiable components only.

The compositions rules for the SU presence state are the following:

- **Uninstantiated:**
  When all components of the SU are in the uninstantiated state the SU is in the uninstantiated presence state.

- **Instantiating:**
  When the first component moves to the instantiating state the SU's presence state becomes instantiating.

- **Instantiated:**
  The SU becomes instantiated when all its pre-instantiable components have reached the instantiated state.
  A non-pre-instantiable SU becomes instantiated when all its components reach the instantiated state.

- **Restarting:**
  The SU presence state becomes restarting when the presence state of all of its already instantiated components becomes restarting. Non-pre-instantiable component that were not instantiated at the time of the restart remain uninstantiated and do not impact the restarting state.

- **Terminating:**
  When the first pre-instantiable component of an already instantiated SU moves to the terminating state, the entire SU moves to the terminating state.
  A non-pre-instantiable SU moves to terminating state with its first component moving to the terminating state.

- **Instantiation-failed:**
  If any component of the SU moves to the instantiation-failed state, the SU also moves to the same state. If there is any component within the SU that already reached the instantiated state AMF terminates it.

- **Termination-failed:**
  If any component of the SU moves to the termination-failed state, the SU's presence state also becomes termination-failed. AMF terminates all the other components in the SU as well.

Table 6.2 summarizes the possible component presences states for each of the SU presence states with respect to the pre-instantiability of the components.

SU Operational State
Just like the presence state, the operational state of a SU is the composition of the operational states of the components composing the SU.

The SU is *enabled* when all its components are in the enabled operational state.

As we have seen at the component operational state this does not mean that the AMF is not aware of any error condition. It may be, however it is still engaged in the restart recovery action the outcome of which shall determine the operational state.

If any of the components within the SU transitions to the disabled state, the SU operational state also becomes *disabled*.

In particular, the SU becomes disabled if any other than the component restart recovery action was triggered or if any of its components transitions to the instantiation-failed or termination-failed states. In addition to the operational state change, these later cases also cause the SU presence state to transition to the instantiation-failed or to termination-failed states respectively.

When a SU becomes disabled the AMF engages in an automatic repair action. As we have seen for components the repair is executed as a restart of the component and since the SU is the composition

---

Table 6.2 Possible combinations of component and service unit presence states

SU presence state	In a pre-instantiable SU		Non-pre-instantiable SU
	Pre-instantiable components	Non-pre-instantiable components	
Uninstantiated	Uninstantiated		
Instantiating	Uninstantiated Instantiating Instantiated	Uninstantiated	Uninstantiated Instantiating Instantiated
Instantiated	Instantiated Restarting	Uninstantiated Instantiated Instantiating[a] Restarting[a] Terminating[a]	Uninstantiated Instantiated Restarting
Restarting	Restarting	Restarting Uninstantiated	Restarting
Terminating	Terminating Instantiated Uninstantiated	Uninstantiated Terminating[a]	Terminating Instantiated Uninstantiated
Instantiation-failed	Instantiation-failed Uninstantiated Instantiating[a] Instantiated[a] Restarting[a] Terminating[a]		
Termination-failed	Termination-failed Instantion-failed Uninstantiated Instantiating[a] Instantiated[a] Restarting[a] Terminating[a]		

[a]Indicates transient component states for the given SU state.

of its components, the SU repair implies the restart of all of its components. If the restart of all the components is successful, the repair is successful and AMF re-enables the SU's operational state.

The automatic repair of an SU may not be appropriate in all cases therefore the configuration allows its disabling for a particular SU or within the SG.

The repair action is disabled for a particular SU usually due to or for the time of some maintenance operations. For example, the software of the components has been upgraded after which a failure is likely due to this upgrade; therefore the restart of the components cannot resolve it and would be a futile. The failure needs to be brought to the attention of the entity performing the upgrade – typically the Software Management Framework (SMF) [49].

Disabling the automatic repair operation in the scope of the SG, on the other hand, is typically related to the service functionality itself, for example, if the restart of a SU may cause some state inconsistency at the service functionality level, or if it requires actions that cannot be controlled by the AMF.

AMF does not repair an SU when the automatic repair action is disabled for the SU or if it is in the instantiation-failed or to termination-failed states and escalation is not possible. In these cases an external intervention is required to recover the SU to the healthy state. Section 'Repaired' on page 184 discusses these repairs.

SU Readiness State

The SU's readiness state determines whether the AMF may select the SU for an assignment. The state values are similar to the component readiness state values, that is, in-service, out-of-service, and stopping.

There are three contributing factors to the decision whether an SU is in-service:

• the SU's environment, whether it is ready so that the SU may take an assignment;
• the SU itself if it is ready to take assignments; and
• whether the SU is allowed to take assignments.

The first condition depends on the AMF node containing the SU. It must be mapped to a CLM member node, which essentially means that there is a running operating system instance which is healthy and capable of executing programs and therefore the software associated with the components of the SU. The AMF node must be enabled, which we will discuss in section 'AMF Node Operational State' on page 164.

If the CLM node is not a member or the AMF node is disabled the SU readiness state is out-of-service.

The second condition depends on the SU operational state and for a pre-instantiable SU on its presence state:

• a non-pre-instantiable SU is always ready to take assignments as long as its operational state is enabled;
• a pre-instantiable SU is only ready if it is enabled and it is in the instantiated or restarting presence states.

The instantiated state implies that all the pre-instantiable components of the SU have been successfully instantiated, which in turn means that all of them have registered directly or indirectly with AMF and therefore indicated their availability.

A SU is in the restarting state if all of its already instantiated components are being restarted and in the restarting state.

An administrator may decide to stop or to take out-of-service some AMF entities. This is controlled by the administrative state of the appropriate entities which will be discussed at length in Section 6.3.5.2.

In any case, if the administrator initiated a shutdown or made ineligible for service the SU itself or any of the compound entities that includes the SU (e.g., its parent SG or hosting AMF node, etc.) then the SU readiness state becomes stopping in case of a shutdown and out-of-service otherwise.

The SU's readiness state reflects back to its components as the readiness of their environment, which is one of the factors taken into consideration when determining the readiness state of a component.

One may perceive the SU readiness state as the synchronization element between the states of its components. The components may become in-service only if their SU reaches the in-service readiness state. Only at this moment AMF may assign CSIs to the SU's components.

When the SU goes out-of-service, all of its components go out-of-service simultaneously and AMF withdraws their assignments.

When the SU moves to the stopping state, all its components move to the same readiness state and AMF assigns the quiescing HA state for their current active assignments. Section 'Shutdown' on page 180 provides more details under what conditions this may occur.

SU HA State

The same way as the component HA state reflects the role the component is taking in servicing a CSI (see Section 6.3.2.3), the SU HA state indicates the role the SU takes on behalf of an entire SI. Again, the values are the same as seen for the component HA state: active, standby, quiesced, and quiescing.

In fact, the HA state assignment at the SU level is accomplished by a set of HA state assignments at the component level. For each CSI of the SI the AMF selects a component of the target SU and assigns the CSI in the desired HA state.

It depends on the redundancy model how AMF distributes the roles on behalf of a given SI among the SUs of the SG. It will try to satisfy first the required active assignments for all the SIs. If this has been satisfied then it proceeds to make the required standby assignments. When AMF selects the components within each of the SUs, it considers their capability model for the type of the CSI it wants to assign to them and their current assignments.

Pre-instantiable components may implement any of the following capability models for each of the component service type they can provide:

- **1_active:** The component can accept only one CSI assignment at a time and only in the active HA state, it cannot take standby assignments.
- **x_active:** The component can accept 'x' CSI assignments at a time and all of them only in the active HA state, it cannot take standby assignments.
- **1_active_or_1_standby:** The component can accept only one CSI assignment at a time either in the active or in the standby HA state.
- **1_active_or_y_standby:** The component can accept either one CSI assignment in the active HA state or up to 'y' simultaneous assignments in the standby HA state. It cannot accept active and standby assignments at the same time.
- **x_active_or_y_standby:** The component can accept simultaneously either up to 'x' CSI assignment in the active HA state or up to 'y' assignments in the standby HA state. It cannot accept active and standby assignments simultaneously.
- **x_active_and_y_standby:** The component can accept up to 'x' CSI assignment in the active HA state and up to 'y' assignments in the standby HA state, all at the same time.

'x' and 'y' are configuration attributes for the different models. The total number of CSI assignments of a given component service type must not exceed the value indicated by the capability model in the selected role. The component may implement a different capability model for each different component service type it supports.

Non-pre-instantiable components implement the **non-pre-instantiable** capability model, which means that the component needs to be instantiated when it becomes associated with a single CSI in the active HA state.

When assigning the active HA state of an SI to a SU, AMF calls back the selected components of the SU to assign the active HA state on behalf of the CSIs of the SI, one component for each CSI of the SI. All these callbacks must be successful for the SU successfully assuming the active HA state.

AMF assigns the standby HA state in a very similar manner, however if it cannot assign the standby HA state of some (or even all) CSIs of the SU because there are no components that can take the assignment due to their capability model, which does not allow standby assignments at all, the assignment is still successful.

To clarify this policy let's consider the following example: there are two service units {SU1, SU2} with two components in each {{SU1.C1, SU1.C2}, {SU2.C1, SU2.C2}}. They provide and protect an SI with two CSIs {CSI1, CSI2}. Components C1 have the 1_active_or_1_standby capability model for the component service type of CSI1. These components implement a database. Components C2 manipulate this database and since all the state information necessary for them is stored in the database they have the 1_active capability model for the component service type of CSI2.

When AMF assigns the active HA state to SU1 on behalf of our SI, it calls back SU1.C1 with the active assignment for CSI1 and SU1.C2 for CSI2.

When AMF assigns the standby HA state to SU2, it only calls back SU2.C1 with the standby assignment for CSI1. AMF does not attempt to assign CSI2 to SU1.C2 as it is not capable of accepting standby assignments.

Note that if there was a second SI to protect and AMF's only option was SU2 for this task, the assignment would not be successful as SU2.C1 can take only one standby assignment.

As explained at the component HA state, the quiesced HA state is used to indicate to the component that its assignment is being switched over and it needs to stop its service and hold the state for the standby to which the assignment is being switched over. The quiesced state has the same meaning for the SU but at the SI level.

Let's develop our example further considering a switch-over.

First AMF assigns the quiesced HA state to the currently active SU, that is, it calls back SU1.C1 with the quiesced assignment for CSI1 and SU1.C2 for CSI2. After the components responded, AMF can proceed with moving the active assignment to SU2 and calls back SU2.C1 with the active assignment for CSI1 and SU2.C2 for CSI2 even though this later one had no standby assignment. When completed, AMF may assign to SU1 the standby assignment for the SI by calling back SU1.C1 with the standby assignment for CSI1 and removing CSI2 from SU1.C2.

One may wonder what the difference is between the 1_active and the non-pre-instantiable capability models. Considering our example we could implement our C2 components as non-pre-instantiable components. This means that the component is not SA-aware and AMF must instantiate it when it needs to be assigned active. Our scenario changes as follows:

- When AMF initially instantiates SU1 and SU2, it instantiates only components C1. When it decides to assign the active HA state to SU1, it calls back SU1.C1 with the assignment for CSI1 and also instantiates SU1.C2. The successful instantiation is equivalent to assigning CSI2 to SU1.C2. To make the standby assignment AMF only assigns CSI1 to SU2.C1.
- To execute the switch-over AMF calls back SU1.C1 with the quiesced HA state for CSI1 and terminates SU1.C2 to prevent it providing service. Then AMF calls back SU2.C1 to change its standby assignment to active for CSI1 and it also instantiates SU2.C2. When completed, AMF changes the quiesced assignment of SU1.C1 to standby for CSI1.

  Note how this scenario can be adapted to integrate existing HA applications with the AMF that do not implement the AMF API.

Finally we need to mention the quiescing HA state, which again have the same meaning at the SI level for the SU as it has at the CSI level for the component. This is straightforward as the state is assigned to the SU by assigning the SI's CSIs in the quiescing HA state to the components currently active for these CSIs. This instructs the components to stop serving new requests, but continue to serve already initiated requests until completion.

Table 6.3 summarizes the possible HA state transitions of a SU on behalf of one given SI and the triggering operations.

The SU may be assigned active or standby as an initial assignment. When we do not want to provide the SI any more we can lock it for immediate stopping of the service or shut it down for graceful termination. Section 6.3.5.1 provides more details on these actions.

AMF Node Operational State

The operational state of the AMF node indicates whether the AMF is aware of any error conditions at the node level. In other words, whenever a recovery action has been triggered by a component failure at the node level, in addition to disabling the operational state of the component, the operational state

Table 6.3 Possible HA state transitions of a service unit

Current state	Next state				
	No assignment	Active	Standby	Quiesced	Quiescing
No assignment	–	Initial assignment	Initial assignment	Never	Never
Active	SI lock	–	Never	Switch-over	Shutdown
Standby	SI lock	Switch-over, fail-over	–	Never	Switch-over, fail-over
Quiesced	SI lock	Failed switch-over	Switch-over	–	Failed switch-over
Quiescing	SI lock	SI unlock	Never	Switch-over, completion of quiescing	–

of the entire node hosting the component also becomes disabled. These recovery actions are the node switch-over, node fail-over, and the node fail-fast. They reflect that the fault causing the error is in the execution environment of the component therefore the recovery is escalated to this fault zone.

In the absence of any of these conditions and after a successful repair the operational state of an AMF node is *enabled;* otherwise it is *disabled.*

Node Switch-Over

To perform the node switch-over operation AMF abruptly terminates the failed component and fails over all the CSIs assigned to the component in the HA active state. The cleanup operation must be successful before AMF can assign the CSI to another component.

If the SU of the component is set to fail-over its components together, then AMF abruptly terminates the entire SU and fails over all SIs assigned in the HA active state to the SU.

In addition AMF switches over all the CSIs assigned in the HA active state to other components hosted on the node, that is, components that belong to other SUs and those that were not terminated abruptly.

Node Fail-Over

AMF fails over all the CSIs assigned in the HA active state to components hosted on the node. Regardless of their operational state the components are abruptly terminated and the cleanup operations must be successful before AMF can assign the CSIs in the HA active state to other components in the cluster.

Node Fail-Fast

Node fail-fast is similar to node fail-over, but rather than staying within the scope of AMF entities, the AMF performs a low level reboot (e.g., at the operating system or at the hardware level) of the disabled node using the PLM. At the same time it fails over all the CSIs assigned in the HA active state to components hosted on the node. At least the termination part of the fail-fast operation must be successful before AMF can assign the CSIs to other components.

When the fail-fast completes successfully, AMF re-enables the operational state of the AMF node as fail-fast operation is also a repair action.

After performing the node switch-over and node fail-over recovery actions the AMF node is still disabled so to repair it AMF may engage in an automatic repair action which means the termination of all remaining components hosted on the node and their re-instantiation as necessary.

This automatic repair action may be disabled for some AMF nodes in which case the repair needs to be performed by an administrator.

SI Assignment State

So far we looked at the states that reflect the status of the service provider entities, but from those it would be difficult to figure out what the status of the services our system is providing. This is the purpose of the SI assignment state which summarizes whether the SI is provided or not, and if it is provided whether AMF could make all the intended assignments for the SI or not.

A SI is *unassigned* if the AMF could not assign it to any of the SUs of the protecting SG in the active or the quiescing HA state. As we have seen earlier these are the only states in which the components (and therefore their SUs) provide the service functionality associated with the CSIs.

If AMF successfully assigned all the intended assignments of a SI (active/quiescing and standby) to the SUs of the SG, the assignment state of this SI is *fully assigned*.

When AMF succeeded with some of the assignments, but not all the intended ones and the assignments includes at least one active or quiescing assignment for the SI, the SI has the *partially assigned* assignment state.

For the purpose of the assignment state, the success of the assignments is evaluated at the SU level. As our example at the SU HA state demonstrated, it is possible that not all CSIs of the SI have a standby assignment when the SI is considered successfully assigned at the SU level. That is, a SI may be in the fully assigned state even though some of its CSI have no standby assignments.

6.3.3.5 Dependencies Among AMF Entities

Throughout the discussion we have seen that there is some implied dependency between certain component type categories. For example, a contained (or proxied) component cannot be instantiated without its container (or proxy) being already instantiated and assigned the container (or proxy) CSI. The AMF is aware of these dependencies and implies them as necessary. For example, if a container component needs to be terminated, AMF will switch-over the services as necessary for its contained components and terminate them first.

However in many cases software and service functionalities have dependencies among each other that AMF is not aware of implicitly and therefore needs to be configured.

As discussed in section 'The Service Unit' on page 152 the SU represents the set of components in the tightest collaboration. This implies that the strictest dependency would occur in this scope. The AMF defines two types of dependencies in the scope of the SU. The first one is the *instantiation level*.

The instantiation level allows the specification of a particular order of instantiation for the components of a SU. The AMF instantiates the SU by instantiating its components in the incremental order of their instantiation levels. Components of a higher instantiation level are instantiated only after all components of the lower levels have been instantiated successfully. Similarly when tearing down a SU the components are terminated in the reverse order of their instantiation level.

Since a component becomes instantiated when it registers with AMF, the instantiation order should cover all the dependencies of the software that needs to be satisfied up to the moment of the registration.

Dependencies of components that need to be satisfied after instantiation and required to provide the functionality associated with a CSI needs to be captured as *CSI dependency* within the SI.

The CSI dependency defines the order in which AMF assigns the active HA state of different CSIs of a SI to the components of a SU. It requires that AMF assigns CSIs of the SI according to their dependency, that is, it assigns first the independent CSIs then their immediate dependents, and so on.

AMF withdraws the active assignments in opposite order of the CSI dependency. If any of the active CSI assignments within this chain of dependency is removed, for example, due to a failure, AMF withdraws immediately all the CSIs assignments depending on this CSI.

Note that the CSI dependency does not apply to the standby HA state assignments since as a result of the standby assignment these components collaborate only within the protection group, which means components of other SUs; they do not collaborate with other components of their own SU.

Finally components may collaborate and therefore may have dependency on each other at the SU boundary. Again, this dependency happens only when the components have the active assignments and is captured as dependency between SIs.

The SI dependency indicates the order in which AMF assigns the active assignment of the different SIs within the cluster. As in case of the CSI dependency, AMF withdraws the assignments in the order opposite to the dependency. But as opposed to CSI dependency this withdrawal is not immediate.

Since the SI dependency reflects a looser coupling, it may tolerate missing some dependency for some short period of time.

Since it is defined at the SI level, SI dependency does not imply the collocation of the components as instantiation level and CSI dependency do. It is possible even for dependent SIs of the same SG that they are provided by different SUs of the SG.

It is important to note that these entity dependencies do not have to reflect only software dependencies. Since they imply ordering, they may be defined purely for such purpose. For example, simultaneous instantiation at system start may put too much burden on the system, so instantiation order can be used for such purpose; however, it applies only within a particular SU. The instantiation of SUs cannot be staged with respect to each other.

A second note is that when one defines dependency between entities that have implicit dependencies, the explicit dependency cannot contradict the implicit dependencies. For example, one must not define that SI1 depends on SI2, if CSI1 in SI1 is provided by a container component and SI2 is by an SU of contained components requiring CSI1 as their container CSI.

Since in case of CSI and SI dependencies, each dependency is given as a partial order between the dependent and its immediate sponsors, one needs to be wary about cyclic dependencies.

It is often confusing and hard to grasp, but the AMF treats SGs of contained components exactly the same way as SGs formed from any other components including those encompassing the container components. The model does not reflect the containment relationship as an additional level in the hierarchy. In the model the relationship can only be tracked down through the container CSI through which the life-cycle of the contained component is mediated by one of the container components.

6.3.4   Redundancy Models

The AMF distributes the assignments for the SIs protected by a SG based on the redundancy model of the SG introduced in section 'The Service Group' on page 153. The redundancy model determines the number of assignments each SI has and how they are distributed among the SUs of the SG. In this section we will take a closer look at each of the redundancy models defined in the AMF specification.

Throughout the section we use an example with the same basic setup: a SG with four SUs protecting two SIs. We show the initial distribution of assignments considering different redundancy models for the SG and the changes performed by AMF after it receives an error report for one of the components with an active assignment.

6.3.4.1   No-Redundancy Redundancy

The simplest redundancy model is the no-redundancy model, which sounds like an oxymoron, but it is not and here it is why: in this model for each SI at most one assignment can be made either in the active or in the quiescing HA state. So from this perspective this model implies no redundancy. However, since in the SG there are many SUs each of which is capable of providing any of the SIs protected by the SG, if one SU fails and it was serving a SI, then AMF can assign this SI to any other SU within the same SG provided there is one available for the assignment. So in this respect this model indeed provides redundancy. This redundancy model suits applications or parts of them that have no service state information to be maintained for service continuity and therefore no need for a standby.

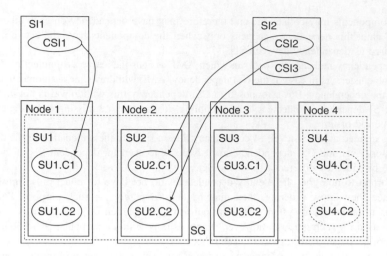

Figure 6.4 No-redundancy redundancy model example: initial assignments.

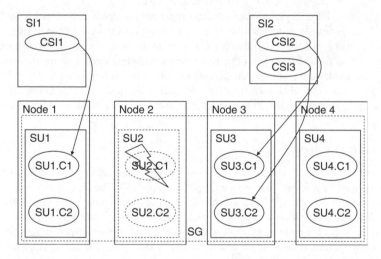

Figure 6.5 No-redundancy redundancy model example: assignments after recovery.

The AMF puts some limitations on SGs with the no-redundancy redundancy model. Namely, in this model an SU may take at most one SI assignment. This requires that in the SG there is at least as many SUs as SIs it protects and at least one additional one to have some redundancy.

Let's assume a SG with four service units {SU1, SU2, SU3, SU4} protecting two service instances {SI1, SI2} according to the no-redundancy redundancy model. The configuration also indicates that three out of the four SUs need to be kept in-service.

First AMF instantiates service units {SU1, SU2, SU3} and assigns the SIs to the first two: SU1 provides SI1 and SU2 provides SI2 as shown in Figure 6.4.

Let's assume that SU2 fails at one point. As a result the AMF performs the following actions:

- It cleans up SU2 and when succeeds AMF assigns SI2 to SU3 as part of the fail-over operation as shown in Figure 6.5. The successful cleanup guarantees that SI2 was indeed removed from SU2.
- At the same time to maintain the number of in-service SUs, AMF instantiates SU4. Since with that the number of in-service SUs is satisfied AMF does not instantiate SU2 for the time being.

2N Redundancy

2N is the best known redundancy model, although it is often referred as 1+1 or active-standby redundancy.

Indeed it implies that for each SI there can be at most two assignments: one in the active (or equivalent) and one in the standby HA states. Furthermore, the active (and equivalent) assignments for all SIs the SG protects are given to one SU in the SG, and all standby assignments are given to another SU. Often the SU with the active assignments is referred as the active SU and the one with the standby assignments is called the standby SU.

This means that the SGs should include at least two SUs, and each of these SUs should be able to accept the assignments for all the SIs the SG protects at the same time.

Of course, there could be more than two SUs in the SG, but at any given moment only (at most) two of them will have actual assignments, and only one will provide the actual service functionality. So the efficiency of this schema is 50% utilization at best.

Let's consider the same scenario as for the no-redundancy model, but assume that the SG protects the two SIs according to the 2N redundancy model. Let's also assume that each of the SUs have two components {C1, C2} and SI1 have one component service instance CSI1, while SI2 has two component service instances {CSI2, CSI3}.

Again, first AMF instantiates the first three service units {SU1, SU2, SU3}. It assigns both SI1 and SI2 in the active HA state to SU1 by assigning CSI1 and CSI2 to C1, and CSI3 to C2. Similarly it assigns both SIs to SU2 in the standby HA state as seen in Figure 6.6.

Let's assume that the AMF receives an error report that SU1.C2 is faulty and needs to be failed over. AMF proceeds as follows:

It cleans up SU1.C2 and when successful it fails over the active assignment for CSI3 to its standby SU2.C2. Meanwhile since SU1's readiness state changes to out-of-service with the termination of SU1.C2, AMF initiates a switch-over for SI1 too assuming SU fail-over is not configured. It puts the

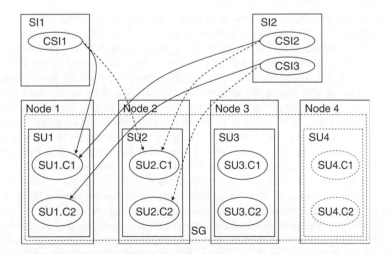

Figure 6.6   2N redundancy model example: initial assignments.

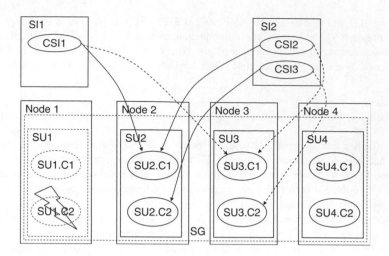

Figure 6.7 2N redundancy model: assignments after recovery.

SU1.C1 into the quiesced state for both CSI1 and CSI2 and when confirmed AMF switches over these assignments to their standby SU2.C1.

Now, AMF can proceed to assign SU3 as the new standby for both SI1 and SI2. In addition to restore the number of in-service SUs it also instantiates SU4 while it terminates the remaining components of the failed SU1. Figure 6.7 shows the assignments after the performed recovery.

Note that even if SU1's readiness state did not change, SI1 would be switched over together with SI2 as all SIs must be assigned to the same SU in the active HA state for the 2N model. Moreover, if the SU was set to fail-over as a single unit, AMF would need to fail-over all the CSIs.

Some AMF implementations could also select to repair SU1 before instantiating SU4. This choice also could be influenced by some other configuration attributes.

6.3.4.2 N+M Redundancy

Better resource utilization can be achieved with the N+M redundancy model. This still implies one active (or equivalent) and one standby assignment for each SI. It also means that a SU may only have active or standby assignments, so the notion of active and standby SUs is still valid. However the number of SUs with assignments is not limited to 2, but as the name suggests there are N active and M standby SUs in the SG. There is no restriction on the relation of N and M, however most often N is greater than M to improve utilization. In particular M = 1 is the most widely used case.

The 2N model can be considered as a special case of the N+M model, where N = M = 1.

The AMF distributes the active assignments for all the SIs of the SG among the N active SUs and the standby assignments among the M standbys. The goal is to maintain all the assignments, that is, have all SIs fully assigned, but this may not always be possible so the next goal is to maintain at least the active and equivalent assignments for all the SIs, that is, have them partially assigned.

The specification does not indicate how the distribution needs to be done. Most importantly it does not currently require an even distribution of the active assignments among the active SUs and therefore AMF implementations may distribute them differently and unevenly.

The specification provides only one attribute to control the maximum number of active assignments an SU may take, which can be used to even out the number of assignments, but if it is used solely for this purpose and not due to capacity limitations, it may also force AMF to drop assignments when it would not be necessary.

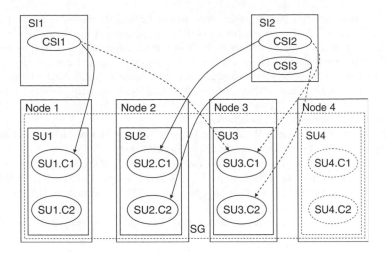

Figure 6.8 N+M redundancy model example: initial assignments.

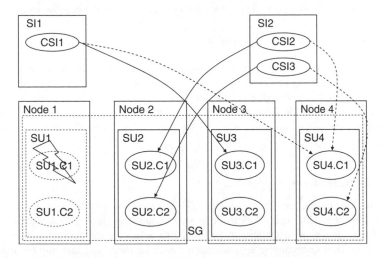

Figure 6.9 N+M redundancy model example: assignments after recovery.

Another issue not regulated by the specification relates to the distribution of the standby assignments: It is not required today that all the standby assignments for SIs assigned to an active SU should be given to the same standby SU. If not, when an active SU fails and AMF fails over its SIs to the SUs having the standby assignments, then suddenly more than one SU need to take over the assignments of the one that failed. From all these SUs all the other standby assignments need to be withdrawn as in this model a SU may only have active or standby assignments.

In general, there is a notion of a preferred distribution of SI assignments within a SG. It is determined through the correlation of different configuration attributes, such as the already mentioned number of active and standby SUs and their maximum number of assignments, but also different ranks that can be given to SUs within an SG and to SIs of the system. The procedure to get back to this ideal distribution is called auto-adjust and controlled also by a configuration attribute.

One should realize however that the assignments cannot be shuffled around at will, for example, to achieve better utilization. AMF uses a series of switch-over procedures to execute an auto-adjust and re-arrange the assignments. In the switch-over the least service impact is guaranteed by providing the time when the current active and the to-be-active components can synchronize the service state. During this synchronization however the service functionality is suspended. Hence AMF gives preference to switch-over to the standby. Thus, if there is a standby, AMF will move the active assignment to the current standby and vice versa. It may take quite a few steps until the intended preferred assignment distribution is reached through such switch-overs.

Let's consider again an example. We have the same SG of four SUs protecting the same two SIs, but according to the N+M redundancy model, N = 2 and M = 1.

Again, first AMF instantiates the first three service units {SU1, SU2, SU3}. It assigns the active HA state for SI1 to SU1 and for SI2 to SU2. In particular CSI1 is assigned to SU1.C1, CSI2 is assigned to SU2.C1 and CSI3 is to SU2.C2. AMF assigns both SIs to SU3 in the standby HA state. The initial assignments are shown in Figure 6.8.

Let's assume that the AMF receives an error report that SU1.C1 is faulty and needs to be failed over. AMF proceeds as follows:

- It cleans up SU1.C1 and in parallel it instantiates SU4. When successful it moves the standby assignments for SI2 from SU3 to SU4, and it fails over the active assignment for CSI1 (i.e., SI1) to its standby SU3.C1. When completed AMF also assigns to SU4 the standby role for SI1 as seen in Figure 6.9.
- AMF also completely terminates the failed SU1.

Note that the exact sequence of actions taken by an AMF implementation may be somewhat different from the above. The key point is though that it must guarantee that the assignments of the failed component are released completely (e.g., there is no process remaining in the system that may still serving any of the CSIs and which is now out of control) before they are assigned to another component. Hence the cleanup operation is part of the fail-over and typically needs to complete before AMF can fail-over the CSIs.

6.3.4.3   N-Way Redundancy

Considering the benefits and the limitations of the N+M redundancy model another schema appeared which is called in the AMF specification the N-way redundancy model. It removes many of the limitations of the other models: while it still implies one active (or equivalent) assignment for each SI, it allows more than one standby assignments for each of them, and this number may differ for each SI protected by the same SG.

It completely removes the notion of active and standby SUs; any SU in the SG may take active and standby assignments at the same time as long as they are not for the same SI.

The specification currently requires that the components used in a SG with the N-way redundancy model should provide their component service types according to the $x\_active\_and\_y\_standby$ component capability model. This seems to imply that when AMF needs to make a standby assignment for a SI, it should be able to make it for all of its CSIs and also that it needs to use the same components in the SU that have active assignments. (One may question whether with these requirements the specification went overboard as (i) whether a standby assignment is required depends on the service functionality as we have seen and not on the redundancy model and (ii) since there is no need that the same component should be assigned active and standby simultaneously, one may achieve the same result at the SU level simply by using more components with a weaker capability model.)

The N-way redundancy model provides a configuration with better control over the distribution of the different assignments. Namely, each SI may have a different affinity to the different SUs of the

SG. This is expressed through the ranking of each SU for each SI. AMF uses this information at runtime and gives the active assignment for a SI to its highest ranking SU among those available for the assignment. The next highest ranking SU will get the standby assignment, and so on.

Again the goal is to first make all SIs at least partially-assigned and then if it is possible then make them fully assigned. During this AMF must follow the SI-SU ranking as a recipe book whether it provides an even distribution of the assignments among the SUs or not. If the configuration provides equal ranking for some or all of the SUs it is implementation specific how AMF distributes the assignments.

To demonstrate the N-way redundancy model let's consider again an SG with four SUs {SU1, SU2, SU3, SU4} protecting two service instances {SI1, SI2}. The SUs again have two components each,

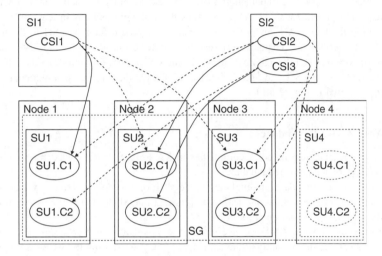

Figure 6.10   N-way redundancy model example: initial assignments.

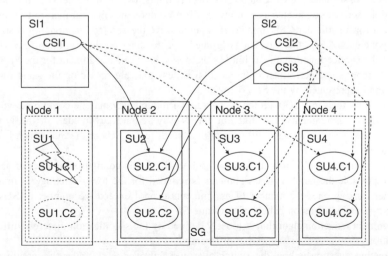

Figure 6.11   N-way redundancy model example: assignments after recovery.

and SI1 has one component service instance CSI1, while SI2 has two CSI2 and CSI3. The preferred number of in-service SUs is three as before.

For each of the SIs we would like to have two standby assignments. Each of the SIs have their own ranking of SUs: for SI1 this defines the order {SU1, SU2, SU3, SU4}. It is the same as the rank of SUs within the SG. For SI2 the ranking defines {SU2, SU3, SU4, SU1}.

At the start AMF instantiates {SU1, SU2, SU3} as they have the highest ranks within the SG. Then it assigns the active HA state on behalf of SI1 to SU1 and on behalf of SI2 to SU2 as these are the top ranked SUs for each of the SIs among those instantiated. AMF also gives the two standby assignments for SI1 to SU2 and SU3, and for SI2 to SU3 and SU1 as indicated by their ranking for the in-service SUs. The initial assignments are shown in Figure 6.10.

Let's assume that at one point AMF receives an error report for the C1 component of SU1 with the recommended recovery of component fail-over. It has currently the active assignment for CSI1 and the standby assignment for CSI2. AMF executes the following actions in reaction to the report:

It cleans up the faulty SU1.C1 component and once completed it fails over SI1 to the first standby, which is SU2. The termination changes SU1 readiness state to out-of-service, so AMF also removes from SU1 the standby HA state assignments for CSI3 and CSI2. AMF instantiates SU4, which becomes the second standby for both SI1 and SI2. AMF also terminates the remaining component of SU1. The new situation is shown in Figure 6.11.

6.3.4.4 N-Way-Active Redundancy

The N-way-active redundancy model accomplishes in SA Forum systems what traditionally referred as active-active or load-sharing redundancy.

It is similar to the no-redundancy redundancy model as it only supports active HA state assignments for the SIs, but it allows more than one active assignment per SI. It also removes the limitation that at a time only one SI can be assigned to an SU.

This means that as far as the AMF is concerned all the SUs that receive an assignment for a given SI provide the same service functionality at the same time. They are sharing the workload represented by the SI. AMF cannot and does not distinguish these assignments; all the components involved receive the same CSI name and same CSI attribute values for a particular CSI of the SI in the same active HA state. As a consequence, AMF cannot give two assignments for the same SI to the same SU.

The multiple active assignments also imply that if a component fails and therefore stops servicing an assignment for a particular CSI, from AMF's perspective the service functionality is still provided by the components and SUs that have an assignment for the same SI. In turn this means that there is no outage of the service functionality as long as there is at least one assignment for the SI. Instead, we can only talk about some performance degradation as less SUs are providing the same functionality, sharing the load represented by the SI.

SIs protected by the same SG with the N-way-active redundancy may be configured with different numbers of active assignments. Just like for the N-way redundancy model, in the N-way-active redundancy model each SI can have different affinity to different SUs of the SG, which is configured through ranking.

The AMF's ultimate goal is to maintain for each of them this configured number of assignments. If this is not possible the number of assignments is reduced for the lower ranking SIs first, but even for the higher ranking ones the number of assignments cannot exceed the number of SUs as an SU cannot take two assignments on behalf of the same SI.

The assignment reduction algorithm is loosely defined by the specification therefore different AMF implementations may execute it somewhat differently.

The interesting question is how the switch-over and fail-over work under the circumstances that AMF does not distinguish the active assignments.

As we mentioned earlier in case of switch-over the components whose assignment is being switched over is quiesced for the CSI for the time the component taking over the assignment confirms the take over. In the CSI assignment AMF also provides the quiesced component name whose assignment is being taken over to the component taking over. Hence if there is a need for state synchronization it is possible even though from AMF's perspective this assignment is identical with assignments that other nonquiesced components have on behalf of the same CSI.

Similarly, when an assignment is failed over AMF indicates which component was active for the assignment being failed over therefore the component taking over the assignment can potentially find out more information about the service function state. However all this is beyond the scope of the AMF and left to the component implementations.

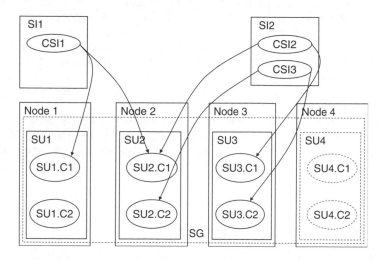

Figure 6.12   N-way-active redundancy model example: initial assignments.

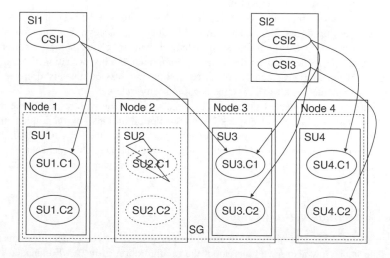

Figure 6.13   N-way-active redundancy model example: assignments after recovery.

The SG of container components always has the N-way-active redundancy model. This is due to the fact that only components with the active assignment provide the service functionality, so standby containers would not have any contained components in them. In addition, the instantiation of contained components is fully controlled by AMF, that is, AMF maintains the state information associated with being a container, and therefore there is no obvious need for a standby.

Using only the N-way-active redundancy model for containers also eliminates the limitations of the no-redundancy redundancy model (that could be another option theoretically) as a SU may serve any number of SIs.

As an example of the N-way-active redundancy model let's consider an SG again with four SUs each composed of two components. Our two service instances again SI1 containing CSI1 and SI2 containing CSI2 and CSI3, and we would like to have two active assignments for each. The number of in-service SUs is three. The ranking of SUs for SI1 defines the order {SU1, SU2, SU3, SU4}, which is the same as the rank of SUs in the SG. For SI2 the ranking defines the {SU2, SU3, SU4, SU1} order.

AMF instantiates the first three SUs {SU1, SU2, SU3} as they rank the highest in the SG. After instantiation AMF assigns the two assignments for SI1 to SU1 and SU2, its highest ranking SUs; and similarly SI2 to SU2 and SU3 as shown in Figure 6.12.

When AMF receives an error report for component C1 of SU2, which has two CSI assignments CSI1 and CSI2, AMF initiates the following actions:

AMF cleans up SU2.C1, and at the same time instantiates SU4. When the cleanup is successful, AMF fails over SI1 to SU3, its next preferred SU, and SI2 to the newly instantiated SU4 by failing over CSI2 to SU4.C1 and switching over CSI3 from SU2.C2 to SU4.C2. When completed AMF also terminates SU2.C2 as shown in Figure 6.13.

6.3.4.5 Remarks on Redundancy Models

We mentioned earlier and to some extent we have seen in the examples, that the AMF has the concept of preferred assignment distribution. This is expressed through the ranking of the SUs within the SG and also for the N-way and N-way-active models ranking them by each SI. SIs themselves are also ranked within the entire system to ensure that more important services have higher priority to resources when there is a shortage. For example, SIs representing the processing of 911 calls should be ranked higher than those representing other calls.

An additional configuration attribute controls whether AMF automatically re-adjusts the assignment distribution after more preferred SUs become available due to a successful repair, for example. To avoid system 'oscillation,' AMF performs such an adjustment only after some probation time has passed and no errors were detected on the repaired components. This probation time applies only to rearrangement of existing assignments. When an assignment was completely dropped and some capacity becomes available for it, AMF assigns it without any delay.

We have mentioned some potential shortcomings of the specification with respect to the redundancy models. They are being looked at as we speak by the SA Forum and also by the implementers of the AMF specification. We included them in our discussion as they are not quite obvious and understanding them help the better understand the overall picture of the AMF and availably management in general. Sometimes describing an error sheds more light on the intricacies than pages of explanations of the correct behavior.

6.3.5 The AMF Administrative Interface

We have seen so far the world of the AMF from the perspective of component developers and the AMF itself. The later view also gave hints about the configuration of an AMF managed cluster and also the information visible about the status of AMF entities and the system as a whole. In this section

Table 6.4 Summary of the AMF administrative operation and their applicability

	Component	Service unit	AMF node	Service instance	Service group	Application	AMF cluster
Unlock-instantiation	–	LI/L	LI/L	–	LI/L	LI/L	LI/L
Lock-instantiation	–	L/LI	L/LI	–	L/LI	L/LI	L/LI
Unlock	–	L, SD/UL	L, SD/UL	L, SD/UL	L, SD/UL	L, SD/UL	L, SD/UL
Lock	–	UL, SD/L	UL, SD/L	UL, SD/L	UL, SD/L	UL, SD/L	UL, SD/L
Shutdown	–	UL/SD	UL/SD	UL/SD	UL/SD	UL/SD	UL/SD
Restart	PC	PC	X	–	–	X	X
SI-swap	–	–	–	X	–	–	–
SG-adjust	–	–	–	–	X	–	–
Repaired	–	D/E	D/E	–	–	–	–
AM-start	X	X	–	–	–	–	–
AM-stop	X	X	–	–	–	–	–

LI – the administrative state locked-instantiation.
L – the administrative state locked.
UL – the administrative state unlocked.
SD – the administrative state shutting-down.
PC – a complete presence state cycle: either instantiated-terminating-unintsantiated-instaniating-instantiated or instantiated-restarting-instantiated.
D – the operational state disabled.
E – the operational state enabled.
X – no state change implications for the target entity.

we take a look at how system administrators can control the system through administrative operations and how the results are reflected in the status information.

6.3.5.1 Administrative Operations

Table 6.4 summarizes the different administrative operations and the AMF entities on which an administrator can issue these operations. Where the operation manipulates any of the states, we also indicated the applicable state transitions as a pair of start and end states separated by slash. In some cases more than one start state is possible, these are listed through comma.

Let's take a more detailed look at each of these administrative operations.

Lock-Instantiation and Unlock-Instantiation
The lock-instantiation and unlock-instantiation operations respectively disallow and allow the instantiation of the target AMF entity and all of its constituents. These operations are applicable to all compound service provider entities as shown in Table 6.5. They change the administrative state of

Table 6.5 Applicability of the lock-instantiation and unlock-instantiation administrative operations

	Component	Service unit	AMF node	Service instance	Service group	Application	AMF cluster
Unlock-instantiation	–	LI/L	LI/L	–	LI/L	LI/L	LI/L
Lock-instantiation	–	L/LI	L/LI	–	L/LI	L/LI	L/LI

LI – the administrative state locked-instantiation.
L – the administrative state locked.

the target entity, which may imply presence state changes of the target entity and some other entities within its scope.

Ultimately AMF instantiates only components, so the instantiation of compound entities is defined as the instantiation of all the components encapsulated by this compound entity. The same applies to the termination of entities.

When the lock-instantiation is applied to a SG, an application or the cluster, none of the constituent components can be instantiated or if they are already in the instantiated state, AMF terminates them as a result of the lock-instantiation.

Without administrative intervention at system startup the AMF selects for each SG of pre-instantiable SUs in the system the set of SUs that makes up the required number of in-service SUs and instantiates them. This selection is based on the SUs' rank within each SG. The administrator issuing the lock-instantiation operation on a SU or an AMF node influences this selection. AMF will not instantiate the target SU or SUs associated with the target node.

Issuing the operation at runtime if the target SU has already been instantiated, AMF will terminate it.

Locking the instantiation of an AMF node will not only terminate all SUs instantiated on the node, but it will also prevent AMF to select this node to host a new SU that was not instantiated before and which is not directly mapped to a node by its configuration. Note however that if such an SU has already been instantiated on the node being locked for instantiation, the SU-node association remains even after the termination of the SU. Currently only cluster restart or reset breaks such an established SU-node association. This restriction may be revised in the future.

The unlock-instantiation operation re-enables the instantiation of the same entities the lock-instantiation prevented.

However when applied to a SU or a node the unlock-instantiation may not immediately result in the instantiation of the target SU or those hosted by the target node. Only if the configuration of the SG an SU belongs requires the SU to be in-service AMF will proceed with the instantiation.

Unlock-instantiation of an AMF node also allows AMF to select the node for hosting new SUs.

Issuing the unlock-instantiation on an SG, an application or the cluster will trigger the instantiation of their constituent entities as performed at system startup. That is, AMF will try to instantiate as many pre-instantiable SUs within each SG of the scope as required to be in-service for the SG.

The lock-instantiation operation is valid only for an entity which is in the *locked* administrative state. It moves the target entity into the locked-instantiation state. Symmetrically the unlock-instantiation operation is only valid when applied to an entity in the locked-instantiation state and it moves the entity to the locked state.

When these operations are issued on an entity in a state different from the above AMF returns an error.

Lock and Unlock

The lock and unlock operations respectively disallow and allow the provisioning of some service. They are applicable to compound AMF entities as shown in Table 6.6. These operation change the administrative state of the target entity and potentially imply assignment, HA, readiness, and presence state changes for the entities within the scope of the target entity.

Once all the required service provider entities are instantiated, the AMF starts the distribution of SI assignments. For each SI it determines the number of active and standby HA assignments needed and selects a SU for each of them within the protecting SG. AMF starts the assignment with the SIs of the highest rank and selects the SUs for each according to their affinity to the in-service SUs of their protecting SG.

The lock administrative operation modifies this selection mechanism.

Table 6.6 Applicability of the lock and unlock administrative operations

	Component	Service unit	AMF node	Service instance	Service group	Application	AMF cluster
Unlock	–	L, SD/UL	L, SD/UL	L, SD/UL	L, SD/UL	L, SD/UL	L, SD/UL
Lock	–	UL, SD/L	UL, SD/L	UL, SD/L	UL, SD/L	UL, SD/L	UL, SD/L

L – the administrative state locked.
UL – the administrative state unlocked.
SD – the administrative state shutting-down.

Locking a SI itself means that AMF will not assign it to any SU and if the SI is currently assigned, it will remove its assignments from all relevant SUs. At the same time the administrative state of the SI changes from unlocked or shutting-down to locked. In addition if the SI was already assigned, then its assignment state becomes unassigned and AMF removes the HA states on behalf of the SI and its CSIs from all related SUs and their components.

On the other hand, locking a service provider entity means preventing that particular entity and its constituents from providing services. It depends on the scope of the target entity whether the operation stops the service provisioning altogether or not.

The operation changes the administrative state of the target entity from unlocked or shutting-down to locked. It also changes the readiness state of all SUs and components within the scope of the target entity, which in turn results in removing any HA state they have on behalf of any SI or CSI. This may result in a change of the assignment state of some of the SI.

In particular, locking the AMF cluster, an application or a SG does mean stopping the provisioning of the SIs in the given scope. Since all components and SUs within the scope go out-of-service none of them may take any assignment and any current HA state is removed. Non-pre-instantiable components and SUs are also terminated. For all SIs within the scope the assignment state changes to unassigned.

If the lock is issued on a SU or an AMF node, similar state changes apply but within the SU or SUs of the node only. Therefore AMF will switch-over the SIs assigned to the SU(s) being directly or indirectly locked to other available SUs. Hence the SIs' assignment states may remain as-is or change to partially assigned or unassigned depending on the availability of the required SUs.

The counter part of the lock administrative operation is the unlock operation. It needs to be issued on the same AMF entity as the lock operation to reverse its effects, that is, to allow the provisioning of some services.

Applying the unlock operation to an entity whose administrative state is locked changes its administrative state to unlocked.

If the unlocked entity is a SI, AMF will also try to make on its behalf the required number of assignments to the SUs of the protecting SG. Depending on whether AMF was able to assign all assignments successfully the assignment state of the SI becomes fully or partially assigned.

If the entity being unlocked is a service provider entity the change of its administrative state allows, but does not necessarily result in changing the readiness state of the entity and all the entities within its scope to in-service.

In particular, if the entity is the AMF cluster, an application or a SG, the unlock operation allows the assignment of the SIs to entities of the scope.

AMF will evaluate the status of the entities within the scope of the unlocked entity and if there are SUs whose readiness state evaluates to in-service, AMF starts to assign the SIs of the scope as it would do at system start. This will result in the appropriate state changes: for the SIs the assignment state moves to partially or fully assigned as appropriate. AMF instantiates the non-pre-instantiable components and their SUs as necessary, and all SUs and their components that receive an assignment change their HA state on behalf of the assigned SIs and CSIs.

If the unlocked entity is a SU or an AMF node, the SU's readiness state or in case of the AMF node the readiness state of the SUs hosted on the node may change to in-service depending on the operational and potentially the presence state of their components and the operational state of the hosting node.

Whether a SU which moves to the in-service state will indeed receive assignments from AMF depends on how the current assignments within its SG match the preferred assignments. For example, if there is an assignment for a SI that AMF was not able to assign, the newly available SU will get the assignment. Also if the unlocked SU is more preferred for an assignment than the SU currently serving it, AMF may rearrange the assignments.

In any case, AMF will re-evaluate the SG and the distribution of assignments against the preferred distribution and if necessary it will make new and redistribute existing assignments. Any adjustment will result in the appropriate state changes of the involved entities.

Note that if a compound entity is locked and therefore all the entities in its scope are out-of-service, it is impossible to unlock only a subset of entities in the scope and therefore put this subset back to in-service. The unlock operation needs to be applied to the same entity, which was locked.

Shutdown

The shutdown administrative operation (Table 6.7) is the graceful version of the lock operation and accordingly its interpretation is that it disallows the service of new requests, however it allows the completion of already initiated requests. This is represented by moving the entity on which the shutdown operation was issued to the shutting-down administrative state first and moving it to the locked state once the operation is completed.

The same way as unlock reverses the effect of the lock operation, it also reverses the effect of the shutdown operation.

When a SI is being shut down, AMF moves the SI's administrative state to shutting-down; however its assignment state does not change. The SUs currently assigned to provide and protect the SI remain assigned, but AMF will change the HA state assignment for those having the active assignment from active to quiescing.

As discussed in Section 6.3.2.3, the components of the SU receiving this assignment change need to reject the service of any new request and complete the service of any ongoing request for their relevant CSI. When no more ongoing requests are left, they need to inform AMF about the completion of the quiescing at which point their HA state for the given CSI transitions to the quiesced state. Once for all CSIs of the SI the components with the quiescing assignments have reached the quiesced HA state, AMF removes all the assignments for the SI and the SI becomes locked. Thus, the SI shutdown operation has completed.

When executing the SI shutdown operation AMF needs to distinguish pre-instantiable and non-pre-instantiable components. One may recall that non-pre-instantiable components are such because they do not implement the AMF API and when started they immediately start providing their service functionality as if they had the active HA state assigned. In other words, their HA state cannot be changed to quiescing and need to be mapped into the active assignment or no assignment.

Table 6.7 Applicability of the shutdown administrative operation

	Component	Service unit	AMF node	Service instance	Service group	Application	AMF cluster
Shutdown	–	UL/SD	UL/SD	UL/SD	UL/SD	UL/SD	UL/SD

UL – the administrative state unlocked.
SD – the administrative state shutting-down.

If there are only non-pre-instantiable components serving the CSIs of the SI, an appropriate mapping is to remove their assignments immediately due to the shutdown operation, that is, to terminate these components. If there are some pre-instantiable components serving some of the CSIs of the SI being shutdown, it is more appropriate to synchronize the termination of non-pre-instantiable components with these pre-instantiable components and therefore terminate the non-pre-instantiable ones only after all pre-instantiable components completed the quiescing.

When issuing the shutdown operation on the AMF cluster, an application or a SG, it has a similar effect as described above on all the entities in the scope. However the administrative state changes to shutting down only for the entity on which the operation was issued.

The SUs and their components having the active HA state for any of the SIs within the scope receive a new quiescing HA state assignment for them, the completion of which they report back to AMF. At completion AMF sets the states and performs the operations as appropriate for the administrative lock.

The SIs of the scope remain partially or fully assigned until the SUs assigned in the active HA state complete the quiescing. After this they become unassigned.

Issuing the shutdown operation on a SU or an AMF node is peculiar and requires attention.

If the target SU or any SU hosted on the node has any active HA state assignments, AMF changes them to the quiescing HA state. If these are the only active assignments for some SIs, this change of the HA state assignment effectively shuts down not only these SUs, but these SIs as well as AMF performs no switch-over at this moment.

As the components of a SU complete the quiescing and report back to AMF this fact, their HA state for their CSI changes to quiesced. When for all CSIs of a SI the components moved to the quiesced state AMF finally switches over the SI to another SU. It will do so because the SI was not the target of the shutdown, it targeted the serving SU only. The SI is in the unlocked administrative state and now with the completion of the quiescing its assignment state becomes unassigned as all its assignments are now quiesced. Thus, if there is another SU available for serving the SI, AMF will assign to that the active HA state on behalf of the SI and the SI will be served again.

As mentioned earlier the unlock administrative operation reverses the effects of the shutdown operation. Once it is issued on an entity, the entity's administrative state changes to unlocked and AMF reverses back the quiescing HA state assignments to active for all the SIs in the scope. Since the quiescing is equivalent to the active assignment this implies no other state changes.

If some entities within the scope have already completed the quiescing, AMF applies the same action as if the unlock operation was reversing the lock operation.

Restart

The administrative operations discussed this far were manipulating the administrative state of the target entities. The other state changes were the result of this manipulation. The restart administrative operation does not impact the administrative state of an entity. The restart administrative operation initiates a presence state cycle on instantiated AMF components and SUs within the scope of the AMF entity on which the restart operation was issued (Table 6.8).

Table 6.8 Applicability of the restart administrative operation

	Component	Service unit	AMF node	Service instance	Service group	Application	AMF cluster
Restart	PC	PC	X	–	–	X	X

PC – a complete presence state cycle: either instantiated-terminating-unintsantiated-instaniating-instantiated or instantiated-restarting-instantiated.
X – no state change implications for the target entity.

As a result of the restart operation, AMF will drive the presence state of the components of the given scope through the instantiated-terminating-uninstantiated-instantiating-instantiated or the instantiated-restarting-instantiated cycles depending on the restartability of the components as discussed in section 'Component Presence State' on page 150. Whether one or the other cycle applies will also determine how the readiness state of the SU is impacted which in turn determines the way the assignments are impacted during this procedure.

The actions AMF will perform to achieve the restart of a component are:

- First AMF will terminate the component via the API or by the terminate CLC-CLI command. If this fails AMF will do the cleanup to achieve at least an abrupt termination.
- If the termination was successful, next AMF will instantiate the component using the CLC-CLI or the API depending on the component category. If successful the component returns to the instantiated presence state.

If the target of the restart administrative operation is a single component and the component is restartable, then the operation triggers the instantiated-restarting-instantiated presence state cycle. The component moving to the restarting state does not change the presence state of its SU, hence it does not trigger a readiness state change and the component can keep all its assignments.

Note however that from the component's perspective – considering an SA-aware component – the procedure is the same whether the component is restartable or not and constitutes the following steps:

- AMF calls back the component removing the assignment;
- AMF calls back to terminate the component;
- AMF instantiates the component via the CLC-CLI (or API for contained components);
- the component registers with AMF;
- AMF assigns an assignment, which for a restartable component will be the same that was removed in the first step.

If the target component is nonrestartable, it needs to go through the instantiated-terminating-uninstantiated-instantiating-instantiated cycle. As soon as the component is terminating, the pre-instantiable SU's presence state changes to terminating as well, which in turn changes its readiness state. Thus, all the assignments of the entire SU need to be switched over to their appropriate standby.

The restart operation has similar effect when its target is a SU. AMF performs the SU restart by restarting all of its components. It is the restartability of the components that determines whether switch over happens or not. During the restart AMF also takes into account the components instantiation order.

When the operation is performed on the AMF cluster, an AMF node or an application, AMF guarantees that first all components of all the SUs within the scope are terminated. It starts their re-instantiation only after the successful termination.

At AMF node restart AMF switches over the SIs as necessary for each of the SUs. For the cluster and for application it performs no switch-over.

As we have seen the restart administrative operation acts only on AMF entities. In particular the administrative restart of the AMF cluster or an AMF node is different from the cluster reset and node failfast recovery actions as AMF executes the later ones at the execution environment level.

Even at the level of the AMF entities, the restart recovery operation involves the abrupt termination of the components while the administrative restart does not. This means that in the second case a component may save some state information that can be picked up by the new incarnation of the same or another component. Hence, the restart administrative operation may not be appropriate for repair.

Table 6.9 Applicability of the SI-swap administrative operation

	Component	Service unit	AMF node	Service instance	Service group	Application	AMF cluster
SI-swap	–	–	–	X	–	–	–

X – no state change implications for the target entity.

SI-Swap

The SI-swap administrative operation can only be initiated on a SI in the partially or fully-assigned states (Table 6.9), which is protected according to a redundancy model that includes active and standby HA state assignments and at least one assignment is available for both HA states. An assignment in the quiescing state is equivalent to an active assignment.

As a result of this operation AMF switches the HA state assignments for the SI between the SUs having the active and the standby roles. Namely, the SU having the standby assignment becomes active and the active becomes the standby after a successful swap.

As a side effect, the assignments of other SIs may also be swapped or re-arranged. In particular, since the 2N and the N+M redundancy models require that an SU may have only assignments in one role, it has to be either active or standby for all the SIs it provides or protects, swapping the assignment for one SI will trigger the swapping of all SIs that are assigned to the active SUs, it may redistribute other assignments.

In case of N-way redundancy the active assignment is swapped with the highest ranking standby assignment. Due to capacity limitations additional rearrangements may happen.

SG-Adjust

The SG-adjust administrative operation gives the administrator the possibility to rearrange the distribution of the SI assignments within a SG in a way that best matches the preferred distribution (Table 6.10).

There are two configuration attributes that define the preferred distribution of assignments:

Within a SG SUs are ranked and this ranking provides AMF with the hint which SUs should be brought into service first when not all configured SUs need to be instantiated. If no other attribute applies, the SU ranking also indicates which SUs should get assignments first. Namely, the highest ranking SUs should receive the active assignments in an SG, followed by those getting the standby assignments if applicable. For example, in a 2N SG the ranking of SUs may make up the following list: {SU1(1), SU2(2), SU4(3), SU3(3)}. In parenthesis we indicated the rank value for each SU; note that they may be equal. AMF will instantiate the highest ranking SUs first up to the configured number of in-service SUs, let's say 3, for example, SU3 remains uninstantiated. The choice between SU3 and SU4 is up to the AMF implementation. Then AMF will try to make the highest ranking SU1 the active SU and the next highest ranking SU2 the standby. The third instantiated SU remains unassigned.

Table 6.10 Applicability of the auto-adjust operation

	Component	Service unit	AMF node	Service instance	Service group	Application	AMF cluster
SG-adjust	–	–	–	–	X	–	–

X – no state change implications for the target entity.

If the administrator locks the node of SU1, for example, for some maintenance purpose AMF will switch-over the active assignments of SU1 to SU2 and make SU4 the new standby. It will also instantiate the last SU3 to have three SUs in-service.

When the administrator unlocks the node where SU1 resides, the assignments remain as above unless the SG is configured for auto-adjust. The administrator can use the SG-adjust operation to return to the initial distribution of assignments. That is when issued, AMF will move first the standby assignment to the now available SU1 and then perform a swap between SU1 and SU2 for all the SIs. It will also terminate SU3 or SU4.

For N-way and N-way-active SGs, in addition to the global ranking of SUs a per SI ranking may be given.

In fact, if the SUs have different ranks within the SG one would like to provide the per SI ranking to distribute the load among the SUs as by default the SU rank would be used by AMF which means that all SIs will prefer the same SUs and AMF will assign all the active assignments to the highest ranking SUs up to the capacity limit. To avoid such an overload one may want to configure the per SI SU ranking with load distribution in mind, for example, as presented in [80].

The operation has no impact on SGs that are configured for automatic adjustment as for these AMF continuously maintains the preferred distribution.

Repaired

The only way AMF attempts to repair its entities is by restarting them. If restart in not a remedy for the fault causing the errors these restarts turn out to be futile and after some configured number of instantiation attempts or if the termination-failed presence state is reached AMF will give up the repair and mark the entity disabled. The configuration may limit the escalation to fail-fast or completely forbid AMF to do any repair attempts, when it is known to be useless or it may even be harmful as discussed in section 'SU Operational State' on page 162.

At this point it is left to the administrator (or some maintenance operations) to repair the entity and re-enable it, so AMF can take over its control again.

The repaired administrative operation allows an administrator to do just that: declare that an AMF entity, more precisely a SU or an AMF node and all its constituents have been repaired so AMF can take them back into its control (see Table 6.11). The administrator needs to ensure that none of the components of the SU or none hosted on the node is running, that is, the presence state of all components in the scope is uninstantiated.

The direct result of the operation is that AMF transitions the operational state of the target entity and any of its constituent entities from disabled to enabled. It also moves the presence states for all the components and SUs in the scope to uninstantiated. AMF does not validate if either of these state changes are indeed appropriate.

These state changes potentially make these entities available for instantiation and assignments. Therefore AMF will evaluate whether any of the newly available SUs need to be instantiated, for example, to make up the configured number of in-service SUs or if it is a preferred SU compared to any of those currently instantiated. Next AMF evaluates if any of the assignments need to be changed

Table 6.11 Applicability of the repaired administrative operation

	Component	Service unit	AMF node	Service instance	Service group	Application	AMF cluster
Repaired	–	D/E	D/E	–	–	–	–

D – the operational state disabled.
E – the operational state enabled.

Table 6.12 Applicability of the EAM-start and EAM-stop administrative operations

	Component	Service unit	AMF node	Service instance	Service group	Application	AMF cluster
EAM-start	X	X	–	–	–	–	–
EAM-stop	X	X	–	–	–	–	–

X – no state change implications for the target entity.

or an unassigned SI can be assigned again with the newly available capacity. If yes, AMF will perform the rearrangements.

EAM-Start and EAM-Stop

The EAM-start and the EAM-stop administrative operations allow the administrator to activate and deactivate the EAM of a particular component or all the components of a SU. That is, those for which the am_start and am_stop CLC-CLI commands exist in the configuration as indicated in Table 6.12.

As mentioned in Section 6.3.2.6 when AMF instantiates a component it also initiates its EAM if the am_start CLC-CLI command is configured for the component. Normally this active monitoring will continue as long as the component is instantiated. This is unlike the healthcheck (also referred sometime as internal active monitoring) or the passive monitoring, which both start and stop as a result of API requests.

In case of EAM the administrator is given the control to stop and start the monitors. AMF maps the EAM-start administrative operation to the am_start CLC-CLI command of the target component or the components of the target SU and executes them. Similarly, it maps the EAM-stop operation to the am_stop CLC-CLI command.

If the EAM-start operation is invoked for a component whose monitor is already running, its status will not change and it remains running. This is important as in the AMF information model there is no object or attribute reflecting the status of external active monitors. In fact, an AMF implementation is not even required to know whether an external monitor it has started at component instantiation is still running. If it has failed, the EAM-start administrative operation provides the means to restart such failed monitors.

6.3.5.2 Administrative State of Compound Entities

As we have seen from the sections on the different administrative operations, the administrative state is defined for all compound AMF entities. The values the administrative state may take and their meaning depends on whether the entity is or includes provider entities or it is on the service side.

On the service provider side the administrative state has the following values:

- **Unlocked** – the entity itself is not blocked administratively from participating in the provisioning or protection of some SIs. AMF may select the entity for an assignment provided all other necessary conditions are also met, that is, the readiness state of components and SUs is in-service.
- **Locked** – the entity and all its constituents are administratively prevented from participating in service provisioning or protection. AMF may not select any SU within the entity's scope for any service assignment. Moreover within the entity's scope AMF has removed any existing assignment. If the locked entity encapsulates all SUs that may provide a particular SI, locking this provider entity implies the suspension of the provisioning of the service represented by the SI.

- **Shutting-down** – the entity and all its constituents are administratively prohibited from taking new SI assignments and asked to remove gracefully any existing assignment. AMF may not select the entity or any provider entity within its scope for any SI assignment. For the existing assignments AMF changes the HA state assignment from active to quiescing by calling back the appropriate components. Shutting down the provider entity may result in the graceful shutdown of the service provisioning for some SIs. However the provisioning may resume after the completion of the shutdown if the entity does not encapsulate all SUs that may provide a particular SI. In other words when the SU becomes locked AMF switches-over its SIs.
- **Locked-instantiation** – the entity and all its constituents are administratively prohibited from instantiation. AMF will not instantiate any component or SU belonging to the scope of the entity and it terminates any currently instantiated entity.

These values are used in the above meaning for SUs, AMF nodes, SGs, applications, and the AMF cluster.

SIs belong to the service side and for them the administrative state has the following valid values and interpretation:

- **Unlocked** – AMF may assign the HA state on behalf of the SI. In fact if there is any in-service SU available for assignment and if all dependencies are met for the SI, AMF will try to assign it in the active and if possible and required in the standby state as well.
- **Locked** – AMF does not assign an HA state on behalf of the SI to any SU and it has withdrawn any existing assignment. AMF has removed all CSI assignments for the SI from all components regardless what HA state assignment they had.
- **Shutting-down** – The service represented by the SI is being gracefully removed therefore AMF has changed the existing CSI HA state assignments for the SI from active to quiescing by calling back the appropriate components. Once the quiescing is completed by a component for a CSI, AMF will remove the CSI assignment completely.

It is important to understand that the effects of an administrative operation do not propagate through the administrative state. The operation changes the administrative state of the target entity only. The effects propagate via the readiness state, which is a cumulative state for all those entities where it is defined (see section 'Component Readiness State' on page 151 for the components and section on the SU readiness state).

This allows a better bookkeeping of the intended administrative states of different entities in the system. Imagine a SU within a SG. If one locks this SG first and then the SU within the group, the two lock operations are kept independent as each represented by its own attribute: one in the SG and one in the SU. The unlock operations can be done independently: unlocking the SG will not unlock the SU. If the effects of the locked administrative state were propagated through the administrative state this would not be possible.

The other nuance that requires some attention is that the administrative state represented by a persistent runtime attribute of the appropriate classes of the AMF information model even though the state changes are triggered by the administrator. This may be different from the approach used in other systems. The persistency of the attribute ensures that even if the system is restarted a completed administrative state change will survive the restart. Using the same mechanism that keeps this value the initial value for the administrative state can be configured when the configuration object are created for the different entities.

AMF sets the administrative state as it executes the actions associated with each administrative operation. It is not mandated by the specification whether an AMF implementation does so at the initiation of the administrative operation or at completion. Therefore it depends on the implementation when the state information is persisted so it will survive a cluster reset, for example.

6.3.6 Interactions Between AMF and Other AIS Services

6.3.6.1 AMF Interaction with PLM

As we have seen the AMF is required to perform the node fail-fast and cluster reset recovery actions on the level of the execution environment. For this purpose AMF should use the PLM [36] administrative API, which provides the appropriate abrupt restart operation and which also has the information how this should be performed for each particular execution environment.

For example, if the execution environment is hosted directly on the physical hardware the abrupt restart is mapped by PLM into a reset of this hosting hardware element. If on the other hand the execution environment is a virtual machine running in a hypervisor, PLM maps the operation into the appropriate operation of the hypervisor.

The identity of the execution environment that needs to be restarted to execute a fail-fast operation AMF needs to obtain from the mapping information for the CLM node hosting the AMF node.

The AMF is not required to have any further interaction with PLM, however ideally, it is aware to some extent of the PLM information model because it provides the information how the AMF nodes map onto the physical nodes providing the platform for the cluster. With the widespread use of virtualization the mapping is not 1 : 1 any more. This means that the failure of a physical node may take out of service a number of AMF nodes. Therefore it is beneficial if AMF can avoid putting all its eggs into the same basket – so to speak. That is assign SUs of the same SG to AMF nodes that map to different physical nodes.

AMF can be helped in this task by a configuration that takes into account this consideration and let's say groups AMF nodes into node groups based on their mapping to hardware elements.

6.3.6.2 AMF Interaction with CLM

The AMF uses the membership information provided by the CLM [37] to determine the scope of the AMF cluster. As described in section 'SU Readiness State' on page 162, one of the pre-requisites to have an SU readiness state in-service is that the AMF node of the SU needs to map to a CLM node which is currently member of the CLM cluster. Moreover, an AMF implementation is required to terminate any component of an AMF node the CLM node of which is not currently member of the cluster (see section 'The AMF Node' on page 155).

All this implies that the AMF needs to obtain the cluster membership information and track its changes to determine which AMF nodes map at any given moment to CLM member nodes.

Through the tracking the CLM provides not only the changes that have occurred already, but since the introduction of the PLM [36] also the planned at the PLM level changes. Such a planned change is provided as a membership change proposal to the AMF that it can evaluate from the perspective whether it would cause any service outage or not considering the current status of the system. If the conclusion is that indeed the change would result in dropping of the active (or equivalent) assignments for some of the SIs, AMF rejects the change. This informs the initiator that service outage can be expected.

At the current state of affairs, AMF does not provide any further information on exactly which SI would be dropped and considering the complexity of the evaluation and that the result partially depends on the AMF implementation another entity cannot replicate the same result.

There is a proposal for AMF to provide this missing information in one way or another as in some cases this can provide a better justification whether to proceed or abandon a particular operation.

6.3.6.3 AMF Interaction with NTF

The AMF uses the Notification service (NTF) [39] to emit alarms and notifications pertinent to the system status. Doing so it also correlates notification ids, that is, if the notification or alarm is generated as a consequence of an event for which AMF is aware of the id of any previously generated alarm or notification AMF puts the correlated parent, root, and any other known notification ids in the appropriate fields of the generated notifications. On request AMF also provides these correlated notification ids through the API.

The alarms generated by AMF indicate emergency situations that may require immediate administrative intervention from AMF perspective as it cannot resolve a problem, which may already impact the availability of some services.

AMF Alarms

The AMF generates alarms for the following situations:

- **Cluster reset triggered by a component failure**

  Any time the AMF receives an error report it contains a recommended recovery action and after verifying the authority of the source of the error report AMF performs the recommended recovery, which may require a cluster reset. This alarm is generated by AMF in such a case.

  AMF clears the alarm automatically after the successful execution of the cluster reset.

  The alarm indicates outage for all services provided by the system.

- **SI unassigned**

  The AMF generates the SI unassigned alarm any time it cannot find the required capacity to assign a SI within the protecting SG or if the provisioning of the SI is prevented administratively. That is, the alarm indicates that the SI assignment state has become unassigned, which means that the associated service functionality is not provided.

  AMF clears the alarm automatically as soon as it can again partially assign the SI to at least one SU.

- **Component cleanup failure**

  AMF has a single attempt to clean up a component: it uses either the cleanup CLC-CLI command or if the component life-cycle management is mediated through a proxy or a container, it asks the mediating component to clean up the component. If the cleanup operation fails the component enters the termination-failed presence and disabled operational states. At the same time the encapsulating SU also moves to the termination-failed and disabled states and it cannot be used any more for service provisioning or protection. Depending on the node configuration AMF may escalate the failure to a node fail-fast operation otherwise it has no other means to terminate the offending component. In either case it informs the administrator about the situation by issuing the component cleanup alarm.

  If the component had an active assignment for any CSI at the moment the failed cleanup was issued, these CSIs and their encapsulating SIs cannot be assigned to any other SU until this alarm condition is cleared. Hence in such a case there is a service outage associated with the alarm.

  The alarm is cleared by the AMF after it receives a repaired administrative operation targeting the disabled SU.

  Alternatively the alarm is automatically cleared after a successful restart of the execution environment the AMF node is mapped to.

- **Component instantiation failure**

  Whenever AMF exhausts all the instantiation attempts permitted by the configuration it moves the component to the instantiation-failed presence state and disables its operational state. As a result the SU encapsulating the component also moves to the instantiation-failed and disabled states and none of its resources can be used for service provisioning or protection. As in case of the termination failure AMF may or may not be allowed to repair the error by initiating a node fail-fast action. In any case it issues the component instantiation alarm to inform the administrator. Depending on the available redundancy, this alarm may or may not indicate service outage.

The alarm is cleared by the AMF after it receives a repaired administrative operation targeting the disabled SU.

Alternatively the alarm is automatically cleared after a successful reboot of the execution environment the AMF node is mapped to.

- **Proxy status of a component changed to unproxied**

  The AMF generates this alarm whenever it could not assign the active assignment for the proxy CSI to a proxy component for an already instantiated proxied component or if all mediation attempts of the existing proxy component have failed. At this point AMF is in the dark regarding the actual status of the proxied component as AMF makes no assumptions about the state of the proxied component resulting from the failure of its proxy component.

  This alarm indicates a service outage with respect to the proxying functionality. However this may or may not indicate service outage with respect to services provided by the proxied component.

  The alarm is cleared by the AMF after it receives a repaired administrative operation targeting the SU of the proxied component, since this implies that the proxied component is in the uninstantiated state.

  Alternatively the alarm is automatically cleared after AMF successfully assigns the proxy CSI to a proxy component that registers on behalf of the proxied component.

AMF Notifications

In addition to updating the AMF information model with state information of all AMF entities, the AMF also generates a notification for any state change. These notifications may provide extra information on the circumstances pertinent to the change. For example, if the SU whose operational state changed to disabled has been upgraded and therefore its maintenance campaign attribute contained a reference to the running upgrade campaign AMF includes the name of the campaign in the notification. Other services such as in this case the SMF can pick up this extra information and take corrective measures as necessary.

Besides the state change notifications AMF also generates a notification each time it engages successfully a new proxy component in proxying a proxied component.

6.3.6.4   AMF Interaction with IMM

The Information Model Management service (IMM) [38] provides the administrative interface for all AIS services including the AMF.

All the administrative operations discussed in Section 6.3.5 are issued using the Object Management interface of IMM (IMM OM-API), which provides a function to invoke administrative operations on different objects of the system information model and delivers these administrative operations to the service implementing those objects in the system.

At system start up the AMF obtains its initial configuration from IMM, which loads the complete system information model from an external source such as an XML (eXtensible Markup Language) [81] configuration file or a repository. This configuration file or repository includes the AMF information model as well.

An AMF implementation also registers with IMM as the object implementer for the AMF information model portion of the system model. It does so in all available roles. From that moment on IMM will deliver to AMF the administrative operations initiated on any of the objects in the AMF information model and also any changes made to the model.

Once the system is up and running the AMF information model can only be changed through IMM.

The same way as IMM exposes the API for administrative operations it also provides the API for populating and changing the information model. Through this API an administrator may create, modify, and delete objects of the classes defined by the specification and implemented as part of the particular AMF implementation of the target system.

The information model (via the IMM OM-API) is also the interface through which the system status can be monitored by an administrator as the AMF – in the role of the runtime owner of the different object of the AMF information model – is required to update the runtime attributes in the model as appropriate.

6.3.6.5   AMF Interaction with SMF

Going through directly the IMM OM-API for all the changes necessary to perform a system upgrade, for example, may be rather cumbersome and error-prone. Therefore a different approach is provided for the SA Forum systems: the configuration changes can be deployed is via upgrade campaigns executed by the SMF [49].

In the background SMF also uses the IMM OM-API to communicate with the AMF. It acts as a management client to AMF driving the administrative operations and model changes necessary for the upgrade through the Object Management interface of IMM.

SMF allows an orchestrated execution of configuration changes through well defined methods that drive toward the reduction of service outages during upgrades. The SMF synchronizes all aspects of an upgrade necessary to make it successful and least intrusive: the administrative operations toward AMF are coordinated with the software installation and removal procedures and with the model changes. All of which is backed up with mechanisms that allows system recovery should any problem occur during execution.

As an input SMF expects an upgrade campaign specification. This whole campaign specification is pre-calculated and provided to SMF, which then ensures the orderly execution of the changes required for the deployment.

For some details on how to come up with an AMF configuration and the way it can be upgraded please consult [82], which uses an example to demonstrate some basic concepts.

Besides the interaction via the IMM OM-API, AMF also provides feedback on potential faults introduced by an ongoing upgrade campaign via the NTF. The SMF subscribes for state change notifications generated by the AMF to detect errors potentially introduced by the upgrade campaign and to initiate the necessary recovery actions.

6.3.7   Open Issues

Throughout the discussion in this chapter we indicated many issues that today are left to an implementation of the AMF because they are not or just partially addressed by the specification.

One of the main issues that we did not mention is the dynamic configuration changes that AMF managed systems need to go through during their long life-cycle. There are regular software updates that need to be deployed, patches need to be applied and also the service functionality changes; it has its own life-cycle. Some of the related issues are addressed in the chapter about the SMF, however there is the AMF side when and how the AMF applies the changes is still open to a great extent at the time of this book being written. This is a somewhat long-term debt of the SA Forum Technical Workgroup, and these issues are being discussed as we speak. They should become part of the next upcoming release.

When it comes to the upgrade of the service functionality it is much less clear whether standardization is possible, to what extent and how much of that should become part of the AMF specification. Maybe the solution is to come up with some patterns like the redundancy models that define some policies that AMF will follow and it becomes to the application designers' responsibility to use the appropriate for the service functionality pattern.

With all that the AMF specification is considered to be one of the most mature within the AIS. Although it is always possible to come up with tricks and tweaks to the specification, these do not change its essence, which can cover the needs of a wide range of applications.

6.3.8   Recommendation

The AMF is the corner stone piece in the SA Forum specification set as it addresses SA itself. It is the answer of the SA Forum for any application that needs to provide services in a highly available fashion.

However, the way AMF is defined it is not restricted to applications that have this need. It can provide life-cycle management for applications that do not implement any of the API functions and not aware of SA at all. At the same time when it can manage applications using sophisticated load sharing schemas, embedded in other execution environments, or not part of the cluster at all.

The flexibility of the AMF information model allows an easy mapping of this wide range of applications and results in a uniform view of the entire system easing system administration and management.

6.4   Conclusion

This chapter introduced the SA Forum AMF.

First we looked at concepts visible through the AMF API and therefore the main concerns for application developers when integrating their design with AMF. These concepts are the component and the CSI. They introduce the logical division of AMF controlled entities into the groups of the service provider entities and the provided service entities.

The AMF performs the availability management based on a model composed of logical entities that represent these two groups. It ensures the HA of the entities representing the provided services through dynamically assigning them for provisioning and protection to the redundant service provider entities in the system. The AMF API is the means AMF uses for this dynamic assignment management. Therefore we looked at the HA states a component can be assigned to on behalf of a CSI; the interpretation of these states and their implications on an implementation of components.

The AMF is capable of managing components that use software implementations that were not developed with the AMF API in mind. AMF can manage them directly as non-SA-aware-nonproxied components or if they at least implement the HA states then using a proxy to mediate the control operations.

Next, to familiarize the logical view that AMF uses for availability management we discussed the AMF information model and its organization. We have seen that components have two types of collaboration and accordingly there are two groupings:

Components that closely collaborate to make up a service functionality desired by the users are grouped into SUs. The service functionality at this level is represented by a SI, which is the combination of the CSIs provided by the single components.

Similar SUs that are redundant in order to protect the service functionality form a SG. Within a SG components of different SUs cooperate to provide availability for their protected service. The role the different component and SUs take in this collaboration is managed by the AMF based on the policies defined for the particular redundancy model implemented by a particular SG.

We compared the main features of the different redundancy models using a simple failure scenario in a relatively simple configuration that still reflected the key points site designers would want to consider when configuring their site.

We also looked at the information model from an administrative perspective as the model contains the status information of the AMF entities in the system and administrative operations are issued on the objects of this model representing these different entities. We have seen how AMF calculates the readiness state of SUs to determine the right candidates for service assignments and how administrative operations impact the readiness state through the administrative and presence states.

The IMM mediates all the administrative operations and information model changes toward AMF.

The chapter also touched upon the interactions of the AMF with other AIS services in the system. Namely it uses PLM for some error recovery operations, tracks the cluster membership changes via the CLM API and generates notifications.

We could not cover all the details and intricacies of the AMF considering that the specification itself is well over 400 pages today. Our goal was to give a different perspective to the specification, which hopefully provides an easier entry point to it, to explain and demonstrate some considerations that drove the solutions of the specification, so one can understand these solutions better and also close the gaps in areas that are left open or not fully defined by the specification.

7

Communication and Synchronization Utilities

Maria Toeroe[1] and Sayandeb Saha[2]

[1]*Ericsson, Town of Mount Royal, Quebec, Canada*
[2]*RedHat Inc., Westford, Massachusetts, USA*

7.1 Introduction

In Chapter 6 we discussed that the Availability Management Framework (AMF) [48] assigns the component service instances (CSIs) to a group of components in different high availability (HA) states to protect the CSI according to the redundancy model. AMF is like the conductor of the orchestra who indicates to the different instruments when they need to enter, take lead, quiet down, but it does not tell how the musician can achieve any of these. Similarly the components receiving the assignments need to figure out what they need to do to fulfill these HA state assignments: Whether they need to save their state, pick up some state information, suspend or resume a service, and so on.

The components in the active and standby roles – the different HA states – on behalf of a CSI need to synchronize the state of the service associated with this CSI so that if the need arises the standby can take over the service provisioning as seamlessly as possible. This requires that the user of the service should not become aware that another component – most likely on another node in the cluster – took over the role of providing the service.

In summary this means that to achieve high Service Availability (SA) there is a general need for utilities that decouple the service provider from the service access point that the service user would bind to as well as for those that allow the replication of any necessary service state information within the cluster so that the service is perceived as continuous.

Some of the Application Interface Specification (AIS) services defined by the SA Forum address exactly these needs. They play an essential role for applications to improve their availability. Applications do not need to use all of these services at the same time but rather they use a subset of these utility services to solve their specific needs in achieving their availability goals.

Service Availability: Principles and Practice, First Edition. Edited by Maria Toeroe and Francis Tam.
© 2012 John Wiley & Sons, Ltd. Published 2012 by John Wiley & Sons, Ltd.

In this chapter we present the three most frequently used utility services. Our emphasis is on what they do, what problems they solve, and how applications can use them to solve their particular availability requirements rather than discussing how they are implemented in the middleware.

One of the simplest communication method that decouples senders from receivers is known as the publish/subscribe messaging pattern [83], which is at the basis of the SA Forum Event service (EVT) [43]. It provides application processes with the ability to publish events on some event channels that in turn can be collected by other application processes subscribing for these channels.

Thus the EVT enables a multipoint-to-multipoint communication in which publishers and subscribers are independent and not aware of each other. The delivery guarantees of the EVT are described as best effort, which definitely leaves room for improvement in the context of SA.

To fill this room the SA Forum also specified the Message service (MSG) [44] which provides delivery guarantees. It enables a message transfer between processes in a multipoint-to-point manner using a priority based message queue serving for its opener as an input queue. Message queues are cluster-wide entities which are accessed using logical names that are location agnostic in the sense that a sender of a message does not have any idea where a receiver is located and conversely the receiver has no sense of where the sender is located unless the sender provides this information. The sender and the receiver are also decoupled that messages are sent to the queue and not to a given process. Anyone opening the message queue becomes the receiver.

Additionally the MSG allows the grouping of Message queues together to form a message queue group, which is addressed the same way as an individual message queue. Thus, message queue groups permit multipoint-to-multipoint communication. Each message queue group distributes the messages sent to the group according to one of the four distribution policies defined in the MSG.

As opposed to these message-based services the SA Forum designed the Checkpoint service (CKPT) [42] for storage-based exchange of information. It allows an active instance of an application to record its critical state data in memory, which then can be accessed by its standby counterpart either periodically or as part of a failover or switchover when the standby transitions to become the active instance.

To protect against data loss, the checkpoint data is typically replicated on several nodes in the cluster. This replication is performed by the CKPT without any involvement of the application.

In this schema application processes are also not aware of each other in the sense that they do not know the identity or the location of their counterpart processes and they may not be aware of the location of the checkpoint replicas either.

In the rest of the chapter we take a detailed look at the features of these three services.

7.2   Event Service

7.2.1   Background: Event Service Issues, Controversies, and Problems

Highly available applications need to communicate amongst their distributed sub-parts as well as with other applications.

One of the forms of such communication is referred to as the publish/subscribe messaging pattern [83], which enables multiple processes – the publishers – to communicate to several other processes – the subscribers – without necessarily knowing where these processes are located. This provides an abstraction much needed in fault tolerant systems to cater a flexible design of application messaging.

It is quite possible that senders and recipients of messages reporting different events in the system are not known from day one when the application design actually starts. It is like the TV channels of the old days, which were established based on the band used for the transmission. A broadcaster would take ownership of such a channel for a particular region where then it could recruit subscriber interested to watch its broadcast.

Indeed, the most common usage issue with recipients is that additional ones can arrive dynamically later in the application design or production life-cycle. A common example is an application that provides Simple Network Management Protocol (SNMP) [53] traps. The application designer builds a SNMP trap generator process which receives the events directly from the various distributed sub-parts of the application, converts them to the proper format, and emits them as SNMP traps. At the time of the design it may not be known who, what manager applications will be the receivers of these traps. The most predictable receiver of course is the system administrator itself.

In this respect, the opposite may also be true that new applications are added to and old ones removed from the system all of which generate SNMP traps and the system administrator or a manager application would be interested to receive all or a particular subset of these traps.

So the question is how to set up the communication between the applications generating these traps and the managers interested in different subsets of the generated traps, all under the circumstances that both of these parties change dynamically over time. Most importantly neither lists are known in advance, that is, at the time the applications (manager and managed) are written.

Furthermore on the managed applications side, they report different issues such as the state of operation, alarm conditions, the temperature they collect from the underlying hardware or just a heartbeat indicating that they are 'alive.' On the other side, not all manager applications are interested in or can deal with all these reports. An availability manager is interested in the heartbeat and the operational state, while the temperature figures may be irrelevant. An alarm is irrelevant in a different way as it may indicate a situation when human intervention is necessary. Hence the system administrator needs to receive all the alarms from all applications and possible other information too, but definitely does not want to see any heartbeats.

Thus, it is necessary that subscribers can tailor their subscription to their interest.

The SA Forum EVT was designed with these requirements in mind.

7.2.2 Overview of the SA Forum Event Service

To address the issues described in the previous section the SA Forum EVT provides a cluster-wide publish/subscribe messaging utility that enables processes within a cluster to communicate without the need for knowing each other's identity or locations.

Processes that send messages are known as *publishers* while processes that receive messages are the *subscribers*. They exchange messages – referred as *events* – over logical message buses called *event channels* as shown in Figure 7.1. Channels are typically created by publishers; however the EVT also allows subscribers to create event channels.

Once an event channel has become available a process may publish events on it. Each event has an associated *pattern*, a kind of label that indicates, for example, the contents of the message and therefore helps in categorizing the event. A process that intends to scoop messages from the channel must subscribe on to the event channel using a subscription *filter*. After this, the subscriber receives an event when its pattern matches the filter of the subscription based on a set of pre-ordained rules.

Each event is normally delivered to all the subscribers whose filters it matches and it is delivered at most once. Due to congestion or failure however events may get lost. So events are also prioritized by the publisher. In case of congestion events of lower priority are dropped before those of higher priority. The events of higher priority are delivered first and in order with respect to a given priority and a given publisher.

In addition to these traditional features, the EVT also supports delayed retrieval of events. That is, if there is a need an event can be retained on a channel for a given period – a retention time – specified at publication and be retrieved by processes subscribing to the channel after the fact of the publication.

With all these features, the SA Forum EVT enables application designers to design applications without needing to know the message end-points from the very start. They can build a future-proof

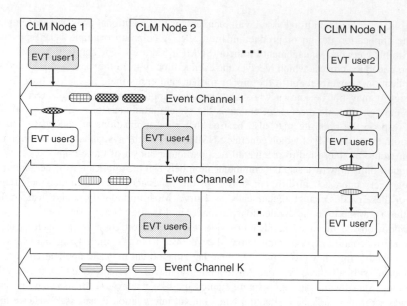

Figure 7.1 Entities of the event service.

application messaging infrastructure based on 'well-known' event channels which has a pre-defined pattern and filter syntax.

Considering our SNMP example we could define an event channel called 'SNMPtrap,' which then would be a cluster-wide entity, meaning that any application running in the cluster can publish events on this channel. Events would represent SNMP traps. Whenever an application generating traps is added to the system it needs to open this channel and whenever it generates a trap, it needs to publish an event containing the trap. It needs to associate this event with the pattern from the agreed upon set that identifies the type of the trap it is publishing. For example, 'amop' may indicate the trap for the operational state, 'amh' for the heartbeat, 'temp' for the temperature, and 'alarm' for the alarms.

On the other hand, the manager applications interested in receiving traps need to subscribe to the 'SNMPtrap' event channel with filters matching the event patterns of their interest. Since the availability manager is interested in the operational state and heartbeats it could set up a prefix filter for patterns starting with 'am.' The administrator would set its filter to match exactly the pattern 'alarm' in its subscription.

From the moment of their subscriptions both the availability manager and the administrator will start to receive events matching their respective filter. But before that the EVT will also deliver any matching event which is being retained in the event channel.

7.2.3 Event Service Architecture and Model

The EVT offers a multipoint-to-multipoint communication mechanism: One or more publishers communicate asynchronously with one or more subscribers by means of events over a cluster-wide event channel.

7.2.3.1 Event

An event is a unit of communication used by the EVT. It carries a piece of information that may be interesting for other processes in the system. It can be an SNMP trap as we discussed in our example,

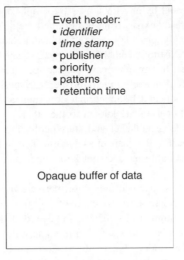

Figure 7.2   The event structure.

an advertisement of some services, or a status update. It can be any piece of information. The most important characteristic of this piece of information is that the processes interested in it are not known to the process, which has the information. It is a 'to whom it may concern' type of announcement where the concerned party needs to do the sorting.

The announcer helps the sorting by adding some labels called the patterns to the information. The purpose of these patterns is to categorize the information so that the interested parties can use filters and receive only the events that are relevant to them. In addition to the patterns the announcer – or the publisher in EVT terminology also assigns a priority and a retention time as well as indicates its own name. The EVT also adds an event id and the publication time to each event shown in italic in Figure 7.2.

The priority and the retention time indicate for the EVT the delivery features of the event. That is, the EVT delivers the events of higher priority before events of lower priority and appropriately when it runs low on resources it will drop events of lower priority first.

The retention time indicates how long the service will keep the event in the channel and deliver it to any new subscriber with a filter, which matches the event's patterns. This implies some persistency for events that are published with nonzero retention time. Indeed in some EVT implementations events may survive even a cluster reboot, but this is not a requirement posed by the specification.

The portion of the event that holds these common attributes characterizing an event is called the event header. While the information portion of the event is referred as the event data.

The EVT does not mandate any structure or format for the data portion. Of course to be able to interpret the event one needs to understand the data format, encoding, and so on. It is outside of the scope of the EVT how the communicating parties establish the common ground. Even for the patterns and their filters only the mechanics are defined in the service; the actual pattern values associated with different event categories are also outside of scope and need to be established by other means. There are no well-known patterns, for example, for SNMP traps.

7.2.3.2   Event Channels

One may think of an event channel as a cluster-wide communication bus that not only delivers messages – well, events – but also stores them. Cluster-wide as it can be accessed from anywhere within the cluster using a unique identifier, the name of the event channel.

An event channel can be created by any EVT user by simply opening it. The service checks if a channel with the given name already exists. If it does not and the opening request allows the creation, EVT creates a new channel with the provided name. If an event channel already exists with the given name, the user is added to the existing channel in the role indicated in the open request even if the creation flag is set in the request. In any other case the service returns an error.

A user can be associated with a channel in the publisher, the subscriber, or both roles. So the same channel may be used for both uni- and bi-directional communication between these two groups of processes. In general neither set of users are known to the other.

A user may open a channel as a publisher and start publishing events on it right away without waiting for any user to attach itself to the channel as a subscriber. In fact using the service application programming interface (API) the publisher does not know and cannot find out whether there are any subscribers at given moment.

As far as the open operation is concerned the subscribers are in a position similar to the publishers'. That is, a subscriber may create an event channel just like any publisher. However after the opening the channel it will not automatically receive the events. It first needs to set up its subscriptions. Each subscription is associated with a set of filters and only events matching these filters are delivered through the given subscription. Each subscription can be viewed as an individual delivery queue for the given subscriber where events are stored for the subscriber until the subscriber collects them.

A user who opened an event channel in the subscriber role may subscribe to this channel any number of times with the same or different sets of filters. The event channel will deliver the events per subscription and not per user. So if two subscriptions of the same user process define overlapping sets of filters, then events matching the intersection will be delivered twice to the process. This is useful when there is a need to modify a subscription as existing subscription cannot be modified. Instead users need to subscribe with the new set of filters and then remove the old obsolete set.

Whenever an event arrives that matches the filters of a given subscription, the EVT calls back the user owning that subscription identifying the subscription and the size of the data. It is the responsibility of the user to collect the event itself. It can collect separately the even header info and the data portion. In any case, once the EVT notified a subscriber about an event, the event is stored for this subscriber as long as the memory allocated for the event was not de-allocated by the subscriber explicitly.

This means that a slow subscriber may pile up a significant backlog and cause the EVT to run out of resources.

Whenever a channel looses an event for a given subscriber it generates a special 'lost event' event and delivers this at the highest priority. Such an event may indicate that more than one event were lost.

Since the service may run out of resources at different phases of the delivery it may not even be able to tell whether the lost event would match a particular subscription, so it may happen that it would deliver a lost event to a subscriber who would not receive the event if it was matched properly with its filters.

Based on these considerations we can say that the EVT provides a best effort delivery. When it delivers events it delivers them at most once per subscription and only complete events are delivered. Events are delivered according to their priority: Higher priority events are delivered prior to lower priority events.

An event channel exists in the cluster until it is explicitly deleted by 'unlinking' it. However the EVT performs the actual deletion only if there is no user process in the cluster that has the event channel open. If there is, the deletion is postponed until the last user closes the channel. At that point, however, the channel is deleted together with any retained event regardless of their retention time.

In contrast, all users closing the event channel does not imply the deletion of the channel. It remains available in the cluster for potential further use.

7.2.3.3   Event Filtering

When a process publishes an event on a channel it specifies a list of patterns for this event. On the other hand when a process subscribes to the same event channel it specifies a list of filters. Based on these the EVT performs a matching procedure: It pairs up the two lists and looks at the patterns and filters as pairs. The first pattern of an event needs to match the first filter of a subscription, the second pattern the second filter, and so on until all patterns are matched up with all filters. An event is delivered to a subscriber only if all patterns associated with the event match all the filters associated with the subscription.

There are four filter options:

- **Prefix match** – an event passes such a filter as long as the pattern starts with the same ordered set of characters as indicated in the filter;
- **Suffix match** – an event passes such a filter as long as the pattern ends with the same ordered set of characters as indicated in the filter;
- **Exact match** – an event passes such a filter if the pattern has the same ordered set of characters as the filter;
- **All match** – any event passes such a filter.

This relatively simple matching technique allows for the design of rather elaborate filtering system for subscribers. The key issue is, however, that the communicating parties need to agree in advance on the conventions in the pattern interpretation and therefore the filter setting. All of which is beyond the scope of the specification.

The patterns and the filters act as addresses and masks that guide the delivery of multicast packages on the Internet, but in contrast to the Internet the EVT does not require location specific addresses. In fact it is completely location agnostic.

7.2.3.4   Conceptual EVT Architecture

From all these considerations we see unfolding a conceptual architecture shown in Figure 7.3.

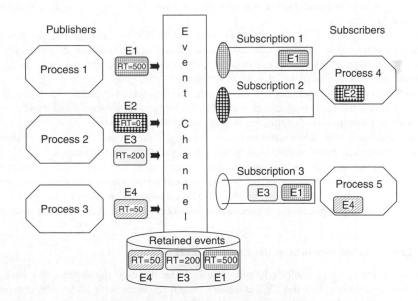

Figure 7.3   Subscriptions to an event channel.

Each event channel implies a highly available storage area where the EVT stores the events published on the channel for the period of their retention time.

When a user subscribes to an event channel the EVT associates a delivery queue with the subscription. This is where the EVT places a copy of each event on the channel event that matches the filters of the subscription and being retained. At the same time the service also informs the owner of the subscription – the user process – about the delivery. Subsequently each time an event is published on the channel the EVT checks if it matches the subscription and if so delivers it.

The event copies remain in the delivery queue until the user collects them or the subscription is cancelled.

When a publisher publishes an event typically it is stored first locally on the publisher's node so that the publisher can free any memory allocated for the event. With that the publishing is completed and successful. Subsequently the EVT transports the event to the main storage area and to the delivery queues of the subscribers whose filters it matches.

When a user unsubscribes from a channel the EVT removes the delivery queue associated with the subscription. When one user unlinks and all other openers close the event channel it is also removed together with its storage area.

7.2.4   User Perspective

The most important aspects of the EVT that users need to be aware are:

- establishing the conventions for an event channel;
- life-cycle management of the channel; and
- event handling and the associated memory management.

7.2.4.1   Event Channel Name and Pattern Convention

To access an event channel the user process needs to know the channel's name as it is used in the API call to open a channel. The method by which the name is established and communicated among the parties involved is not discussed in the EVT specification. Users are left to their imagination to find a suitable mechanism.

The situation is similar with the pattern conventions: Publishers need to use patterns that the subscribers are aware of so that they can install appropriate filters. Otherwise they can only use the filters that let through all events. These may be used to discover the patterns used on a channel; however, the bigger challenge is to interpret the event data.

The publishers and subscribers need to use the same conventions with respect to the event data so that they interpret it the same way at both ends. The EVT offers no solution for this either therefore it requires some planning at the application design and development time.

All these also mean that unless some well-known names, patterns, and formats are established for the service the event channels will be used within the scope of a given application.

An application may use other AIS services to communicate at least partially the needed information. In particular, the SA Forum Naming service [46] was defined for providing a directory type of service. One may also rely on the functionality of the SA Forum Information Model Management service (IMM) [38] by defining application configuration information provided at deployment time.

7.2.4.2   Life-Cycle Management of the Event Channel

Event channels are managed completely and exclusively by the user applications. We have seen in the conceptual architecture that the EVT associates certain resources with each event channel and it relies on the users that they use these resources efficiently.

To avoid accidental creation of event channels just by attempting to open them, it is important that only processes that are responsible for the creation of the channel set the creation flag in the open call. Also it is essential that once an application does not need a channel it unlinks the channel to let the EVT free its resources.

A service implementation also needs to be prepared to handle failure cases of the application managing the channel life-cycle.

7.2.4.3 Event Handling

When it comes to event handling memory management is the most important issue that user applications need to take care of.

To publish an event a process allocates the data structure associated with the event header using the provided event allocation function. This header part can only be manipulated using the appropriate API calls. This portion can also be reused by the process as many times as needed since the memory remains allocated until the user calls the API to free the previous allocation.

At publication in addition to the header the user also needs to provide the data buffer and its size. This portion is handled completely by the user.

When the user has finished constructing both parts, it can publish the event at which time the EVT creates a copy of the entire event; it takes complete charge of the event content so that the user can free any or all memory allocations associated with it or reuse any part of it to construct a new event. The publication is confirmed by the service issuing an event id associated with the event. This id may be used subsequently to clear the event's retention time, which also removes the event from the event channel (i.e., from its main storage area, but the event remains in any delivery queue until consumption).

Symmetrically when an event is delivered the EVT allocates some memory for the event header and informs the user of the data size of the event. For this latter portion the user needs to allocate an appropriate buffer and pass it in the call at the time it wants to retrieve the event data. For the header portion again the user needs to free the allocation using the provided API after it has collected all the needed information.

Since the subscriber can obtain the event id from the event header, it can also clear the event's retention time and therefore delete the event from the event channel.

7.2.5 Administrative and Management Aspects

At the time of writing, the specification defines no configuration or administrative operations for the EVT itself or its event channels. The service also generates no notifications or alarms.

However the EVT reflects the presence of event channels in the system as runtime objects in the information model. These objects expose a set of read-only run-time attributes that can be accessed via the SA Forum IMM if desired. This provides some statistics about the use of the channel such as the number of openers, the number of retained and lost events, and so on.

7.2.6 Service Interactions

7.2.6.1 Interactions with CLM

The SA Forum EVT provides its services only to users within the cluster for which it needs to be up to date about the cluster membership. This requires that the service implementation tracks the membership information through the SA Forum Cluster Membership service (CLM) [37]. Whenever CLM reports that a node left the cluster the EVT stops serving user processes on that node. This implies some housekeeping on EVT's side as with the node leaving the cluster all the subscriptions of processes residing on that node needs to be cancelled and the related resources freed including any associations between these processes and any event channel they opened.

However any event channel created will remain in the system as long as it is not unlinked.

Note that the EVT is not allowed to loose event channels as a consequence of nodes leaving the cluster. It may however loose events that were not secured yet for safe keeping.

7.2.6.2 Interactions with IMM

The only interaction between EVT and IMM is related to the event channels that the EVT exposes as runtime objects in the information model. For this the service needs to use the Object Implementer (OI) API of IMM. Accordingly it registers with IMM as the implementer for the class representing event channels (SaEvtChannel) and provides the necessary callback functions.

Whenever a new channel is created by a user the EVT creates an object of this class using the IMM OI-API. The name and the creation time are set at creation. All other attributes are non-cached meaning that the EVT does not update them unless someone attempts to read their values. If this happens IMM calls back EVT using the callback function EVT provided at registration as an OI and ask for the requested attributes.

7.2.7 Open Issues and Recommendations

The EVT is one of the services that were defined in the early stage of the specifications. It was also among the first ones implemented in the different middleware implementations. As a result it is mature and from the standardization perspective there are no open issues.

The SA Forum EVT is a very good fit when asynchronous communication is needed amongst processes that could reside in various nodes of the cluster. It is particularly suited for (multi)point-to-multipoint communication when the set of senders and the receivers not known in advance or dynamically changes over time.

Its biggest drawback is that it does not provide the delivery guaranties that applications may require in the HA environment. It should not be used in applications where reliable message delivery is absolutely essential. In particular it is not suitable for notifications and alarms for which the SA Forum Notification service (NTF) should be used instead. In the next section we will also look at the SA Forum MSG, which was designed for reliable communication.

The concept of retention time associated with an event goes beyond the basic publish/subscribe paradigm. It enables the communication between processes that need not exist at the same time which is quite powerful and can be used to preserve state even between incarnations. The possibility of clearing of the retention time allows one to remove the event from the channel as soon as it has been taken care of or become obsolete, which eliminates the guessing at the publisher side.

Users may experience some ambiguity regarding the possible preservation of events and the implied persistency of channels over cluster reboots. It may come as a surprise for an application that after a cluster reboot it will find events in an event channel from before the reboot. Considering that these events may convey state information, they may also save faulty states that might have been contributing to the cluster reboot on the first place. So the possibility to control the behavior is desirable, but in the current situation at least applications need to be aware that event preservation may occur.

7.3 Message Service

7.3.1 Need for Reliability and Load Distribution

Maintaining SA requires cooperation and coordination amongst applications within a cluster as well as within the various distributed subparts of an application, which in turn requires a reliable messaging service that can be used to communicate important events in a timely fashion.

Considering the redundancy used to achieve fault tolerance it is easy to see that it is useful if this communication does not require the knowledge of the exact location of the end-points. In a highly available environment the end-points may change dynamically as the active and standby instances of an application change roles due to failovers and switchovers. This dynamism calls for providing an abstraction between message senders and receivers that allows for the senders to send messages without knowing the exact identity and location of message receivers.

Although the EVT provides the decoupling of publishers and subscribers it does not guarantee reliable communication which is essential for many HA applications.

Load distribution to avoid overloads is another important requirement in highly available systems. We would like to share the load amongst multiple processes that may even be located on different nodes of the cluster. For example, the traffic served by a web-server may experience wild swings during the course of a day. During peak hours we may want to be able to deploy two, three times as many instances to serve the traffic as we would need when the traffic is low. In this case there are two issues: One is that we would like to distribute the traffic load at any given moment of the day among all the server instances. Secondly we would like to be able to increase and decrease the number of instances sharing the load as the traffic changes. All this we want to be able to pull off without any support or even awareness of the users accessing our web-server.

So there is a need to be able to submit requests or jobs to a service access point, which then could distribute them somehow magically among all the processes that are capable of processing those jobs without overloading any of them. The clients submitting these jobs should not be involved and even aware of any aspect of this distribution.

The SA Forum has defined the magic bullet to address these in the form of the MSG specification. It describes a reliable messaging service which enables multipoint-to-point as well as multipoint-to-multipoint communication. It also supports load-sharing amongst multiple processes within a cluster.

7.3.2 Overview of the SA Forum Message Service

The SA Forum MSG specification defines a reliable buffered message passing facility that has all the features expected from a true carrier class messaging service. It includes capabilities that can be leveraged to build applications that provide continuous availability under the most demanding conditions.

The main concept defined in the MSG is the message queue, which is a cluster-wide entity with a unique name as shown in Figure 7.4. One may think of it as an input queue of a process, but instead of addressing the process itself this input or message queue is used as the destination point of the messages. This approach makes the process using the input queue interchangeable.

A process which needs such a queue opens a message queue and can start receiving messages sent to the queue. The MSG message queue can be opened by at most one process at a time as usual for input queues. However when the first process closes the message queue another process on the same or from a different node may open it and continue processing the messages. Meanwhile the message queue acts as a buffer for the messages sent to the queue.

There is no restriction on who may send messages to a given message queue therefore a single message queue allows for a multipoint-to-point communication.

A process sending messages to a message queue is unaware of the process receiving it. The latter may change any time, for example, due to a failover, and the sender of the messages will have no idea that the original receiver process has been replaced with another one possibly on a different node of the cluster. This abstraction provides a location agnostic communication with seamless handling of failovers and switchovers which is a necessity in HA systems.

For example, we can associate a message queue with an Internet protocol (IP) address and send all requests addressed to this address to the messages queue. The process of the application serving

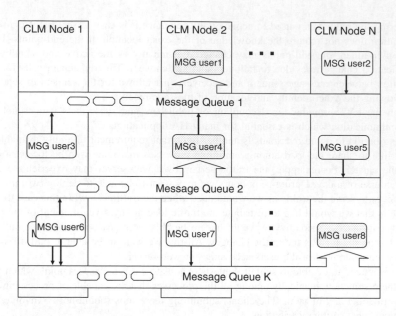

Figure 7.4 Message queues with single receiver and multiple sender processes.

the requests associated with the address such as a web-server then can sit on any of the nodes of the cluster, it will receive the requests even if one node fails and a new process is assigned the task on another node. The new process just needs to open the same message queue.

The MSG preserves all the messages sent to the queue that have not been consumed yet. It does so as long as the queue exists even when the queue is not open by any process. That is, the life-cycle of an application process that has opened a message queue is not tied to the life-cycle of the message queue itself.

The MSG implementation may store the message queues on the disk and therefore its contents may survive a cluster reboot, however this is not a requirement imposed by the specification not even for persistent queues as we discuss later. The implementation choice depends on the targeted communication performance as such feature may hamper performance and within the suite of AIS services the MSG is intended for high-performance communication.

To address the need of load distribution the MSG offers the concept of message queue groups. It is a set of message queues, which can be addressed by the same unique name in the cluster and has an associated distribution policy. With message queue groups the MSG enables multipoint-to-multipoint communication.

From the perspective of a sender a message queue group is very much like an individual message queue, but each message is delivered to one selected message queue in the group (unicast) or to all the queues in the group (multicast).

Considering the web-server application this means that if we associate the IP address with a message queue group instead of a particular message queue then we can install several application processes each serving one message queue of those grouped together in the message queue group. The MSG will take care of the distribution of the requests incoming to the message queue group. It will apply the distribution policy selected for the queue group and deliver the messages to the message queues of the group accordingly.

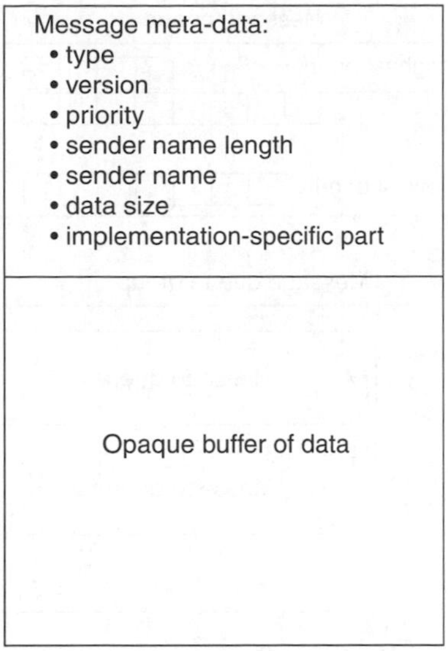

```
┌─────────────────────────────────────┐
│ Message meta-data:                   │
│   • type                             │
│   • version                          │
│   • priority                         │
│   • sender name length               │
│   • sender name                      │
│   • data size                        │
│   • implementation-specific part     │
├─────────────────────────────────────┤
│                                      │
│                                      │
│                                      │
│                                      │
│           Opaque buffer of data      │
│                                      │
│                                      │
│                                      │
│                                      │
│                                      │
└─────────────────────────────────────┘
```

Figure 7.5   The message data structure.

7.3.3   Message Service Architecture and Model

7.3.3.1   Message

A unit of information carried by the MSG is called a message. As in case of events, messages may carry any type of information in their data portion while the specification defines the message header, which is called the message meta-data pictured in Figure 7.5.

A standard set of message attributes and a fixed size implementation-specific portion comprise this meta-data. The standard attributes describe the type, the version, the priority, the sender name and its length, and the data size. The implementation-specific portion is handled by the service implementation exclusively.

The message data is an opaque for the MSG array of bytes.

The specification defines the data structure only for the standard attributes portion of the message meta-data, which is used at both ends of the communication. This provides for some negotiation between the senders and the receiver with respect to the interpretation of the message data content. However again, the specification only provides the tools (i.e., the type and version fields) the actual interpretation of these fields and any negotiation procedure are outside of the scope of the MSG specification.

7.3.3.2   Message Queue

A message queue is a buffer to store the messages. It is divided into separate data areas of different priorities as shown in Figure 7.6. A message queue is identified by a unique name in the cluster and it is global – cluster-wide – entity.

Message queues are dynamically created using the MSG API. Once a message queue has been created any process can send messages to the queue. To receive messages sent to the queue a process needs to open it.

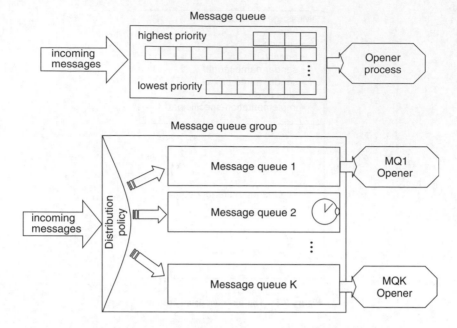

Figure 7.6   The message queue and message queue group.

The process that has opened a message queue successfully becomes the sole receiver of the messages arriving to the queue until this relation is terminated by the process failing or closing the queue.

The MSG delivers the message to the destination message queue and from that moment on the opener process of this queue may collect the message. Depending on the selected option the MSG may call back the opener process whenever a new message arrives to queue.

When the opener process collects the messages the MSG returns the messages of higher priority before those of lower priority. Therefore this delivery mechanism may reorder the messages of the same sender. Each message of the queue delivered at most once after which it is deleted from the queue.

Messages that have not been collected yet do not expire from the message queue.

Closing the message queue does not delete the queue or its contents. It terminates the 'receiver' relation established by the opener process, which then allows another process to attach itself to the same message queue. This allows for a graceful switchover between processes handling the messages of a queue. For example, when the standby instance takes over the servicing of incoming requests it can just continue with the next message available in the queue.

However when a new process opens the message queue it has the option of flushing its content. Since a MSG implementation may preserve the contents of message queues over cluster reboot, a process may make sure that no faulty information is preserved from the previous run by flushing it first.

Messages are also deleted when the queue ceases. A message queue may be persistent or non-persistent. For nonpersistent queues the closure also starts a timer – the retention timer. If no other process opens the queue before its expiration then the queue is deleted from the cluster by the MSG. Persistent message queues need to be deleted explicitly by an API user.

7.3.3.3   Message Queue Groups

The MSG offers load distribution among message queues. To apply a distribution policy among a set of message queues they need to be grouped into a message queue group as shown in Figure 7.6.

Senders send messages to a message queue group exactly the same way as to a message queue – that is, by indicating the name of the message queue group. From the sender's perspective it is a single global entity addressed by a name. The sender does not need to be aware whether it uses a name of a queue or a queue group.[1]

A message queue group is created empty – no queue associated with it. They are added subsequently. The same message queue can be inserted into any number of message queue groups. It will remain the member of a message queue group as long as it exists, the message queue group exits, or it is explicitly removed from the group. When a message queue ceases to exist it is removed from all message queue groups where it was a member.

The deletion of the message queue group has no impact on the existence of the message queues themselves. It only disassociates them, meaning that they stop participating the distribution scheme defined for the message queue group.

The specification defines three unicast and one multicast distribution policies for message queue groups:

- Unicast policies:
  - **Equal load distribution:** Each message sent to the queue group is delivered to one message queue within the group which is selected in a round-robin fashion. This policy is mandatory and should be supported by any MSG implementation.
  - **Local equal load distribution:** The MSG delivers each message to a single queue according to round-robin selection, but priority is given to queues that are opened by processes on the same node as the sender process. That is, queues which are opened by processes on other nodes are taken into account only if there is no queue with a local opener.
  - **Local best queue:** Again preference is given to message queues opened locally and among them to those that have the largest space available.
- Multicast policy:
  - **Broadcast:** The message is delivered to all member message queues that have enough space to hold the entire message (i.e., to preserve message integrity).

In case of the unicast policies if the selected message queue cannot receive the message (e.g., it is full) a MSG implementation may choose between delivering an error to the sender or selecting another queue from the group by applying the same policy. In case of multicast the sender receives an error only if the message could not be delivered to any message queue.

There is no restriction on who can manage a message queue group. Any MSG API user may do so and these manipulations have immediate effect on the group. Processes interested in the changes of the membership of a particular queue group can track the queue group and be notified at any change.

7.3.4 User Perspective

With respect to the MSG an API user need to resolve similar issues as with respect to the EVT. Namely, the naming and other conventions associated with message queues; the life-cycle handling of message queues and groups; and the handling of messages.

In addition the management of the message queue group membership also rests in the hands of API users since the MSG also does not expose any administrative operations or configuration objects in the system information model.

---

[1] Since the RDN attribute type is different for message queues and message queue groups the DN indicates which is the case.

7.3.4.1 Naming and Other Conventions

Sender processes need to know about message queues and message queue groups they need to send their messages. The MSG offers no discovery option.

They also need to be in agreement with the receiver processes about the format, the encoding and the interpretation of the data portion of the messages. The MSG facilitates this task by including in the meta-data portion of a message the type and version fields to characterize the message content, but the specification does not define the usage of these attributes. So their values and the interpretation are again completely open for applications. This is an advantage and a disadvantage at the same time as applications may establish any conventions without any restriction, and they must establish their conventions before being able to use these facilities.

7.3.4.2 Message Queue Management

The MSG also puts the control of the life-cycle of message queues in the hand of the API users.

When a process attempts to open a message queue, the MSG checks whether the queue exists.

If there is no message queue in the cluster by the name indicated in the call then the MSG creates a new queue using the creation attributes of the call and the process opening the queue becomes its sole opener.

If the creation attributes are not provided or there is already a message queue by the same name but its attributes differ from those provided, the opening operation fails.

Otherwise if the queue exists and no other process has it open then the operation is successful and the opener process can start receiving messages sent to the queue for as long as it has the queue open.

If the opener process fails or closes the message queue, it becomes available for other processes to open it, but only one at a time. The MSG enforces this exclusivity therefore no special conflict resolution or coordination is needed on the application's side.

The new opener has the option of flushing or preserving the queue's current content. It can also indicate whether it wants to be called back whenever messages arrive or it will collect the messages at its own pace.

In contrast to event channels, a message sender cannot create a message queue. In fact senders do not need to open the message queue to which they want to send messages. Of course this means that they cannot send messages until a receiver becomes available in the cluster and creates the message queue.

Senders also cannot detect the existence of the message queue in any other way than trying to send messages to it.

Among the creation attributes, the user process indicates whether the message queue is persistent or not and provides a retention time for the later case. The persistency of the message queue indicates who can delete it. Once closed nonpersistent queues remain in the system for the period of the retention time after which the service deletes the queue. Or a user may unlink the queue any time to delete it as soon as it has no opener. Persistent queues can only be removed by explicitly unlinking them.

Unlinking of the queue is not coupled with the opener/receiver role. Any process may delete a message queue using the API. This may happen while another process has the queue opened. In this case the deletion will proceed at the moment the process having the message queue open closes it.

The MSG allows for a very precise management of the service resources. At the creation of the message queue the creator indicates to the service the requested size of each priority area in bytes. In addition the API provides users with functions to check the status of message queues and to set the thresholds for the different priority areas.

There are two watermarks associated with respect to each priority area. The high-water mark indicates the capacity at which the priority area reaches the critical state and may not be able to accept any more messages. The low-water mark indicates that there is enough capacity available in the priority area so that its state can return to noncritical again.

The API function allows the user to set the high- and low-water marks for each priority queue separately.

Whenever the high-water mark is reached for all priority areas of a message queue the MSG generates a state change notification indicating the critical status of the queue. It also generates a notification clearing this condition when at least in one priority area the usage drops below the low-water mark.

To control and calculate the usage is the reason why the API exposes even such detail as the size of the implementation-specific portion of the meta-data.

7.3.4.3   Message Queue Group Management

With respect to message queue groups the specification defines no dedicated role aspects or restrictions. Any process may manage the life-cycle and the membership of a message queue group whether it has any relation to the queues in the group or not.

Anyone or even different processes can configure and reconfigure a message queue group any time during the existence of a message queue group. These changes determine the load-distribution among the processes associated with the message queues and take effect immediately.

When the message queue group is created it is empty. Message queues are inserted subsequently.

While the membership of a queue group can be modified its distribution policy cannot be changed. This may be an issue on the long run because if the initially selected policy needs to be changed, there is no easy way to make the change without the notice of the sender processes.

Processes may track the membership changes. Using the track API a user may find out not only the current members of the message queue group and the subsequent changes, but also the distribution policy associated with the queue.

Tracking can only be started on existing message queue groups, but since anyone may create a message queue group, this is hardly an issue. As a result a want-to-be-sender process is in a better position when it comes to message queue groups. It may create the group and start tracking it so it can send its message as soon as a queue is added to the group.

The MSG provides a callback at the deletion of the message queue group.

7.3.4.4   Message Handling

The guarantees of the message delivery depend on the option chosen by the sender at sending each message. There are four options to choose from:

A sender may decide to send a message asynchronously.

In this case it may or may not ask for any delivery confirmation. Such a call will not wait for the service delivering the message to the message queue. It will return as soon as the MSG 'collects' the message from the sender and if the sender asked for it the MSG will provide delivery confirmation once the message has been placed into the message queue.

A sender may send a message synchronously.

In this case, the call returns only when the message has been delivered to the destination message queue. If it is full, it does not exist, or the message cannot be delivered by any other reasons, the call returns with an error. This call provides the same delivery guarantee as the asynchronous version with delivery confirmation. In neither case is it guaranteed, however, that a process will receive the message.

So alternatively a sender may send its message synchronously and request a reply from the receiver all within the same call. In this case the MSG will deliver the message to the destination queue and expect a reply from the process collecting the message from the queue. The MSG delivers this reply at the return of the initiating call.

To ensure that any of the synchronous calls do not block forever the sender needs to provide a timeout.

On the receiving end the MSG calls back the process which has the message queue open provided it has asked for such notification at the opening of the queue. The callback however is not a prerequisite and the opener process may collect the messages any time from the queue. The MSG keeps messages in the queue until a process – the current opener – successfully retrieves them. If there is no message in the queue the get operation blocks for a period chosen by the retriever.

Message handling also involves some similar memory handling issues that were discussed in Section 7.2.4.3 in relation to event handling. Users need to pay attention to these to make sure that the service resources are used efficiently.

7.3.5   Administrative and Management Aspects

The MSG does not expose any administrative operations or configuration options through the IMM. The management of all MSG entities is done via the service user API, so management applications may use this if they need to manage message queue groups or delete message queues. They can use the track API for monitoring these entities.

The MSG does expose runtime objects via the IMM. Through these the status of message queue groups, message queues, and their priority areas can be obtained. For message queues and their priority areas the status information includes among others the allocated and used buffer size, the number of messages in the different queues. For message queue groups it lists the member queues.

The MSG also generates state change notifications via the NTF that management applications or the administrator may want to monitor. However they typically would not be able to act on these notifications as it is the process that has the message queue open, which should retrieve the messages from the queue to free up resources when the queue is in a critical capacity state so even the user applications may want to subscribe for these notifications.

7.3.6   Service Interaction

7.3.6.1   Interactions with CLM

The SA Forum MSG provides its services only to users within the cluster and an MSG implementation determines the cluster membership using the CLM track API [37].

If CLM reports that a node left the cluster the MSG stops serving user processes on that node. If any process on that node has a message queue opened that queue implicitly becomes closed and any other process from within the cluster may reopen it.

By the same note, processes residing on the node that left the cluster will not be able to send messages to any of the message queues or queue groups.

With respect to the message queues and queue groups, the MSG has the responsibility to maintain them in spite of failures. Note, however, that the specification allows the MSG to loose messages from a message queue when it is implemented as node local resource.

7.3.6.2   Interactions with NTF

The MSG generates no alarms, but it generates state change notifications that a MSG implementation shall propagate using the SA Forum NTF.

The MSG specification defines two states for a priority area based on the utilization of its buffer capacity. A priority area enters the critical capacity state when the used space of its buffer exceeds the high-water mark set for it. When all its priority areas are in critical capacity state the message queue is in critical capacity state and the MSG generates the state change notification for the message queue.

When all message queues of a message queue group are in the critical capacity state the message queue group also enters the critical capacity state and the service generates also a state change notification for the entire message queue group.

As soon as in one of the priority areas of any of the message queues of the queue group there is enough space – that is, its used area drops below the low-water mark set for that priority area the state of the message queue to which the priority area belongs and of the entire message queue group returns to noncritical. The MSG again generates a notification about this state change for both the message queue and the message queue group.

To free enough space in a priority area so that its usage drops below the low-water mark, the process which has the message queue open needs to retrieve enough messages from the queue. This means that a process which opens a message queue may also want to subscribe with the NTF to receive the notifications whenever the message queue reaches critical capacity state.

When a message queue is created the high- and low-water marks are equal to the size of the priority area. That is, the critical capacity notification is generated when the queue has already run out of space. To receive the notifications while there is still some capacity available in the queue the opener process needs set the threshold values for the different priority areas as it is most appropriate for itself.

7.3.6.3 Interactions with IMM

Since the MSG does not expose administrative operations or configuration objects, the only interaction between MSG and IMM is to reflect its different entities as runtime objects in the information model. Accordingly, a MSG implementation needs to register with IMM as the OI for these classes and use the OI-API to create, delete, and modify the objects.

Most of the attributes of the MSG object classes are cached, which means that it is the service implementation's responsibility to update these attributes in the information model. For IMM to be able to obtain from MSG the values of noncached runtime attributes, the service implementation needs to provide the callback functions at the initialization with IMM.

7.3.7 Open Issues and Recommendations

The SA Forum MSG specification is also one of the first AIS services defined as the cluster being a distributed system the need for reliable, location independent communication service that can be used between cluster entities is obvious. We can make a step further, the need is for service interaction rather than the interaction between the service provider entities as the latter may fail at any time and be replaced by other providers momentarily, but this should not reset the service interaction.

The SA Forum MSG addresses these needs and provides a rich set of APIs that application developers can use to build highly available applications. It provides synchronous as well as asynchronous APIs for sending messages thus accommodating a variety of programming styles and providing different delivery guarantees.

One of the key differentiator of the SA Forum MSG is the decoupling of the life-cycles of the message queue and the process that created and opened it. This provides an isolation that enables the survival of messages in the queue that has not been consumed yet and that can be consumed by another receiver thus effectively continuing the service interaction.

Since these features typically hamper performance it would have been nice if the specification also considered options that target high performance and low latency cases. This way, users could use the same programming paradigm for all their communication needs.

The MSG also addresses the load distribution among service provider entities. It allows for grouping of message queues together in to message queue groups and defining a distribution policy – one of the four policies defined by the specification – which then will be applied by the MSG implementation to distribute messages among the queues of the group. The configuration of the message queue group can be changed dynamically at runtime by adding and removing message queues and therefore adding or removing processing capacity according to the actual needs. These changes also can be tracked by interested entities.

A somewhat controversial point of the MSG specification is the statement that it permits the loss of messages if the message queue is implemented as a node local resource. Unfortunately an application cannot determine whether this is the case or not, and it also contradicts to the intention of the MSG providing reliable communication facilities as well as message queues being cluster-wide entities.

As a result if an application targets any MSG implementation it may decide to prepare for the worst, which hampers significantly performance while in most cases this is completely unnecessary. MSG implementations typically do provide the reliable communication facilities as intended by the specification as this is the main distinction between the MSG and traditional communication facilities and also from the EVT.

Using the MSG an active components and its standby may reliably exchange their state information so that they stay in sync with each other and the standby can take over the service provisioning whenever the active fails.

Alternatively the two parties can receive the same message simultaneously and, for example, the active may react by providing the service, while the standby only uses the message to update its state, so it can take over the service provisioning on a short notice.

So the question is whether with MSG we have the perfect solution for the communication and state synchronization needs to provide SA.

7.4 Checkpoint Service

7.4.1 Background: Why Checkpoints

Applications that need to be highly available are often state-full. In other words the active instance of such an application often has state information, which is critical for service continuity and needs to be shared with the standby in order to enable the standby instance to take over the active's responsibilities within a bounded time-frame. In Chapter 6 we presented the stopwatch service as an example for which the elapsed time is such a state information that enables another process to resume the service in case of the failure of the first process.

There are different possibilities to propagate this information. The first option is that the process which actively provides the stopwatch service regularly sends updates to its standby process about the elapsed time it has measured.

In general when a message-based state synchronization is used, the active instance packages, marshals, and transmits to the standby the data that describes its state. The standby instance receives, unmarshals, and seeds its data structures to mirror the state of the active. To keep the standby in sync with the active subsequently only the changes need to be sent as updates.

Such message-based synchronization can be implemented using either the EVT or the MSG presented in the previous sections. It provides a fast update of the standby instance's state, but this scheme may require complicated application logic to keep the state information consistent between the active and the standby instances. The standby may need to receive all messages in correct order. In addition if the standby fails the new instance needs to receive from the active instance all the state information again creating an overhead that may hamper the performance of the regular workload.

A storage-based synchronization scheme could eliminate these issues therefore the SA Forum defined the CKPT.

Since the writing to the checkpoint is performed normally by the active while serving its clients the operation needs to be efficient. Some applications may have performance constraints that can be fulfilled only by local write to memory, which is volatile.

Considering our stopwatch example or any other case it is essential in that if the active instance fails due to a node crash, the state information, for example, the record of the elapsed time survives this node crash. So another expectation towards the CKPT is that it ensures that checkpoints survive node failures. This typically involves multiple replicas of the data on different nodes.

This creates the third expectation that the CKPT maintains all the replicas of a checkpoint without the involvement of the applications.

Depending on the number of replicas this operation can be rather heavy therefore we would also like to be able to tune the update operations so that one can achieve an optimal trade off between reliability and performance.

In our example the elapsed time can be recorded as a single string or an integer, but applications may have elaborated data structures describing their state. It is vital that not the entire data content needs to be exchanged at each update or at each read operations, but that the changes can be done incrementally for both write and read operations.

The SA Forum CKPT addresses all these requirements and therefore enables applications to keep their standby instances current with the state of the active instances.

Let us see in more details what features and options the CKPT offers to its users.

7.4.2   Overview of the SA Forum Checkpoint Service

In the CKPT the SA Forum provided a set of APIs that allow application processes to record some data incrementally that can be accessed by the same or other processes. As a result these data records – if storing in them some state information – can be used to protect an application against failures by using the data to restore the state before the failure.

For example, for an application that processes voice or video calls it is important that the information about ongoing calls that the active instance is handling – the state information for this application – is replicated in the standby instance so that it is ready to step in to replace the active at any point in time. For this purpose the application can use the CKPT: The active instance would create a data structure – a *checkpoint* – using the service API to store all the information needed about the ongoing calls and update it as some of the calls complete and new ones are initiated. The standby can access this checkpoint also using the CKPT API regardless of its location. That is, checkpoints are global entities as reflected in Figure 7.7.

If and when the active instance fails the standby is told by the availability manager – for example, the SA Forum AMF – to assume the active state. Since the standby already has or can obtain from the checkpoint an accurate record of the state of the ongoing calls it has a detailed knowledge to assume the active state seamlessly.

For performance reasons the specification requires that checkpoints are stored in the main memory rather than on disk. That is, this data is not persisted. Nevertheless the CKPT implementation will make all possible precautions against data loss. This means that the service implementation maintains multiple copies or *replicas* for each checkpoint in the system on different nodes so that a node failure causes no data loss.

Accordingly, checkpoints do not survive cluster restart or reboot, not even the restart of the CKPT itself. The CKPT is not suitable for data that needs to be kept through these global operations.

The CKPT synchronizes the checkpoint replicas without any user assistance. The user may create a checkpoint with either synchronous or asynchronous synchronization mode. In case of synchronous updates the write operations initiated on the checkpoint return only when all the replicas have been updated and the caller is blocked for the entire operation.

In case of asynchronous updates the write operation returns as soon as one replica – the primary or *active replica* – has been updated. Thus, the process that initiated the write operation gets back the control and can continue with its job while the CKPT takes care of the update of all other replicas of the written checkpoint.

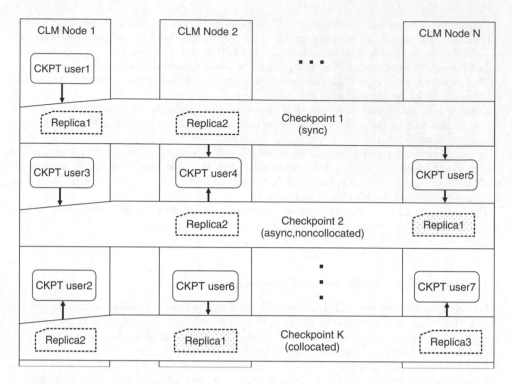

Figure 7.7 Checkpoint are cluster-wide entities maintained as a set of replicas located on different nodes.

This way the application may achieve a better performance at the expense that a given moment in time the replicas may not be identical, which imposes some risk as the active replica may be lost due to a node failure, for example, before the service had the chance to propagate the update to other replicas. Applications need to consider this risk.

The performance of asynchronous updates can be further improved by using collocated checkpoints.

Normally it is the CKPT implementation, which decides on the number and location of the replicas. In case of a collocated checkpoint this decision is given to the application, which can choose when and where to create checkpoint replicas and which one of them is the active at any given moment in time.

Similarly to the nonpersistent message queues of the MSG checkpoints also have retention time as well as close and unlink operations. When the checkpoint has been closed by all users the CKPT keeps the checkpoint for the period of the retention time. Once this expires the CKPT automatically deletes the checkpoint with all its replicas. A user may also ask the deletion of the checkpoint explicitly with the unlink operation in which case as soon as all users close the checkpoint it is deleted by the service. That is, no coordination is needed on the application side.

While the Event and the MSGs of the SA Forum can be used for the communication between processes of the same application and different applications, the CKPT is primarily intended to be used those of the same application. It is a cluster-wide equivalence of processes communicating through shared memory. It is geared toward the needs of state synchronization between active and standby instances of the same application.

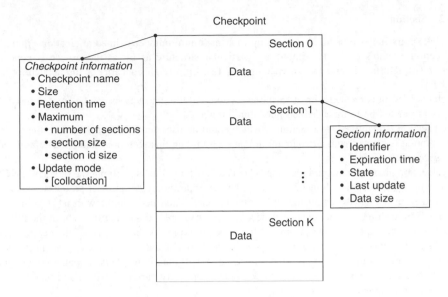

Figure 7.8   Checkpoint organization.

7.4.3   Checkpoint Service Model

7.4.3.1   Checkpoint

A checkpoint is a cluster-wide entity managed by the CKPT. It has a unique name in the cluster, which is used by the different processes to access the checkpoint from anywhere within the cluster. Processes store data in a checkpoint, which in turn can be read by same or other processes as necessary. The same or different processes may have the same checkpoint open many times simultaneously.

For performance reasons checkpoints are stored in memory therefore to protect against node failures the CKPT typically maintains multiple copies of the checkpoint data on different nodes of the cluster. These copies are called replicas of the checkpoint and at most one of them may exist on each node.

Each checkpoint is characterized by a set of attributes specifying the checkpoint's name, size, creation flags, retention duration, and the maximum size and number of sections it may contain Figure 7.8.

The creation flags indicate whether the replicas are updated synchronously or asynchronously and in the later case whether atomicity is guaranteed and/or if the checkpoint is collocated, that is, the user controls the replica creation.

The atomicity guarantee means that when an update operation returns successfully it is guaranteed that the active replica contains all the changes proposed by the operation. If it fails then no change is made to the active replica.

The CKPT maintains the status of each checkpoint. Whenever any of the maximums defined for the checkpoint is reached it becomes 'exhausted' otherwise it is 'available.'

The user data is stored in a checkpoint in sections discussed next.

7.4.3.2 Section

Each checkpoint has one or more sections up to a maximum number defined at creation Figure 7.8.

A section contains raw data without any particular structure as far as the service is concerned. Again any structuring, interpretation, encoding, and so on, are all concerns of the application using the CKPT.

Each checkpoint is created with at least one section. Additional sections can be created and deleted by users as needed up to their defined maximum number.

Each section is identified by a section id which could be user defined or generated by the CKPT. If the checkpoint is created with a maximum of one section then this section is called the default section and has a predefined id.

Each section also has an expiration time. This is an absolute time specifying the moment the CKPT should remove the section automatically. It is like the expiration date on meat products, which must be cleared from the selves 'automatically' when the date is reached. Unlike with meat products, the section expiration time can be updated if needed – for example, each time the data in the section is updated – or even set to the 'end of time' value to never expire like in the case of the default section.

Sections also have a state. In most cases it is valid, but if the checkpoint is asynchronous with weak writes (i.e., the atomicity of updates is not guaranteed) then it may happen that a write operation returns with an error while the sections to be updated are only partially written and therefore those that have been changed become corrupt.

Finally the CKPT also keeps track of the last update of each section.

After creation sections can be updated individually by overwriting the section or as part of a write operation performed on the checkpoint that updates multiple sections identified by their ids.

Since sections are dynamically created and deleted throughout the checkpoint's life time one may discover them by iterating through them.

7.4.3.3 Replicas

As we mentioned already to protect against data loss the CKPT maintains multiple replicas of each checkpoint on different nodes in the cluster as shown in Figure 7.7. On a given node there is always at most one replica of a given checkpoint.

Replicas may be managed by the CKPT implementation completely (i.e., without any user participation) in which case the service determines the number of replicas maintained in the cluster and their location. The user has no control over them. This may not be an issue when all replicas need to be in synch all the time and therefore the checkpoint is created in the synchronous update mode. Synchronous update guarantees that write operations return only when all replicas were successfully updated or the operation failed and therefore none of them has changed. To maintain consistency and atomicity the CKPT implementation would need to use special protocols known from the database technology (e.g., two-phase commit protocol [84]), which require significant time.

Alternatively to improve performance, a checkpoint can be created with asynchronous updates. In this case there is the notion of the active replica, a particular replica on which the write operations are performed. A write operation returns as soon as it has updated the active replica or the operation failed. The CKPT implementation updates the remaining replicas asynchronously from the write operation while the process that updated the replica can carry on with its task.

This means that even read operations need to target the active replica it is the only one which is guaranteed to be up-to-date. So the replicas are there really for the protection of the checkpoint data and not, for example, to provide local access. Other replicas may lag behind as the service implementation propagates the changes to all of them.

For an asynchronous checkpoint the CKPT still manages the number of replicas, their location, and also the selection of the active replica. This option suites the case when multiple processes write to the checkpoint simultaneously from different nodes, so there is no favored location for the active replica or the replicas in general. For example, in a service group using the N-way-active redundancy model all active components of a CSI may need to update the same checkpoint.

When there is a particular process or component which does the updates of the checkpoint the update operation can be optimized by making sure that the active replica and this updater process or component are hosted by the same node and therefore the update is a local operation.

To achieve this, the CKPT has the notion of collocated checkpoints. For example, components protecting a CSI according to the N-way redundancy model may take advantage of this update mode as it is always the active component which writes the checkpoint and the standbys read it. Even though the read operation is remote in this case because the standbys do not provide the service like the active does the delay is less likely to impact service performance.

The replicas of a collocated checkpoint are managed by the users of the checkpoint. Whenever a user opens such a checkpoint the service implementation makes sure that there is a local replica of the checkpoint on the node this opener is running. It is also the users who decide which of these replicas is the current active. If there is no active replica at a given moment the checkpoint cannot be read or updated.

In case of collocated checkpoints, it is also the users' responsibility to create enough replicas so that there is enough protection against data loss. They may want to ensure also that there are not too many replicas to hamper performance by the extra replica update operations.

Note however that for collocated checkpoints the opening and closing operations are not symmetric. While the service ensures a local replica for each open the opposite is not true for closure, that is, the service is not required to remove the replica if there is no other local opener. In general, the specification does not define how long replicas exist in the system and it is left to the discretion of the service implementation.

7.4.4   User Perspective

To access and interpret the checkpoint data application processes also need to agree on the name of the checkpoint and the data organization and interpretation. Since checkpoints are typically used among processes or components of the same application these conventions are less of a concern that they were for the EVT and MSG.

In case of the CKPT the different life-cycle management and data access are the main focus of the user applications. Depending on the update mode selected at the checkpoint creation time the user application may need to manage the life-cycle of the sections and the checkpoint itself and in addition it may have the responsibility of managing the replicas as well.

7.4.4.1   Checkpoint Creation and Opening

After initializing the CKPT to be able to use a checkpoint a user process needs to open it.

At opening a checkpoint the service implementation checks first if a checkpoint by the given name exists already or it needs to be created. If it does not exist the creation attributes indicate the checkpoint's features such as the update mode, the different size limits, and the retention time. To successfully create the checkpoint the user also needs to permit the creation by setting a flag, which protects against accidental creation. If this is not set then the user can open only an existing checkpoint for reading, writing, or both. If the creation flag is set and the checkpoint exists then the creation attributes are compared with the features of the existing checkpoint and the call succeeds only if they are the same. All these checks allow seamless creation and opening of checkpoints without particular synchronization among the user processes accessing the checkpoint potentially simultaneously while preventing accidental creations or duplications.

Note that a checkpoint can be opened using a synchronous call, which blocks the caller for the time of the operation, or through an asynchronous, nonblocking call. These are not to be confused with the update mode of the replicas, which depend on the setting of the creation flags within the creation attributes.

A user can check these attributes of an existing checkpoint by getting the status of the checkpoint. This will also include the current number of sections and the memory used to store checkpoint data.

7.4.4.2 Write Access

After opening the checkpoint the user will be able to perform on it only the operations that were indicated in the open flags of the open call, which is the standard way for databases. Comparing to databases, however, the CKPT does not provide concurrency control such as exclusive access to the checkpoint or its portion. As mentioned earlier, many users may open a checkpoint simultaneously and they are able to read or write it simultaneously. If concurrent access is an issue the users need to resolve it among themselves by other means, for example, by using another AIS service: the Lock service [45].

The reason why the CKPT does not provide concurrency control is the consequence of its intended use: It was designed for efficient checkpointing of some entity's state related to a job it is performing so that another entity, a standby, can stay in sync and take over the job in case the first fails. That is, the assumption is that there is only one entity (e.g., process, component) writing its state and one or more others reading this information. Even if there is more than one entity that writes the same checkpoint, they are assumed to write to different sections.

Under these assumptions concurrency control would add an unnecessary overhead and slow down a CKPT implementation. We mention these assumptions as they limit the CKPT's applicability.

7.4.4.3 Section Management

Users who opened a checkpoint for writing can create new sections in the checkpoint and fill these sections with data. Each of these sections has an expiration time – an absolute time – when the data contained in them becomes obsolete. The nice feature of the CKPT is that it does the clean up for its users based on the expiration time. When the expiration time is reached for a section the CKPT automatically deletes the section from the checkpoint. Based on this feature a user may come up with different strategies: It may create a section and overwrite it as necessary, or add a new section each time with the new data 'a la EVT.' All depends on the needs of the given application.

Existing sections can be written individually or simultaneously respectively using the overwrite or the write operations. In both cases the section(s) to be written is identified by the section id given at the section creation. The overwrite operation as its name indicates overwrites the selected section completely and the new data size becomes the size of the section. The write operation on the other hand writes portions of different sections starting from the given offset. Many section and multiple portions of the same section can be written with a single write operation.

If the checkpoint is updated synchronously when either of the write operations return successfully all replicas have been updated with the new data.

If it is a checkpoint with asynchronous update then only the active replica was updated and the CKPT will propagate the change to the other replicas in due course. This however implies some risk as if the active replica gets lost to a failure or the node leaves the cluster before this happens then the changes will be lost. It is also not necessary that all changes are propagated to all replicas at the same time. Service implementations are free to use any suitable update protocol.

Further risk is introduced by the weak update mode of asynchronous checkpoints. This means that when any of the write operations return with an error, some of the changes might have been applied to some sections and therefore corrupt their data. From a service implementation perspective this means that when updating such a checkpoint the data can be changed directly in the active replica. There is no promise of the 'all or nothing' atomicity and therefore the operation can be faster at the price of a greater risk.

7.4.4.4 Replica Management

With respect to the asynchronous checkpoints we also need to mention the collocated option as in this case the user application has to take the responsibility of managing the replicas.

First of all replicas are created only due to user actions: When a user process opens a checkpoint on a node that does not host a replica for it, the service will create a new replica locally. This means that to protect the checkpoint data adequately the users need to open the checkpoint from different nodes so that the replication occurs.

Considering that in a cluster the user processes may not be in control of their location this seems to be an odd requirement toward applications considering the goal of the SA Forum.

In reality this checkpoint option assumes that the user processes are part of components managed by AMF that form a protection group. In this case, AMF will make sure that these components are running on different nodes and therefore they only need to coordinate their management of the checkpoint replicas with the assignments they receive from AMF for the associated with the checkpoint CSI. So the checkpoint collocation mechanism actually makes sure that mechanism replicating a checkpoint and the processes accessing this checkpoint and its replicas are not controlled by different tunes of different managers, but they are synchronized under a single manager, which is AMF.

7.4.4.5 Access for Read

To read a checkpoint a process needs to open it with the read flag set. In this case the creation attributes and flags do not need to be set.

Once a process has successfully opened a checkpoint for read, it may collect any stored information any time provided it is aware of the checkpoint structure. The process needs to know at least the section ids so that it can indicate in the read request the sections it wants to read, they need to be named explicitly.

The process may or may not provide buffer space for the actual data. In the latter case the CKPT will make the memory allocations that the process will need to free once it processed the collected data.

If the process is not aware the actual contents of the checkpoint it can use the checkpoint status get and the section iteration API to collect the information needed to formulate a valid read call.

For the interpretation of the checkpoint data again user processes need to follow some conventions; however as we have mentioned, since these are assumed to be processes of the same application the issue should not be critical.

The concurrent access may also require considerations at read access. Since a user may read a checkpoint while another user is writing it. As the CKPT is not required to provide concurrency control, the data obtained by the reader may be inconsistent, that is, partially written at the time of the read. Service implementations, however, may provide stronger guarantees.

As opposed to the EVT and the MSG, the CKPT does not provide any API through which it would inform its user processes that opened a particular checkpoint for reading that the checkpoint has changed. This is probably the greatest issue with the CKPT specification available at the time of writing.

The reader processes may use different strategies to keep up to date with the changes. This may range from reading the checkpoint at given intervals to relying on the service assignments performed by AMF or using the CKPT with EVT or MSG in tandem and communicate the fact of the updates through them.

7.4.4.6   Removal of Checkpoints

Application processes indicate to the CKPT that they stopped using the checkpoint by closing it. This does not remove the checkpoint from the system as other processes may have the same checkpoint open. Furthermore, even if there is no other opener at a given moment the checkpoint remains in the system for the retention time. This allows a second dimension of user processes to exchange data using checkpoint. That is, a subsequent incarnation of a component may read the stored state of a previous incarnation and therefore become capable of continuing some task from where its predecessor left of.

To remove a checkpoint a user process needs to unlink it. This deletes the checkpoint and all its replicas immediately if there is no other opener in the cluster. If there is, the service maintains the checkpoint until no opener remains. During this existing openers can use the checkpoint without any change, while new open requests of a checkpoint with the same name will create a new checkpoint.

7.4.5   Administrative and Management Aspects

The CKPT maintains an information model as part of the system information model. This reflects the checkpoints and their replicas existing at any given moment in the system. Both checkpoints and their replicas are represented by runtime objects created and deleted by the service as these entities are created and destroyed dynamically in the system.

For a checkpoint the representing object contains the creation attributes and creation flags discussed in Section 7.4.3.1 and related statistics such as the total number of openers, and the readers and writers among them; the number of replicas and the number of sections and how many of them are corrupted.

For each replica the only attributes reflected in the model are the replica name and whether it is the active one. An object representing a replica is a child of the object representing the associated checkpoint. From this information one cannot deduce the location of a replica and, for example, an administrator may accidentally lock two cluster nodes simultaneously that host the only replicas for a checkpoint.

Since each CLM node may host only one replica for a given checkpoint the replica object can be interpreted as an association object between a checkpoint and a CLM node. With this interpretation and the rules presented in Chapter 4 the replica Relative Distinguished Name (RDN) would need to be the Distinguished Name (DN) of the hosting CLM node, which combined with the checkpoint's DN creates the replica's DN. As a result the information model can reflect not only the number of checkpoint replicas existing in the system, but also their location. This usage has been discussed by the Technical Workgroup of the SA Forum and accepted for a future – at the time of the writing – release of the Checkpoint specification.

The CKPT specification defines no administrative operations or configuration objects. But a service implementation should generate notifications related to the checkpoint status discussed in the next section.

7.4.6   Service Interaction

7.4.6.1   Interactions with CLM

In an SA Forum system the CKPT is only provided to user processes residing on CLM member nodes therefore the service implementation needs to track the membership information using the CLM API [37]. Whenever the CLM reports that a node left the cluster, the CKPT implementation needs to stop serving user processes on that node. It must also consider any lost checkpoint replica that was stored on that node, which means that it may need to restore the missing replica on a node within the cluster.

The service implementation is responsible for the replication of noncollocated checkpoints; therefore for them this loss automatically triggers the replication process.

In case of collocated checkpoints it is the user's responsibility to restore the redundancy if needed. If the process with whom the replica was collocated was managed by AMF then AMF will apply the required recovery actions, which should trigger this restoration if the components coordinate their assignment with the checkpoint access properly.

Since the CKPT exposes the checkpoints and their replicas through the information model it may need to adjust the picture to the CLM changes.

If more than one node leaves suddenly the cluster it is also possible that the CKPT loses all the replicas for some of the checkpoints. This is acceptable according to the CKPT specification, but considered a rare event and therefore the higher performance achieved by storing replicas in memory outweighs the risk of data loss.

7.4.6.2   Interactions with NTF

Even though it might be significant for some the loss of all replicas for some checkpoints the CKPT specification at the time of writing defines no alarms for such a case. It only specifies state change notifications that a CKPT implementation shall generate for a checkpoint when the resources usage exceeds the level defined for the checkpoint at its creation. To propagate these notification CKPT implementations shall use the SA Forum NTF [39].

The CKPT generates the checkpoint section resources exhausted notification in two cases:

- If the maximum number of sections defined for the checkpoint has been reached; or
- If the maximum checkpoint size has been reached, that is, the total amount of data stored in the different sections is equal to the maximum size defined at creation.

Notice that neither of these indicates error conditions as the checkpoint can be still updated by overwriting the existing contents. So if a checkpoint is permanently in this state that may just mean a very efficient use of the service resource.

The reason to worry comes only when a write access fails with the SA_AIS_ERR_NO_RESOURCES error code.

As soon as both resources show availability again the service generates the checkpoint section resources available notification.

7.4.6.3   Interactions with IMM

As discussed the CKPT exposes in the information model the checkpoints and their replicas as runtime objects. That is, an implementation needs to register with IMM as the OI for the classes of these objects

(SaCkptCheckpoint and SaCkptReplica) and use the IMM OI-API. Since the service offers no administrative operations the only callback it needs to provide is related to the non-cached attributes of the SaCkptCheckpoint class. Since these are not cached they are updated on user request, which IMM receives at its Object Manager API and forwards to the service implementation.

As checkpoints and their replicas are dynamically created and removed from the system the CKPT implementation reflects them in the information model. IMM and the information model is the only window the CKPT provides for its internal behavior that management applications can rely on across service implementations.

7.4.7 Open Issues

The main issue with the current CKPT specification is that it does not provide a push or callback mechanism for reader processes that a checkpoint has been updated. Interested parties need to pull the information or use other mechanism, other services to propagate the information.

The issue has been discussed in the Technical Workgroup of the SA Forum and a tracking interface was proposed to resolve the issue. It follows the track API defined in the programming model (Chapter 11) and existing in other services like CLM or AMF.

This solution facilitates the implementation of the so-called hot standby without the use of additional services as the standby component gets the notifications and can obtain in a timely manner any change the active component makes available through the checkpoint.

A second issue that may qualify as open is whether the CKPT should generate an alarm when all replicas of a checkpoint are lost. The reason the specification does not require this is that alarms are intended to be delivered to the system administration for resolution but this is not possible as the CKPT exposes no administrative operations or configuration options as discussed in Section 7.4.5.

Nevertheless the reason behind the loss could be an administrative action performed on the cluster membership, for example, locking all nodes that hold replicas for a given checkpoint. Since the objects representing the replicas in the information model do not reflect their location this is quite possible. Using a naming schema suggested in Section 7.4.3.3 provides a solution, but it also implies that when nodes leave the cluster the objects representing the checkpoint replicas must be removed and recreated as the service restores them, which might generate extra load toward IMM.

Some consider that replicas should not be reflected in the information model at all as their dynamism creates to much load on IMM already.

There were also considerations to provide options for API users and/or system administration to configure the replication aspects of the service. For example, to allow a user process creating a checkpoint also to identify the CLM nodes it desires to host the replicas even in case of noncollocated checkpoints.

Practice shall show whether these issues are indeed important and need to be addressed. At the moment the real demand only supports the introduction of the track API.

7.4.8 Recommendation

As we discussed in Section 7.4.4 the concept of collocated checkpoints is suited for the use of the CKPT by components of applications managed by AMF. However the processes of the component interacting with the CKPT need to act almost like proxies between the AMF and the CKPT to achieve the desired behavior.

It may seem that this option is putting a lot of responsibility on the user application it requires very little extra compared to just using the CKPT with its built in mechanisms: The application only needs to manage the active replica selection according to AMF's orchestration of the HA active state and coordinate the access to the checkpoint with the assignments. These measures provide automatically the protection level appropriate for the redundancy model and the performance optimization for the write access. In all other cases applications may consider the use the built in mechanisms of the CKPT.

Since the CKPT provides data replication and stores the replicas in memory. It may be looked at as a high performance alternative for databases for more transient application data which is incomplete or irrelevant at cold start such as application or cluster restart, but which improves performance at runtime. In particularly data that supports the service recovery aspects of an application. To allow recovery after cold start the relevant information needs to be stored persistently on a disk and it is usually referred as backing up the application data and it is typically a demanding operation.

Checkpoints are viable alternatives that are much less demanding and that can be used in combination with, for example, logging to provide the needed protection between two backups. The backup and the log together provide protection for the case of cold restarts while the checkpoints allow for quick failovers and switchovers.

Compared to real databases the CKPT does not provide concurrency control, so if this is necessary the Lock service or other mutex mechanism can be used.

7.5 Conclusion

7.5.1 Common Issue: Entity Names

Before concluding this chapter we need to mention a reoccurring issue that we did not discuss at the particular services as it exists for all of them. It is the issue already mentioned in Chapter 4 at the discussion of the naming of runtime objects.

The problem is that all three services of this chapter reflect their service entities in the information model as runtime objects. The DN of these objects is given to the service implementation by the user application, which may be written at the time when the actual information model of the system is not known. Since it is a DN which is given, it implies the DNs of all the ancestor objects for the runtime object to be created.

In other words the name of an event channel, a message queue, a message queue group, or a checkpoint determines where the runtime object representing the entity is inserted into the information model. This implies that at naming of these entities one needs to take into consideration of the persistency and the life-cycle of all the parent objects.

As a result it might be tempting to put these objects directly at the root, but not desirable to do so. Instead one option is to insert these objects under the object representing the appropriate service itself. Alternatively a process which is part of an AMF component can find out using the AMF API about the component name and as a result the application's name (since it is part of the component name). In addition such a process can find out about the CSI names assigned to the component, which are most likely associated with the use of these service entities.

7.5.2 Conclusion

In this chapter we presented the three most often used utility services of the AIS defined by the SA Forum. They allow different ways of communication between application processes among others for the purpose of synchronizing their states when collaborating in the provisioning and protection of some services. That is, at providing SA, which is the aspect we are most interested in this book.

These services are the EVT, the MSG, and the CKPT. All three services offer solutions that decouple the communicating parties that need to share some information. For the purpose of this communication the parties synchronize through the respective service entities, the names of which are the minimum required shared knowledge among the involved parties.

Beyond this information the presented services require no or minimal synchronization from the user application even with respect to the life cycle of these service entities. It is actually an application choice whether this life-cycle synchronization is expected or not. That is, both parties the senders or the information owners and the receivers – the information user may create the service entity through which the information exchange is conducted.

Most importantly these service entities are cluster-wide and therefore location transparent allowing for failure recoveries transparent for the application service users and application service providers alike.

The SA Forum EVT provides a publish/subscribe based communication mechanism that can be used among multiple anonymous processes of the cluster. Publishers and subscribers may join and leave 'the club' – represented by an event channel – as they please. Publishers do not know about the receivers of their communication. On the other hand receivers have enough knowledge about the expected communication so that they can subscribe to them even if they do not know the exact identity of the publishers. Event channels have some 'memory' as they can retain events if needed for potential late comer subscribers, but overall the service provides best effort delivery.

The SA Forum MSG provides a communication mechanism, which can be tuned for greater performance or more reliability. The basic service entity is the message queue, which is an input queue for a set of processes collaborating in the protection of the service represented or requested through the messages delivered by the queue. Any number of anonymous senders may send messages to the queue processed by a single receiver at a time. In case of the failure of this receiver the message queue acts as a buffer until the new receiver takes up the message processing task.

When message queues are combined into message queue groups the MSG also provides a load distribution mechanism among multiple message processors.

If not for the creation of message queues, the message senders may not be aware whether they send their messages to an individual message queue or message queue group. The method of communication is somewhat alike to data streaming where if part of the data is missing the whole stream may need to be retransmitted.

Finally the SA Forum CKPT provides a cluster-wide equivalence to shared memory based communication. The service entity representing this 'shared memory' is the checkpoint which is accessible from anywhere in the cluster for processes needing to write or read part or the entire data. It conveniently holds the complete picture of the information to share at any moment in time like series of snapshot that follow one another without the need to keep the history of changes.

Since checkpoints are stored in memory the service allows for fast dissemination of any information and it is the reader's choice whether it reads the entire checkpoint or just the relevant or updated parts of it.

The CKPT offers options particularly tailored for applications managed by the AMF so that the data replication performed by the CKPT is in complete synchronization with the redundancy of service provider entities under AMF control.

As none of these services impose a format on the units of communications used with their service entities (i.e., events, messages, and sections) application designers have a lot of freedom in defining the layout that is most suitable for their application. But this also imposes the limitation as for successful communication the sender and the receiver need to be in an agreement about the interpretation of these data structures.

Therefore the scope within which this agreement exists determines the scope within which the services can be used. As a result without further standardization of at least the negotiation mechanisms all three services discussed are most suitable for intra-application communication. In combination with existing standards (e.g., higher level communication protocols, other AIS services) however they can be extended beyond the application limits and suddenly the MSG can be used as a load balancer front end for web services or the EVT can be used as the carrier of alarms and notifications as defined by NTF.

8

Services Needed for System Management

Maria Toeroe

Ericsson, Town of Mount Royal, Quebec, Canada

8.1 Introduction

System management is the cornerstone of maintaining high service availability (SA). It allows the configuration and reconfiguration of the system, monitoring its operation and it is essential for fault management. We have seen that the Availability Management Framework (AMF) [48] is the first line of defense when an error is detected. It uses redundancy and deploys recovery and repair actions to recover the system's services in the shortest possible time. However this may not be enough to prevent future problems. As we discussed in Chapters 1 and 6 an error is only the detected manifestation of a fault, recovering from and repairing the error may not repair its root-cause – the underlying fault.

To prevent future problems we need to go further and analyze the situation to find this root-cause of the detected error or errors, which requires more information from the system. Therefore systems built for SA systematically collect data that may reveal any information about any faults in the system.

Within the SA Forum Application Interface Specification (AIS) two services provide such capabilities. These are the Log service (LOG) [40] and the Notification service (NTF) [39].

The LOG provides applications with an interface to record any data that they consider important among others for fault management. Applications may create new or open existing log streams each of which is associated with one or more log files where the data inserted by the application is preserved for some period of time according to the policy defined for the log stream. The contents of these files may then be harvested periodically for offline analysis of the system or particular applications. Most importantly neither the LOG nor the other SA Forum specifications define the contents or the format of the log files. They are tailored as needed. The formatting information for each log stream is available for potential readers as a separate file, which needs to be collected together with the log files.

The LOG is suitable for collecting large amount of data from the system typically used in offline analysis.

Service Availability: Principles and Practice, First Edition. Edited by Maria Toeroe and Francis Tam.
© 2012 John Wiley & Sons, Ltd. Published 2012 by John Wiley & Sons, Ltd.

For cases when a prompt reaction may be necessary to the events occurring in the system the AIS includes the NTF.

The NTF defines different categories of notifications, their contents, and format each indicating some kind of incident. Management entities in the system subscribe to the category and class of notifications they are capable of handling. Applications as well as system services generate the notifications whenever a significant event occurs and provide the contents as required via the NTF interface.

Most of the notifications are intended for the consumption of management entities within the system. We will see such an example with the Software Management Framework (SMF) [49], which is such a management entity and correlates an ongoing software upgrade with the errors reported within the system.

However not all situations can be handled within the system. Notifications that report situations where external intervention or administrative attention is required comprise the categories of alarms and security alarms. From an operator's perspective in telecommunication systems an alarm signals a situation requiring human attention or even intervention and therefore such an alarm often results in turning on an alarm sound signal and/or dispatching a call, an SMS to the person on duty in addition to logging it in the system logs and reflecting it on the management interface.

So the system administrator when receiving an alarm needs to act on it. This is the point when the third AIS service, the Information Model Management service (IMM) [38] comes into the picture. As mentioned several times in this book it stores the information model for the different services and combines them into a consistent system information model. The information model reflects the state of the system and it also acts as a management interface. Accordingly the IMM exposes an object management (OM) interface through which the managed objects of the information model can be manipulated. The administrator can create, delete, and modify configuration objects and issue administrative operations on both runtime and configuration object as appropriate. The OM-API, application programming interface is often referred to as the northbound or management interface of a system.

As discussed at the introduction of the system information model and at each of the services the objects of the information model reflect entities or logical concepts meaningful only for the service or application for which they have been defined. Therefore the manipulations carried out by the administrator on the managed object needs to be delivered to the service in charge of the object. In IMM terms this service or application is the object implementer (OI) deploying the managed object in the system and for this interaction the specification defines the OI-API.

Each time an object is manipulated via the OM-API, IMM delivers the actions to the OI associated with the object via the OI-API and returns any received outcome to the manager via the OM-API. IMM only acts as a mediator; the whole transaction is opaque for IMM.

In this chapter we will take a closer look at each of these AIS services starting with the LOG then moving onto the NTF and finally concluding with the IMM.

8.2 Log Service

8.2.1 Background: Data, Data, and More Data

As we pointed out in the introduction the main purpose of the LOG is to collect data from the system that can be used for root-cause analysis to diagnose problems that occurred during operation so that similar situations can be prevented in the future. Probably the most difficult question to answer in this respect is what information is useful for this purpose and what would be superfluous.

To answer this question we need to look at who is the user of the collected information. The users of the data collected by the LOG are the system or network administrators or tools at their level. In any case they have a general overview of the system, but may not be aware of the specifics of each particular application running in the system.

The information required by this audience is different from the software developer, for example, who has an intimate knowledge of the application details. The Log specification distinguishes this application specific low level information as trace data. While it can provide beneficial information for a more generic audience the amount of such information generated in a system would be so enormous that its collection on a continuous basis would create too much overhead in the system.

Instead the information collected through logging needs to be suitable for system level trouble shooting by personnel or tools possessing some high-level overview of the system.

Even with such higher level information volume remains the main challenge of logging. This can be tackled from two main directions: the organization of the data so it can be efficiently collected, stored, and retrieved when necessary; and the efficient purging of data when it becomes obsolete. In addition it is beneficial if the service provides filtering capabilities so that the amount of collected data can be adjusted to the particular circumstances, system requirements, and application needs, if it can take off the burden of record formatting from the applications, so they do not need to be aware if the output format changes due to requirement or locality changes.

All these issues are addressed by the SA Forum LOG with which we begin our exploration of the AIS management infrastructure services.

8.2.2 Overview of the SA Forum Solution

The SA Forum LOG offers its users cluster-wide logical entities called log streams that represent a flow of records, which is stored persistently as shown in Figure 8.1. Any number of users may open the same log stream simultaneously and insert log records into this flow. The log stream delivers these records to an associated file where the record is saved according to a format also associated with the log stream. The user writing to a log stream does not need to be aware of the format associated with the stream. It only needs to be aware of the pieces of information that it needs to provide in the API call for each write operation. Based on the provided data and the formatting associated with the stream the LOG saves the record in a file linked at the time of the saving operation with the log stream.

Since more than one API users may write to the same log stream simultaneously the LOG also serializes these records before saving them in the destination file. The specification does not stipulate

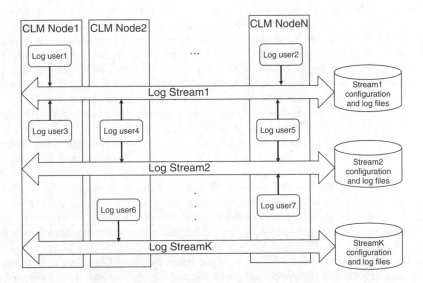

Figure 8.1 The overall view of the Log service.

how this serialization occurs and therefore different implementations of the LOG may use different strategies. For example, the specification does not require the ordering of the records according to the time stamp provided in the API call. One implementation may use this time stamp, while the other may not.

For each log stream the file formatting information and other properties are contained in a separate configuration file. The formatting information is essential for anyone wanting to read the log file as this defines the different fields, their length, and format within each saved record. The specification defines the different format tokens and their use.

In addition the configuration file also explains the maximum size of the log files generated and policy used by the LOG when this maximum size is reached. There are three strategies that may be defined:

- If the **halt option** was defined the LOG stops saving further records for the given log streams and records are lost until more space becomes available. To prevent the loss of records a threshold is defined for such a log stream and the LOG generates an alarm when this threshold is reached. It generates a second alarm if the file size still reaches its maximum.
- The **wrap option** instructs the LOG to continue saving the log records in the same log file even when it reaches full capacity by overwriting the oldest records existing in the file – that is, wrapping around. Hence doing so also may cause the loss of some information, although in this case from the log file itself, it has been proposed to add a similar threshold attribute to this type of log streams as well and generate an alarm whenever the threshold is reached.
- The third, the **rotate option** is a special case of the wrap option. The difference is that rather than wrapping around in a single file it uses a set of files. Whenever the maximum file size is reached a new file is created until the maximum number of files is reached. At this point the oldest file is replaced by the LOG. Setting the maximum number of files to the maximum possible value results in continuous creation of new files as they fill up, which may in turn result in running out of system resources.

Whenever the LOG creates a new log file it indicates the creation time in the file name. When the file is full and the LOG closes it, it is renamed by appending also the closing time.

The question is how log streams come to existence. The specification defines two classes of log streams: those that are configured by an administrator and those created by an application at runtime using the LOG API. Configured log streams exist from the moment they are added to the configuration up until they are removed from it. On the other hand application-created runtime log streams exist as long as there is at least one LOG user in the cluster who uses them. When the last user closes a stream (or stops using it due to the failure of the opener process, or the node leaving the cluster, etc.) the LOG closes the files associated with the log stream and removes the stream.

There are three configured log streams mandatory to implement by a LOG implementation. They are well-known – their name is defined by the LOG specification. These are the alarm log stream recording alarms and security alarms; the notification log stream recording all other categories of notifications; and the system log stream that can be used by applications and services alike to record situations that affect their service or the system.

As one may conclude the first two well-known streams are used for notifications carried by the SA Forum NTF, which will be discussed in Section 8.3. The reason to distinguish these two streams is that the LOG also allows the filtering of log records; however it cannot be applied to the alarm and notification log streams. The specification mandates that all alarms and notifications need to be logged in the system.

As we pointed out in the Section 8.2.1 one of the issues that the LOG needs to deal with is the amount of data generated in the system. Of course the code written for the applications and system components cannot be changed just because one would like to have more or less information generated.

With a significant number of such software in the system it also becomes unimaginable of configuring each one of them separately.

Instead the LOG offers the possibility to define the level of severity at which logging should be performed.

Whenever a user writes a log record to a log stream it provides the severity level associated with the record. This information is then used by the LOG to filter out records that are below the currently set severity level. In addition the LOG also informs log writers not to produce records below this set level.

8.2.3   The LOG Information Model

The information model defined for the LOG is rather simple and specifies two similar object classes: One for configuration log streams and a second for runtime log streams.

Configuration log streams[1] are created – just like any other configuration object of any other AIS service – by system administrators using the IMM object management API. These log streams exists as long as the configuration object exists. Users may open these log streams by referring to their names, that is, to the distinguished name (DN) of the configuration object representing the log stream.

All the characteristics of the associated log streams are determined by the attribute values of the configuration object including the name of the log file, its path, maximum size, the policy applicable when the file is full as well as the record format.

Most of these attributes are also writable, which means that if an attribute value, which configures in some way the log file associated with the log stream, changes then the LOG needs to close the current log file and create a new one, which satisfies the configuration defined by the new set of attribute values.

Attribute changes that do not define the log file features are applied immediately. For example, changing the severity level changes the filtering policy for the log stream across the cluster. The LOG applies this change right away.

Since these logs streams can exists in the system as log as the LOG implementation is present in the system the suggestion is to create the representing configuration objects as child objects to the configuration object representing the LOG itself.

Log streams can also be created by applications using the LOG API. These log streams are represented in the LOG information model as objects of the runtime log stream class. The configuration attributes for such log streams are provided by the creator process at creation time and therefore they remain in effect for the entire period of existence of the log stream.

The only attribute that can be manipulated and – being a runtime attribute – only through administrative operations is the severity level.

The name of the log stream provided by the creator is used by the LOG as the DN of the runtime object it creates to represent the log stream in the LOG information model. This means that the log stream name needs to be valid with respect to the actual system information model. More precisely the parent object to the log stream object needs to exist otherwise the LOG cannot insert the log stream object into the model and the creation operation fails.

Since the LOG API users generally do not need to be aware of the system information model this constraint may require some considerations from application designers who wish to use the LOG API in such a manner particularly if the application needs to be portable between systems.

There are a few potential parent objects that an application may know about without any knowledge of the current system information model:

- The well-known objects representing the different AIS services and in particular the one representing the LOG itself. As we mentioned the suggestion is to use it as the parent object of the configuration log stream objects.

---

[1] At the time of writing configuration log streams for applications are part of the upcoming release.

- The system root is a special well-known object; however, one would quickly realize that this option is the least favored by operators and system administrators as it flattens – some would even say pollutes – the information model.
- In case of applications managed by the AMF a process
  - can obtain the name of the AMF component it belongs to using the AMF API;
  - learns about component service instances names as CSIs are assigned to the component the process belongs to.

In the later cases the IMM may not allow the removal of the AMF component or CSI unless the runtime log stream object is removed first. So their ancestors may be more suitable.

A runtime log stream exists as long as there is at least one opener associated with it in the cluster. When the last opener closes the stream the LOG closes the files associated with the stream and deletes the log stream. Accordingly it also removes from the information model the runtime object representing the now closed log stream. If there are several writers to the stream it may be difficult to achieve this in sync with the manipulation of the AMF objects.

There is no way to find out about closed runtime log streams from the information model, therefore the LOG also generates notifications in association with runtime log streams. It generates one at the creation of such a log stream and one at the deletion. Both notification types include the information necessary for collecting data from the log files.

8.2.4   User Perspective

The API defined for the LOG is simple and straightforward. The same way as for other AIS services, processes that wish to use the LOG need to link its library and initialize the API. At initialization three callbacks may be provided among which the most interesting is the one associated with the severity level. If this callback is provided at initialization the LOG is able to call back the process whenever the severity level is changed for any of the log streams opened by the process. Thus, the process can operate more efficiently by not generating log records that would be discarded by the LOG anyway.

Before the user process can start to write log records to a log stream it needs to open the stream. If the stream already exists in the system – it is a configuration log stream or it has been already created – then the new opener is added to the list of openers for the particular stream maintained by the LOG. This list has particular significance for runtime log streams as they exists as long as there is at least one opener process associated with the stream. When the last process closes the log stream it ceases to exist, the LOG closes the associated log files and removes the representing runtime object.

If the log stream does not yet exist, when a user process tries to open it the LOG creates the stream provided that the opener process has provided all the attributes necessary for configuring the log stream.

A special case is when two processes both assuming that they are the first openers – and therefore the creators of the log stream – and provide the configuration attributes for the log stream. To resolve the potential conflict the LOG creates the log stream based on the attributes of the request it receives first. If the second opener-creator defined the same attributes than there is no conflict as it would have created the same log stream. If however any of the attributes are different, the operation fails and the second user will receive an error that the log stream already exists. It can open the stream successfully only after removing the attributes associated with the log stream creation. (Well, it could also try to match the attributes of the existing log stream; however, that requires more efforts, for example, the use of the IMM OM-API to read the runtime object.)

Once the log stream has been opened by a process it can write log records to the stream. For the write operation the user needs to provide the log header, the time stamp, and the record data as an opaque buffer.

If the record is intended for the alarm or the notification log stream then the specially defined notification log header is used, however applications typically do not use these streams. For all other cases the generic log header is used, which includes the severity level, which is used at filtering as we have seen.

The user may provide an actual value for the time stamp or ask the LOG to provide it by setting its value to unknown.

A user process may invoke both the opening and the writing operations in a blocking or in an asynchronous nonblocking manner. The latter case requires that the process provides the appropriate callbacks at initialization.

After the initialization of the LOG the same process may open as many log streams as required; however implementations may limit the total number of log stream in the system.

8.2.5   Administrative and Management Aspects

The LOG offers different management capabilities depending on the type of the log stream.

As we already discussed in Section 8.2.3 administrators can create, delete, and reconfigure configuration log streams:

- Creating a configuration log stream means that the associated files are created and user processes may open the stream and write to it from anywhere in the cluster.
- Deleting a log stream means that the LOG will close all the files associated with the log stream and the configuration object is removed from the information model. Users who had the log stream open will receive an error next time they attempt to write to the stream.
- Modifying the configuration of such a stream most often results in the LOG closing the current log file and creating a new according to the new configuration. The exception is the modification of the severity level, which changes the filtering policy for the log stream.

Note that configuration log streams of applications will be first defined in the third release of the specification. Until that only the well-known log streams exist as configuration log streams and the LOG creates them. That is to say an administrator cannot create or delete any of these well-known streams; however, their reconfiguration is still possible except for the filtering of the alarm and notification streams, which are also prohibited by the service.

For runtime log streams a single administrative operation is available that changes the filtering criteria for the stream. It is equivalent to setting the severity level of a configuration log stream. The administrator has no other direct control over runtime log streams.

8.2.6   Service Interaction

8.2.6.1   Interaction with CLM

The LOG uses the membership information provided by the Cluster Membership service (CLM) [37] to determine the scope of the cluster. It provides the LOG API only on nodes that are members of the CLM cluster.

The reason behind this limitation is that log streams are cluster-wide objects accessible simultaneously from any node of the cluster. When a node is not member of the cluster it and its processes have no access to cluster resources such as the log streams. This implies that circumstances and events that need to be logged while a node is not a member need to be logged by other means, for example, most operating systems provide some local logging facilities. Any integration with such facilities is beyond the scope of the LOG specification and is implementation specific.

This also means that a LOG implementation is required to finalize all associations with processes residing on a node that drops out of the membership and any initialization of the service should fail as long as the node remains outside of the cluster. To obtain the cluster membership information the LOG needs to be a client of the CLM and track the membership changes.

8.2.6.2 Interaction with IMM

The IMM [38] maintains the information model of all AIS services including the LOG. It also provides the administrative interface toward it.

At system start up the LOG obtains the set of configuration log streams it needs to implement from IMM, which loads them as part of the complete system information model. A LOG implementation also registers with IMM as the OI for the LOG information model portion of the system model. It does so in all available roles.

From this moment on IMM will deliver to the LOG the information model changes initiated on configuration log streams and also the administrative commands issued on runtime log streams.

As the OI, the LOG implementation uses the IMM OI-API to create, update, and delete objects representing the runtime log streams existing within the cluster. It also updates any runtime attributes of configuration log streams as appropriate.

The IMM OM-API is the interface through which an administrator can observe and also to some extent control the behavior of the LOG in the system.

8.2.6.3 Interaction with NTF

According to the upcoming release, the LOG generates two types of alarms and three types of notifications and uses the NTF API to do so.

The two alarms announce situations when information may be lost either because it cannot be logged or it is about to be overwritten.

In particular the capacity alarm is generated when the threshold set for a log stream has been reached and again when the maximum capacity is reached. Depending on the policy set for the log stream at the reach of maximum capacity the logging cannot continue (halt) or it will result in overwriting of earlier records (wrap and rotate). The administrator needs to figure out which is the correct course of actions – preferable after the first alarm. In case of halt policy, the only option is really to collect the data from the log file, so that logging can continue. In case of wrap and rotate overwriting earlier records may well be the intention in which case no action is necessary. Otherwise again the data needs to be collected to avoid any loss.

The second alarm indicates that the LOG is not able to create a new log file as it is required by the configuration for the log stream. This may occur due to misconfiguration, but also due to running out of resources in the system. However, the alarm is generated only if the LOG is unable to communicate the error by other means. For example, if the error can be returned via the service API in response to the creation attempt of a runtime log stream, the alarm will not be generated.

Whatever the reason of the file creation alarm, it signals that the LOG is not able to save log records as required and they are being lost. The administrator needs to find out the actual reason to remedy the situation.

As the LOG will reattempt the operation that triggered the alarm it will detect that the problem has been corrected and clear the alarm automatically.

The notifications the LOG generates provide to the interested parties the information necessary to collect the data so that they do not need to monitor the LOG information model.

The LOG generates notifications at the creation and deletion of runtime log streams to warn the administrator or anyone interested and subscribing to these notifications that collecting the logged data may be due.

The LOG also generates notifications that a configuration log stream has been reconfigured and therefore the access to the logged information might have changed.

Besides the LOG using the NTF to produce alarms and notifications the NTF may also use the LOG. As we pointed out at the discussion of the well-known log streams, two of them are intended for alarms and notifications respectively. It is not specified who writes these streams and there are several possibilities: The first approach is that the NTF writes both streams. This is the intention and it is most efficient; however, the NTF specification does not require it. Since it is the LOG specification, which requires these streams a second approach is that the LOG also writes them. For this it needs to become also a subscriber to the NTF as discussed in Section 8.3.3.1. Again this is not spelled out in the specification, so an implementation may not do so. In this case finally, a dedicated application may subscribe to NTF and write to the log streams by that bridging the two services.

8.2.6.4   Interaction with AIS Services

It is recommended that all AIS service implementations log their actions and that their use for this the SA Forum LOG. In these relations the LOG is a provider for the other services.

8.2.7   Open Issues and Recommendations

After the release of the third version of the specification there will be only few open issues with respect to the Log specification. Probably the least clear from the specifications is the relation of the Log and the NTFs. This is primarily due to the NTF reader API discussed later in this chapter in Section 8.3.3.3 and the Log requirement of providing the well-known alarm and notification log streams. They seem to overlap.

Indeed it is possible for the NTF to use the LOG to persist the notifications with the purpose of reading them back whenever someone asks for it. Whether it should use the pre-defined alarm and notification log streams is another issue. The problem with using these well-known streams to support the NTF reader API is that they are reconfigurable by an administrator including their logging format. This means that the new format may or may not satisfy the requirements of the NTF also that its 'reader module' would need to adjust to the new format again and again. This is an unnecessary complication for the NTF implementation as it could create its own runtime log stream for which the format is locked in exactly the way it is needed for optimal use.

By the same logic we can define a rule of thumb when to use runtime log stream as opposed to configuration log streams.

When the log files are consumed by external users it is beneficial to use configuration log streams as their output format can be readily adjusted as required over time. Having the same type of adjustment of a runtime stream would be rather cumbersome as the stream needs to be recreated with the new attributes, which could require significant coordination if many users use the stream, not to mention that the new format would need to be configured somehow anyway. This was actually the rational why configuration log streams were introduced for applications as well.

On the other hand, when the logs are consumed by tools and applications optimized for a particular formatting the use of runtime log streams can preserve the format provided at the creation of the log stream for its entire existence.

Now returning back to the relation between LOG and NTF: All these consideration about the reader API do not prevent an NTF implementation to write the alarm and notification streams provided by Log. It is the intention. However, since it is not mandated yet another approach to resolve this relation is that the LOG or a dedicated application subscribes to alarms and notifications and writes them to the well-known log streams as appropriate. The critical issue here is to make sure that there is no duplication of the recordings. This is actually an issue in general for these well-known streams as the specification does not indicate the expected writes to the different streams. Hence an application designer or implementer reading the specification may be compelled to write the notifications the application generates by the application itself to the appropriate streams, which results in duplication if the middleware does so already.

The point is that applications producing alarms and notifications typically do not need to worry about them being logged. The middleware is expected to take care of it.

In summary we can say that the LOG provides a standard way to record situations, circumstances, and events that may be useful for – primarily – offline analysis of the system behavior whether this data is generated by system components or by applications. It is capable of collecting great amount of data organized into different streams of information while releasing the data producers from the burden of formatting and organization of these records.

The collected information is very beneficial in fault management; however, in high availability and SA systems there is a need for faster reaction to certain situations. The LOG is not intended to trigger these reactions; it is intended to provide the context to events that require thorough analysis.

The service addressing the need of reactiveness is the NTF that we will discuss next.

8.3 Notification Service

8.3.1 Background: Issues, Controversies, and Problems

We have seen that the AMF will try to repair components by restarting them and if that is unsuccessful then escalating it to the next fault zone, then potentially to the next, and so on. Obviously if the cause of the problem is a corruption of the file containing the executable code or a malfunctioning hardware, a simple restart at any level will not solve the issue. The administrator needs to get involved and fix the problem; therefore AMF needs a way to alert the administrator.

The same is true for other services. When a problem is beyond their self-healing capabilities they need to inform the administrator that the problem they are facing is persistent and their remedies failed. Of course we could define for each service an interface for this purpose, but if each of the services defined their own interface with their own methods of communicating the problem the combined system management would become quickly a nightmare for the system administrators.

Instead we would like to have a common communication channel and method through which the system entities (including applications) can signal an emergency situation that external intervention is needed.

In addition this channel may also be used to provide those external 'forces' – or for that matter anyone else who may be interested – with regular updates about significant events and circumstances to provide some context to any potentially upcoming emergency situations.

At the same time it is also important that different listeners of this 'emergency broadcast channel' are not overwhelmed with data they cannot digest anyway. We would like them to receive only the information relevant to them and that they receive all of that information so they can make sound decisions should an emergency situation occur.

This means that the 'broadcasted' data needs to be organized such that anyone can read it easily as well as it can be classified efficiently so that the filtering and correlation of different pieces are both possible. Correlation may become necessary at a later point when an error occurs and its context becomes important to make an educated decision.

From the perspective of the entities reporting the emergency situations it is also important that they can produce the notifications efficiently as they are often in the critical path of dealing with some incident such as the emergency situation they are about to report. In case of AMF, for example, while it is busy in orchestrating the fail-over of the service instances served from a failed node (reported by Platform Management service (PLM) via (CLM)) to a healthy one any additional work of reporting and logging is undesirable, so we would like to keep it at the minimum.

Notice also that in the above scenario not only AMF, but the complete service stack, that is, PLM and CLM will also report their view of the issue. Later, when the administrator fixes the failed node (e.g., replaces the hardware) this action tends to clear the problem for all AIS services and not only for PLM. There is no need to deal separately with the CLM issue unless there is indeed some connectivity problem too. So it is just as much important that the system can tell whether an emergency situation has been dealt with or need further actions as reporting the emergency situation.

The SA Forum NTF was defined with these goals in mind. It was also aligned with the existing related specification of the International Telecommunication Union, the ITU-T X.7xx recommendations [72, 85, 86], and [87].

It provides a single communication channel dedicated to signaling significant changes and situations in the system, most importantly circumstances, when the problem persists and the system cannot resolve it permanently or at all; and therefore external resolution is required. It defines the format of these incident reports so that all interested parties can read the information and also it can be organized properly.

Let's see in more details how these are achieved.

8.3.2 Overview of the SA Forum Notification Service

Maybe the best everyday-life analogy to the NTF is the 911 emergency number. People call it when they find themselves in a situation they are unable of handling. It is a single point of access to all emergency services: The Police, the Ambulance, and the Fire Department.

The SA Forum Notification is a similar single well-known point of access through which system components can report issues they cannot deal with or 'consider' important. Any service or application may use the NTF for reporting situations of significance. In terms of the NTF these services and applications are the *producers* of the incident reports or *notifications* as shown in Figure 8.2.

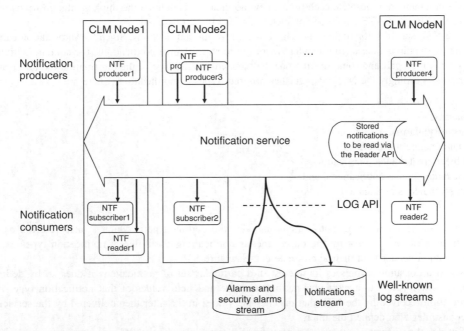

Figure 8.2 The overall view of the Notification service.

Most AIS services are producers of notifications and the service specification defines the 'situations of significance' that an implementation of the service shall 'consider' important and therefore should report.

On the other side, there are entities inside and outside of the system, whose task is to listen to the different reports and if a problem is reported then to resolve it. It is like the Fire Department, the Ambulance, or the Police at the other end of the 911 phone line. The dispatcher takes the report of an emergency situation and alerts the appropriate service or services, which then in turn take over the handling of the situation based on the report they have received. They are the consumers of the emergency reports; hence in the NTF there is the notion of the *consumer* of notifications.

The notification producers typically do not know who or which entity is going to be the consumer of their notifications. Neither does the NTF, at least not in advance. System entities need to express their interest in the different kinds of notifications to the NTF, so it can deliver them the notifications accordingly. This *subscription* mechanism replaces the 911 dispatcher and allows the NTF to deliver the right notifications to the right consumer. The consumer may be internal (e.g., another service) or external (e.g., system administrator) to the system. It is also possible that multiple consumers listen to the same notifications, or that nobody listens to some of them.

To make sure that even though at a particular moment there is nobody listening to some notifications these notifications are not lost – after all they are reporting some important situations – the NTF implementation has the obligation to store the notifications persistently. Any consumer that misses some notifications that are of significance can *read* them back from the NTF, from these stored notifications.

The question still remains how consumers express their interest in a certain set of notifications and how the NTF can sort the received notifications to match the subscribers' interest.

The AIS specifications define the different notifications the AIS services produce, so a possibility would be to list them based on the different specifications. However this approach would render a NTF implementation obsolete each time a new notification is defined the implementation needs to be updated. Also, the NTF would be completely useless for applications.

The NTF specification defines the data structure for the notifications. Since the information carried in the notifications ranges from incident reporting to state change notifications, the content is divided into a common part and some specific one that depends on the type of the information the notification carries. Accordingly the NTF distinguishes different types of notifications. There are

- alarms,
- security alarms,
- state change notifications,
- object create/delete notifications,
- attribute change notifications, and
- miscellaneous notifications.

The data structure of all notifications includes the common part, which is also referred as the notification header. Except for the miscellaneous notifications the different notification types extend the common header with their appropriate data structure.

When a consumer expresses its interest in a particular set of notifications it does so by defining a *filter* which refers to a particular notification type and field values of that notification type. Any notification received by the NTF that matches the values in the filter then delivered by the service to the consumer who defined the filter.

Besides the different contents, these notification types also imply different importance. An alarm is generally perceived as the indication of some real urgency, while an attribute change notification is usually considered less important. The idea is that alarms – both security and regular – reflect incidents that the system cannot deal with and to resolve them human intervention is required. Strictly speaking these are the true 911 cases, while the other notifications are typically consumed internally by other services of the system or provide context for the alarms.

The NTF functionality linked with the importance of notifications is the notification *suppression*. When in the system there is no consumer for certain notifications it may be a reason not to produce them on the first place as producing notification takes away processing time, bandwidth, and other system resources. Other times even though there would be consumers, but some shortage of resources – due to extreme high traffic, for example – might require that less important notifications are turned off. The suppression mechanism allows for this type of tuning of the system at runtime by the administrator.

The NTF specification requires that its implementations guarantee the delivery of at least the alarms and security alarms. They also must store alarms persistently. We have already seen in Section 8.2.2 that the LOG provides a well-known alarm log stream for this purpose, which cannot be filtered either.

An implementation may however provide lower delivery guarantees for other types of notifications and also may not store them persistently. The LOG specification defines a second well-known stream also for notifications.

For all notifications however it is required that all NTF implementations deliver only complete notifications and that each subscriber receives each notification at most once. Moreover with respect to the same producer subscribers receive notifications in order, that is, in the same order as they were produced.

Finally the NTF also supports *localization* to the extent that it provides the API call to get the localized message associated with a notification. The specification does not cover the method the localized information is provided to the service; it is left to each service implementation. The appendix of the specification only outlines at high level the general idea behind this localized message catalog and its mapping to notification parameters.

8.3.3   User Perspective

The NTF specification defines three sets of API functions. They are for the notifications producers, subscribers, and readers. Regardless which portion of the API a user wants to use it needs to link the NTF library.

8.3.3.1   Producer API

Producing a notification means creating and filling in the required data structure and submitting it to the NTF.

The most important moment about producing notifications is that it often happens during some critical operations, considering, for example, the AMF when it is handling a fail-over. According to the specification AMF needs to produce a series of notifications with the fail-over as the different entities change their states. This also means that AMF needs to produce many similar notifications as, for example, at a service unit (SU) fail-over each of its components and the SU itself will change several of their state (i.e., the presence, the readiness, and the high availability (HA) states at least, but the operational state may also change).

The NTF producer API offers some optimization possibilities:

First of all, an NTF user that wants to produce some notifications may pre-allocate a notification identifier. Normally the notification identifier, which is a cluster-wide unique identification number, is allocated during the send operation. The NTF returns the notification id at the completion of the send operation and a user may reference it in subsequent notifications that are related to the sent one. The need to reference related notifications serializes the send operation as well as any invocations requiring the notification id, which may take some time. However if there is a pre-allocated notification identifier the user may use it right away in any related notifications and invocations. Since there is no ordering requirement associated with the notification id, NTF implementations may use different strategies for the allocation of this cluster-wide unique number that may also shorten the send time.

The second optimization possibility that users may consider is the reuse of the allocated data structure. Again allocations may take some time and also use additional resources. If more than one notification of the same type needs to be sent (e.g., state change notifications for all components of an SU failing over its services) they will use the same data structure. Therefore the user may allocate this data structure once and reuse it as many times as needed by updating it with the values appropriate for each new notification. Of course, this again serializes the operations as each data setting needs to be sent before the data structure could be used for the next.

Finally, if a producer provided the appropriate callback at initialization time, the NTF will inform it whenever some notifications are suppressed in the system. Thus, the producer can stop generating them all together potentially saving valuable time and resources.

Toward the producer NTF passes the suppression as a bitmap of the event types, for which the producer should stop generating any notifications. The event type is one of the common attributes of all notification types.

8.3.3.2   Subscriber API

When it comes to the consumer API, the first task a 'want to be subscriber' or a 'want to be reader' needs to do is to set up the filters that describe the notifications it would like to receive or read.

The filter setting requires the setting of a similar data structure as the notification data structure itself as the filter conditions are set for this very same data structure. Even though this subscription setup is usually not part of critical operations, for symmetry the NTF API supports data structure reuse between different subscriptions.

With the different filters in place the user is ready to initiate the subscription from which moment on it will receive a callback if a notification matching any of the filters is broadcast in the system. The notification is delivered within the callback.

For each notification instance delivered the NTF consumes some of its resources that are freed only when the subscriber has processed the notification. In some cases when a subscriber does not consume the notifications fast enough, NTF may run out of its resources. In this case, the NTF cannot deliver any more notifications, instead it informs the subscriber that it dropped some notifications and provides the required information for the subscriber to retrieve these discarded notifications using the Reader API.

8.3.3.3   Reader API

If the notification consumer wants to read persisted notifications in addition to setting up the filters it also needs to define a search criterion to find the first notification to be read. To understand this we need to indulge a bit more in the way notifications are persisted:

Since there are many notification producers in the system, they may try to send notifications simultaneously. The NTF serializes these notifications by that creating a single sequence of notifications which are then written to some persistent storage, such as a file on a disk. Alternatively the NTF may use the already discussed LOG to record the notifications in which case the serialization is effectively performed by the LOG.

In any case, the result is an ordered list of notifications within which, however, the order may not exactly be the same as the time stamp associated with each of the notifications. When reading the notifications this recorded order prevails and the time stamp can only be used as a search criterion to find a first notification to read. From this first notification, however, the reader may navigate forward or backward in the list selecting the direction on each read operation.

An easier way to identify the first notification to be read – provided it is known – is its notification identifier.

As we described in Section 8.3.3.2, under certain circumstances the NTF may not be able to deliver a notification to a particular subscriber (e.g., the subscriber is consuming the notifications it is receiving too slowly). In this case it will still deliver a callback that it had to discard a notification with a particular notification identifier. The subscriber can use the reader API to retrieve the notification using the provided notification identifier.

8.3.4   Correlation of Notifications

We mentioned in the previous section that it is desirable that a notification producer references related notifications in each generated notification. This is referred as correlation of notifications.

The correlation of notification is particularly important at root-cause analysis as an error may trigger multiple reactions as the error or its consequences reach the different layers and subsystem. The notifications they generate can be organized into a tree.

For example, if a hardware node goes down at least the PLM, the CLM and the AMF react as each handles its own entities that resided on the node:

1. PLM detects it as a failure of the hardware elements and the execution environments mapped to this hardware node and needs to adjust their states for which it generates a set of notifications. It reports this starting with the entity closest to the error and proceeds toward the impacted dependent entities. When reporting the state change of a dependent entity in a notification it references the notification identifier of the state change notification of the entity that triggered the state change being reported in this notification.
   In addition PLM also informs all its interested tracking users about the change. In this track callback again PLM provides the notification identifier of the notification reporting the triggering state change.
2. CLM sees the event at least as one or more nodes suddenly leaving the cluster. It may detect this on its own or since it tracks the execution environments to which the cluster nodes map, it also receives the track callback from PLM together with the notification identifier. CLM produces its own notifications about the nodes left. In them CLM references the notifications produced by the PLM using the notification ids received in the PLM track callback.
   Again, the same way as PLM reports the failure through the track API, CLM does the same also including the notification identifiers as appropriate.
3. AMF tracks the cluster membership, hence it receives the callback from CLM. It perceives the disappearance of a node as a simultaneous failure of all those components hosted on the nodes that left the cluster, so AMF needs to recover all the services they might have been providing. This means failing over all the CSIs from the node that left to the appropriate standby components – which equates to a lot of state change notifications. Again AMF reports these state changes following the entity dependencies as applicable in a similar manner as the PLM did in step #1.

In these state change notifications AMF references the notification identifier associated with the triggering event. However, the AMF does not propagate the notification identifiers automatically within its callbacks. The user process requiring it can obtain the ids by calling the appropriate function and providing the invocation identifier of the callback.

These events trigger the following set of notifications. By #n we indicated the notification id here, but it also reflects a possible sequence of the generated notifications. In parenthesis we also indicate the root and the parent notification ids referenced as correlated:

```
#1: PLM HE alarm ( ,   )
 #2: PLM HE operational state change notification (1,   )
  #3: PLM HE presence state change notification (1, 2)
   #4: PLM HE readiness state change notification (1, 3)
    #5: PLM EE presence state change notification (1, 4)
     #6: PLM EE readiness state change notification (1, 5)
      #7: CLM member node exit notification (1, 6)
       #8: AMF node operational state change notification (1, 7)
        #9: AMF SU readiness state change notification (1, 8)
         #10: AMF component readiness state change notification (1, 9)
          #11: AMF component presence state change notification (1, 10)
           #12: AMF SU presence state change notification (1, 11)
```

We see that there is a sequencing involved here. The assumption is that CLM receives the track callback from PLM before it detects itself that a node left the cluster and the same is true for AMF. Only in this case the generated notifications can reflect the reality and only if they are correlated properly. Even though the failures and error are definitely related and compose a chain of reactions the timing of the different error reporting and detection mechanisms may show them unrelated or related.

It is possible that a PLM implementation generates notifications #2-#4 as direct consequences of #1, but not each other. Same may go for notifications #5-#6, and at the AMF level for #9-#12. This would change the notification tree, but still put the hardware failure as the root-cause of the entire tree:

```
#1: PLM HE alarm ( ,   )
 #2: PLM HE operational state change notification (1,   )
 #3: PLM HE presence state change notification (1,   )
 #4: PLM HE readiness state change notification (1,   )
  #5: PLM EE presence state change notification (1, 4)
  #6: PLM EE readiness state change notification (1, 4)
   #7: CLM member node exit notification (1, 6)
    #8: AMF node operational state change notification (1, 7)
     #9: AMF SU readiness state change notification (1, 8)
     #10: AMF component readiness state change notification (1, 8)
     #11: AMF component presence state change notification (1, 8)
     #12: AMF SU presence state change notification (1, 8)
```

However if CLM detects the node leaving the cluster before PLM reports the readiness status change via the tracking API, the two incidents become unrelated. This is reflected in the correlation of the notifications:

```
#1: PLM HE alarm ( ,   )
 #2: PLM HE operational state change notification (1,   )
 #3: PLM HE presence state change notification (1,   )
 #4: PLM HE readiness state change notification (1,   )
  #5: PLM EE presence state change notification (1, 4)
  #6: PLM EE readiness state change notification (1, 4)
#7: CLM member node exit notification ( ,   )
 #8: AMF node operational state change notification (7,   )
  #9: AMF SU readiness state change notification (7, 8)
  #10: AMF component readiness state change notification (7, 8)
  #11: AMF component presence state change notification (7, 8)
  #12: AMF SU presence state change notification (7, 8)
```

Seeing all these events as related, partially related, or unrelated may make all the difference whether a management entity observing the notifications takes the right action.

Our example used a scenario of some AIS services, however any notification producer process generating notifications should follow the same principles. If there is dependency between the generated notification and previously reported triggering events the process should reference the related notification identifiers.

Within the AIS services APIs that report some kind of status changes expect the user to provide some notification identifiers when calling the API. A typical example is the error report API of AMF, which assumes that whoever reports the error toward AMF also produced a notification toward NTF. Passing the notification ids in the API call allows AMF to correlate its own notifications with that produced by the error reporter.

There is also a set of notifications which come in pairs. A typical case is the alarm: An alarm is reported when the emergency situation arises and a second alarm clearing the emergency is reported when the situation has been resolved. The second alarm always references the first to indicate what situation has been cleared and it references it in a third position.

```
#22: PLM HE alarm (20, 21,  )
...
#27: PLM HE alarm cleared (20, 21, 22)
```

From all this we can see that it is the responsibility of the NTF user to correlate the different notifications, the NTF only provides the data structure to do so.

There could be several correlated notifications. Considering the tree of correlated notifications discussed above one may see that it is typically rooted in a single incident, which was reported in a single notification. This we refer as the root notification and NTF distinguishes it from other correlated notifications. In addition there is usually an immediate 'parent' node in the notification tree reporting a change that triggered one or more subsequent notifications. Not surprisingly this notification is referred as the parent notification and its identifier also distinguished in the data structure provided by NTF.

No additional notification is distinguished, but any number of them can be reported in a notification just by listing them. For example, the clearing of an alarm references the alarm being cleared in this nondistinguished position.

8.3.5 Administrative and Management Aspects

The administration of the NTF is about controlling the suppression of notifications generated in a system. It consists of two steps:

- First the administrator sets up the static filters for notifications that may need to be suppressed. These filters are configured as objects of the NTF information model. The NTF configuration classes mirror the data structure of the different notification types and allow the administrator to set for them filter expressions.
- Subsequently the administrator may activate and deactivate these filters as necessary through administrative operations.

There could be different reasons for the suppression of notifications:

Among programmers there is a tendency to provide as much information as possible about the behavior of the application or system for the case when there is a problem its root-cause can easily be identified. The line between the sufficient and too much is not well defined and different designers may draw it differently. In an open system where applications may come from different vendors this causes heterogeneity. Of course the operator of the system on which these applications are deployed also has its own ideas and goals, so through notification suppression the system can be tuned to generate only the information one is interested in.

There is however another more significant reason: Notification suppression allows one to free up some resources, which in an overloaded system may become critical in service provisioning.

The opposite is also true: When a problem persists less suppression can provide more information about the context and help in the resolution of the problem.

The precondition of efficient suppression is that the notification producers are prepared for suppression. Whenever the administrator activates a filter, the NTF informs its users, who provided the appropriate callback that they shall not produce the particular set of notifications. If so the producers will not waste resources as if a user still tries to produce a notification matching any of the active filters the NTF will drop the notification right at the source not to waste further any resources for something not wanted.

Note however that there is a difference between the suppression filtering criteria specified by the administrator and the information propagated toward the producers. In the callback to the producers only the event types can be specified while the administrator may specify additional criteria in the filter configuration. That is, to the producers NTF can indicate only the event types that are suppressed all together. If only some notifications of an event type are suppressed, the NTF itself needs to do the filtering, which means some wasting of resources.

As mentioned earlier alarms and security alarms cannot be suppressed. There is no filter class defined for these notification types. This means that an application designer needs to take special care of deciding when an alarm is really necessary. To make this decision easier it is worth mentioning that an operator would map alarms to emergency calls to or to paging the maintenance personnel to attend the situation. So even if the situation seems to be critical for the application, but it is another entity within the system that needs to take care of it then there is no need for an alarm, but a notification should suffice. An example within the AIS services is the upgrade related errors detected during an upgrade campaign. AMF generates state change notifications only which are intended to be intercepted by the SMF, which then decides when to suspend the campaign execution.

Before leaving the subject it is probably worthwhile to compare dynamic filtering and static suppression of notifications. They might be confusing at first glance as both mechanisms use filters but in somewhat different ways:

- Static suppression sets the filter at the producer end and stops the generation of the notifications matching the filter completely in the system. It is configured and controlled by the administrator. It cannot be applied to alarms and security alarms.
- Dynamic notification filtering sets the filters at the consumer end of the communication channel and it is specific to a particular subscriber. That is, the subscriber using the NTF API restricts what notifications are delivered by indicating what notifications it is interested in. So as opposed to the suppression, notifications that match the filter are delivered. All notification types can be filtered this way including alarms.

8.3.6   Service Interaction

By its nature the NTF is expected to interact with most services and applications in the system. More specifically, all entities in the system should use the NTF to provide at runtime information useful for fault management. In this role the NTF provides the means for this common functionality. The specification of each of the AIS services contains a section defining the alarms and notifications the service generates following the common data structures laid out in the NTF specification itself.

In the rest of this section we look at those service interactions where the NTF appears in the role of the service user.

8.3.6.1   Interaction with CLM

The NTF, just like other AIS services, uses the membership information provided by the CLM [37] to determine the scope of the cluster and provides the NTF API only on nodes that are members of the CLM cluster.

This single common communication channel provided by NTF is a cluster-wide entity accessible simultaneously from any node but only within the cluster. It is not provided to processes residing on nonmember nodes even if it is a configured node. Just like in case of the LOG if there are incidents that would need to be reported while the node is not a member yet, NTF cannot be used and some other mechanism needs to be provided by the platform or middleware solution.

This also means that an NTF implementation is required to finalize all associations with processes residing on a node that drops out of the membership and any initialization of the service should fail as long as the node remains outside of the cluster.

To obtain the cluster membership information the NTF needs to be a client of the CLM and track the membership changes.

This situation is, however, somewhat contradictory as both the PLM and the CLMs define notifications and even alarms. The PLM is below CLM and does not require membership at all. Similarly the CLM reports incidents and state changes related to member and also nonmember nodes.

The resolution comes from the understanding that the NTF provides a cluster-wide collection and distribution of the notifications. As long as the producer's node is within the cluster it can a generate notifications even if those report incidents and changes were outside of the cluster. That is, the node on which the CLM generates the notification about a cluster change needs to be member, but it does not – it cannot be the one which has left, for example.

Similarly the dissemination of the information happens within the cluster boundaries only, but this does not limit the use of agents to provide external connections and delivery of this information beyond the cluster.

The point is that the integrity of the service is guaranteed within the cluster boundaries only, beyond that the respective producers and consumers are responsible for it, hence the restriction.

8.3.6.2 Interaction with IMM

As for all AIS services for NTF also the IMM [38] maintains the information model and provides the administrative interface.

At system start up the NTF implementation obtains the configuration of suppression filters and deploys the active ones in the system as appropriate. It also registers with IMM as the OI for the NTF information model portion of the system model. It does so in all available roles.

From this moment on IMM will deliver to the NTF implementation any change in the filter settings and also the administrative commands issued on them.

The actual setting and the activation status of the different filters are exposed to the external world through the IMM OM-API.

The NTF has no runtime objects to implement. It exposes no information through the information model about its producers, consumers, or even outstanding alarms.

8.3.6.3 Interaction with LOG

When it comes to the interaction between the NTF and the LOG we need to look at two different aspects and also that the specification does not mandate the use of the LOG by NTF; it only recommends so.

Whether to use the LOG or any other mechanism is really a question with respect to the NTF reader API and not so much with respect to the well-known configuration log streams defined by the LOG. We have mentioned that the LOG provides two distinct well-known configuration log streams: one for the alarms and security alarms and a second for other notifications.

According to the reader API, the NTF needs to persistently record the notifications and particularly all the alarms generated in the system so they can be read back for an NTF user should it request so.

One possible option for an NTF implementation is to use the SA Forum LOG for this persistent recording. This seems to be particularly suited as the LOG is also expected to provide two configuration log streams dedicated to NTF content.

However, as we pointed out in earlier discussions, these log streams are configuration streams and therefore may be reconfigured 'at the whim' of an administrator. By that information essential for NTF reader API may be lost or the format may be inefficient for the particular NTF implementation. These streams also separate alarms from other notifications, which complicates the sequential reading as it is provided by the reader API.

For these reasons if using the LOG the use of a runtime stream set up by and tailored for the NTF implementation itself is more suitable. However since the use of LOG is optional, an NTF implementation may come up with its own solution.

The main benefit of the alarm and notification configuration log streams is that they are well-known streams, which makes them easily locatable for everyone in an SA Forum system for reading as well as for writing. This raises the question who writes to these streams.

The intention is that in the presence of a LOG implementation an NTF implementation writes alarms and security alarms to the alarm log stream and all other notifications to the notifications stream.

Alternatively the LOG implementation may subscribe to all notifications and write to the two streams appropriately. Yet another option is to bridge the services at system integration time as it is easy to come up with a dedicated application that subscribes to the NTF for the missing notifications and writes them to the appropriate stream.

In any case to avoid any duplication, applications should not log their alarms and notifications themselves but leave this task to the middleware.

8.3.7   Open Issues and Recommendation

The NTF is also one of the more mature services within the AIS therefore relatively few open issues remain after the release of its fourth version. However, many find the interface rather complex and therefore difficult to understand and use. The disliked features are mostly related to the way the API addresses efficiency with respect to data structure reuse. It is debatable whether the efficiency concern should be addressed by the API or can be left to compilers, which today are capable of doing an excellent job of code optimization. It is hard to debug and fix complex code, so a simpler API is something to consider in the future.

A more difficult and unresolved issue at the time of writing is the different service interactions: Namely, the interaction with services that do not require cluster membership (see Section 8.3.6.1), and the interaction between the NTF and the LOG (discussed in Section 8.3.6.3). These are currently left to the middleware implementation and therefore applications should not make assumptions.

We have provided some guidelines regarding the use of alarms versus notification, and the conclusion was that if the information is consumed internally by the system, notifications are more appropriate. In this case however APIs may also be a solution. The question again is which to use and when:

An API can provide a prompt reaction to an issue, but on a one-to-one basis. We could observe this in the AIS track APIs. They provide similar state change information as the state change notifications, but via an API, so that a receiver can immediately act and react to the information. In fact it can even provide a response. Note however that services providing the track APIs also produce notification as the information is important for fault management purposes.

Notifications, on the other hand, are like broadcasts, all interested parties receive the information, and they are also anonymous at least to the party producing the information. The reaction to this information is typically not well-defined; it may require 'other means' such as manual intervention, or it may not be time critical. An example of this is the use of AMF notifications by the SMF discussed in Chapter 9. In this case time is not as critical, AMF is also not aware of the reacting party. For AMF it is a 'to whom it may concern' type of announcement to which some type of administrative action is expected.

To summarize the NTF it provides a common communication mechanism primarily for fault-related incidents as well as changes that give some context to these incidents and therefore allowing for more educated reaction to emergency situations. Depending on their contents the reports may fall into different categories that require different handling: they may be correlated, raised and cleared, and so on. While NTF allows for these different functionalities it is actually quite oblivious to the notifications it delivers. It really provides the means for the notification delivery and expects the users, notification producers to use this infrastructure meaningfully.

On the consumer side it provides both push and pull mechanisms of dissemination. It pushes the notifications to its interested subscribers in a timely fashion and for those who missed any information or do not subscribe it also provides a pull interface through which the same information can be retrieved. In both cases the consumers express their interest by defining dynamic filters and therefore they receive only the information they need or capable of handling.

Static filters allow system administrators to suppress the generation of notifications and therefore limit the resource usage in overload situations, for example.

In short the NTF allows the system to report its state and solicit reaction from when the situation requires it. Next we will look at what methods the reacting party – a management application or a system administrator can use to express its reactions of handling the situation. We will continue our investigation with the IMM.

8.4   Information Model Management Service

8.4.1   Background: Issues, Controversies, and Problems

Once an alarm has been received the administrator needs to take actions to deal with the emergency situation. This may range from issuing administrative operations on different entities in the system to reconfiguring part of the system including the AIS services and also applications running in the system.

In the discussion of the different AIS services we have already seen that most of them define an information model with the idea of providing manageability. Services are configured using objects of the configuration classes of their information model, they expose the status of their entities and themselves through runtime attributes and objects, and finally administrative operations target also these objects. So the question was posed how to present all these in a coherent way toward system management: Should each AIS service expose separately the management mechanisms it defines? Or should there be a common management access point for the entire system through which the individual services can be reached too?

There was a consensus within the SA Forum Technical Workgroup that it is beneficial to provide a single entry point for all management access. A number of existing system management standards were examined to facilitate integration with existing system management applications through the north-bound interface. An initial proposal to use the Common Information Model Web Based Enterprise Management (CIM/WBEM) from the Distributed Management Task Force (DMTF) was made. However in the process of constructing the model for the services by specializing the standard object classes from the CIM schema it was found that the resulting model would be too large and complex for many practical applications. The Simple Network Management Protocol from the Internet Engineering Task Force (IETF) was also considered. While it provided a simpler interface to management systems the naming and navigation between objects exposed in the model made it awkward to reflect the natural hierarchical structure of the AMF's system model. It was finally decided that there was no suitable standard that could be used 'as is.' The team finally agreed to base the solution on the well known Lightweight Directory Access Protocol from the IETF [60] incorporating the eXtensible Markup Language (XML) [81] approach of WBEM [94]. It provided a convenient hierarchical naming structure together with a flexible object model. Thus the idea of the IMM was born. This meant the following requirements toward IMM:

It needed to be able to integrate the information model of the different AIS services, those that have been already defined at the time this decision was made and also those that would come later. This required some flexibility, which in turn would allow applications to use the same means for their own information models and management needs.

Thus such a solution could provide common management not only for the AIS services, but also for the applications running in an SA Forum compliant system.

IMM then would expose the entire integrated model to any management application or agent that requires management access to the system. This would provide a uniform view of the model and a uniform handling of its objects, their classes, and operations. The management entity would not need to be aware to which service or application process implements a particular object and how to reach this service or application. Instead the IMM becomes a mediator that dispatches management request to the service or application whose portion of the information model is being manipulated.

IMM needed to offer an easy way to integrate with the different AIS service implementations and also with applications that expose their information model through IMM. Considering the HA environment this also meant that this integration had to allow for easy fail-over and switch-over of roles between different processes of these implementations even if that occurs during the execution of a particular operation.

AIS services and applications exposing their model through IMM needed to be able to obtain their configuration described in their information model. They should also receive any modification that occurs and the administrative operations issued on the model objects.

With this respect the main issue was the preservation of the integrity of the information model in spite of the fact that multiple managers may have access simultaneously to the model. Considering a bank account to which many people have access it is like trying to keep the account in the black while these people depositing and withdrawing from the account simultaneously. It may be challenging.

Even if it is a single manager that makes modifications to the configuration, the changes may impact more than one service and/or application in the system. There is often a dependency among different configuration objects, which means that changes need coordination.

The problem with this is that a management entity may only be aware of a new configuration it wants to achieve, but it may not know all the dependencies that need to be taken into account. Moreover transitioning a live system may impose further dependencies such as the ordering of the modifications.

Let's assume again the bank account with a current balance of zero and a single person accessing it. He or she would like to transfer some amount from the account and of course to deposit the amount to cover the transaction. As long as the deposit covers the transferred amount the account is in a good the standing. To guarantee this common sense dictates that the two transactions need to be in a specific order: first the deposit and then the transfer.

Since we are talking about the same account it is easy to see the issue and resolve the problem. The situation may be similar in the information model: Dependencies may exist among attributes of the same or different objects of the configuration. The manager would need to know and take into account all dependencies to perform any intended change.

This may require a deep knowledge of the system and therefore be challenging. It would be easier if the ordering could be left to the system to resolve, that is, to bundle the transfer and the deposit as a single transaction and let the account manager figure out in which order they need to be applied to avoid penalty charges.

To leave the task to the system makes sense as services and applications in general need to be prepared to check the validity at least of their own configuration. Cross checking also easier and more consistent if it is done by the system than if it is left to an administrator who may not know all the issues; and of course the IMM needed to support all this.

In response to management operations, but also due to operational changes within the system the status of the different entities in the system may change. On the one hand side services and applications should be able to communicate and expose these changes. On the other hand, the managers should be able to collect this information efficiently. As opposed to configuration changes and administrative operations, which are considered being relatively infrequent in a system, reading the information model and particularly the status information is required frequently. Therefore the IMM needed to be bias toward the read access.

Finally, IMM needed to provide a way of inputting the information model to provide the system with an initial configuration and also to store any current configuration persistently so that when the system is restarted it can reuse it as needed. It can also be used at fault recovery to restore a known correct configuration.

In the rest of this chapter we will look at how the SA Forum solution addresses these different requirements and needs.

8.4.2 Overview of the SA Forum IMM Solution

The IMM maintains the integrated information model for the entire SA Forum system. It is a repository for the model objects.

IMM was designed with the AIS services in mind; however this does not restrict its use. Applications may use it for the same purpose as long as their needs align with the basic assumption of IMM, that is, it is primarily for read access, and modifications (i.e., write operations) happen infrequently.

By managing the information model the IMM has to interact with two groups of entities in the system:

- It interacts with the management side (also referred as northbound interface), that is, with management applications, management agents, or the system administrator. In IMM terminology they are the *object managers* (OM).
- IMM also interacts with the services and applications implementing the entities in the system that are represented by the managed objects in the model, hence they are called *object implementers* (OI).

Accordingly the IMM specification defines an API for each of these sides as shown in Figure 8.3. An object manager has the following functions:

- defining the information model by defining its object classes; this includes configuration as well as runtime object classes;
- creating and modifying the information model by creating, removing, and modifying configuration objects; all these changes prescribe for the related OIs the configuration they expected to implement in the system;
- applying administrative operations to both configuration and runtime objects;
- controlling the administrative ownership of objects; only the OM holding the administrative ownership may manipulate an object, for example, modify a configuration object and issue an administrative operation on any object;
- monitoring the system status through reading and accessing part or the entire contents of the information model.

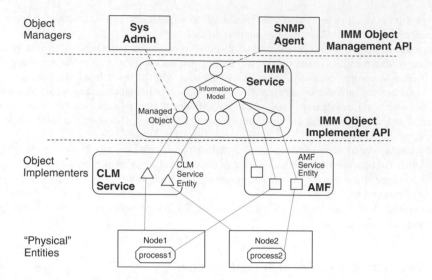

Figure 8.3   The overall view of the Information Model Management service.

On the OI side the IMM distinguishes three different roles:

The *runtime owner* is the entity in the system implementing the managed object. All objects need to have a runtime owner for deployment and each can have at most one at any given time. The runtime owner's responsibilities are:

- to update the relevant runtime information in the model: this includes the creation, deletion, and maintenance of owned runtime objects; and the update of the runtime attributes of owned configuration objects;
- to carry out the administrative operations issued by an OM on an owned object.

Besides the runtime owner configuration objects may also have *CCB validators* and *CCB appliers*. CCB stands for configuration change bundle, which is a container for *transactions* of configuration changes put together by an object manager.

These roles (CCB validator and CCB applier) are associated with the validation and deployment of such a transaction of configuration changes in the system.

Whenever an object manager wants to manipulate the system configuration it constructs the list of configuration changes that describe the difference between the current and the intended configuration of the system. Such a list composes a given content of a CCB. When it is complete the OM applies the whole CCB content as a single transaction of configuration changes. In turn IMM delivers these changes to the OI side, first to the CCB validators and subsequently to the CCB appliers if applicable.

Thus the IMM invokes first all the CCB validators of the objects involved in the given CCB transaction. The validators check whether the new-proposed-configuration is valid (e.g., the changes are consistent and complete) and whether the application of the CCB transaction is possible (e.g., if all the resources are available). CCB validators also ensure that if they accept the proposed changes then the deployment of the modifications cannot fail. To ensure this they may need to make some resource reservations or other preparations. If any of the CCB validators rejects the transaction IMM aborts the CCB transaction and the system remains in the same configuration as before and this also makes sure that all the reservations are released and preparations are reverted.

If the validation is successful, next IMM invokes the CCB appliers to carry out the transaction. They deploy all the changes necessary to implement the new configuration.

The IMM invokes the CCB appliers in a predefined order, which can be configured.

Let us consider again our bank account example. Considering that the account can be accessed by many people aka many object managers, the one that wants to carry out a transaction needs to obtain the administrative ownership of the account for the time of the transaction. Once it has done so it can go ahead and define the CCB transaction consisting of the money transfer and the deposit. It does not need to worry about the order of the operations whether the transfer or the deposit comes first. Next, it can apply the CCB transaction.

Accordingly the IMM first validates the transaction with the CCB validator, which checks that at the end of the transaction the account remains in good standing. If it is the case IMM invokes the CCB appliers to deploy the transaction. In our case we have two of them: one dealing with credit and another performing debit transactions.

To guarantee that an account can be debited it needs to be credited first; hence the applier dealing with credits is configured to be invoked first. Once the credits have been processed the applier responsible for debits can execute its task and complete the operation.

It is possible that the same entity performs in all three OI roles for a given object or even a group of objects within the information model. In fact for most of the AIS services it is the service implementation which is expected to take all three roles. The role of the CCB applier is also suitable for those entities that are interested in being alerted when configuration changes happen in the system regardless whether they need to adapt to the change or not.

After this high level overview of the IMM let's look in some details at its APIs to get a deeper understanding.

8.4.3   The Object Manager API

The IMM uses a 'minimalist' approach. The Unified Modeling Language (UML) representation used by the AIS services for their information model is significantly simplified and the IMM provides a toolbox only for the essential features necessary for system management. The IMM relies extensively on the knowledge of the OIs and OMs. It assumes that the object management side does not need to specify anything that can be known by the OI side. It also means that an OM needs to be aware of the information model. For example, considering the information model defined for the AMF, both the AMF implementation and an OM (e.g., an administrator) are expected to know the hierarchy of AMF entities: That a component can be defined as a child of a SU, which in turn needs to be a child of a service group, and so on. In other words the IMM representation of the information does not reflect class relations. Similarly, it does not reflect administrative operations, so one cannot specify in the IMM representation, for example, on which objects one may issue an administrative lock. The OM and AMF are expected to have this knowledge.

Thus, management applications having no previous knowledge about the information model typically will not be able to obtain it from the IMM to be able to manage the system meaningfully. This obviously imposes limitations on the possible use of the IMM.

8.4.3.1   Object Class Management API

On the positive side this means that the information model can be easily populated with new classes and their objects. There is no need for a complex API.

Indeed IMM provides the interface to create new object classes for the information model. To do so, an object manager indicates for each new object class the name, the category (configuration or runtime), and the attributes with their features. The attribute features are limited to:

- the type – one of the value types defined in the IMM representation, which are those common to all AIS services, if the required type is not among these common types it needs to be mapped into one;
- a set of flags – indicating the multiplicity and the handling of the attribute (e.g., configuration, runtime, cached, etc.); and
- an optional default value.

This part of the class definition can also be inquired using the OM class management API.

IMM complements each defined class with an appropriate set of service attributes that are necessary for IMM itself. These contain information such as the different OIs associated with the objects of the class, the administrative owner, and so on. Note that we used the word 'complements' as IMM also does not support inheritance even though 'classes' and 'objects' suggest an object-orientated approach. Accordingly no inheritance can be defined between object classes either.

The A.03.01 version of the IMM specification, the latest at the time of writing, allows only the deletion of object classes. It does not permit their modification, not even the extension with additional attributes, which is often allowed by similar services and therefore it is desirable to remove this limitation.

Once a class has been defined one can populate the information model with the objects of the class. Objects of configuration object classes are created, deleted, and modified via CCBs. Objects of runtime object classes are handled exclusively by the OIs.

An OM may delete an object class only if it is not used in the system, that is, it has no objects in the model.

8.4.3.2 Administrative Ownership

The administrative ownership is a kind of locking mechanism provided by the IMM that implies some exclusiveness for accessing and manipulating IMM objects.

One could think of this mechanism as a rudimentary password mechanism. An object manager wanting to manipulate some objects in the information model creates a password that represents an administrative owner. Then it asks IMM to flag the target objects with this password. As long as the requested objects have no other administrative owner associated with them, the operation is successful. From this moment on only object managers initializing the same password aka having the same administrative owner name are able to manipulate these objects.

If an object manager keeps this information for itself it will have exclusive access to the reserved objects until it releases the administrative ownership. If the first object manager shares the information with other object managers then IMM will allow all of them to manipulate the reserved objects and the resolution of any type of concurrency between these managers sharing the administrative ownership is left to themselves.

Holding the administrative ownership is required to perform administrative operations and configuration changes. It does not prevent read access, which means a reader may obtain an inconsistent view across objects if some of them are being manipulated at the same time.

8.4.3.3 Configuration Changes

To perform any type of configuration changes an administrator needs to create a CCB first (see Figure 8.4). It is a container for a set of configuration changes, which is associated with an administrative owner and an identifier. Only objects owned by this associated administrative owner can be manipulated within the CCB.

The OM may use the same CCB to compose and apply subsequent configuration change transactions. Each such transaction is a set of operations that create new configuration objects, modify, and

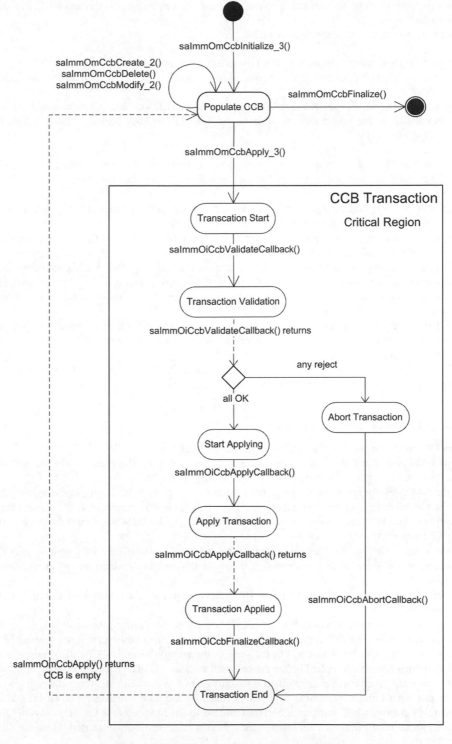

Figure 8.4 CCB action diagram. (Based on [38].)

delete existing ones. The administrative owner of a newly created object is the administrative owner of the CCB.

To create an object one needs to indicate its

- parent object indicating where the object is inserted in the information model;
- object class given by its name; the creator needs to know this name as it cannot be inquired from IMM;
- attribute values – at least the Relative Distinguished Name (RDN) value and those attribute flagged as 'initialized' in the object class need to be provided. To identify these, the class definition is obtainable from IMM.

To delete an existing object it is enough to indicate its DN. The object may have no child objects at the time of the deletion. We will reflect on the considerations that this restriction requires when the child object is a runtime object.

To modify a configuration object again the object's DN is required as well as the modification type (add, delete, and replace) and the attribute values need to be provided.

The set of changes defined in a CCB is applied to the information model as a single transaction meaning that either all the changes succeed or none of them. The IMM serializes the application of CCBs to avoid concurrent modifications of the information model potentially leading to inconsistency. As a result one may consider that the IMM specification is overly cautious and it certainly may cause performance limitations when it comes to configuration updates. Thus this restriction is based on the assumption of infrequent model changes.

Once the content of the CCB has been applied it is cleared regardless whether the application was successful or not. The CCB as a container can be reused by the same OM to construct a new configuration change transaction.

While constructing a configuration change transaction the CCB identifier can be used to read the information model as if the changes have been applied already.

8.4.3.4   Reading the Information Model

For an OM there are two ways to read the information model:

If an OM knows the DNs of the objects it wants to read it can use the accessor API and get all the attributes with their values for each of those objects using their DN.

If the OM does not know the names of the objects it is interested in, it can use the search API to let IMM find all objects in the information model that match the search criteria. If the search criterion is empty, all objects of the information model will match it and the IMM will return each one of them as the OM iterates through the model.

If the search criterion specifies the attribute name or even its value then only objects owning an attribute with the given name are returned and only if the specified value matches at least one of its values.

Depending on the specified search options IMM may return only the object names, some, or all of their attributes.

When initializing either of these APIs with a CCB identifier, IMM returns the result as if the referred CCB has been already applied to the information model. At this time however it is not guaranteed that after applying the CCB the picture will be the same as there could be other CCBs that were created in parallel and that have been applied before the one with the given id. The exception is the period when the CCB with the given id is being validated since only one CCB can be in this stage at any given time.

Note that even though these APIs have been defined on the object manager side an OI may also use them. Of course, when doing so it becomes an OM itself. In fact this is the only way an OI can obtain its initial configuration. This Janus-faced behavior is needed only at startup. Once the OI has obtained its initial configuration it does not need to use the OM-API any more and may completely rely on the OI-API as we will discuss it later.

OMs are not aware whether the objects they are manipulating have OIs at the time of the manipulation. They however may indicate to the IMM whether the configuration changes should be applied in the absence of the relevant OIs by initializing the CCB appropriately. They can indicate whether any CCB validator, any CCB applier or both may be absent at the application of the CCB transactions.

8.4.3.5   Administrative Operations

Administrative operations are applicable to both configuration and runtime objects. They also require the administrative ownership of the targeted object; however, they cannot be part of a CCB and an object cannot be targeted by an administrative operation and a configuration change transaction at the same time.

We point this out as configuration changes may require that the target object – or more appropriately, the entity it represents – is in a particular administrative state, which can only be achieved by issuing an appropriate administrative operation. That is, in such cases the OM needs to synchronize the administrative operation with the configuration change transaction.

A typical example of this synchronization case is AMF, which requires that before one is able to remove a SU and its components from the configuration the entities need to be terminated. This an administrator can only achieve by locking them for instantiation, which in turn requires a lock operation preceding the removal. Thus, the lock operation(s) needs to be issued before the delete change request is added to the CCB.

Each administrative operation requires a target object. An operation may have additional options, however very few AIS services use this possibility. When the OM issues the administrative operation toward the IMM, IMM forwards the request to the runtime owner of the target object. Only the runtime owner interprets the requested operation, executes it, and returns the result to IMM, which the forwards it to the OM.

The peculiarity of the API provided by IMM for administrative operation is related to the potentially long execution time of some operations and the HA nature of the system. If an object manager process requesting an administrative operation dies while the operation is in progress, IMM will not know to whom to return the result of the operation at the time it becomes available. Due to the HA nature however failures are handled by failovers, which means that there is normally a new process replacing the failed one that would like to receive the result. To do so, OMs initiating an administrative operation may provide an identifier with it. This identifier serves then as a token to obtain the result should the original process die during the execution of the operation. Of course the initiator process needs to share this information with its standby.

8.4.4   The Object Implementer API

The OI side of the 'IMM world' is more complex than the OM side. This is due to the intention of relieving the OMs of some of the tasks and knowledge that can be embedded in the system or more precisely in the OIs. So the intention is to put most of the burden on the OIs and not on the OMs.

8.4.4.1   Object Implementer Roles

As we mentioned earlier one of the major issue in the IMM is maintaining the consistency of the information model in the context of concurrent and interdependent changes. The third version of the IMM specification put most of the efforts into resolving this issue. Accordingly, it has introduced the three different OIs roles: the runtime owner, the CCB validators, and CCB appliers.

The need for the CCB applier role is most straightforward. As we have seen on the OM side, IMM provides no alert mechanism that a configuration object has been modified in the information model. A CCB applier is a process, which needs such an alert. The reason a process would like to receive an alert is usually because it needs to adapt somehow to the change. Hence it applies some of the changes introduced by the CCB to the system – it is a CCB applier.

Now if there are some constraints to what extent or in which cases this imaginary process can adapt to the change then it would want to make sure that the changes remain within those constraints. To achieve this it needs to validate first the proposed changes, that is, it needs to be a CCB validator. In the CCB validator role a process gets a 'voting right' whether the change is acceptable or not. It also gets the obligation to prepare for the change so if it gets accepted by all voters, that is, all CCB validators the CCB appliers indeed can go through and deploy the change.

The runtime owner is the process which maintains the runtime information presented in the information model. This reflects toward OMs the runtime status of the deployment. The runtime owner also carries out the administrative operations issued by an OM on a managed object.

The runtime owner typically would need to be a CCB validator and a CCB applier as well. However IMM does not imply this. OIs need to register with IMM for the roles they want to take up with respect to a given configuration object or an entire configuration object class.

In case of runtime objects the situation is much simpler: They only have runtime owners and it is the one who creates the object.

8.4.4.2   Object Implementer Role Registration

Because we are looking at HA systems we need to consider fail-overs and switch-overs with respect to the OI roles. To allow these the IMM specification defines an indirect registration process:

First of all we select an OI name to represent a process that will need to fulfill different OI roles for a set of objects. Then this name is set for these different objects in the intended OI roles. This can be done by setting the name for individual objects or for entire object classes.

Subsequently when a process capable fulfilling these OI roles wants to take up the roles for all those objects associated with the name, it only needs to registers for the name and with this registration it obtains all the appropriate OI roles for all the associated objects.

If the process fails unexpectedly the new process replacing the failed one needs to do the same: register for the OI name. If a process needs to turn the roles over to another process it can clear its own registration for the OI name after which a new process can register itself for the name and thus obtain the roles the OI name represents.

When an error occurs that an object has no OI that typically means that there is no process associated with the appropriate OI name. It is less likely that the OI names have not been set as the IMM attributes storing this information are persistent runtime attributes for all persistent objects (i.e., configuration and persistent runtime objects).

8.4.4.3   CCB Validator and CCB Applier API

When the OM constructing a configuration change transaction added all the required changes to the CCB it applies the CCB. This is the moment that the CCB validators and CCB appliers step in.

First the IMM calls back the CCB validators for all the objects involved in the CCB. This means the objects being manipulated, and if necessary their parent objects as well. For example, when an object is created the CCB validator for the parent object is invoked as there could be limitation on what type and what number of children it may have.

Each CCB validator receives only one callback per CCB transaction providing the CCB identifier only. The CCB validator needs to find out itself how many of its objects and in what type of changes are involved in the current transaction.

To find out this information the IMM provides a CCB iterator API on the OI side. For a given CCB id this API is available only from the moment the object **manager** applies the CCB and only until this function call returns.

The API is similar to the search API discussed on the object management side except it provides access only to the model changes of the given CCB transaction (i.e., the objects created, deleted, and modified).

The CCB validator needs to evaluate all the changes for all objects it is responsible for and if it finds them appropriate it needs to make all the preparation to ensure that the CCB can be applied. This may mean cross checking with other objects in the model or entities in the system and may go as far as resource reservations.

Once the CCB validator made sure that there is nothing on its side that would prevent the deployment of the CCB transaction, it can accept the CCB by returning an OK. Otherwise it returns an error and the CCB cannot proceed.

If any of the CCB validators responds with an error or does not respond within the expected period, the CCB transaction is aborted by the IMM. This means that the IMM invokes again all involved CCB validators to inform them about the abortion so that they can release any state or resources held for the perspective of the deployment of the transaction.

When the IMM receives an OK from all validators invoked for a CCB transaction, it proceeds by invoking the CCB appliers to deploy the configuration changes. IMM invokes the CCB appliers in the order of their configured ranks starting with the highest ranking appliers and invoking the next rank only after the first group has completed the operation and returned from the call or had enough time to do so.

Even if an applier does not respond to the callback (i.e., the associated timer expires before the callback completes or the applier process dies), the IMM proceeds with the deployment. The CCB transaction cannot fail after the CCB validators accepted it. All changes are applied at least to the information model and an applier may deploy them later when it becomes available again.

8.4.4.4   Runtime Owner API

Runtime owners maintain the runtime objects and attributes of the information model and also execute the administrative operations issued by OMs.

For the first task IMM provides an API that allows the creation and deletion of runtime objects; and the update of runtime attributes of any object let it be configuration or runtime. These are similar to the creation, deletion, and modification of configuration objects except that they are not bundled into transactions since only the runtime owner has write access to runtime objects and attributes and there is always at most one runtime owner for each object.

The nuance of the attribute update is that it may be performed in two ways: Either the runtime owner updates the attribute each time its value changes – these are the cached attributes – or the IMM asks the runtime owner to perform an update to satisfy an OM's request of obtaining the value of a non-cached runtime attribute. Whether a runtime attribute is cached or not is defined in the object class definition.

The IMM also calls back the runtime owner of an object if an OM issues an administrative operation on the object. Since the information model does not have any information on the applicability of administrative operations, it is the runtime owner which verifies whether the operation is valid for the particular object or not. If so, it executes the operations. In either case the runtime owner provides the IMM with result that in turn the IMM passes on to the OM, which requested the operation (or asked for the result).

The 'result' that the specification allows to pass through is simply an error code indicating whether the operation was successful or not. It does not allow the passing of any additional values, which in some cases could be beneficial.

That is, the IMM is quite ignorant about the administrative operations and only provides a very simple pass through mechanism between the OMs and the OIs.

However, even this simple mechanism allows for the switch or fail-over of the OM role requesting an administrative operation by providing an operation identifier, which can be used later by another process to obtain the result of the requested administrative operation.

8.4.5 IMM XML File

The IMM specification defines the OM-API and OI-API as part of its user interface. In addition it provides a standard XML schema, which defines the XML format to represent the content of the system information model. Any time the IMM is asked to export the content of the information model it stores the appropriate data in a file using this format. Such an XML file can also be given to IMM as an initial configuration of the system.

The schema defines two main schema elements: One for class definitions and another for object definitions. The sub-elements of the class definition describe the features of the class attributes; while sub-elements of the object definition assign values to the class attributes.

8.4.6 Administrative and Management Aspects

The IMM itself uses two configuration object classes: One for configuring the service itself and the second for configuring the CCB appliers. The single object of the first class allows an administrator to define:

- whether at startup IMM should use its internal repository or an XML file containing an initial configuration for the system;
- the timeout period to be used with CCB validators and CCB appliers.

In addition the object also provides different status information such as:

- the time of the last configuration update in the model;
- the current number of initialized CCBs, OIs, and administratively owned objects;
- the uniform resource identifier (URI) for the XML file where the IMM content was exported.

The IMM specification defines only one administrative operation, which is applicable to this same object. As a result of this administrative operation an IMM implementation exports the persistent content of the information model to a file with the path name given as a parameter to the administrative operation. Obviously, this also updates the URI status attribute mentioned above.

The second configuration object class is to configure CCB appliers by their name and rank. The rank determines the order in which IMM invokes them when applying a CCB transaction as we discussed in Section 8.4.4.3. This configuration is not exhaustive, that is, it is not necessary to list all CCB appliers in the system. CCB appliers that were not configured explicitly are invoked last after all configured CCB appliers.

8.4.7 Service Interaction

All the AIS services that have an information model use the IMM for storing their respective information model and to expose their administrative API, which allow the manipulation of the information model.

Each of the AIS services acts as the OI for its part of the system information model. That is, it registers with IMM as the OI for the object classes representing its entities.

Typically an AIS service takes all three OI roles for its configuration objects. Accordingly IMM calls the service back in the CCB validator role when any of its objects is part of a CCB that is being applied by an object manager. If the validation was successful, IMM also invokes the service in the CCB applier role. The application of the CCB may also result in status updates of the object(s), which the AIS service performs in the runtime owner role.

As a runtime owner a service is responsible to update all the runtime attributes and also create and delete runtime objects; as well as execute the administrative operations invoked on its objects. Runtime objects have only runtime owners.

However before any of the OIs can implement its configured objects it needs to take up the role of an OM to obtain the initial configuration as only the OM-API provides read access as we have seem in Section 8.4.3.

8.4.7.1 Interaction with CLM

The same ways as other cluster-wide services the IMM APIs are also available only on cluster member nodes. This means that an IMM implementation must use the CLM API to determine the scope of the cluster at any given moment in time and process user requests accordingly. That is, it needs to use the CLM track API.

On the flipside, the CLM as well as the PLM use IMM to store their respective information model. This creates a chicken and egg situation. The IMM specification leaves it to the implementations to resolve the issue and does not mandate any particular solution. One possibility though is to store some of the PLM and CLM configuration locally and use it during cluster formation until the node becomes part of the cluster. Once the cluster has formed, IMM becomes available and the implementation can adjust to the most up-to-date configuration available in the information model.

8.4.7.2 Interaction with NTF

The IMM does not generate any alarms, but it does generate notifications and uses NTF to communicate them to all interested parties.

Rather than the individual AIS services, it is the IMM which generates the notifications to indicate the start and the end of administrative manipulations such as administrative operations and configuration changes. This centralization is also geared toward the consistent handling of these aspects of the SA Forum system management.

Whenever an object manager invokes an administrative operation on an object in the information model, IMM generates the associated notification. It does the same when it returns to the OM the result received from the OI.

In both notifications IMM provides the name of the object targeted by the operation, the id, and the parameters of the administrative operation, and any correlated notification id as appropriate. On that note, an administrative operation end notification is always correlated at least with the administrative operation start notification.

IMM also generates notifications for CCB transactions: It generates a configuration update start notification when an OM applies a CCB; and it generates the corresponding configuration update end notification when the apply call returns. At this point the CCB might have been deployed successfully or aborted. In any case the notification contains the CCB identifier and the error code returned to the OM.

It is important to realize that the CCB identifier used in the notification is more of a session identifier and it does not uniquely identify the particular CCB transaction. On the other hand, the two notification ids generated by NTF at the beginning and at the end of the CCB transaction are unique for the CCB transaction. Moreover IMM returns to the OM the notification identifier generated at the end of the transaction as an out parameter to the OM's apply call. In addition the same way as for the administrative operations, IMM correlates the configuration update start and end notifications.

8.4.8   Open Issues

The IMM specification is a relative latecomer to the set of AIS specifications. The version in effect at the writing is A.03, which means that there were only a few revisions so far. In particular, the first two versions of the specification did not distinguish between the different OI roles. They were introduced by the A.03 version, which means that this most complicated portion of the API had no revision yet. Some question its maturity.

However the open issues are not limited to this newest portion of the specification, but also touch other portions.

One of the most debated issues probably is the separation between the north- and the southbound interfaces – in IMM terms between the OM-API and the OI-API.

As we have seen in Section 8.4.4 the OI-API does not provide the capability of reading the content of the information model. Any OI that needs to obtain its initial configuration has to become at least temporarily an OM to read its configuration.

The SA Forum Technical Workgroup came to the conclusion that it is mostly the question of taste whether complete separation exists between the two interfaces as it introduces the duplication of the functionality. In the attempt of keeping the APIs simpler for the time being the choice was not to repeat essentially the same API on both (OM and OI) sides if it was not necessary.

Other issues are also stem from this minimalist approach. Namely the class management API allows the creation and deletion of classes, but not for the modification. This means that whenever a class of the information model is upgraded all its objects need to be deleted first to be able to delete their class. Only after this the new version of the class can be created and repopulated with the earlier deleted objects now adjusted to the new class structure.

Many see this procedure as rather cumbersome and avoidable particularly when the new version of the class only extends the existing class with some new attributes. IMM implementations could handle such an update easily and efficiently even in a transactional manner, if it was required by the specification. However this implies additional class handling functions not present in the current API. Considering the life cycle of HA systems this issue demands a rather urgent resolution.

Another issue that needs short term resolution is the restriction that objects cannot be deleted if they have child objects. The reason for the urgency is that this means that applications effectively are able to 'freeze' the configuration by appending runtime objects as children to configuration objects. The IMM will not allow an administrator to delete any of such configuration objects unless the runtime objects are removed first. However there may not be a way or it may not be obvious how to make the OI to remove those runtime objects.

This shows that application designers need to pay particular attention when designing the information model of their application. Most of the AIS services give a free hand of inserting the runtime objects as most appropriate for the application rather than defining the relevant insertion point.

The issue has been discussed by the SA Forum Technical Workgroup and there is an agreement to allow an OM (e.g., the administrator) to delete from the information model configuration objects even if they are parents to other objects. This change will allow – albeit indirectly – the administrative deletion of runtime objects such as runtime log streams which contradicts with the current definition of the 'runtime' category and it is a paradigm change.

At the time of the writing it has not been decided yet how this will be handled on the OI side. The problem is that none of the AIS services that define runtime object classes is prepared for the situation that such an object is deleted by an administrator. Currently they are in full control of their runtime objects. This means that neither the IMM OI-API nor the different AIS service APIs define the appropriate functions. Considering a runtime log stream, for example, it is a big change in the service philosophy that such a log stream can cease to exist while being in use by any of the service users.

The last issue that we would like to address in our discussion is related to the serialization of CCB transactions. Namely, that IMM performs the validation and application of CCB transaction one at a time.

This issue is related to the basic assumption made at the outset of the definition of the IMM according to which IMM is used primarily for read operations as configuration updates occur infrequently. Changing this paradigm requires significant changes to the whole service definition, which is not desirable at this time. It requires more justification supported by deployment experience.

Obviously the bigger the information model an IMM implementation manages more restrictive this limitation may become. It may become necessary considering that the different CCBs may manipulate unrelated parts of the information model.

However today there is no definition what 'unrelated' means for CCBs. The problem is that even though an object manager may be manipulating only object X in the model, the manipulation may be based on reading objects Y, Z, and W. There is no way to know this relation. The traditional database handling paradigm would require a lock on all four objects accessed in such a case. Within IMM a similar effect can be achieved currently by the object manager obtaining the administrative ownership of all the additional objects. However IMM does not require this and another CCB may manipulate any of Y, Z, or W compared to the time of the reading. Even the serialization does not help in this case. It would be the CCB validators' task to catch such an inconsistency.

Note that since the CCB would include only object X, an OI (CCB validator or applier) would need to use the OM-API to check these objects, but due to the serialization during the validation phase the immutability is guaranteed.

What may become more critical is that some systems may need to handle embedded CCBs. That is, considering the previous example the modification of X included in a CCB would require not just the reading, but the modification of Y, Z, and W, which are not part of the current CCB. Typically this would be solved by nesting a new CCB into the current one, and it also requires that the current CCB should succeed only if the nested CCB succeeds. While obtaining the administrative ownership of Y, Z, and W can guarantee that they can be modified in a subsequent CCB as necessary, this mechanism cannot guarantee the rollback of both CCBs should any of them fail. This would require substantial effort on the CCB validator of object X.

8.4.9 Recommendation

Considering all the aspects of the IMM the most important recommendation is to use it only for cases when the basic assumption is true, that is, the model is mainly accessed for read and write operations happen relatively seldom.

Application designers using the AIS service APIs that trigger runtime object creation need to consider the consequences of inserting these objects at different places in the information model. The tendency may be to attach them to the configuration object they are in relation with. This may be appropriate if the administrative manipulation of the configuration object controls the lifecycle of the runtime objects. For example, if the administrative locking of the AMF SU, which triggers the removal of any CSI assignment from its components also removes the runtime objects. Therefore changing the configuration of components within the SU becomes possible as they will not have attached such runtime objects, which is one of the typical reasons why a lock operation would be issued, for example, during software upgrades.

One of the reason why using the component as a rooting point for runtime objects becomes tempting is that its name can be obtained via the AMF API at runtime. In reality, however, the related CSI name can also be collected from the API callbacks even if it cannot be obtained by solicitation.

A CSI may serve as a better insertion point for service related runtime objects, such as a checkpoint associated with the service state of the served CSI. Although this may not be suitable for all cases.

Sometime the runtime objects' life cycle cannot be linked to any other object than the object representing the service or the application itself. This is the path the AIS information model recommends for many service specific configuration classes. For example, it is recommended that configuration log streams are rooted in the well-known AMF application object representing the LOG itself.

When no other option is suitable, objects can be placed at the root of the information model. However this is least recommended as it 'pollutes' the information model.

8.5 Conclusion

In this chapter we reviewed the AIS services that allow the control, management, and monitoring of an SA Forum compliant system.

In particular we looked at the IMM, which combines the information models of the different AIS services and applications into a single model to expose it to management entities (e.g., system administrators, management applications) in a coherent and consistent way. Using the IMM OM-API these management entities may monitor, configure, and otherwise manipulate the SA Forum system through monitoring and manipulating the objects in the information model. The IMM dispatches these actions to the OIs registered as responsible for the manipulated object. Object implementers are the implementations of the different AIS services and applications that expose their objects through the IMM.

As a result the IMM is the primary management interface to an SA Forum system and its applications. Its specification defines the OM-API and OI-API together with the information model that configures the IMM implementation itself. In addition the SA Forum also defined the XML schema to describe the content of the IMM information model, which can be used to provide the system with an initial configuration or to export the current configuration of an SA Forum system.

Through exposing the information model, the IMM provides an up-to-date view of the system, however it does not collect data over the system life-cycle that would suitable for the offline analysis of the system behavior, nor it is capable of alerting the system management whenever an urgency requires so.

To address these issues in addition to the IMM, the AIS provides API definitions for the LOG, and the NTF.

The LOG is tailored for collecting information about the system behavior over relatively long periods of time, even its lifetime. This means that the information gathered by the LOG is higher level information suitable for system management for the analysis at the system level rather than trace data that would allow the detailed analysis and debugging of specific applications and software components.

The LOG defines an API through which users write their log data to the appropriate log stream without needing to be aware of the formatting and other needs of the output file or files. The output formatting and other configuration features of the log files defined and managed separately, for example, via the IMM and the LOG deploys these changes transparently for the users using the impacted log streams.

The LOG provides API only to collect information; it does not provide API to access the collected data. The log files need to be harvested together with their associated configuration files for correct interpretation. On the other hand, using these means anyone shall be able to interpret the collected information within or without the system.

As oppose to the LOG's long-term data gathering, the NTF allows its users to dispatch alerts about significant changes in the system that potentially require immediate reaction from parties other than the dispatcher. This receiver could be another SA Forum service, an application, but most importantly an administrator – a person responsible for the system maintenance who may need to perform some manual intervention. For cases when such manual intervention may be necessary the NTF defines the categories of alarms and security alarms, while for cases that are expected to be handled by other components of the system or serve informational purpose the specification provides different categories of notifications such as state change, attribute change, object create and delete, and miscellaneous notifications.

Accordingly for entities dispatching notifications and alarms the specification defines the producer API, while entities monitoring these reports use the reader and/or subscriber API of the NTF.

Both the LOG and the NTF provide different filtering possibilities that allow users and administrators to protect against overload situations and also to customize the information collected or the delivered to their needs.

In summary the three services presented in this chapter define a consistent and coherent management view and access to an SA Forum system, its services as well as its applications using these services.

When the systems' expected in-service performance is five '9s' or above system management becomes a deal breaker. However managing such complex systems in an open and expandable way is a complex task by itself. Therefore the uniform handling of different system management aspects allows one to focus on the actual issues at hand. While this feature is essential in HA systems, it may also be beneficial in other complex systems with less stringent requirements.

9

Model-Based Software Management: The Software Management Framework

Maria Toeroe
Ericsson, Town of Mount Royal, Quebec, Canada

9.1 Introduction

The key expectations toward a system providing Service Availability (SA) is that its services are available to their users virtually any time, be it at 3 a.m. on a Sunday night or 3 p.m. on Thursday afternoon. While a couple of decades ago only few services had to satisfy this requirement, today with the Internet becoming the infrastructure for all aspects of life the possibility of shutting down the system for maintenance and software upgrades has long gone. Moreover with the Internet's global reach even the notion of slow business hours is fading away quickly: 3 a.m. in North-America means 3 p.m. at the other side of the globe with all its urgency that is not localized any more.

Nevertheless the maintenance and upgrades still have to be carried out somehow at some point and the Software Management Framework (SMF) [49] of the SA Forum offers an approach to resolve this conflict.

As we have discussed in Chapter 6 the Availability Management Framework (AMF) [48] is already designed so that it is capable of maintaining SA in the presence of a failure. The solution is based on the redundancy introduced to the system so that services provided by a failing component can be failed over to another healthy component standing by for exactly this purpose. Luckily just because we introduced this redundancy our component should not fail more often, so this redundant standby is there for most of the time 'idling,' waiting for something to happen that requires it to take over the service provisioning. It is also clear that it does not have to be a failure that causes the take over. AMF also provides administrative operations that can be used to rearrange the assignments. For example, if we would like to perform some maintenance on or upgrade the active component we can take advantage of this redundancy as well and use administrative control for coordination. Thus, we do not need a completely new solution; we only need to complement the existing one – AMF – so

Service Availability: Principles and Practice, First Edition. Edited by Maria Toeroe and Francis Tam.
© 2012 John Wiley & Sons, Ltd. Published 2012 by John Wiley & Sons, Ltd.

that we can handle the additional requirements of coordinating upgrades and reconfigurations. The SA Forum Technical Workgroup has defined the SMF exactly with this intention in mind.

In this chapter we look at the solution offered by the SMF, which achieves its task in tight collaboration with AMF and Information Model Management service (IMM).

The first version of the SMF specification revolves mainly around two aspects of software management that we are going to discuss in details: The software inventory and the upgrade campaign.

The chapter introduces the entity type which is a key concept related to the software inventory. We have already mentioned at different Application Interface Specification (AIS) services that define entities that are represented in the information model. We also indicated that these entities often have types that reflect a grouping of these entities based on their similarities. The primary source of these similarities is that these entities run the same software, which leads us to the domain of software management. We will look in more details at the notions and requirements SMF introduces toward software vendors to describe their software intended for a SA Forum compliant system and how this information can be used to derive the different entity types.

The chapter also introduces the method SMF utilizes to deploy new software and new configurations. Its key notion is the upgrade campaign that the SMF specification defined to be able to describe for an SMF implementation the target of the upgrade and the procedures to be used to achieve this target. It is reflected in the information model to provide a way to monitor the execution.

The discussion covers the failure handling mechanisms required by the specification during upgrades, their basic concepts, and failure case coverage. It discusses the additional measures that may need to be taken by applications and a middleware implementation.

Finally we indicate the areas of software management not covered by the first version of the SMF specification that need to be dealt with in future releases or by other means.

9.2 Background

The first question is what we mean by software management in the context of SA. Is it any different from software management in general? Even those who have an idea about software management in this context may still wonder what it has to do with some application programing interface (API) specification such as the SA Forum AIS.

There might be different interpretations, but software management typically covers the software life-cycle from the moment of the completion of its development – software delivery – to the moment it is deployed in a target system. That is, it normally deals with the packaging, the versioning, and the distribution of the software product on the software vendor side; and on the customer side, with the installation, verification, configuration, and finally the deployment of the software on the target site where it is used for the services it can provide to its users.

There are a number of solutions even standards covering the first part (i.e., packaging, versioning, and distribution). Many of these solutions address to different extent even the second part.

From the perspective of SA this second part is the critical one as to deploy a new version of some software usually means that the instance running currently the old version of the software is terminated and replaced with an instance running the new software version. Obviously, during this replacement the instance terminated cannot provide any services. The provisioning needs to be switched to another redundant instance in a similar manner as in the case of failures discussed at the AMF in Chapter 6.

The redundancy introduced to protect against failures in systems built for SA can be used in other context as well. It does not matter from this perspective whether a service provider entity is taken out of service for the time of its upgrade or it went out of service due to an error and it needs to be repaired before it can take assignments again.

Yet, the problem is not exactly the same.

The notable difference between failure handling and software upgrade is that while at repair the same software is restarted, in case of an upgrade (or downgrade) a different, a new version of the software is being restarted and in case of redundancy this new software needs to work together with the old version at least for some time.

So the issue of compatibility comes into the picture.

Compatibility issues are of course not completely new and unique to systems providing SA. Today's systems are built from software components coming from different vendors. System integration can be a challenging task as all these software and hardware components need to be compatible and work with each other flawlessly to compose a system with the intended characteristics and features.

Unfortunately it is virtually impossible to guarantee this flawlessness as the essence of software solutions is their flexibility with which comes the huge state space a running instance of the software can be in at any moment of time. Putting together two or more such instances explodes the state space very quickly and only a finite number of combinations can be tested in finite time.

An attempt for the solution of compatibility is the standardization of interfaces through which different software pieces interact. In any case, it creates a dependency as if an interface changes both parties collaborating through that interface need to align. This means that if one piece of software is upgraded to the new interface, the other one needs to follow otherwise they will not be able to collaborate. This in turn means that the scope of the upgrade may need to be widened to resolve all compatibility issues and instead of the upgrade of a single software piece the complete vertical software stack needs to be upgraded.

With redundant applications and systems the compatibility issue expands yet into another dimension. Now software components that are peers and working together to protect a service (i.e., the same functionality) may run different software versions yet need to be compatible to collaborate properly. It is as if the horizontal dimension of these peers was added to the vertical dimension of the stack of software pieces collaborating to achieve certain functionality.

Another aspect that needs to be taken into account at deployment is that different pieces of software may impact each other just by being collocated, because they use the same resources. All of us using computers are familiar with the popup window that tells us that the upgrade will take place only after a system restart. Well, this means that not only the services provided by the software entities targeted by the upgrade, but all service of the entire computer will be impacted by the restart operation and only because their software is executing on the same computer.

Yet the biggest challenge in case of SA is that during upgrades and system reconfigurations there are so many things that may go wrong. Murphy's Law applies many many times. There are synchronizations that need to be fulfilled because of dependencies in addition to those normally present between the peers protecting the services. There could be errors, bugs in the system introduced by the upgrade itself or happening as part of the 'normal' operation. Issues may be the result of given combination of software that might have eluded testing, which is not that difficult after all considering the combined state space of the system.

In short, the software management solution for SA system needs to be able to recover from conditions that are beyond normal operation; and it needs to be able to do so with the least service impact.

Finally to top all these challenges is the fact that upgrades are very software specific: Meaning that different pieces of software may require different solutions for optimal result. There are a lot of specificities at the application level that need application level coordination, that require application level knowledge, and therefore hard or impossible to handle from the system or middleware level being transparent to these lower layers.

The first version of the SA Forum SMF specification could not possibly address all these issues in a limited time with limited resources. Instead the SA Forum Technical Workgroup defined a limited scope that would enable the addition, the removal, and the upgrade of AMF applications and in particular the addition or removal of partial or complete applications as well the upgrade of the redundant logical entities of running applications.

In spite of this limited scope the concepts defined in the specification were designed so that it will be possible to expand them to the entire system potentially supporting even hardware upgrades and reconfigurations.

In the rest of the chapter we present the software management solution offered in the SA Forum SMF specification and we also list the issues that this first version does not address.

9.3 Software Management a la Carte

9.3.1 Overview of the SA Forum Solution

The SMF is one of the latest additions to the SA Forum AIS. As the previous section showed there were great challenges – a big shoe to fill. So the decision was to proceed stepwise and the primary goal of the first version of the specification was to facilitate the addition, removal, upgrade, and reconfiguration of applications while maintaining SA.

Since in SA Forum systems the AMF is responsible for maintaining SA and applications providing such services are expected to be AMF applications, the solution offered in the current SMF also takes into consideration for SA only AMF applications. As a result it heavily relies on the AMF itself and the AMF concepts in general. SMF builds around the AMF information model, the features associated with the represented AMF entities and the AMF administrative API.

However, since in the long run the solution should not be limited to AMF entities only the main concepts of the SMF were defined at a higher abstraction level and therefore they are more generic.

The SMF specification defines the concept of the *deployment configuration*, which is the collection of all the software deployed in a system as well as the logical entities configured through the information model. Accordingly it identifies two areas of software management:

The first area is concerned with the software inventory, the delivery of new software to the system, the removal of any obsolete one, and the description of the features, capabilities of the software, and its configuration options. The SMF specification identifies this area as the *software delivery phase* and the related information model as the *software catalog*.

The second area of concern is the act of deploying a new configuration of entities in a live system which may remove, replace existing, or add some new entities or just rearrange, reconfigure existing ones, and any combination of these options. The specification refers to this as the *deployment phase*.

Note that when existing applications are being reconfigured and rearranged in the deployment phase there may not be always a need for a preceding delivery phase if there is no software involved (e.g., reconfiguration or removal of entities); or it may require only the installation and/or removal of the software on certain nodes within the cluster using an image already delivered to the system and stored in its *software repository*.

The assumption is that logically there is only one software repository in an SA Forum compliant system and the images in this repository can be used at any time for software installation, validation, repair, and removal as necessary. The 'unit' of these images is the *software bundle*. The validity, consistency and integrity of software bundles are checked when the bundles are delivered to the repository.

However the first version of SMF does not define operations for this check, delivery, or any other operation concerned with the repository. Neither does it define the format or the exact content of software bundles. The reason for this is that these are platform and therefore implementation specific details and most platforms already have gone through the definition of them such as the widely used for Linux systems RPM Package Manager (RPM) format (formerly known as Redhat Package Manager) and related utilities standardized in the Linux Standard Base specifications [88]. It would have been a huge effort to define these and it would have been just a re-inventing of the wheel.

So the focus was shifted to the issues that are specific to SA. In this respect there were only one item identified in the area of software delivery, and that was the description of the software capabilities and configuration options. For this purpose the SMF specification includes an XML (eXtensible Markup Language schema [81, 89] to be used by software vendors to describe their products intended for SA Forum compliant systems and in particular for software to be managed by AMF. We will look at the *entity types file* (ETF) [90] and its role in more detail in Section 9.3.2.

While the software inventory is purely the concern of the SMF, the logical entities that need to be configured and upgraded are of the concern of other AIS services such as the AMF or the Platform Management service (PLM), or even user applications – considering that the IMM [38] is not restricted to AIS services only.

The expectation is that all these entities are somehow reflected in the information model as (managed) objects, which define the configuration of the entities that the system should implement.

This is a key point: The SMF focuses only on the configured content of the system information model, which is manipulated during upgrades and reconfigurations. The runtime content should follow this as the consequence of the configuration changes being deployed.

One may also realize that even though in the information model there might be seemingly independent objects that are implemented by different AIS services and applications, they are not necessarily independent. They may reflect different aspects of the same logical entities in the system.

For example, an AMF application is described through components and service units (SUs), and so on, for the purpose of the availability management performed by AMF. In the actual system these components may be processes that are started and stopped by AMF using the component life-cycle commands and the AMF API as described in Chapter 6. These same processes may communicate using message queues of the Message service [44], which in turn implements the queues themselves. However their creation and deletion is controlled by the application processes (started as AMF components) and they may potentially use some application specific configuration information, which can be part of the system information model as configuration objects the same way as the AMF components and SUs.

Obviously, changes in the software the components are running may require changes in this application specific part of the configuration as well, and all these changes need to be made in a synchronized way during an upgrade while maintaining SA during and of course after the upgrade.

To deal with the problem at hand the SMF specification defines the concept of the *upgrade campaign*, which is composed of *upgrade procedures* that may or may not be ordered Figure 9.1. Each of the upgrade procedures consists of one or more *upgrade step*s typically executed in sequence. Each of the steps upgrades and/or reconfigures a set of entities by performing the necessary software installations and removals and the information model changes in a synchronized way. We will go into more details on the upgrade campaign specification in Section 9.3.3 and its execution in Section 9.3.4.

To perform an upgrade or reconfiguration an SMF implementation expects an upgrade campaign specification. It is an XML file that follows the *upgrade campaign specification* (UCS) XML schema [90] accompanying the SMF specification.

The upgrade campaign specification relies on templates as shorthand to identify existing target entities in the system. At the same time they also allow for the specification of generic upgrade campaigns applicable to different deployments that an SMF implementation can tailor to the actual configuration of the system at runtime.

When executing an upgrade campaign the system administrator would like to see a completely different perspective: He or she would like to know exactly what is being upgraded, how the upgrade is progressing and if something fails where it has happened in the sequence of operations of the upgrade campaign and physically in the system.

There is also a need to recover from failures as much as possible automatically and with as little as possible service outage and loss of any kind.

To address these needs the SMF specification defines also an information model and associated state models for the upgrade campaign that reflects the status of the campaign execution.

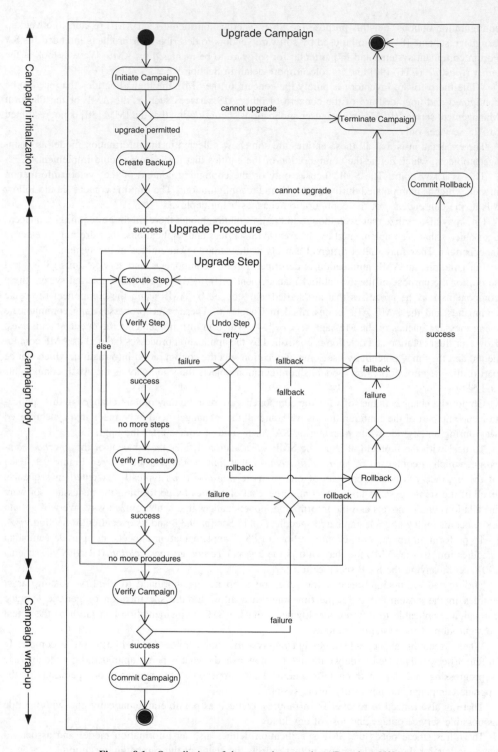

Figure 9.1 Overall view of the upgrade campaign. (Based on [49].)

Through the state models, SMF also identifies failure scenarios that may occur during campaign execution and their handling options that an administrator can choose from should an error be detected. Section 9.3.4 goes into all these details at some length.

9.3.2   Entity Types File: Is It Eaten or Drunk by SMF?

Vendors are expected to deliver their software intended for SA Forum compliant systems as software bundles and describe the content of each software bundle in an associated ETF. It is an XML file following the standardized XML schema that allows the characterization of the software in AIS terms. That is, according to the features the different AIS services define for their entities. The first release of the specification focused only on the AMF entities and therefore the schema also contains only elements for them; however, the approach is extendable to other services and even applications as needed.

We have seen in Chapters 5 and 6 that both PLM and AMF define types in their information model that characterize groups of entities in the system, which run the same software and have similar configuration. As a result these types serve as a single point of control for these groups of entities from the perspective of AMF and PLM. It also means that these types fully define all configuration features required for the group of entities they characterize. The main concepts of this view defined by the SMF basic model presented in Figure 9.2.

When it comes to the software itself, however, it is usually more flexible and can be deployed in a variety of ways. For example, we have seen that the AMF needs to know how many component service instances (CSIs) it can assign to a component in the active and the standby states so for the component type we can configure that all components of the type can accept two active assignments or five standbys. Configuring two actives or five standbys in the AMF information model does not necessarily imply that this is all that the software is capable of. It may only mean that this is the configuration that AMF needs to apply at runtime due to other limitations. One may restrict a configuration to limit the load, to synchronize the collaboration between different pieces and for many other reasons. A software that can handle multiple active and standby assignments at the same time may also be used in configurations that require only active assignments (e.g., N-way-active redundancy model), or either active or standby assignments, but never both at the same time (e.g., 2N or N+M redundancy model).

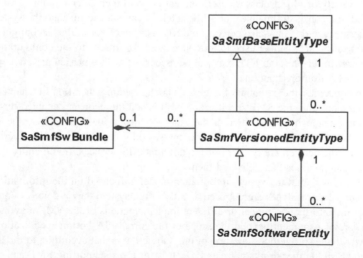

Figure 9.2   The SMF basic information model [49].

To characterize all such possible (AMF) types may be impossible; instead the SMF introduces the concept of the *software entity prototype*. A prototype does not have to be fully specified like the entity types of the AMF information model, for example; it only needs to indicate the limitations built into the software implementation so that proper types can be derived from them.

If a feature is not specified in the prototype description, it means that the software has no limitations in that respect. For example, the prototype of the above component may not have limits for the number of active and standby assignments at all, which would mean that the software implementation can handle any number of assignments; or the prototype may limit the active assignments as a maximum of 5 and the standbys in 10 as it is limited in the implementing code.

The ETF is the description of these prototypes that a system integrator can use to create a valid configuration in which the derived types describe all and the exact configuration attributes appropriate for the particular deployment configuration.

The AMF also defines compound entities (SUs, service groups, etc.) and their types. These characterize the collaboration of entities in the system.

Again if we look at their limitations we can see that some software implementations may put assumptions or restrict the way they collaborate with other software pieces. Others may have no limitations at all. A component prototype, for example, may require another and this sponsoring component prototype may limit that an instance of it may be able to collaborate with at most 10 component instances. Such restrictions can also be described in the ETF through the appropriate compound entity prototype.

Since these restrictions are optional, there are mandatory and optional prototypes in the ETF – at least when an ETF is used to describe the content of a software bundle. The mandatory prototypes are those that describe the basic building blocks delivered by the software bundle. In case of AMF, these building blocks are the component prototype and the component service prototype. If there is no restriction on the combination of these in a deployment configuration, there is no need to describe any further prototypes. Otherwise the specified compound entity prototypes describe the ways that these elementary prototypes can be combined at deployment.

Note that the notion of software description with respect to its potential configurations opens the door to the automation of configuration generation. Indeed creating a valid configuration can be a challenging task for systems that deploy many different software pieces collaborating with each other and also configured for SA. To appreciate the challenge one just needs to recall the AMF information model with its variety of object classes describing entities and their inter-relations. All these need to be fully defined in a consistent manner for AMF before it can start to manage the system.

From this perspective one may also perceive the ETF as a set of constraints a configuration needs to satisfy, which means that it can be used to introduce these constraints to the configuration generation method. Of course in this case an ETF may not be associated with a particular software bundle or it may be associated with more than one.

We will explore further the potential of automatic configuration generation in Chapter 16. Here we simply make the point that the SMF specification has opened the door and enabled this possibility.

The SMF information model in force at the time of writing does not include the object classes describing the different entity prototypes. It encompasses only the object classes describing the software bundles and the derived entity types of the different AIS services. Some entity types reference directly the software bundles that delivered them.

The reason for this is that the system information model is focused on the information essential at runtime, and since the prototypes are not used directly in the system they are not essential. They may become part of the information model or different implementations of the SMF may include them to automate software management beyond the scope of the current SA Forum specifications.

So the answer to the 'proverbial' question of this subsection is that according to the first version of the specification SMF neither drinks nor eats ETF. It is for tools supporting SMF and the SA Forum compliant system in general, which may become part of the system in the future.

9.3.3   The Upgrade Campaign and Its Specification

An UCS is also an XML file – like ETF – and it describes for SMF implementations the parameterization and customization of the sequence of standard upgrade actions to perform to carry out the upgrade.

In other words, the SMF specification defines a standard process for upgrades which can be customized depending on the target. Then the upgrade campaign specification focuses only on this customization, it describes the object of the different standard actions and also if needed defines the appropriate custom actions.

As opposed to ETF, an SMF implementation requires an UCS file as an input for an upgrade (which in reality may be downgrade or 'just' a reconfiguration).

The specification defines three main parts of an upgrade campaign: the initialization, the campaign body, and the campaign wrap-up.

9.3.3.1   Campaign Initialization

During campaign initialization the SMF prepares the system for the upgrade campaign. This includes checking the different prerequisites that

1. The system is in a state that ensures – as much as possible – the successful execution:
   This includes the verification that the SMF implementation is healthy and the software repository is accessible, all software bundles needed (new and old) are available in the repository, the upgrade campaign is applicable – e.g., the system configuration has not been updated compared to the referenced one, there are enough resources in the system to perform the upgrade such as memory, disk space, and so on, and that no other upgrade campaign is running.
   If necessary SMF may even verify through API callbacks that the applications are ready for an upgrade as we will see in Section 9.3.6.
2. The system is in a state that does not jeopardize SA unnecessarily during the execution of the campaign:
   Upgrades may be performed for different purposes. Sometimes they add new services, remove old ones, or improve their quality, but not in a way that would justify an outage. In these cases if the system does not have the required redundancy, for example, one would not want to initiate the campaign execution. Other times the goal of the upgrade may actually be the repair of the degraded system, the improvement of its security or stability. In these cases – particularly if it is a repair – the system may not have a chance to recover and may further degrade without the execution of the campaign, which makes it urgent regardless the outage it may cause. Yet another issue is that services may be protected at different levels by the system. Dropping an active assignment for a service instance (SI) that has three standby assignments normally is less permissible than losing one which is configured to have a single active assignment and no standbys at all.
   To be able to decide which case it is, the campaign specification includes a section that indicates the permitted outage during execution in terms of AMF SIs. While typically no outage is allowed, an urgent campaign would allow any outage – all SIs may go unassigned. In other cases the configuration may be such that dropping some less important SIs is necessary and therefore allowed.
3. The system can prepare for the upgrade so that it can recover from failures occurring during execution:
   The most important preparation is the creation of a *backup* that can be used to restore the current consistent system state from scratch should everything else fail. If such a backup is not possible, the campaign execution is not allowed to start. In less drastic cases SMF may be asked to gracefully undo the changes made by the upgrade for which it needs to reserve the required resources – for

example, history file(s), checkpoints – as the SMF implementation requires. SMF is also required to log all the upgrade process, so it needs to prepare for this too.

In addition, since the system continues to provide services even during the upgrade and therefore its state is changing continuously, application may also want to log their application specific changes. This needs to be done in such a way that, should the backup be restored at any point during the upgrade, using these logs the applications will be able to recover the changes performed from the moment of the backup till the moment the system failed so that this backup was restored. This recovery process is usually referred as roll-forward and the first version of the specification leaves the responsibility with the applications whether it is needed or not.

Interested applications may want to receive a callback from SMF that indicates that they should create a backup and anything associated with this operation.

As part of the standard operations of the campaign initialization SMF adds the new entity types to the system information model.

The campaign initialization can be tailored as mentioned by

- Referencing a configuration base to which the campaign applies;
- The list of new software bundles, which need to be available in the software repository;
- The description of the new entity types to be added to the information model;
- The parameterization of the standard callbacks to registered users.

In addition, one may specify any administrative operations, information model changes, custom callbacks, and command line interface (CLI) operations.

The upgrade campaign specification contains only the customizations of the standard operations and additional custom operations. It does not contain the standard operations themselves.

9.3.3.2 Campaign Body

The upgrade campaign body is dedicated to the actual upgrade procedures that deploy the changes in the system. The first version of the specification distinguishes two types of upgrade procedures based on the method they use for the upgrade.

- **Single-step procedure**
  The single-step procedure – not surprisingly – consists of a single upgrade step. It is geared toward upgrades during which SA is not a concern. Typically these are the cases when new entities are added to the system or when old entities are removed. In the former case SA is not a concern as the services the new entities will provide are not provided yet or if provided they are provided by other entities already existing in the system. Similarly in the latter case, when we remove entities from the system it also means that we do not count on them any more for SA. Their services are either removed or provided by other entities.

- **Rolling upgrade procedure**
  The second procedure type uses the rolling upgrade method. The idea behind the rolling method is to utilize the redundancy existing in the system for the purpose of fault tolerance also for the purpose of the upgrade. That is, we split up the group of entities protecting some services into subgroups of redundant entities so that taking out of service one such subgroup does not cause any service outage. Then we take each of these subgroups one at a time out of service and upgrade their entities within an upgrade step. The upgrade completes once all subgroups have been upgraded – when the upgrade steps rolled over all the subgroups. Since the entities in the different subgroups are redundant they are expected to be similar, which means they can be characterized

through a template describing their common features. For AMF entities, this may be as simple as having the same component type or SU type or belonging to the same service group, which typically implies having the same SU type.[1] This similarity also means that these entities are upgradable using the same upgrade step.

What is an Upgrade Step?

The specification defines an upgrade step as a standard sequence of actions performed in order to upgrade a set of software entities. This sequence of upgrade actions is (we first give the actions then explain them one by one).

1. Online installation of new software
2. Lock deactivation unit
3. Terminate deactivation unit
4. Offline removal of old software
5. Modification of the information model
6. Offline installation of the new software
7. Instantiation of the activation unit
8. Unlock of the activation unit
9. Online removal of old software

Installation and Removal of the Software

The installation of the software means the creation of an executable image of the software within the SA Forum system using the software bundle in the software repository as a source. The removal operation removes such a previously installed software image from the system. The installation and the removal operations are typically invoked by CLI commands associated with the particular software bundle.

The executable software image in question is associated with a particular PLM execution environment (EE) within the system on which it is installed meaning that entities that need to execute in that EE are instantiated by executing this particular image.

Why are we so complicated in describing this? The reason is that often in these systems nodes do not have dedicated hard drives. The image may be created remotely on a common file system. Moreover it is also possible that the same image may be associated with more than one node. All these are implementation specific details that the SMF implementation should resolve for the particular system it is executing on. From a 'user perspective' the installation and removal CLI commands are issued on the node, in the PLM EE for which the executable image is being created or removed.

Offline and Online Operations

In the context of the SMF the online–offline distinction is related to the impact the operation has on the entities of the system. An online operation may not have any impact on any entity running or potentially running in the system at the moment of or after its execution all the way until the related information model modifications are made. For example, the software image may be installed for a new component type beforehand. However this image is not used and does not impact in any way the component to be upgraded to this new type until the object representing the component in the information model is modified to refer to this new type. This no impact includes even the case of the component restart due to a failure, for example. That is, such a restart should still use the software

---

[1] SUs of a service group that are hosted on the same node group are identical except for the time of an upgrade during which – due to the nature of the operation – belonging to the same service group does not imply the same SU type.

image associated with the component's current type and the new installation should not have any impact on it.

An offline operation has an associated scope of impact, that is, a set of entities that may be impacted if they are running or starting at the time of or after the execution of the operation. In order to avoid any unplanned impact, these entities need to be taken offline. They need to be kept offline until the upgrade of the related entities has completed and they are ready for deployment. Note that once the impacted entities have been taken offline the operation in fact becomes an online operation with respect to the rest of the system as it must not affect any other remaining entity.

Deactivation and Activation Units

The deactivation unit is the set of entities that is taken offline for the time of some offline operations (installation, removal or both). The activation unit is the collection of entities that is taken back online once the offline operations have completed.

The reason of distinguishing the activation and the deactivation units is that when we want to remove some entities from the configurations they will not be taken back online, and when we add new entities they do not exist yet to take them offline. At the extreme, an upgrade step may have an empty activation unit if all the entities taken offline are also removed from the system; and the deactivation unit is empty if all the entities to be taken offline for the operations are newly introduced within the step.

To take the entities offline typically one needs to administratively lock them first. This, as appropriate, will also trigger a switch-over of the services they may be providing. After this they can be terminated without any service impact.

To bring the entities online first they are instantiated and if it is successful they can be unlocked to allow them to take service assignments as necessary.

The assumption is that the new entities are all created in the locked administrative state so that SMF can control their release for service. This is done only when everything (the entities themselves and their environment) is ready for them to take assignments.

Since the SMF uses the administrative operations to take the entities offline and online, the impacted by an operation set of entities needs to be adjusted (e.g., expanded) to the set for which these required administrative operations have been defined. For example, AMF components cannot be locked or terminated by themselves as the lock operation is not defined for them. So if a component needs to be taken offline the deactivation unit needs to refer to its encapsulating SU.

Another way of looking at the activation and deactivation units is that these are the entities to which SMF applies the administrative operations defined by the upgrade step.

Modifications of the Information Model

In this action the SMF applies the information model changes to the objects representing the entities targeted by the upgrade step. As we have seen these entities may not be the same as the activation and deactivation units but they are usually within their scope. Hence SMF interprets them this way too unless they are explicitly listed.

Explicit listing is unavoidable in case of new entities. For SMF to be able to add a new entity to the configuration it needs the complete configuration of the entity, that is, it requires the specification of the configuration object representing the entity in the information model with all its mandatory attributes.

In case of existing entities the template approach can be used to identify a set of similar targeted entities and also to describe the changes in their configuration. For example, components of a given component type can be selected using this component type as the template. Their upgrade may only require the configuration change that their object representation references the new component type, which can be given as a single attribute modification that SMF can apply for each object representing a targeted component.

Reduced Upgrade Step

Looking at the standard upgrade step described at the beginning of section 'What is an Upgrade Step?' on page 275, if there is no need for any offline operation within an upgrade step (action #4 and #6 would be empty) the lock operation (actions #2) can be skipped completely and instead of the separate termination and instantiation (in actions #3 and #7) a restart operation can be used once the information model changes have been applied in action #5. Since no lock is performed there is no need to unlock (action #8) either. We are left with the following sequence:

1. Online installation of new software
2. Modification of the information model
3. Restart of the activation unit
4. Online removal of old software

This is referred as the reduced upgrade step. It is ideal from the perspective of SA as it upgrades entities while they may be providing services. In the context of AMF this is the case when a component is restartable (see Chapter 6 Section 6.3.2.4) and also it requires no offline installation or offline removal operations, that is, it impacts no other entity and even the component is not impacted until the information model has been modified.

Considering the PLM – which was out of scope of the first version of the SMF specification – when upgrading an operating system one may have no other option than to use this reduced step as to be able to communicate with the EE and control its upgrade SMF would need to use the old operating system. Only when the installation and configuration of the new operating system has been completed using the old environment, SMF can restart the EE with the new operating system and if it is successful the old can be removed again using the new environment.

To summarize, to specify an upgrade procedure one needs to identify:

- which upgrade method to use: single-step or rolling;
- the activation and deactivation units, which are always template based for rolling upgrades;
- the software bundles that need to be installed on and removed from the nodes;
- the configuration changes for the entities targeted by the upgrade;
- whether the standard or the reduced upgrade step applies; and
- the execution level of the procedure, which determines the execution order of procedures within the upgrade campaign.

SMF initiates procedures of a higher execution level only after it has completed all procedures of the lower levels. SMF may execute procedures of the same execution level simultaneously or in any order.

In addition to the upgrade step specification that we discussed in some details, each procedure also has an initialization and a wrap-up portion similar to the upgrade campaign.

The most important information given in the initialization portion is the permitted outage information, which is actually used during the campaign initialization for the prerequisite check. The reason it is listed at the procedure is that to be able to determine the outage one needs to know the deactivation unit of the upgrade steps of the procedure.

Upgrade Procedure Customization

The rest of the initialization and the procedure wrap-up allow for adding customized action to prepare for a procedure, to verify its outcome and/or to wrap it up.

A typical use would be the case when the upgrade is performed to provide some new services by the new version of the software. Since we are upgrading existing entities we would do that using a rolling upgrade procedure. This, however, does not allow us to add the new SIs. Instead of defining a separate single-step just for the addition of these entities the right place to add the new SIs to the information model is the procedure wrap-up when all the SUs protecting them have been upgraded. Symmetrically, if the new version does not support some obsolete SIs they need to be removed in the procedure initialization.

In the procedure initialization and wrap-up customized actions may contain information model changes, administrative operations, CLI commands and also so-called customized callbacks, which are API calls to some registered users.

We will have a closer look at the callback mechanism in Section 9.3.6. Here the important point is that SMF can transfer the control of the upgrade execution to some other entity for a certain time period. During this period this other entity may perform additional actions necessary for the upgrade in a synchronized manner with the upgrade procedure.

These customized callbacks may be used also for the customization of the upgrade steps. The UCS schema defines a set of insertion points where SMF would provide customized callbacks within particular upgrade steps of an upgrade procedure. The particular upgrade step on which a callback is made may be the first, the last, or the step halfway of the procedure; it may also be made on all steps.

The hook-up points for the callbacks within an upgrade step are:

- before lock,
- before termination,
- after IMM changes,
- after instantiation, and
- after unlock.

9.3.3.3 Campaign Wrap-Up

The most important task of the campaign wrap-up is to verify that the system is working correctly after the completion of the upgrade, and when this has been established to free all the resources that were allocated to ensure the success of the upgrade and to enable system recovery in case of a failure. It also provides an opportunity to clean up the information model by removing objects representing entity types and related software bundles that have become obsolete.

Again there are some standard operations that the SMF specification requires from implementations to perform. Most importantly there is an observation period during which SMF is in an 'observation' mode: It listens whether anyone reports any errors that it can correlate with the just completed upgrade campaign.

The upgrade campaign specification establishes the time period for which SMF performs such monitoring. This period of 'waiting-to-commit' the campaign starts at the moment the campaign has completed successfully including all the upgrade procedures and the wrap-up actions required for the completion.

At the end of the waiting period SMF prompts the administrator to make the decision whether to commit the campaign or to revert the configuration changes gracefully. This is the last chance for such a graceful rollback as after the administrator issuing the commit operation SMF frees all the resources enabling it.

However if necessary it still remains possible-albeit at some losses-to recover the system state in effect before the upgrade campaign by restoring the backup made at the initialization. For this a second waiting period defines during which SMF blocks the initiation of a new campaign and keeps the backup intact. Once this timer expires there is no guarantee that the state before the upgrade can be restored.

Besides the timers just as for campaign initialization and in the campaign body one may define administrative operations, information model changes, CLI commands, and custom callbacks at two points of the wrap-up.

- At the completion of the campaign (i.e., after the completion of its last procedure) – the purpose of these actions is to complete the upgrade campaign and to verify its success, for example, at the application level. If specified, the waiting-to-commit timer starts once these operations have completed.
- At commit – these actions are related to committing the campaign and also to the release of the different resources dedicated to the campaign.

The upgrade campaign specification contains the parameterization of the timers, the optional actions and the list of entity types and software bundles that SMF needs to remove from the information model and from the system's software repository. It does not include the standard actions as described here and in the SMF specification. The upgrade campaign specification indicates only the modifications that need to be made to the standard upgrade campaign process to customize it for the upgrade of the targeted entities.

9.3.4   Upgrade Campaign Execution Status and Failure Handling

One of the main requirements of manageability is the capability of monitoring the system's behavior. This need is even greater during upgrades and configuration changes. In case of the SMF, we would like to be able to monitor the progress of an upgrade campaign, which entities have been upgraded, which were taken out of service and then back. Most importantly we would like to obtain as much information as possible when something goes not quite as planned.

Everyone is familiar with the progress bar when upgrading some software on their computer. Indeed this was a suggestion, but how to translate an upgrade campaign into a simple progress bar? What does 57% mean for an upgrade campaign? Is that information useful if an error should occur?

Instead of trying to interpret percentages, the SMF specification defines an information model presented in Figure 9.3 together with a state model for the concepts that we have already seen – the upgrade campaign, its upgrade procedures and their upgrade steps. This way during execution the progress of the campaign can be observed through the SMF information model in which the objects of the relevant classes reflect – among others – the state information for each step and procedure of the campaign.

An upgrade campaign is represented by an object of a configuration object class and it reflects the executions of the upgrade campaigns. Under each upgrade campaign object SMF creates a tree of runtime objects that represents the procedures and their upgrade steps as they apply to the system configuration when the campaign is being executed.

At the upgrade procedure level this means that SMF creates exactly as many objects as there are procedures defined in the campaign. However, under each rolling upgrade procedure it creates as many upgrade steps objects as the number of objects in the system configuration that satisfy the template defined for the rolling upgrade procedure.

In the tree, each upgrade step object has its associated activation and/or deactivation units, each of which is associated with some nodes where software is installed or removed during the execution of the upgrade step.

This tree, which is rooted in the upgrade campaign object, is essentially the unfolding of the information contained in the upgrade campaign specification as it applies to a particular system configuration. Meaning that SMF applies the templates of the upgrade campaign specification to

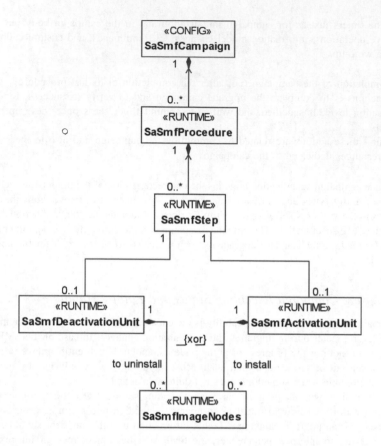

Figure 9.3 The SMF information model for upgrade campaigns [49].

the current system configuration to identify the entities that compose the activation and deactivation units and that are targeted by the configuration changes.

For example, if the activation unit template of a rolling procedure refers to an entity type X and this template is applied to configuration A, which has five such entities then it defines five upgrade steps with their activation and deactivation units. If the same template is applied to configuration B which only has two entities of type X, then only two steps will appear in the information model of system B.

This mechanism allows for the specification of generic upgrade campaigns that are adjusted automatically by SMF to the different deployment configurations.

SMF performs this unfolding of the campaign specification at the campaign initialization the latest.

Once the campaign execution starts SMF reflects its progress by setting the states defined in the specification for upgrade campaigns, upgrade procedures, and upgrade steps in the objects representing these logical concepts.

The SMF state model combines the state machines defined at each level (e.g., step, procedure, and campaign). Each state machine reflects the execution stages of a step, procedure, or the campaign including the potential error handling and the administrative operations. The resulting state model is rather complex and all its details can be found in the specification [49]. Here we provide only a summary that we believe will help to grasp the main ideas behind.

9.3.4.1 State Model Overview

The campaign, the procedure, and the step state machines all start from an initial state and the intention is to reach the (execution) completed state.[2] If this cannot be achieved due to a failure or an administrative intervention we would like at least to complete the campaign execution in a state which is equivalent to the system's configuration at the initiation of the upgrade campaign.[3] We would like to achieve this with no more service interruption than the campaign would allow us.

This means that when we start the execution of an upgrade campaign all objects in its tree have their state attribute set to 'initial.' When the campaign successfully completes all the objects in the tree should have their state attribute set to the appropriate 'completed' or 'execution completed' state.

If at any point during the execution one decides to revert back to the campaign's initial configuration – a rollback is initiated – then those upgrade steps and procedures that moved away from the 'initial' state need to reach their appropriate 'rolled back'[4] state, while those that stayed in the initial state remain so.

This means an action-by-action step-by-step undoing of the already executed portion of the upgrade campaign in order to return to the configuration that was in effect at the initiation of the campaign.

When all upgrade steps and procedures of a campaign are in either the 'initial,' 'undone,' or 'rolled back' states the system state is considered to be consistent and the configuration is equivalent to the one in effect at the campaign initiation.[5]

Accordingly, there are two parts of each state machine: one for the upgrade or forward path and one for the rollback or reverse path.

The execution starts in the forward path and remains so until a rollback is initiated. When this happens, the execution switches to the second part: to the rollback path. It remains there until the rollback either completes or fails.

Failure of an Upgrade Action

Whether the execution is in the forward or in the reverse path, it can only proceed as longs as the executed upgrade actions are successful and therefore the system state is consistent and known. If an action fails the SMF is typically permitted to retry the upgrade step.

Retry means that first the SMF performs in reverse order the undo action – as defined in the SMF specification – for each of the already executed upgrade actions within the failed step. This should bring the configuration back to the one it was at the beginning of the step.

Subsequently if the failed step is undone successfully and the retry is permitted, SMF makes another attempt to execute the same upgrade step. If successful, the campaign execution can proceed as if no error occurred.

If the retry is not allowed or too many attempts have been made, the campaign cannot complete in the forward path any more. In this case the only possible remaining goal is to roll back the entire system to the configuration initial for the campaign.

If the actions of the failed step cannot be undone successfully, the entire upgrade campaign has to fail because the system state has become unknown therefore consistency cannot be claimed. To bring the system into a consistent state it needs to be built up from scratch by performing a so-called fallback.

This is when the backup created at the campaign initialization becomes important: The backup stores all the information necessary to restore the same consistent system state that was in effect when

---

[2] Note that the 'completed' state is not a final state in the sense as the final state is defined for finite state machines.
[3] For the purpose of the current discussion we will refer to this state as the campaign's initial configuration. It is not to be confused with the initial configuration in effect at system startup time.
[4] For the upgrade step the 'undone' and the 'rolled back' states are equivalent. 'Undone' is a final state of the upgrade step automaton causing a rollback, while other steps that reached the 'completed' state will move to the 'rolled back' state.
[5] Note that we distinguish the system configuration from the system state here. We elaborate on this distinction in section.

the backup was created but from scratch. The restoration of such a saved system state after a complete system restart is referred as a fallback.

Fallback is the last resort of the SMF to recover the system after some failure. The first version of the specification leaves it implementation specific and it does not require that an SMF implementation performs the fallback operation automatically. Instead it expects that the administrator makes this decision.

Obviously failures may occur also during the rollback of the upgrade steps in the reverse path. In such cases the SMF attempts a retry in a similar manner as during the forward path. The main difference is that since during rollback the system is already on a recovery path the only remaining recovery option is the fallback whether it is due to another failure or exceeding the permitted number of retry attempts.

Whether the campaign execution reaches the 'execution completed' state or the 'rollback completed' state, the SMF returns the control to the administrator for the final decision to commit the upgrade campaign or its rollback.

As mentioned earlier at the campaign wrap-up, committing the campaign implies that SMF releases the resources used to protect the campaign execution and after this the system cannot return gracefully to the state before the campaign initiation. A fallback still remains possible for some time if the upgrade campaign specification set such a waiting time as discussed in Section 9.3.3.3.

Asynchronous Failures

The problem with upgrades is that the success of the upgrade action does not imply that the newly deployed configuration and/or software indeed functions as intended. In addition these functional mishaps may not even show up right away when a new or the upgraded entity is put back into service. This is partially because the function may not be exercised immediately or – considering an AMF entity – it may not get an assignment immediately. It is up to the AMF to decide when to instantiate such an entity and when to give it an assignment. As a result correlating an error detected in the system with the upgrade action that led to the problem is virtually impossible.

To complicate the situation further, once the entity is under the control of AMF, the error detection mechanisms and recoveries defined for AMF apply. This means that either AMF will detect the error or it will be reported to AMF so it performs the appropriate recovery and repair actions. However, if, for example, a bug or corruption causing the error was introduced by the upgrade campaign, AMF will not be able to remedy the situation through a simple restart of the entity, not even by the reboot of the node which it uses as a repair action.

The upgrade campaign needs to be rolled back as soon as possible to recover the earlier healthy state. To make this possible the information about the problem needs to be funneled to SMF. At the same time it is also desirable that AMF stops all futile repair actions after the isolation of the faulty entity and most importantly it does not escalate the error to a wider fault zone and with that taking even more entities out of service.

While the situation is similar at the upgrade of entities of other services and even applications, the SMF specification focuses on a solution for the AMF entities since in an SA Forum compliant system these are the entities considered for SA.

The solution offered by the first version of the SMF specification is the following:

SMF while executing the upgrade campaign marks those SUs that are altered in any way by the upgrade campaign by setting their maintenance campaign attribute to the distinguished name (DN) of the upgrade campaign object.

This setting disables the normal repair actions that AMF would apply in case of a failure. AMF only isolates the SU and recovers any service it was providing at the time of the failure, but AMF does not attempt to repair it. Instead AMF sets the operational state of the SU to be disabled and reports this in a state change notification indicating also that the SU was under maintenance and provides the DN of the object representing the campaign.

From the beginning of the campaign execution the SMF subscribes to these state change notifications. When SMF detects a notification referencing the DN of the upgrade campaign in progress it interprets it as a potential indication of an asynchronous error. Therefore it suspends the campaign execution and returns the control to the administrator to decide whether the error is campaign related.

Depending on whether the campaign is on the forward or the reverse path, the administrator has several options and may decide that:

- the error is not campaign related or it is related, but can be corrected; in either case the administrator needs to deal with error and clear it so that the entity can be returned to AMF's control; in this case the campaign can proceed on its current path;
- the error is campaign related.
  - In this case if the rollback option is still available, the administrator may order a rollback in which case the SMF implementation deals with the problem implicitly by reverting the altered entities including the one on which the error was reported to their earlier configuration and clears the error condition for AMF this way.
  - If rollback is not possible or the error situation is more severe the more drastic recovery, the fallback restores the backed up state of the system and clears the error condition for AMF at the same time.

Fallback vs. Rollback

It is worth spending a few words on explaining better the difference between rollback and fallback.

If we consider the system state it encompasses:

- the system configuration as defined by the configuration objects in the system information model; and also
- the software image associated with this system configuration including system and application software;
- the runtime information reflecting the actual status of the represented entities; and
- the application data.

At normal operation only the last two items change: The application data changes according to the services the system provides; and the runtime information changes depending on the changing status of the entities in the system.

When an upgrade campaign is being executed it manipulates the first two items: The system configuration and the associated software image. But since the system continues to provide its services even during the execution of an upgrade campaign – as required for SA – the runtime information and the application data also continue to change.

The changes in the application data are critical for SA as they reflect in some application specific way the services that the system provides and protects.

When it comes to the runtime status information we need to distinguish two categories: The persistent and the nonpersistent data.

The persistent runtime objects and attributes of IMM [38] play a similar role as the configuration objects and attributes – that is, they must survive system and IMM restart – except that they are created, maintained, and destroyed by an object implementer (e.g., applications, AIS service) and not by an administrator. Hence they require a similar handling as application data.

Non-persistent objects and attributes by definition do not survive system or IMM restart as they have relevance to the current incarnation of the system.

Chart A of Figure 9.4 provides a graphical representation of these changes in normal – no-error – conditions. The upgrade changes shown in dotted line refer to the configuration and software

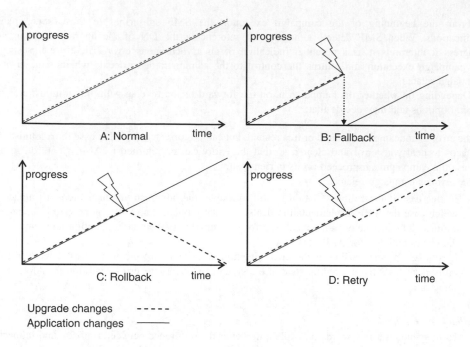

Figure 9.4 Comparison of the fallback, rollback, and retry operations.

image changes, while the application changes are reflected with the solid line and refer to the persistent runtime data and application data changes. During upgrade as time progresses both upgrade and application changes are made progressively.

A tangible example would be the upgrade of the online banking service application from version 1 to 2 while a bank customer makes a deposit to his account. The change of the software version is the upgrade change while the deposit to the account is the application data change.

At the initiation of the upgrade campaign when the backup is created it includes the information for all four parts. This means that if a fallback is used to restore the state stored in this backup it clears all the upgrade and application changes made from the beginning of the upgrade as it is shown in chart B of Figure 9.4. Obviously this is not an ideal situation for our bank customer as this means that if we consider the previous deposit scenario then the system would lose the information about his deposit.

We must note that the reason for a fallback may be anything not only an upgrade. In high availability systems backups are performed on a regular basis so that if the system occurs in an unstable or unknown state for which no other known remedy exists a stable healthy state is restored by a fallback to the latest backup where SA can be guaranteed again.

The SMF also mentions the concept of rolling forward although currently the specification only recommends it to applications. A roll-forward is the process of re-applying of some logged transactions after a fallback in order to restore the state changes committed from the moment of taking the backup up until the moment of the fallback occurred.

For example, if the deposit transaction was logged by the banking system and this log survived the fallback then it can be used to restore the correct bank account information of our bank customer.

Since these transactions are application specific the first version of the SMF specification left the roll-forward to the discretion of applications.

When a rollback is performed it does not impact the service provisioning. In chart C Figure 9.4 we see that while the upgrade changes regress back to the same point as at the beginning of the upgrade the application changes continue to progress. For our deposit example this means that while the software is changed back through the rollback to the original version 1 the information about our bank customer's deposit remains intact.

For comparison purposes chart D of Figure 9.4 also shows the retry scenario. Here the system rolls back the upgrade changes to the one at the beginning of the upgrade step, but then in the retry it performs the same changes again while the application changes proceed without any interruption. As a result the retry if successful only introduces some delay in the completion of the upgrade campaign.

9.3.5   Administrative and Management Aspects

The SMF allows administrators to create and delete software bundle and upgrade campaign objects; and to control the execution of upgrade campaigns by issuing administrative operations on the upgrade campaign object representing the appropriate upgrade campaign.

9.3.5.1   SMF Information Model Management

Software Bundle Objects

Software bundles delivered to the system's software repository are represented in the SMF information model, which defines a configuration object class for this purpose (see Figure 9.2).

A software bundle object contains the information necessary to install and remove the represented software bundle within the SA Forum system. These are the CLI commands and command line arguments listed separately for online and offline installation and removal. The attribute values need to be set exactly in the way the SMF implementation should issue these commands in the EE in which the software bundle needs to be installed or removed from. The upgrade specification only refers to the software bundle object from which the appropriate CLI command and its arguments are fetched when the upgrade step calls for the operations.

Additional attributes indicate the default timer for the CLI operations and the scope of impact for each of the offline operations as it applies to the particular system. The first helps to determine the failure of an operation. The second attribute provides a further hint on the activation and deactivation units, a given implementation may or may not use.

Due to this tailoring of the attribute values to the concrete target system, the same software bundle may be represented by objects with different settings in different systems reflecting the system specifics such as the needs or assumptions of SMF implementation, the file system solution used, or its organization.

In any case, the presence on the software bundle object in the SMF information model indicates that the software bundle is available in the software repository and can be accessed during upgrades as required.

The exact way it is presented or the way it is imported is not specified by the first version of the SMF specification. All these details and the image management – beyond the mentioned installation and removal CLI commands – are completely left open for the SMF implementations to decide. Different solutions may extend the currently defined object class with additional attributes such as the path to the software bundle in the repository or the required package handler utility if several of them are supported.

An implementation may also decide to link the software bundle object creation and deletion operations with the import and removal of the software bundle to/from the software repository and by that implying some image handling operations. Although this is not required it is a natural interpretation of the current specification.

Upgrade Campaign Objects

As discussed in Section 9.3.4, upgrade campaigns are represented in the SMF information model by configuration objects. They link the XML upgrade campaign specification file with its execution in the system. The configuration object reflects the execution state of the upgrade campaign specification it represents.

When the administrator creates an upgrade campaign object SMF takes the associated XML file as an input and applies it to the current configuration of the system. It may perform some checks (e.g., whether the campaign is applicable to the system) and create the entire sub-tree of upgrade procedures and steps right away.

Once the upgrade campaign object has been created the administrator may issue administrative operations on it to initiate the execution, suspension, rollback, fallback, or to commit the campaign.

Since the campaign object is a configuration object it will remain in the information model until the administrator deletes it. In a way, these objects keep record of the system evolution.

The rest of the sub-tree consists of runtime objects and in their case it is left to the discretion of the SMF implementation their life span. The specification only requires their presence for the time of the execution of the campaign as they reflect the status of the execution. So an implementation may remove them when the campaign has been committed or keep them until someone deletes the campaign object.

9.3.5.2 SMF Administrative Operations

The first version of the SMF specification defines the following four SMF administrative operations applicable to upgrade campaign objects:

- execute,
- suspend,
- rollback, and
- commit.

The administrator issues them on the object representing the upgrade campaign to be performed.

Although the specification does not define the exact format and semantics of the fallback administrative operation yet, it requires that an administrator should be able to initiate a fallback. The reason of not being defined in the specification is that the fallback is not limited to the scope of software management only. It is used to recover the system from any type of situations when no other remedy can be used. Similarly the backup operation has not been defined by the SMF specification, it only requires that at the beginning of the campaign it is performed in some system specific way and the SMF implementation knows how to trigger it.

It has not been decided yet which AIS service shall offer the backup and fallback operations. One potential candidate is the PLM, the other one is of course the SMF, or they may deserve the definition of their own entirely new service.

After this small detour let us see the administrative operations defined by the first version of SMF.

Execute

The execute administrative operation initiates the execution or the continuation of the upgrade campaign represented by the campaign object on which the operation is issued. It is applicable to campaigns in the initial or suspended states. It can be issued only for one upgrade campaign at a time. That is, simultaneous execution of upgrade campaigns is not supported.

If the sub-tree of runtime objects of the upgrade campaign has not been created yet, SMF creates it when it receives the execute command and checks the campaign pre-requisites. If successful, SMF starts the execution of the upgrade procedures following their execution level starting with those at the lowest.

The execute operation returns when either:

- the campaign has completed successfully by completing all procedures and the wrap-up actions verifying the success;
- the campaign has been suspended;
- the campaign reached a failure state.

The execute command is effectively a resume operation when used in the suspended state regardless of the reason of the suspension. If the suspension is due to an asynchronous error, issuing the operation implies that the error condition has been dealt with.

If the campaign is in a failure state (an error other than an asynchronous has been detected) the execute operation does not apply any more and returns an error. It depends on the failure state whether rollback may still apply or a fallback needs to be performed.

Suspend

The administrator may suspend the execution of an upgrade campaign any time whether it is executing in the forward path or rolling back.

Suspending the campaign means that SMF carries on with any ongoing upgrade step until it is completed, undone, or rolled back at which point the parent procedure becomes suspended. So at the procedure level SMF suspends ongoing procedures at the next upgrade step boundary. When all ongoing procedures reach such a suspended state, the entire campaign reaches the suspended state.

Procedures in their verification stage cannot be suspended and similarly the final verification phase of the campaign cannot be suspended either.

In the suspended state the administrator decides whether to continue the execution of the campaign, to roll it back, or to perform a fallback. To resume a campaign suspended in its forward path the already mentioned execute operation is used. Alternatively a rollback may be initiated or resumed using the rollback operation discussed next. As we mentioned earlier there is no standard fallback operation defined currently, but it needs to be available in some way for a system to be SA Forum compliant.

The suspend operation is applicable while the campaign is in the executing or in the rolling back states; and it returns as soon as the ongoing campaign reaches a suspended or a failure state.

Rollback

The rollback operation applies only in a suspended state or when the campaign execution has completed successfully. In this case and if the campaign was suspended on the forward path, the rollback command initiates the graceful undoing of the configuration changes performed by the campaign thus far.

When issued on a suspended campaign which is already rolling back, the operation resumes the rollback process.

Similarly to the execute operation the rollback returns either when:

- the rollback completed successfully;
- the rollback was suspended; or
- the rollback has failed.

In the last case from SMF perspective the only available administrative operation is a fallback.

Commit

With the commit operation the administrator confirms that the campaign or its rollback was successful and the SMF may release all resources but the backup allocated for the execution and the protection of the upgrade campaign. As a result it applies only in one of the completed states.

After committing a campaign a rollback cannot be performed any more, but the fallback operation may still return the system to the configuration before the campaign provided that the backup created at campaign initiation is still available. As mentioned earlier the upgrade campaign specification may specify a period between subsequent upgrade campaigns and guarantee that the backup and with that the fallback are available at least for this period.

9.3.6 User Perspective

When supporting upgrades at the middleware level the challenge is that the middleware cannot be aware of special application features and needs such as synchronization procedures or switching between functionalities or versions. These are only known at the application level.

For example, a new version of the software may implement a new synchronization method between the peer components protecting the same services. Since the currently running version does not understand this new method, it cannot be used as long as there may be old components in the protection group. But when all components have been upgraded to the new version it is desirable to switch them to this new synchronization method. One way to do it is to implement a negotiation in the new version of the software through which peers decide when the new method can be used.

The potential problem with this is that successful negotiation requires that all the parties impacted participate in the negotiation. Considering the AMF, it instantiates only the required number of components which may not mean all components. For example, if in the service group only five SUs need to be instantiated at a time out of 10, the uninstantiated ones cannot participate in the negotiation and may still have the old software. If the instantiated five have been upgraded and switched to the new method when one of them fails and therefore one of the old ones starts to instantiate it cannot effectively collaborate with the others.

The SMF would be aware that there are 10 nodes to upgrade with the new software to complete the upgrade of the service group even if only five of its SUs are instantiated. And since it controls the upgrade, it also knows when the last node of the 10 is upgraded. What it does not know is that this is a significant event for the application. If the application could indicate somehow to SMF that it would like to know when the last node was upgraded and SMF could signal this then the application could perform the change to the new synchronization method at the right time.

The SMF specification introduces the concept of upgrade-awareness. It refers to SMF user processes that need to synchronize 'application level'[6] actions during upgrades and therefore need to know about their initiation and progress.

At the discussion of the customization of the upgrade step and procedure in Section 9.3.3.2, we mentioned that the upgrade campaign schema allows one to define customized callbacks to user processes for certain stages of the campaign execution. These stages are the campaign and procedure initialization and wrap-up; within upgrade steps at particular actions of a particular upgrade step (or all steps) of an upgrade procedure; and also at rollback initiation.

These stages were defined with upgrade-aware processes in mind. The solution works as follows:

On one side, in the upgrade campaign specification there are these hook-up points at which one can define callbacks. The callback definition includes a callback label and a set of parameters. The label is like a keyword identifying some kind of actions at the application level. The optional parameters associated may provide further information if necessary. For SMF they are all completely transparent.

---

[6] Being 'application level' is determined from SMF perspective and not from the typical system stack. It means that it is outside of the scope SMF is aware of and designed to handle.

On the other hand, any process which needs to synchronize certain actions with an ongoing upgrade needs to register with the SMF to be able to receive a callback and also to define a filter for the callback labels that matches its interest.

When during the execution of the upgrade campaign SMF reaches a point for which the upgrade campaign specification defines a callback, SMF checks if there is a registered process which defined a filter that the specified callback label satisfies. If so, SMF calls back the interested process. In the callback SMF implementation provides the callback label and parameters as defined in the upgrade campaign specification together with the DN of the SMF information model object representing the upgrade campaign, procedure, or step within which the call is being made and whether the actions need to be executed in the forward path or the reverse path.

For each callback, the upgrade campaign specification also indicates whether the SMF implementation needs to wait for the response of the process and for how long.

If the UCS specifies a waiting time, SMF essentially hands over the control to the application process for this period at most. When the process completes the actions associated with the callback label, it needs to report back the outcome to SMF, so SMF can decide whether the campaign can proceed or it encountered an error.

If the process reports a failure within the timeout period, it is treated as any other action failure of the campaign. That is, if it is within an upgrade step, the step can be retried otherwise a rollback or a fallback may be required.

If no wait timeout is specified or the timer expires before the response is received, SMF proceeds with the campaign regardless of the outcome.

To help the work of upgrade campaign designers to define the callbacks within an upgrade campaign, in the ETF the component prototype descriptor has a section in which the software vendor can describe the callbacks the implementation expects or able to interpret. It consists of the labels and the conditions these labels indicate for the software implementation.

For example, in the ETF description of a component prototype the vendor may say that the components derived from the prototype expect a callback with the label BACKUP at campaign initialization at the time the system backup is created. Such a component would like to receive the callback whenever a new upgrade campaign starts, so the appropriate callback needs to be included in all campaigns targeting the system.

Another case is our earlier example on the new synchronization method for which the vendor may indicate that on the last step of the upgrade procedure (or in the procedure wrap up) the components expect a callback with the AppSyncSwitch label. Such a component would only be interested in campaigns that upgrade its own base type, so the callback needs to be included only for such upgrades.

9.3.7 Service Interaction

9.3.7.1 SMF Interaction with IMM

As we discussed in Sections 9.3.4.1 and 9.3.5, the inventory of software bundles and upgrade campaigns are represented as objects in the SMF part of the information model. The administrator interacts with the SMF by manipulating these objects and issuing administrative operations using the IMM OM-API (object management application programing interface).

Accordingly, an SMF implementation needs to register with the IMM as the implementer of these objects using the IMM OI-API (the object implementer API of the IMM). From that moment on it will receive a callback from IMM whenever these objects change or when the administrator issues an administrative operation on any of them. When the operation completes SMF reports the result to IMM, which in turn delivers it to the administrator.

This part of the SMF interaction with IMM is the same as for all the AIS services.

However to perform upgrade campaigns the SMF also need to act as a management client to the AIS services whose entities are being manipulated within a campaign. In IMM terms SMF needs to act as an object manger to perform this task and it needs to use the IMM OM-API. As a result it can issue administrative operations and change the system configuration by manipulating the configuration objects of the system information model. The prerequisite to this is that SMF is able to obtain the administrative ownership of the objects targeted by the upgrade campaign. In IMM the administrative ownership represents the right to manipulate an object (Chapter 8).

In particular, the standard actions of the upgrade step that are defined on the activation and deactivation units map into administrative operations on the entities of these units. We even gave this alternative definition of the activation and deactivation units in Section 9.3.3.2.

For example, in case of AMF, the action 'terminate deactivation unit' means that SMF needs to issue a lock-instantiation administrative operation on each of the entities that match the deactivation unit template.

The first version of the SMF specification provides the mapping of the upgrade actions to the AMF administrative operations only. It is not straightforward for other AIS services and other mapping may be required. For example, Cluster Membership service (CLM) does not provide separate administrative operations that could be mapped into the lock and the termination upgrade actions, so when the scope of the SMF is extended to CLM and other services these differences need to be dealt with.

In addition to administrative operations, SMF also maps the configuration changes that are defined in the upgrade campaign specification into configuration change bundles (or IMM CCBs). This part of the upgrade campaign specification schema was defined so that this mapping is more or less straightforward.

Another part of the upgrade campaign that SMF maps into CCBs is the addition and removal of entity types and the customized upgrade action of campaign and procedure initialization and wrap up that indicate IMM operations.

Finally, SMF is also required to verify that the software bundles listed in the upgrade campaign specification are indeed in the software repository. The assumption here is that if the bundle is in the repository then there is an object in IMM representing it. However since this is related to image management, which is not covered by the first version of the specification, the specification also allows for implementation specific solutions.

9.3.7.2 SMF Interaction with AMF

From the previous section it follows that the SMF is not in direct interaction with the AMF. Instead this interaction is mediated through the IMM when it comes to the administrative operations. SMF changes the AMF configuration also through IMM by applying CCBs to the AMF objects that match the template specification for the entities targeted by the upgrade.

As we noted earlier SMF does not control how AMF distributes the assignments during an upgrade. It cannot do so exactly because of the way it interacts with AMF.

When SMF unlocks some entities after they were taken out for an upgrade and returns them to AMF's control AMF may or may not give them an assignment right away. It depends on the configuration and the current conditions in the system.

This means that any fault introduced by the upgrade may not manifest right away. There is no guarantee for this even in those cases when they would get an assignment.

As mentioned in section 'Asynchronous Failures' on page 282, if an error was introduced during the upgrade the AMF recovery and repair mechanisms cannot resolve the resulting problem, therefore we need to prevent the normally triggered AMF repair and fault escalation mechanism. Instead the AMF specification provides a way to feed back the error to SMF as it is a potential upgrade failure.

To implement this feature the SMF evaluates the entities targeted by the configuration changes (i.e., those that are part of CCBs) and determines the enclosing AMF SUs. SMF sets the maintenance campaign attribute of these SUs to the upgrade campaign, which in turn disables the AMF repair attempts and also AMF can provide this information as a feedback to SMF in the notifications it generates.

Obviously, SMF needs to listen to these notifications so that it can react to them in a timely manner. Currently it suspends the upgrade campaign and reports the problem to the administrator for deliberation.

Once the campaign has been committed SMF clears the maintenance campaign attribute setting so that AMF can apply the repair actions again.

9.3.7.3 SMF Interaction with NTF

As we have seen in the previous session part of the interaction between AMF and SMF is mediated by the Notification service (NTF) [39]. To receive the appropriate AMF notifications, the SMF needs to subscribe to the operational state change notification generated by AMF for its SUs. In particular it needs to look for those that indicate in their additional information field the DN of upgrade campaign being executed.

As a notification producer, the SMF generates no alarms at all, which may be a surprise to some readers. The reason behind this decision is that the upgrade campaign is controlled through administrative operations and all the errors and failures are reported through this interface. There is no particular need to replicate the same information in the form of alarms.

The first version of SMF assumes a rather tight administrative control during the upgrade campaign execution and leaves all significant decisions to the administrator even in those cases when the state model would suggest only a single operation as a possibility. Under these circumstances there is no reason to alert the administrator by additional alarms.

With respect to notifications, a SMF implementation is expected to generate state change notifications for step, procedure, and campaign state changes. This provides an alternative – a push based – way of monitoring of the upgrade campaign in progress.

9.3.7.4 SMF Interaction with LOG

Even though no details are given in terms of the level, the content or the format, the SMF specification requires that an SMF implementation logs the progress of the upgrade campaign execution. It is recommended that SMF uses the AIS Log service [40].

9.3.8   Open Issues

Considering the scope of software management and that at the moment of writing only the first release of the SMF specification is available, it is not a surprise that the list of open issues is rather long.

The goal of the first release was to address the issues that are specific to SA and set the direction of software management for SA Forum compliant systems. As a result the scope of the first SMF specification was limited to:

- the addition and removal of AMF entities; and
- the upgrade of redundant AMF entities provided that the new version is compatible with the old.

No upgrade method has been defined for incompatible upgrades and for nonredundant entities such as services. These issues remain for future releases.

The area most people expect to be covered in a software management related specification is the software packaging and image management. As discussed in Section 9.3.1 in this area there are already a number of solutions, recommendations, and standards, which actually made superfluous to address the topic. Yet, it would be beneficial to define the integration between the SA Forum SMF and the existing package handling utilities and image management solutions. We have seen already that the fact of representing software bundles as objects in the SMF information model offers the opportunity of linking the object creation and deletion with the package import and removal operations.

Other aspects – like the installation and removal of software within the cluster – due to the variety of clustered solutions do not lend such an easy resolution. It is difficult to define the required operations without making assumptions about the underlying architecture, which in turn may put unnecessary limitations on implementations and therefore undesirable.

The situation is similar and related in case of the backup and the fallback operations. The associated roll-forward operation adds complexity with the diversity of application. Yet, the subject of SA calls for a solution for these operations.

Obviously in the future the SMF has to address the entire AIS stack, and in particular the upgrade of the entities of the PLM [36], which covers the middleware, the operating system, the virtualization facilities, and even firmware.

The current SMF specification ultimately leaves to the administrator to decide whether the execution or a rollback of an upgrade campaign was successful or not. Considering the complexity of these systems it is a question of whether it is feasible to expect the administrator to make such a decision or based on the information available in the system the decision should be automatic. Experience with implementations shall show whether an automatic decision is preferred or the administrator should always be able to rule over the system's decision.

Since this is the section to let our imagination soar we can imagine a software management solution that would know when a new hardware node is plugged into the system what software it requires and install it automatically before the associated cluster nodes may join the cluster. The potential for such a solution is in the SA Forum architecture; however, the first release of specification does not cover this area.

Another desired scenario that many would like a software management solution to resolve is the following: One would like to be able to provide a new desired configuration of the system and leave it up to the SMF to figure out how to migrate the system to this new configuration. After a short pondering about the task one may realize that to be able to do this for all possible cases would require a rather elaborate solution hence the SA Forum Technical Workgroup decided not to require this capability from SMF implementations and has come up with the definition of the upgrade campaign. This allowed to focus on the runtime aspects of the problem of software upgrades and also brought the potential of implementing the SMF specification closer.

It also means that the problem is left for solutions that provide additional value to SA Forum compliant implementations. Note also that since the upgrade campaign has been standardized, the provider of such a solution can be anyone that follows this specification. Of course the challenge is to be able to come up with a solution that can generate an upgrade campaign that migrates the system from its current configuration to any new desired configuration with no or minimal service outage.

9.3.9 Recommendation

To provide guarantees with respect to SA in SA Forum compliant systems at least the upgrade of AMF entities that are in the scope of the first release of the SMF specification should be handled according to the specification. The reason behind this is that in such systems only these entities are considered for SA.

Therefore following the specification should provide the same SA guarantees across a variety of systems regardless of the actual SMF implementation used as all of them should align on at least the requirements of the standard. Hence application and system designers can rely on them when designing a new system, a new application or its new version.

As mentioned in earlier sections, the design of the SMF specification implies some assumptions beyond those of the AMF specification when it comes to SA during upgrades and application and system designers need to satisfy these assumptions.

So what is needed to be able to provide SA during upgrades?

Probably the most important requirement applications need to satisfy is the peer compatibility within a protection group as shown in Figure 9.5. That is, the new version of the software needs to be able to collaborate with the old version during the process of the rolling upgrade when the active and standby components may be running different versions of the software yet need to be able to exchange state information to participate in a protection group.

From this perspective one needs to consider not only the upgrade, but also the rollback or downgrade scenario. If anything goes wrong during the upgrade – and since Mr Murphy never sleeps this may happen – it is important that the rolling back of the campaign does not cause service outages either.

This backward compatibility requirement means different things for different applications depending on whether they are stateless or stateful and in what way.

The vertical compatibility between components of different types collaborating within a SU (which is most people consider when the subject of compatibility comes up) determines the scope of each upgrade step. Whether a component can be upgraded by itself, or the entire SU or hosting node needs to be upgraded together.

As presented in Section 9.3.6 the SMF specification caters for application level coordination. This also needs to be considered for the case of rollback: If there is a customized callback on the forward pass the SMF implementation will make a callback on the rollback pass symmetrically to the forward pass and will indicate the direction it is executing the campaign specification, which needs to be interpreted by the application properly.

One needs to proceed with caution, however, as this part of the specification is least tested in practice at the moment of writing; particularly when it comes to the upgrade campaign creation.

From Section 9.3.8 it is also clear that there is no standard way of upgrading the complete stack of the SA Forum compliant systems, which means that different SMF implementations may treat differently the upgrade of non-AMF entities.

The main vision of the SMF specification is, however, that software management should be automated. This vision is reflected in Figure 9.6.

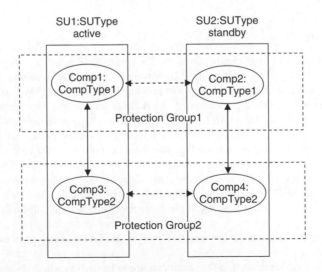

Figure 9.5 Compatibility types.

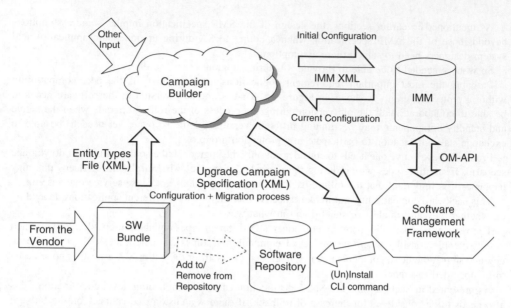

Figure 9.6 The information flow envisioned in the SMF specification [49].

Accordingly vendors would describe their software in ETFs, which are delivered with the software bundle. The ETF information is then used by a campaign builder tool to create a new configuration that uses the new software to provide the required services (given as 'other input') on the cluster the configuration of which is fetched from IMM. For a new system it may be enough to provide the new configuration via IMM. However for a running system the new configuration is processed to generate an upgrade campaign that would migrate the system from its current configuration to this targeted new one. In both cases the software bundles are delivered to the system's software repository so that SMF can use them to install the new software.

9.4 Conclusion

Among the existing standards and specifications, the SA Forum SMF makes a leap forward in the field of software management. This is because the efforts were focused in the area where no specification or standard has provided guidelines yet. This was possible because of the AMF specification, which provided a conceptual basis that the SMF specification could rely on and develop further.

Of course since the problem has existed ever since high-availability systems were used people have been dealing with it in different ways, so there were a variety of solutions available. However most of these solutions were proprietary. The terminology and the concepts used in the field varied enormously, so even when using the same term people would mean different things. So the first achievement of the SMF specification was to define a common terminology appropriate for the problem domain.

As we demonstrated throughout the chapter the SMF specification also provided, indeed, a framework to deal with high availability and SA in the context of software upgrades and system reconfigurations.

Even more, it set the direction for the area to follow in the future as it further evolves addressing the still open issues.

The concepts, the methods defined in the specification can be used in a much wider context and not only for SA Forum systems. It is true that certain features are highly specialized for these systems, but the ideas are not limited to them and can be extended easily to the software entities of platform or even to hardware entities. Grid and cloud computing are candidates for such extensions.

In some cases the current specification is overly cautious since it targets systems at the highest extreme of the scale. These precautions can be easily removed in systems of lower requirements. Even for SA systems some of the precautions may need to be removed in the future as it is extremely difficult for a human to make decisions about the consistency and the correctness of such systems. These breakpoints are really there to be able to feed in the results and reports of other tools that may help the administrator.

As discussed in Section 9.3.8, future releases should address the upgrade of the entire SA Forum stack and also the integration with existing utilities that complement the scope of the SMF.

10

Combining the Services

Maria Toeroe
Ericsson, Town of Mount Royal, Quebec, Canada

10.1 Introduction

In Part Two of the book we started out with the architecture of the Service Availability (SA) Forum interface specifications and the information model used in systems compliant to the specifications. These chapters together painted the overall picture of such systems and provided the basics for the discussions to follow.

In subsequent chapters we looked at the different service specifications in details starting from the platform level up toward the applications and the management functionalities.

Our suspicion is that these chapters even in this format presented quite a bit of challenge for readers not particularly familiar with the domain. Nevertheless we can assure these readers that reading the specifications themselves would have been a more daunting task.

In any case, we have done the decomposition and the analysis of our architecture and this is the point where we need to pull everything we learnt about our system together and do the synthesis.

We can look at our system from the perspective of the designers and the developers of an application, who need to build an application that will be able to provide its services in a highly available manner; and for that purpose who want to take advantage of the SA Forum specifications and their implementations by using the SA Forum services instead of developing the same functionality from scratch.

The other perspective is of the system architects and designers who need to come up with the platform on which such applications can be deployed. Their primary goal is to come up with a flexible platform that can accommodate the anticipated range of applications effectively and also support the operational and maintenance activities that are inevitable throughout the system life-cycle.

Finally we also take a quick look from the perspective of system administrators who have the task of managing and maintaining the system. They would like to see a rather homogeneous picture in spite of all the diversity that may exist in the actual system.

10.2 Application Design and Development

After reading Part Two of the book, probably the question one ponders about is: Do I really need to deal with all these services and that kind of complexity?

Service Availability: Principles and Practice, First Edition. Edited by Maria Toeroe and Francis Tam.
© 2012 John Wiley & Sons, Ltd. Published 2012 by John Wiley & Sons, Ltd.

We are happy to say that no, application designers and developers do not need to deal with all the different services. They can focus only on those that provide functionality needed in their application implementation.

Application designers of course need to understand the SA Forum services enough to be able to decide which one of them offers the functionality they need in their applications and if there are more than one, which one of them is the most suitable. Here we provide some hints on how to go about these choices.

Unless an application manages or monitors some hardware in the system it does not need to use the Hardware Platform Interface (HPI) [35].

If it does, for example, the application service is actually provided using some hardware elements then the use of the HPI interface becomes handy as it allows the application to manage this piece of hardware in an implementation independent way and therefore the application can control these hardware elements on any system providing the HPI interface.

If the application does not manage the hardware, but still needs to know its status or it needs to influence certain administrative actions applicable to platform elements then using the Platform Management service (PLM) [36] may be enough. Through the PLM API (application programming interface) applications are able to track the status changes and vote – so to speak – whether the removal or the locking of a platform element impacts their services. The prerequisite of this tracking is that the piece of platform (hardware or software) the application is interested in is reflected in the PLM configuration.

Even though we emphasized at each service that the Cluster Membership service (CLM) [37] is the authority on the membership, this information is more important for the Application Interface Specification (AIS) services themselves than for applications. Applications that are managed by the Availability Management Framework (AMF) [48] will always run on member nodes, so using the CLM API is not needed to determine membership. But it can provide some additional information by passing through the node related PLM tracking information. If this is needed the application has the choice to listen to CLM or go directly to PLM.

Using these APIs and being managed by AMF may actually create confusion as both the application and AMF may receive the same validation requests resulting in contradictive reactions. The application needs to coordinate its actions with the AMF high-availability (HA) state assignments and provide feedback toward AMF using the HA readiness state settings.

Considering that the subject of this book is SA we expect that application designers and developers interested in the subject will want to use the AMF and therefore its APIs.

As we have pointed out already in the Chapter 6, for a developer there is no need to understand all the details of AMF. After all AMF can manage an application even if it does not implement any of its APIs (see non-proxied-non-SA-aware components and proxied components in Section 6.3.2.5. Obviously the SA aspect may not result in the best possible user experience, but for some applications these might be a perfectly viable options as we will see in Part III of the book where we present different integration options.

Applications that need to communicate within or without the cluster in a location independent way that also decouples the senders and the receivers can benefit from the described in Chapter 7 Event service (EVT) [43] and Message service (MSG) [44].

If the communication is intra-cluster and related to some state synchronization among different application instances, the Checkpoint service [42] becomes an option to alleviate some of the burden the developer needs to face.

We have to emphasize that none of these or any other AIS Utility Service is mandatory by any means. So, for example, if high performance is an requirement for the application communication, the use of other methods may be more appropriate than either EVT or MSG. However such choice also means that the developer will have to resolve at least some of the challenges we presented for these particular services.

Besides the EVT, Messaging, and Checkpoint services the SA Forum also standardized – and an application designer may want to consider – the Lock [40], the Naming [46], and the Timer services [47]. Chapter 3 presented a brief overview of each of these AIS Utility Services.

The AIS Management Services are there to unify the system management. As a result most applications that need to be integrated with such systems will need to use at least the Notification service (NTF) [39] for their notifications and alarms to provide the system administration with some online and offline status information about their operation. Such data becomes essential in the root cause analysis of failures and outages. Since notifications and alarms are logged by the middleware the application needs to use the Log service (LOG) [40] only if additional data needs to be recorded.

If the application requires some configuration and it exposes some administrative control, it may be beneficial to integrate it with the Information Model Management service (IMM) [38] as object implementer for the configuration object classes defined as an extension to the SA Forum information model. The objects of these configuration classes can offer a configuration interface and also expose the application status. The AMF and PLM are good examples that application designers may study.

Alternatively an application may chose to expose its status through runtime object classes in a similar way as we have seen in case of the AIS Utility Services.

Finally applications that need to coordinate their actions with their own and system upgrades may also use the Software Management Framework (SMF) [49] API. The anticipation though is that it will be mostly management applications, system components and such that have the task of coordinating different activities in the system that would benefit from the possibility offered by the SMF API. For example, if an application needs to backup its application level data or deploy additional safety measures for the duration of an upgrade the application designer may consider the callback mechanism offered by the SMF.

To facilitate the application developers' task in Part Three of the book we review the basics of the programming model for both the C and Java APIs. As mentioned already we will present an example of integrating a relatively simple application with some AIS services. We will also look at legacy applications how they can utilize the SA Forum services in the hope that these examples convince everyone that the SA Forum specification did indeed make life easier for many application developers.

10.3 Application Platform Design

While the application developers can pick and choose from the SA Forum Services as connoisseurs from a smorgasbord, system architects who design the platform to support these applications need to follow the dependencies existing among the SA Forum services at the specification level, as well as at the implementation level and also the needs of the targeted applications.

In Chapter 3 we have pointed out that modularity is the basis of the architectural design of the Service Availability Interface (SAI) specification. This means that in a particular system not all the services need to be present all the time. That is, application platform designers also have some choices, but instead of the smorgasbord allowing any choice it is more like an 'à la carte' menu where complete entrées are defined.

The SAI specification defines some dependencies among its services. These need to interact and rely on each other for proper operation and therefore they need to be included in the application platform together.

An important assumption about these service interactions is that a dependent service should be able to obtain everything it needs from the required sponsoring service using the standard APIs as defined in the specification. That is, generally there is no need to extend the service interface to enable service interactions, which provides the necessary condition for interchangeable service implementations. Of course, it depends on the particular implementation whether this remains to be the case or not.

At the discussion of each of the services we dedicated a separate section to present their interaction with other SA Forum services. Therefore here we only focus on the global picture and we will not go into details.

Among the service interactions the most important one is the dependency of the AIS services on the CLM with the exception of the PLM and the Timer services. The result is that for all practical reasons we can assume that CLM is part of the middleware providing AIS services.

But the CLM specification does not define its dependency on the PLM as mandatory, nor does PLM's dependency on the HPI. Service implementers are free to make their own choices.

Since our focus is on SA our target applications require availability management and therefore the AMF becomes a must. Note that this does not require that all managed applications are SA-aware. AMF can manage applications without them implementing the AMF APIs.

AMF receives its configuration and exposes its administrative API via the IMM which needs to be included in the application platform.

Considering the HA requirement – approximately five minutes down time in a year – it is easy to see that there is no opportunity for shutting down the system for maintenance or upgrade. This also makes the SMF an essential part of the system. During upgrades SMF introduces the AMF configuration changes and the administrative control again using IMM.

The same applies to all AIS services which define an information model or expose administrative control. Therefore IMM becomes just as important in the application platform as CLM.

Throughout Part Two the other recurring service in the service interaction sections was the NTF. The different SA Forum services use NTF as the carrier of their notifications and alarms. The expectation is the same toward applications developed to be deployed in SA Forum clusters.

Finally all SA Forum services are expected to use the LOG to record significant events that may facilitate system analysis and in particular root cause analysis. This interaction we did not mention in the service interaction sections as the specifications – except for a few cases – do not elaborate on the information that needs to be collected. The middleware implementation should LOG at least the notifications and alarms generated in the system by the system and also by applications. Therefore LOG is also considered as a fundamental application platform service.

From the perspective of applications besides availability management they need communication and synchronization facilities as we presented in Chapter 7. However from the perspective of the SAI architecture all these services are optional. The question is whether any and which one of them needs to be included in the application platform.

The good news is that since the specifications do not define dependency between these services, they could be added at any time when a new application requires them provided that the service implementations follow the specifications.

This brings us to the perspective of the middleware implementation. Different middleware implementations may define their own requirements on how the service implementations are coupled. For example, in the OpenSAF [91] implementation of the SA Forum specifications, some AIS service implementations rely on the AMF implementation for managing their HA. This means that even if the target applications do not require AMF, the middleware itself does and needs to be included in the application platform.

It actually makes complete sense as the middleware services are expected to be highly available which means that one way or another, their availability management needs to be implemented. So it is natural that they rely on one of their own, which was designed for exactly this purpose, it is 'the professional' for the job. This also means that the functionality is not duplicated in each service creating potentially significant overhead. This comes with the added benefit that applications can also use the services of AMF, reducing the overhead even further.

In Part Three we will also look at more details and considerations of the system level design again in the context of migrating complex applications requiring complete solutions.

To provide some insight to the middleware itself we will have a peek in Chapter 13 at OpenHPI [71] and OpenSAF [91] and their architectural solutions to implement the SA Forum services. This chapter is only a teaser for these implementations matured to be carrier grade.

10.4 Operation and Maintenance

If the SA Forum did its job well and the application designers aligned with the concepts correctly we should have a homogeneous administrative interface to our system. The administrative control is exposed through the IMM object management API, while status information is available also via the NTF. The LOG can collect additional data delivered in terms of log files.

The main question though is who are the expected consumers of all this information?

Looking at the content of the SA Forum information model it is hard to imagine a human administrator tweaking the system by adjusting dozens of attribute values in hundreds of configuration objects. In fact many have the view that most of these attributes should not be exposed to humans as human errors are a, if not the major, source of outages. Complex and overwhelming management data backfires easily.

That is to say that most of the management information exposed by IMM is there for management applications rather than for the human operator. A very good example of a management application is the SMF presented in Chapter 9, which expects an upgrade campaign as an input and uses it to deploy all the configuration and other changes in the system in a systematic manner. By keeping track of all the operations performed and the possible recovery actions which it can deploy, SMF is able to perform systematically an upgrade or a reconfiguration. As we mentioned already to protect the system from human errors SMF is considered the only way to modify system configurations except for complete system initialization from a new configuration file.

Now looking at the information model as the source of the system configuration and the upgrade campaigns as scripts to manipulate this configuration one still may not be at ease. Someone somehow needs to come up with these artifacts. The only reassurance the SA Forum specifications can provide is that they can be designed and developed offline and tested just like one would do with software to be deployed in the system.

The idea of using scripts and automating these administrative tasks is not completely new. However the systematic end-to-end approach taken by the SA Forum is quite revolutionary in our opinion. Some of the details of the solution have been presented already in Chapter 9 and we will take another look in Chapter 16 of Part Three from the perspective of how software engineering techniques known better in the field of software development can be deployed here too for these tasks to further increase SA.

With that let us embark on our adventure and get a taste of the use of the SA Forum specifications. Let see them in action.

Part Three

SA Forum Middleware in Action

11

SA Forum Programming Model and API Conventions

Francis Tam
Nokia Research Center, Helsinki, Finland

11.1 Introduction

In Part Two of the book, we have discussed the founding principles, motivation behind the design, usage scenarios, and limitations of the Service Availability (SA) Forum services and frameworks. In this chapter, we turn our attention to how we can use these services and frameworks in practice.

For developers, one of the first places to look for implementation related information is the service's Application Programming Interface (API) definition sections of the specification. In the SA Forum specifications, all the service and administrative APIs are defined using the C programming language [92]. The question typically asked by people has been: why C? It was a decision based on the member companies at the time when the specifications were developed. This represented a common ground for the participating companies, one of the key characteristics of open standards development. This decision was chiefly backed by the evaluation of the language's suitability from a technical perspective. C is a well-known and popular candidate system implementation language for operating systems and embedded system applications, primarily due to its efficiency. It has been recognized for code portability across many different platforms. Its flexibility such as ease of access to hardware and relatively low demand on system resources at runtime add to the list of advantages for implementing the SA Forum middleware. The fact that the C++ programming language is source and link compatible with C offers another edge that the services can also be used by an object oriented language.

As we have seen, the whole SA Forum specification has been divided into a number of services. This was due to the complexity of the entire system to achieve modularity, as well as a way to develop the ecosystem such that different participating member companies could take part in just a subset of the specifications and their implementation, instead of the all-or-nothing or monolithic approach. This called for a need to have a consistent naming conventions and coherent behavioral patterns in similar functions across all the services. As a consequence, the usage model established a generic interaction between a service user process and a library for a service area. It is important to note that this division is a logical one. The implementation of service areas could be physically combined into a single library.

Service Availability: Principles and Practice, First Edition. Edited by Maria Toeroe and Francis Tam.
© 2012 John Wiley & Sons, Ltd. Published 2012 by John Wiley & Sons, Ltd.

In the SA Forum specifications, the definition of a process is based on that of Portable Operating System Interface for Unix (POSIX) [73]. However, this does not mandate the middleware be implemented in a POSIX system, although at the time when the specifications were developed, POSIX was the choice of implementation platform amongst most member companies. This explains why many interfaces have this influence behind their designs.

We begin this chapter by describing the programming model, which is applicable across all the Application Interface Specification (AIS) services. This includes the interface relationships between service users and implementers, naming conventions and type definitions, real-time support, the usage model, and the tracking capability. As the API specifications are typically the first place where developers would look for programming related information, we outline what one can expect from the service and administrative API sections of the specifications. Finally we discuss a number of topics and issues that are frequently considered by developers in practice. These include the interaction with POSIX, memory management, the handling of pointers, finding out the implementation limits, the availability of area servers, and the concern of backward compatibility.

11.2 Programming Model

In this section, we explain the general C programming model of the SA Forum specifications [63], which is applicable to all the AIS services. This includes the AIS area service interfaces, naming conventions, predefined types and constants, real-time support, library life-cycle, and the concept of tracking.

11.2.1 AIS Area Service Interfaces

The SA Forum specifications have been divided into a number of services according to their functions. As we have seen in Part Two, each service provides the corresponding functions through standardized interfaces. Each of these services can therefore be regarded as an area of its own. In the rest of this chapter, we use the term area service to refer generically a SA Forum AIS service, without specifically naming the service.

Besides the consistent naming conventions and type definitions, interactions between a service user process and the area service must also be consistent. Figure 11.1 shows the interface relationships among a user process, the area service interface implementation library, and an area server. It must be noted that all area service packages have the same logical structure.

The SA Forum Application Interface for an area has been defined to be between a process and an area library that implements the area service interface. On behalf of the user process, an area library invokes the appropriate functions at an area server that carries out the actual service functionality as described in the corresponding AIS service specification. It is important to point out that this interaction is a logical one. This is due to the freedom an implementer has to construct a separate physical module for each area server or combine a few into a single physical one; and the choices of centralized or distributed implementation of an area server. It must also be noted that the method of communication between an area library and area server has not been specified, therefore, it can either be local or remote, depending on how an area server has been implemented.

11.2.2 Real-Time Support

The SA Forum interface has been designed to accommodate both unthreaded and threaded models. In a single-threaded program, if the main execution thread blocks on a long-running task, the entire application can appear to be frozen, thus giving the impression that the service is not being provided

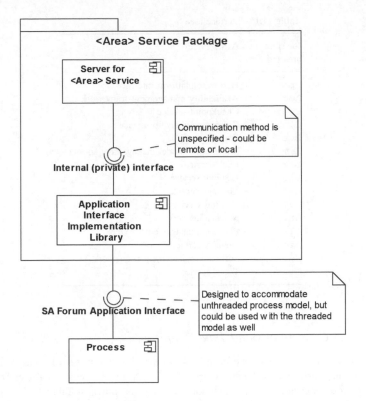

Figure 11.1 Interface relationships.

and the application does not exhibit the characteristic of guaranteeing a response in real-time. A multi-threaded programming model allows for multiple threads of execution within a process. The threads share the same address space, state information, and other resources within a process. Context switching between threads within a process is typically faster than that of processes. By moving long-running tasks to a thread that runs concurrently with the main execution thread, it is possible for the application to remain responsive to user input while executing tasks in the background.

Multi-threading is not the only way to keep program responsive, nonblocking interactions can also be used to achieve the same result. This is one of the reasons why the SA Forum interface supports both synchronous and asynchronous programming models. A synchronous programming model is easy to use by developers. An application invokes an interface in the implementation library and blocks until the call returns. In an asynchronous programming model, an application invokes an interface in the implementation library, but control returns to the caller immediately. The caller has to specify a callback function that will be invoked when the operation is completed. On the middleware side, there is also a need to use the callback mechanism to initiate interactions to an application. There are also operations that have an asynchronous nature, for example, a change in the cluster membership.

Since the C programming language only offers a single-threaded control flow, support is therefore needed by the underlying operating system to provide the asynchronous programming model. The notion of a selection object (see Section 11.2.4) has been introduced in the programming model such that the operating system dependency is hidden in the area service implementation library. This leaves the user application to be as portable as possible across other AIS implementation platforms.

Table 11.1 Interface area tags

Interface area tag	AIS services
Hpi	Hardware platform interface
Amf	Availability management framework
Ckpt	Checkpoint service
Clm	Cluster membership service
Evt	Event service
Imm	Information model management service
Lck	Lock service
Log	Logging service
Msg	Message service
Nam	Naming service
Ntf	Notification service
Plm	Platform management service
Sec	Security service
Smf	Software management framework
Tmr	Timer service

11.2.3 Naming Conventions and Type Definitions

Since the SA Forum specifications have been broken down into a number of areas corresponding to the AIS services, an interface tag is therefore needed to identify an individual service. The assignment used to tag each area interface to its corresponding services is shown in Table 11.1.

For the AIS services, a function declaration is expressed as:

```
type sa<Area><Object><Action><Tag>(<arguments>);
```

where

$\quad\quad\quad\quad\quad\quad$ sa — prefix for 'service availability'
$\quad\quad\quad\quad$ <Area> — interface area
$\quad\quad\quad$ <Object> — name or abbreviation of object, or service
$\quad\quad\quad$ <Action> — name or abbreviation of action
$\quad\quad\quad\quad\quad$ <Tag> — tag for the function such as Async or Callback

An example is:

```
SaAisErrorT saClmClusterNodeGetAsync(
    SaClmHandleT clmHandle,
    SaInvocationT invocation,
    SaClmNodeIdT nodeId
);
```

where

$\quad\quad\quad$ <Area> = Clm for the Cluster Membership service,
$\quad\quad$ <Object> = ClusterNode,
$\quad\quad$ <Action> = Get, and
$\quad\quad\quad$ <Tag> = Async.

Other declarations are expressed as:

- Type

```
typedef <...> Sa<Area><TypeName>T;
```

- Macro

```
#define SA_<Area>_<MACRO NAME> <macro definition>
```

- Enumeration type

```
typedef enum {
        SA_<Area>_<ENUMERATION_NAME1> [= <value>],
        SA_<Area>_<ENUMERATION_NAME2> [= <value>],
        ....
        SA_<Area>_<ENUMERATION_NAMEn> [= <value>]
} <enumeration type name>;
```

Table 11.2 lists all the standard and predefined types and constants for the AIS services.

11.2.4 Usage Model and Library Life Cycle

The logical interface relationships introduced in Section 11.2.1 has the premise that a user process must associate itself with an area service library for the duration when the area service functions can be used. This duration is expected to be between an area service library is initialized and finalized by the service user process. Figure 11.2 shows an example interaction sequence among a process, application interface implementation, and an area server. It is important to stress that this is just an example scenario and it should not be interpreted as the only way to implement an area service library and its interactions with the area server.

Actions 1 and 2 show the sequence of an area service library initialization. As part of the library association step, the callback functions that the user process has implemented and can be invoked by the library are passed as parameters. The returned handle is used to represent the association between the user process and the library in subsequent interactions. It is worth pointing out that a user process is allowed to call the library initialization multiple times. Each call would return a different handle for a different instance of the same library.

Actions 3 and 4 illustrate the sequence of obtaining a selection object for making the callback mechanism usable in C. A selection object is an operating system dependent entity that allows for a process to wait until there are callbacks pending. In the POSIX environment, the selection object is a file descriptor that is used with either a select() or poll() system call to detect incoming callbacks. It is important to point out that a process may be able to obtain different selection objects for the same library handle. This is used to support the multi-threaded dispatching of callbacks.

The motivation behind the introduction of a selection object concept is to hide this operating system dependency in the library implementation. For an AIS implementation on a platform other than POSIX, the equivalent function of a selection object to support the callback mechanism in C must be implemented in the library.

Assuming that the wait is over for the user process (Actions 5, 6, and 7), Action 8 signals to the library that the process is ready to process the pending callback. Here a user process has the options to choose how to handle the pending callbacks according to the supplied dispatch flags in the dispatch call. The flags have the following behaviors and provide for flexibility to the user process when combined with using the multi-threaded model:

- SA_DISPATCH_ONE – Invoke a single pending callback and then return.
- SA_DISPATCH_ALL – Invoke all the pending callbacks before returning.

- SA_DISPATCH_BLOCKING – One or more threads execute callbacks as they become pending. The threads remains in the dispatch and stay until a finalize call is made by one of the threads in the process.

According to the user-selected dispatch flags, a callback function is invoked by the library (Action 9) in order for the user process to execute the command (Action 10). The appropriate response generated by the user process as a result is then returned to the library (Action 11), which relays it back to the server of the area service (Action 12).

Finally, the user process enters the finalization stage when it does not require the use of the area service library by calling sa<Area>Finalize() in Action 13. This results in disassociating

Table 11.2 Standard and predefined types and constants

Description	Types	Remarks
Boolean	SaBoolT	SA_TRUE, SA_FALSE
Integer	SaInt8T, SaUint8T SaInt16T, SaUint16T SaInt32T, SaUint32T SaInt64T, SaUint64T	Signed and unsigned integers of 8, 16, 32, and 64 bits
Floating point	SaFloatT SaDoubleT	IEEE 754 32-bit single-precision IEEE 754 64-bit double-precision
String	SaStringT	typedef char * SaStringT; Terminated with the null character ('\0')
Size	SaSizeT	typedef SaUint64T SaSizeT; Used to specify the sizes of objects
Offset	SaOffsetT	typedef SaUint64T SaOffsetT; Used to specify offsets in data areas
Time	SaTimeT	Positive number in nanoseconds as either an absolute timestamp or time duration

Sequence	SaAnyT	typedef struct {
		    SaSizeT bufferSize;
		    SaUint8T *bufferAddr;
		} SaAnyT;
		Used to define a set of an arbitrary octets

Name	SaNameT	#define SA_MAX_NAME_LENGTH 256
		typedef struct {
		    SaUint16T length;
		    SaUint8T value[SA_MAX_NAME_LENGTH];
		} SaNameT;

Service	SaServicesT	An enumeration type for all the defined SA Forum services

Version	SaVersionT	typedef struct {
		    SaUint8T releaseCode;
		    SaUint8T majorVersion;
		    SaUint8T minorVersion;
		} SaVersionT;

Table 11.2   *(continued)*

Description	Types	Remarks
Track flags	Constants	SA_TRACK_CURRENT SA_TRACK_CHANGES SA_TRACK_CHANGES_ONLY SA_TRACK_LOCAL SA_TRACK_START_STEP SA_TRACK_VALIDATE_STEP

Dispatch flags	SaDispatchFlags	typedef enum {
		    SA_DISPATCH_ONE = 1,
		    SA_DISPATCH_ALL = 2,
		    SA_DISPATCH_BLOCKING = 3
		} SaDispatchFlagsT;

Selection object	SaSelectionObjectT	typedef SaUint64T SaSelectionObjectT;
Invocation	SaInvocationT	typedef SaUint64T SaInvocationT;

Limits	SaLimitValueT	typedef union {
		    SaInt64T int64Value;
		    SaUint64T uint64Value;
		    SaTimeT timeValue;
		    SaFloatT floatValue;
		    SaDoubleT doubleValue;
		} SaLimitValueT;

Error codes	SaAisErrorT	An enumeration type for all the defined error codes for AIS services

the user process from the library. It must be noted that multiple handles obtained by calling the sa<Area>Initialize() multiple times must be finalized separately.

11.2.5   *Tracking*

Tracking is a useful capability that enables a service user process to follow the changes that are of interest. An application requests a tracking operation to be performed on an entity, or a group of entities. A track callback is then invoked when the requested changes on the entity or entities have occurred, or about to occur. There are many scenarios in which AIS services require the support of tracking, for example, when the cluster membership changes in Cluster Membership service (CLM) (see Section 5.3.4), protection group changes in Availability Management Framework (AMF) (Section 6.3.2.4), and readiness status changes in the Platform Management service (PLM) (see Section 5.4.4).

The format of a tracking function is:

sa<Area><Xxx>Track[<Func>]()

where

<Xxxx> – is the kind of changes to be tracked and
<Func> – is the tracking function to perform:

- start tracking an entity or an entity group;
- stop a previously initiated tracking on an entity or entity group;

- callback notification;
- respond to a callback notification.

A service user process can also specify the different tracking behaviors according to the different tracking flags as shown in Table 11.3.

When the area service informs a service user process that the changes requested to be tracked have occurred, the information about the changes as specified by the tracking flags are returned in a callback function. The required memory for this information is allocated by the area service library. It is the responsibility of the called back process to invoke the corresponding `sa<Area><Xxx>NotificationFree()` function to free the memory allocated by the area service library.

In some use cases, a service user process needs be able to accept or reject a request to change the status of a tracked entity, or to be able to perform actions prior to the change takes effect. This allows for graceful reaction to certain administrative operations and is supported by the enhanced tracking.

The enhanced tracking can occur in three different ways. As shown in Figure 11.3, the first one begins at the top with the following steps:

1. **Validate** – A subscribed process may choose to accept or reject the proposed change. A response is needed by the subscribed process. If at least one process has rejected the proposed change, the AIS service invokes the track callback indicating that the pending action has been aborted. No change happens as a consequence of the aborted status. Note that no response from a process is counted as an 'accept.'
2. **Start** – All the subscribed processes have accepted the proposed change and with this callback the AIS service requests them to carry out the change. Processes respond as soon as they completed the actions required for the change.
3. **Completed** – When all subscribed processes involved in the start step reported that they have completed their actions, the AIS service also performs the actions required to complete the change. When the change is completed, the subscribed processes are notified by the AIS service.

The second way begins at the start step. This happens when a subscribed process does not subscribe for the validate step or if the pending change cannot be rejected. Therefore, such processes are directly notified at the start step without the validate step. They then continue to the completed step as described in the first way above.

The third one occurs at the completed step, which means that this is the same as the non-enhanced tracking as described in the beginning of this section, that is, processes are notified after the change has taken effect.

It is useful to note that the use of the three different paths of enhanced tracking can be generally thought of as:

1. **Validate-start-completed** – for tracking changes caused by operations that may be rejected.
2. **Start-completed** – for a graceful execution of the operation.
3. **Completed** – for the abrupt execution of an operation or a notification about an unsolicited change such as a failure.

Both PLM and CLM offer the enhanced tracking mechanism to track their corresponding entities. Interested readers are referred to Sections 5.3.4 and 5.4.4 for more details.

11.3 Making Sense of the API Specifications

What is in the API sections for each defined AIS service? In this section, we explain the API conventions for AIS services and administrative operations.

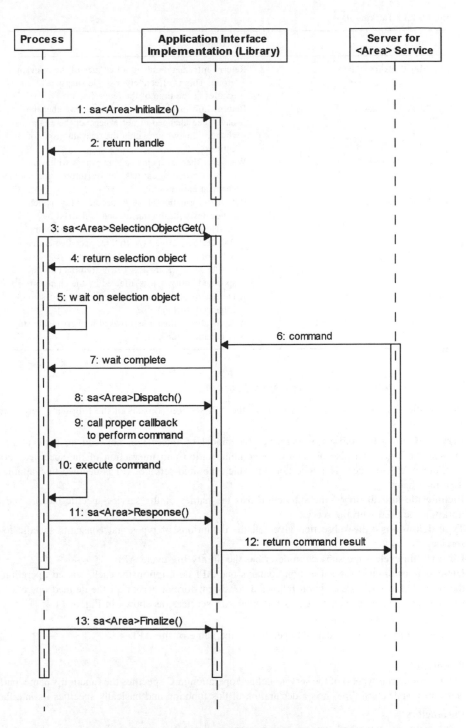

Figure 11.2 Example interaction sequence.

Table 11.3 Tracking flags

Tracking flag	Description
SA_TRACK_CURRENT	Returns information about all entities of the relevant group. This is effectively a single snapshot of all entities at the time of the request
SA_TRACK_CHANGES	Returns information about all entities, whether they have been changed or not, whenever there is a change somewhere within the relevant group of entities
SA_TRACK_CHANGES_ONLY	Whenever there is a change somewhere within the relevant group of entities, returns information about changed entities only
SA_TRACK_LOCAL	Returns information about a specific entity, which collocates with the tracker and is defined by individual area services. This flag is used with SA_TRACK_CURRENT (returns specified entity only) or SA_TRACK_CHANGES, or SA_TRACK_CHANGES_ONLY (returns only the specified entity if it is affected by the changes)
SA_TRACK_START_STEP	A notification callback is invoked in the start step of enhanced tracking
SA_TRACK_VALIDATE_STEP	A notification callback is invoked in the validate step of enhanced tracking

11.3.1   Structure of Service API Specification

In the specification of each of the AIS services the API section consists of the following subsections:

- **Service model** – the subsection describes the model of interaction between an application and the specific service. It also discusses the communication and interaction of the specified service with other AIS services. The conditions in which the area service API is deemed unavailable are described here.
- **Include file and library** – the subsection lists the names of the service-specific include file and library to be used with the service.
- **Type definitions** – the subsection gives all the definitions of types and constants specific to the service.
- **Library life cycle** – the subsection describes the library life-cycle APIs.
- **Defined operations** – the subsection contains one API description for each defined operation for the service. Each API description follows a consistent format across all the defined operations in the specific service, as well as across the other AIS services, as shown in Figure 11.4.

At the top of each service-defined operation is the name of the API.

- **Prototype**
  The function prototype(s) of the service-defined operation in C specifies the function's name, parameter types, and return type. It is a declaration of this function and basically specifies its interface.
- **Parameters**
  Apart from the name and type of each parameter, the AIS service API pages use the notations [in], [out], and [in/out] in the parameters to give further information regarding how these parameters should be treated.

[in] indicates that the parameter contains input from the caller to the called function. Also, the called function does not modify this parameter and can therefore be considered as using the pass by value method.

[out] indicates that the parameter contains output from the called function to the caller. The caller does not supply any further information via this parameter, which is essentially a pointer to a memory area where the information is to be provided. This can be considered as using the pass by reference method.

[in/out] indicates that the parameter is used for both input and output. The caller places further information via this parameter, which is essentially a pointer to a memory area where the called function provides the result. This can be considered as using the pass by reference method.

- **Description**
  This describes what the service-defined operation does and its usage.

- **Return values**
  This provides a list of all the expected error codes for the service-defined operation, together with an explanation of the likely cause of such an error.

- **See also**
  This refers the readers to other related APIs.

11.3.2   Administration API

For some AIS services, there are administrative operations and possible states defined and exposed by the Information Model Management service (IMM) to the system management applications. The administration API for each such AIS service has the following subsections in the specification:

- Administration API model covering the administrative states and operations for the specific AIS service.
- Include file and library name – the subsection lists the names of the service-specific include file and library to be used with the administration API. Since for all AIS services IMM exposes the administration API, they are the include and library files for IMM.
- Type definitions – the subsection gives all the definitions of types required by the administration API.
- Defined operations – each of which consists of an administrative API definition following the format, as shown in Figure 11.5.

At the top of each service-defined administrative operation is the enum value for the corresponding administrative operation as defined by the particular AIS service. It is passed as a parameter to the IMM OM-API (information model management service object management application programming interface) function defined to invoke an administrative operation. Since it is always the same IMM API – a variant of the `saImmOmAdminOperationInvoke()` – which is used to invoke an administrative operation the function prototype is not given. It has been discussed in Section 8.4.3.5.

Parameters
  These include the administrative operation identification and required parameters.
Description
  This describes what the service-defined administrative operation does.
Return values
  This provides a list of all the expected error codes for the service-defined administrative operation, together with an explanation of the likely cause of such an error.
See also
  This refers the readers to other related service APIs and/or administrative operations.

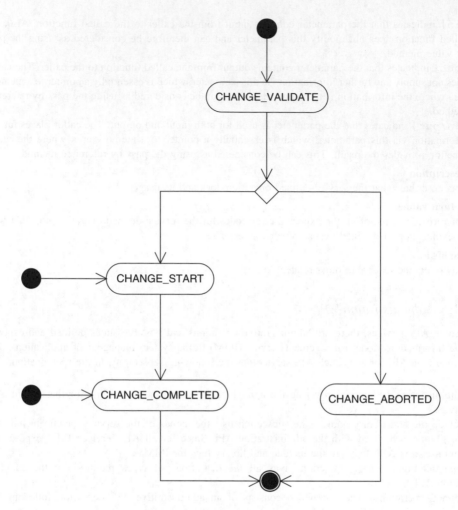

Figure 11.3 Enhanced tracking steps.

11.4 Practical Topics

In this section we look at a number of topics that are related to the use of the SA Forum middleware in practice. These include short notes on interacting with the POSIX environment, memory management, pointers handling, finding out the limits of an implementation, conditions in which an area service is deemed unavailable, and the issue of backward compatibility.

11.4.1 Interacting with POSIX

For those applications that are developed in a POSIX environment, greater application portability from one AIS implementation to another can be achieved by following the SA Forum guidelines for developing applications related to the use of signals and threads.

In connection to the asynchronous programming model, signals in a POSIX environment are widely used by applications as a mechanism to be notified when an event is occurring in the system. Examples

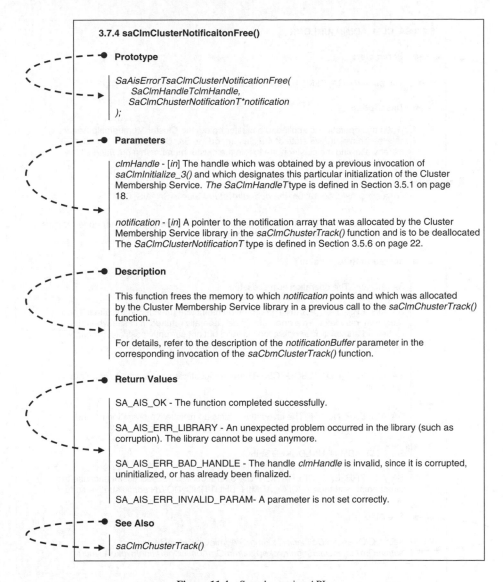

3.7.4 saClmClusterNotificaitonFree()

Prototype

SaAisErrorTsaClmClusterNotificationFree(
 SaClmHandleTclmHandle,
 SaClmChusterNotificationT*notification
);

Parameters

clmHandle - [*in*] The handle which was obtained by a previous invocation of
saClmInitialize_3() and which designates this particular initialization of the Cluster
Membership Service. *The SaClmHandleT*type is defined in Section 3.5.1 on page
18.

notification - [*in*] A pointer to the notification array that was allocated by the Cluster
Membership Service library in the *saClmChusterTrack()* function and is to be deallocated
The *SaClmClusterNotificationT* type is defined in Section 3.5.6 on page 22.

Description

This function frees the memory to which *notification* points and which was allocated
by the Cluster Membership Service library in a previous call to the *saClmChusterTrack()*
function.

For details, refer to the description of the *notificationBuffer* parameter in the
corresponding invocation of the *saCbmClusterTrack()* function.

Return Values

SA_AIS_OK - The function completed successfully.

SA_AIS_ERR_LIBRARY - An unexpected problem occurred in the library (such as
corruption). The library cannot be used anymore.

SA_AIS_ERR_BAD_HANDLE - The handle *clmHandle* is invalid, since it is corrupted,
uninitialized, or has already been finalized.

SA_AIS_ERR_INVALID_PARAM- A parameter is not set correctly.

See Also

saClmChusterTrack()

Figure 11.4 Sample service API.

of such events include detection of hardware exceptions, timer expirations, terminal activity, and
user-defined signals. Upon receiving a signal, an application responds by taking an appropriate action,
which can be a default one as specified by the system, ignore the signal or invoke a user-specified
function known as a signal handler to deal with the event.

Since the SA Forum specifications do not require that the AIS functions can be safely invoked
from a signal handler, the following should be observed:

- Avoid using any SA Forum API from a signal handler.
- Do not assume that SA Forum APIs are interruptible by signals.

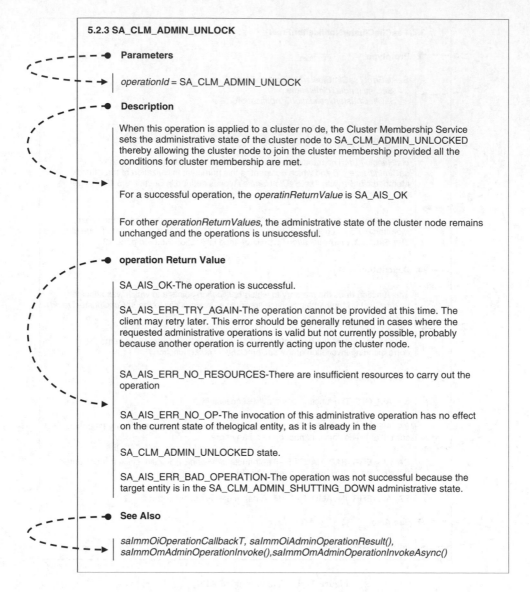

5.2.3 SA_CLM_ADMIN_UNLOCK

Parameters

operationId = SA_CLM_ADMIN_UNLOCK

Description

When this operation is applied to a cluster no de, the Cluster Membership Service sets the administrative state of the cluster node to SA_CLM_ADMIN_UNLOCKED thereby allowing the cluster node to join the cluster membership provided all the conditions for cluster membership are met.

For a successful operation, the *operatinReturnValue* is SA_AIS_OK

For other *operationReturnValues*, the administrative state of the cluster node remains unchanged and the operations is unsuccessful.

operation Return Value

SA_AIS_OK-The operation is successful.

SA_AIS_ERR_TRY_AGAIN-The operation cannot be provided at this time. The client may retry later. This error should be generally retuned in cases where the requested administrative operations is valid but not currently possible, probably because another operation is currently acting upon the cluster node.

SA_AIS_ERR_NO_RESOURCES-There are insufficient resources to carry out the operation

SA_AIS_ERR_NO_OP-The invocation of this administrative operation has no effect on the current state of thelogical entity, as it is already in the

SA_CLM_ADMIN_UNLOCKED state.

SA_AIS_ERR_BAD_OPERATION-The operation was not successful because the target entity is in the SA_CLM_ADMIN_SHUTTING_DOWN administrative state.

See Also

saImmOiOperationCallbackT, saImmOiAdminOperationResult(), saImmOmAdminOperationInvoke(),saImmOmAdminOperationInvokeAsync()

Figure 11.5   Sample administrative API.

As stated earlier, SA Forum APIs have been designed to support both the single-threaded and multi-threaded models. Therefore, the implementation of area service libraries must be thread-safe for those applications that may have different threads of a process concurrently calling the area service APIs. Thread-safety is a property that attempts to eliminate the risk of interference among threads and unwanted modifications to another thread's data elements.

One of the many mechanisms of ensuring thread-safety when terminating a thread is to carry out a thread cancellation operation in a controlled manner. This involves the thread to be cancelled is allowed to hold cancellation requests pending until it reaches a thread cancellation point. At which point, the cancellation request is acted upon and the thread to be terminated is allowed to perform application-specific cleanup processing. In other words, thread cancellation may only occur at these cancellation points. Many POSIX functions have been specified as thread cancellation points. However, it must be pointed out that the SA Forum APIs cannot be assumed to be thread cancellation points.

There is a general problem of making `fork()` work in a multi-threaded process, which boils down to deciding what to do with all the threads in the newly created process. One approach is to copy all the threads to the new process. This however must prevent those threads that are suspended on system calls or just about to execute systems calls from executing in the new process. An alternative is to just copy the thread that calls `fork()` but this might create a situation in which a thread that holds a resource will not be present in the new process, resulting in the resource is never released. The guidance from POSIX [73] to this problem is to use an appropriate thread creation call when a new thread of control is needed, and only use `fork()` to create a new process to run a different program. Since we cannot assume that the AIS functions are fork-safe, if a process using AIS functions forks a child process in which AIS functions will also be called, the child process should therefore `exec()` a new program immediately.

11.4.2 Allocating and Freeing Memory

When the amount of memory required for a program cannot be determined beforehand, dynamic memory allocation at run time is needed. The allocated memory must be freed and returned to the memory pool for re-use, otherwise a memory leak may exhaust the available memory in a system. In the context of the SA Forum APIs, where there are interactions between a user process and an area service library, either party may be responsible for the dynamic memory allocation.

The SA Forum API specifications have a ground rule which states that memory dynamically allocated by one entity, either the user process or area service library, must be freed by the same entity. However, it is sometimes simpler or even better to have the area service library allocating the memory and the service user freeing the memory afterwards. For example, it is difficult to provide a buffer of the appropriate size by the invoking process when it is problematic to determine how much memory is required for an operation in advance. There are also situations where excessive copying can be avoided if this approach is followed, as a result, the performance is improved. In either case, each area service that provides a function to dynamically allocate memory for a user process also provides a function to be called by the user to free the memory. A user process must therefore call these memory free functions to avoid any memory leaks.

This type of usage is clearly documented in the SA Forum APIs. Table 11.4 lists the memory allocation calls and their corresponding memory free calls for those relevant area services.

11.4.3 Handling Pointers

When parameters of the type of pointers are passed between a service user process and an area service library, the lifetime of these pointers must be taken into account in the program for both synchronous and asynchronous APIs.

In synchronous APIs where pointers are passed as parameters by a service user process to the area service, the area service must not keep any pointer beyond the function has returned.

In asynchronous APIs where pointers are passed as parameters by a service user process to the area service, the area service must not keep any pointer after the area service has invoked the corresponding asynchronous callback function. On the user process side, it must not use that pointer after the callback function has returned. If the user process needs to keep the information passed by the pointer beyond its lifetime, a copy of the information must be made locally before the callback function returns.

Table 11.4 Per service memory allocation and free calls

Area	Function to allocate memory	Function to free memory
AMF	saAmfProtectionGroupTrack()	saAmfProtectionGroupNotificationFree()
CKPT	saCkptSectionCreate() saCkptCheckpointRead()	saCkptSectionIdFree() saCkptIOVectorElementDataFree()
CLM	saClmClusterTrack()	saClmClusterNotificationFree()
EVT	saEvtEventAllocate(), saEvtEventDeliverCallback() saEvtEventAttributesGet()	saEvtEventFree() saEvtEventPatternFree()
IMM	saImmOmClassDescritionGet()	saImmOmClassDescriptionMemoryFree()
MSG	saMsgQueueGroupTrack() saMsgMessageGet(), saMsgMessageSendReceive()	saMsgQueueGroupNotificationFree() saMsgMessageDataFree()
NTF	saNtf\<t\>NotificationAllocate() Where \<t\> is the type of notification saNtfLocalizedMessageGet() saNtf\<t\>NotificationFilter Allocate() where \<t\> is the type of filter	saNtfNotificationFree() saNtfLocalizedMessageFree() saNtfNotificationFilterFree()
PLM	saPlmReadinessTrack()	saPlmReadinessNotificationFree()

11.4.4 Finding Out Implementation Limits

An area service implementation usually has limits, for example, the maximum number of supported entities of a certain type, the maximum size of certain objects, and so on. It is important for developers to find out what these limits are, so that they can be taken into account of the overall application design. Some AIS services offer a function to a user process to obtain these implementation limits, each of which is defined by an enumeration type Sa\<Area\>LimitIdT:

```
typedef enum {
     SA_<Area>_<NAME1>_ID = 1,
     ...
     SA_<Area>_<NAMEn>_ID = n
} Sa<Area>LimitIdT;
```

Table 11.5 lists the AIS services that have defined implementation limits.

A process can obtain at runtime the current value of a particular implementation limit by invoking the sa\<Area\>LimitGet() function:

```
SaAisErrorT sa<Area>LimitGet(
     Sa<Area>HandleT <Area>Handle,
     Sa<Area>LimitIdT limitId,
     SaLimitValueT *limitValue
);
```

The handle \<Area\>Handle is the handle which identifies this particular initialization of the \<Area\> Service, which was obtained by a previous invocation of the sa\<Area\>Initialize() function.

Table 11.5   Per service implementation limits

Area	Name of implementation limit
Event	SA_EVT_MAX_NUM_CHANNELS_ID
	SA_EVT_MAX_EVT_SIZE_ID
	SA_EVT_MAX_PATTERN_SIZE_ID
	SA_EVT_MAX_NUM_PATTERNS_ID
	SA_EVT_MAX_RETENTION_DURATION_ID
Lock	SA_LCK_MAX_NUM_LOCKS_ID
Log	SA_LOG_MAX_NUM_CLUSTER_APP_LOG_STREAMS_ID
Message	SA_MSG_MAX_PRIORITY_AREA_SIZE_ID
	SA_MSG_MAX_QUEUE_SIZE_ID
	SA_MSG_MAX_NUM_QUEUES_ID
	SA_MSG_MAX_NUM_QUEUE_GROUP_ID
	SA_MSG_MAX_NUM_QUEUES_PER_GROUP_ID
	SA_MSG_MAX_MESSAGE_SIZE_ID
	SA_MSG_MAX_REPLY_SIZE_ID
Naming	SA_NAM_MAX_NUM_CLUSTER_CONTEXT_ID
	SA_NAM_MAX_NUM_NODE_CONTEXT_ID

limitId identifies the limit and the <Area> service returns the implementation-specific value in the memory area to which limitValue points.

These limits typically refer to the maximum number or size of entities of a certain type that the implementation can support, however, other limits such as thresholds can also be used.

11.4.5   When an Area Service is Unavailable

Since we are dealing SA applications, it is therefore important for developers to understand when the underlying availability middleware may not be available. There are primarily two conditions under which an area service is not available to a service user process. The first one, which is signaled by the returned error code SA_AIS_ERR_UNAVAILABLE, indicates that the service user process is not allowed to use the operations offered by the area service. The second one, which is indicated by the returned error code SA_AIS_ERR_TRY_AGAIN, gives the information that the provider is currently experiencing some difficulties in delivering the service, although the service user process is allowed to use the operations in this area service. The reasons may be due to transport, communications, or other issues but they tend to be of a temporary nature.

In order to understand why a service user process is not allowed to use the operations in an area service, one has to go back to the premise of the SA Forum solution relating to a cluster. As defined in the CLM (see Section 5.4), a cluster is formed by a set of cluster nodes. Each cluster node is the logical representation of an execution environment that has been administratively configured into the cluster configuration. A member node is a cluster node that CLM has recognized to be healthy and well-connected to be used for deploying high availability applications and services. CLM is the authority that decides when a cluster node can be transitioned into a member node.

Another important distinction related to a cluster is whether a logical entity is cluster-wide or node-local. The former indicates that an entity spans one or more cluster nodes and its name is unique in the entire cluster, while the latter defines an entity on a cluster node and accessible by processes on this node only. A checkpoint in the Checkpoint service is an example of a cluster-wide entity whereas a timer of the Timer service is a node-local entity example.

Operations of an area service that are meant for cluster-wide entities are in general not allowed for processes running on cluster nodes that are not in the cluster membership. Accordingly the membership status reported by CLM determines whether the SA_AIS_ERR_UNAVAILABLE or the

`SA_AIS_ERR_TRY_AGAIN` code is returned at the time the area service is unavailable. Exceptions are those operations that enable or detect the formation of the cluster membership.

For those operations that are intended for node-local entities, the cluster node does not need to be a member node at the time when the operation is performed.

Guidelines for service implementers include the use of CLM to determine the membership status of a node before returning the appropriate error code, and specify the behavior of the service API functions when they are placed under various conditions that cause the service to be unavailable within the scope of a node.

11.4.6   Backward Compatibility

Backward compatibility can generally be defined as the ability of a newer version of a system to interoperate with an older version of the same system. For the SA Forum specifications, this is an appropriate concern as the specifications evolve over time. In the context of the SA Forum specifications development, backward compatibility is achieved by following a set of rules when inevitable changes must be made in the newer releases.

A ground rule is that a function or type definition never changes for a specific SA Forum release. If changes are unavoidable in a function or type definition, a new function or type name is defined in the new version. This new name must be built from the original name in the previous version with a suffix that indicates the version where the change is made. This new name then replaces the old one in the new version, that is, there is only one and only entity providing the function or type definition. As an example, `saAmfComponentErrorReport_4()` was introduced in the AMF's B.04.01 release. The function replaces an earlier version of the component error report function `saAmfComponentErrorReport()` in previous releases. As a result, there is only one such function, the latest one in the new version of the specification. This newest version of the function interoperates with all the other existing functions that did not change in the library.

There is one exception though. The type name does not need to be changed if the size of the enum, flag, or union type does not change when new enum values, flag values, or union fields are added to an existing enum, flag, or union types.

It is important to note that the library can provide more than one version of a service implementation. In this case, different versions can be used by different service user processes according to the corresponding version request in the library initialization call. AIS implementers must also ensure that they respect the version numbers provided by the application when the library is initialized:

- New enum values must not be exposed to applications using older versions.
- New or modified error codes that only apply to newer functions are not exposed to applications using older versions.
- Only functions for the specific version are accessible by applications.

11.5   Concluding Remarks

In this chapter we have focused on the aspects related to using the SA Forum specifications from a developer's perspective. We have described the generic programming model for AIS services, in particular the interface relationships between service users and service implementers, naming conventions and type definitions, real-time support, its usage model and library life cycle, and the capability of tracking. We have outlined the type of information one can expect from the API sections for each service in the SA Forum specifications. A number of issues associated with using the SA Forum APIs in practice have been discussed, which included POSIX interactions, memory management, pointers handling, obtaining implementation limits, availability of area service libraries, and the concern of backward compatibility.

The C APIs are the definitive specifications of the programming interfaces in the SA Forum's services and frameworks. However, there is a need for other programming languages as well in practice. Java is one good example and the SA Forum has responded by defining a Java language mapping standard. In the next chapter we will be looking at the Java language mapping specification, its design rationale and usage example.

12

SA Forum Java Mappings: Specifications, Usage, and Experience

Robert Hyerle[1] and Jens Jensen[2]

[1]*Hewlett-Packard, Grenoble, France*
[2]*Ericsson, Stockholm, Sweden*

12.1 Introduction

This chapter first introduces the reader to the history, rationale, and the architecture driving the Java mapping specifications: in particular, their relation to the existing C language specifications.

Subsequently, the chapter details the conventions and patterns/idioms used when developing the Java mapping specifications. With this knowledge, the application programmer will be able to more easily understand and use the Java mappings. The Java mappings are based on being able to understand their semantics by referring to the original C specification. Understanding the underlying mapping logic helps immensely in 'connecting' the Java mappings with the original specifications (see [61, 93] and [63]).

The examples of usage that follow increase understanding as well as shed light on various practical concerns.

12.2 Background

12.2.1 Early Exploration of Java Mappings in Hewlett–Packard

Hewlett-Packard (HP) and the Telecom Infrastructure Division (TID) in particular, has long been involved with the Service Availability (SA) Forum. In 2005 we turned to the service specifications for a particular use quite different from what was originally envisioned. The HP server divisions were investing in, and supporting, Web-Based Enterprise Management (WBEM) [92] as an alternative to Simple Network Management Protocol (SNMP) [53] for both hardware and software management. We in the TID investigated how to interface our software components with the WBEM agent: the Common Information Model Object Manager (CIMOM).

Service Availability: Principles and Practice, First Edition. Edited by Maria Toeroe and Francis Tam.
© 2012 John Wiley & Sons, Ltd. Published 2012 by John Wiley & Sons, Ltd.

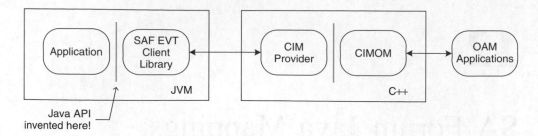

Figure 12.1 An early Java mapping.

In the HP implementation, a CIMOM management agent resided on each operating system host. That implementation was written in C/C++. The standard method of attaching a component with the CIMOM was with a plug-in called a provider as shown in Figure 12.1. The component specific provider, running as part of the CIMOM, communicates with the component. Hence, the communication between the component and the provider is private. If the component is provided with a library that both implements a programmatic interface and the communication scheme with the associated provider, then the component (in this case, our telecom software) would be isolated from the library/provider communication.

Our primary interest at this point was implementing WBEM indications, roughly equivalent to SNMP traps. The component would signal an event of interest via the library Application Programming Interface (API). The provider would receive the event and invoke the appropriate CIMOM function such that an indication would be emitted.

The communication scheme was relatively simple: it was local to the host and it had one producer and one consumer. We used Unix named pipes for the channel.

We could have invented an API for our components to use, but we realized that the Application Interface Specification (AIS) Event Service (EVT) [43] defined an interface that suited our purpose: the event semantics and the SA Forum pattern of 'client library API with encapsulated/hidden communication to a server' fit. We would implement the interface but not the EVT itself.

The missing piece was Java. Some of components were written in Java and the EVT, at the time, had only a C language binding. Over the course of a few weeks, we investigated and developed a Java Language mapping for the EVT. We resisted defining a subset or usage specific mapping. The Java mapping might be useful in the future as a real EVT. As well, defining the task as simply mapping the existing standard provided focus and eliminated the need for any problem specific analysis.

The pleasant surprise during this mapping exercise was that the EVT API, specified for C, was object oriented. Using a few mapping conventions, a Java API emerges almost directly.

Few of the mapping conventions have changed since this first mapping of the EVT. HP went on to define – for internal use – mappings for other SA Forum Services using the same conventions. In 2008, this work was contributed to the SA Forum and provided the greatest part of the basis for the official Java language mappings. These conventions have been used by others to contribute Java mappings.

This chapter details the mapping conventions. They are the key to understanding the Java API.

12.2.2   Java in Ericsson

More than 10 years ago Ericsson started to develop products implemented in Java. During the last six years the main focus has been on Java Enterprise Edition (Java EE) [97] technology, that is, the implemented applications are designed to be deployed on application servers.

We needed to support several deployment scenarios. In some scenarios our Java applications should execute in a heterogeneous environment together with legacy native code but yet be managed in a unified way. For these scenarios the SA Forum specified services provided an appropriate way to achieve the unified management. In other scenarios our Java applications should execute in a homogeneous environment where the management support provided by the application server was adequate.

We therefore wanted the applications to be agnostic about the frameworks used to manage and to control them. We also thought that it was essential to use existing Java standards as far as possible in order to gain wide acceptance in the Java community and to be able to incorporate third party applications in a seamless way. At the same time it should be possible to plug in different back-ends in order to support the wanted deployment scenarios.

The resulting approach was applications using standard Java APIs and plug-ins mapping these APIs to the SA Forum services. The mapping was quite straightforward for the Log service (LOG) and the Notification service (NTF). We published the SA Forum whitepapers [95, 96] describing how the Java Logging API should be mapped to the LOG [40] and how the Java Management Extensions (JMX) [97, 98] notifications should be mapped to the NTF [39].

Mapping the Information Model Management service (IMM) [38] was more complicated. JMX is the Java standard for management. It is mainly used as the interface for management systems and for applications to expose configuration and runtime information to the management systems. However, it is to a lesser extent used for applications to access their configuration. For this purpose there exists a number of different approaches in the Java community – all of them with serious limitations. We have used JMX as the API for applications to access their configuration, but we have also tried JavaBeans [99] for this purpose, as this approach results in concise, straightforward and easy-to-read code. There is obviously a need for further Java standardization in this area.

No Java standard for availability management existed. We therefore implemented enhancements in the application server in order to be able to plug in different availability managers such as the SA Forum Availability Management Framework (AMF) [48]. The applications did not need any modifications. Based on this experience we started the standardization of Availability Management for Java, JSR 319 [100], which now has finished.

Mapping Java standard APIs to SA Forum services implies compromises and some features of the SA Forum services are not possible to use. However, our conclusions were that these compromises are quite acceptable for most Java applications.

12.2.3   The SA Forum Java Mapping Initiative

The pace of introduction of new services continues to increase. Being able to leverage a large community of developers and tools as well as third party products and frameworks is critical to rapid development. The Java community is large as is the rich Java software collection.

In the telecom as well as other domains, carrier grade high-availability (HA) and fault tolerance is the norm: it needs to be present when Java is employed. Important problems to solve are how to provide HA and Operations, Administration, and Maintenance (OAM) integration to Java applications in a mixed (Java plus other programming languages/frameworks) application environment without imposing an intrusive technology.

The availability support that exists for Java EE today covers only the Java parts and does not include availability support for hardware and non-Java applications. It is desirable to have a framework that can provide availability for mixed applications in the same clustered system. The same problem exists for configuration and fault management in mixed technology clusters.

One way to solve these problems is to use a standardized, global framework for availability, and configuration/fault management as the base for both Java and non-Java applications. However, this runs counter to nonintrusiveness. For wholly new components, this framework might be fine. But again, the assumption is that applications and services today are largely built from pre-existing pieces.

A central idea when starting the work on the Java mappings for SA Forum services was to define one set of Java interfaces that is a rather straightforward mapping of the SA Forum interfaces. These interfaces can then be used directly, or, be used for integration with existing solutions within the Java community.

The SA Forum would specify the primary mapping, and then the integration with existing Java standards could then be described in white papers. For the parts where Java lacks a standard, these could be defined within the Java Community Process (JCP).

Not all of the services defined in the SA Forum standard are of interest for Java and Java EE. Some of the services already exist with the same or similar functionality standardized for Java, and there is no need for an interaction between Java and non-Java applications.

Other SA Forum services, for reasons of integration, benefit in having a Java mapping:

- **OAM alignment.** Java applications and native applications will be hosted in the same environment. One example of native applications that are already colocated with Java is databases. It is highly desirable that both types of applications be managed the same way.
- **SA Forum enabled third party products.** When the ecosystem around SA Forum takes off, there will be a more and more SA Forum enabled components available. It is desirable that these products can be used together when mixed in with Java and Java EE.
- **High availability.** When Java EE is to be used in a mixed environment with native applications it is important that one framework controls both the Java and non-Java application redundancy and distribution. A new Java API has been standardized for this: Availability Management for Java [100].

12.3 Understanding the Java Mappings

12.3.1 Java Application Integration Architecture

A programmer or analyst, seeking to use the SA Forum AIS services with Java applications or components, would naturally turn to the Java mapping specifications. These specifications are published as JavaDoc [101]; a format familiar to all Java programmers.

However, that programmer would quickly discover that the JavaDoc alone is insufficient for understanding the API. It is so sparse, that one might conclude that it was written by a junior programmer unschooled in producing meaningful documentation comments.

Why is this? The fundamental reason rests in two design choices adopted by the SA Forum when producing these specifications:

First, the Java API should be on the same level and have the same semantics as the C API while being expressed with appropriate Java constructs and conventions. One could imagine Java interfaces (APIs) to SA Forum services in many forms: for example, encapsulating frameworks, virtualized resources, mapping to existing Java objects such as Streams or MBeans, and so on. A 'higher level' interface might fit better with current Java style, might hide more implementation details, might be more flexible, or might be easier to approach for Java programmers. A 'lower level' interface – some sort of direct mapping via the Java Native Interface – might simplify the specification process and enable a straightforward way to use existing SA Forum service implementations in a Java environment. Any choice of a mapping approach has drawbacks of course. The higher level interfaces would have required significant specification and implementation development plus a period of experimentation and refinement. The lower level interfaces would have exposed much of the irrelevant C implementation details. The Java Native Interface [102] adds risk to the integrity of the Java runtime and is not appropriate for general application programming. We felt that a low level interface would often be wrapped in more appropriate Java by users – creating a multitude of unique and nonstandard interfaces.

The 'middle ground' approach was a choice of practicality. It seeks to remove as many C artifacts as possible without losing a correspondence to the underlying C specifications. This is done by using regular rules or patterns to transform the C specification into Java as opposed to ad hoc interface redesign. These rules are described below.

It is hoped that this mapping approach will result in Java interfaces that have no need to be wrapped unless semantics different from the C specification are to be introduced. At the same time, if necessary, the standardized Java mappings can be used as a foundation and be completely hidden when building a higher level interface. Such higher-level interfaces have, in fact, been developed including some that have been standardized [100].

Secondly, there should be a single specification of SA Forum services. With a choice of producing a Java mapping close to the C language specification meant that there was little to gain by producing a separate, stand-alone specification. As well, with limited resources, this work would have significantly delayed the appearance of such a specification. Finally, with a single specification, there was less risk of errors introduced by duplication of information, and much less risk of inconsistencies between specifications.

Hence, to use the Java mapping specification, one must start by reading and understanding the C language specification. Only then will the Java mapping be understandable.

The Java mapping in JavaDoc is suitable as a reference but rather unsuitable for front to back reading. As a reference, it of course provides class and method signatures as well as package structure. It also provides references into the C language specifications. So, having the C specification immediately available as an additional reference is generally necessary.

While one can follow the references into the C language specifications to understand the Java mapping, it is often simpler to understand the conventions used to produce the mapping. The remainder of this section explains those conventions. Armed with an understanding of the original specification and these conventions, the programmer can effectively use the terse JavaDoc.

12.3.2 Naming

The Java API uses commonly accepted naming conventions [103]. The C language APIs fortunately use constant, variable, typedefs, parameter, and function naming conventions that are roughly compatible in terms of capitalization and noun/verb usage.

On the other hand, the C language APIs rely on identifier (ID) naming to simulate namespaces. The common pattern is to use prefixes to reflect a (missing) package mechanism in C. Since Java has packages, these are used directly. This removes the need for prefixes. The package structure used in the Java mapping is described below.

In addition, the C language API convention of terminating all type definition names with a capital 'T' is not employed in the Java mapping.

With these conventions, it is generally possible to recognize the correspondence between the C and Java names as in this example from the NTF: the typedef definition `SaNtfSpecificProblemT` in C corresponds to the class definition SpecificProblem in Java within the `org.saforum.ais.ntf` package.

There are some exceptions to increase readability and decrease confusion:

1. Exception classes corresponding to `SaAisErrorT` values are all defined in the `org.saforum.ais` package. The 'Ais' prefix is retained. For example, `SA_AIS_ERR_TIMEOUT` in C becomes `AisTimeoutException` in Java.
2. The commonly used 'handle' idiom is embellished with the handle type in its name; for example, `AmfHandle` instead of `Handle`.
3. Certain SA Forum specific constants retain this reference in their names: for example, `SA_TIME_END` appears in both the C and Java mappings as the 'SA' is used as an adjective and not a namespace designation.

12.3.3 Package Structure

The Java mapping uses packages to organize the different service APIs. There are three levels in the package name hierarchy.

At the top level of naming there is a package meant to be included – in whole or in part – by all packages using SA Forum services in Java: `org.saforum.ais`. This package roughly corresponds to definitions found in the SA Forum AIS Overview and C Programming Model specifications (see also: the `ais.h` header file) [93]. This package factors out definitions common to all AIS services. Additionally, it contains library handle interfaces providing lifecycle control and the infrastructure for asynchronous communication between the AIS services and the client code. Finally, it adds a generic factory class to provide the 'bridge' from the client to the SA Forum service implementation.

Below the global level are the various service packages corresponding to the C language specifications. These names are of the form `org.saforum.ais.<service>`. Each service package is named by its abbreviation: for example, 'amf,' 'clm,' and so on. Some services are defined using only a single package.

Other services (e.g., 'ntf') have subpackages at a third naming level based on the 'roles' supported by the service. These names are in the form of `org.saforum.ais.<service>.<role>`. These subpackages contain definitions specific to roles: here, for example, the NTF [39] has 'producer' and 'consumer' packages; the IMM [38] has 'om' (object manager) and 'oi' (object implementer) packages. When subpackages are present, the second level package contains definitions common to the roles.

12.3.4 The Underlying Objects

The naming conventions and package organization provide a framework for organizing the various service specifications and translating IDs. They do not address the correspondence between the C language definitions and Java objects. In particular, they do not guarantee that the C definitions can be cast into an object-oriented, or any other, style.

Fortunately, the C language service specifications are designed using an object-oriented style. This style is not well explained in the specifications, but is easily discovered with careful reading. Once understood, the translation to Java objects is straightforward.

The C language specifications use a pattern of obtaining an 'object reference' via a 'factory,' followed by repeated 'method invocations,' followed eventually by 'object destruction.' The object references are called handles in the C specifications. The factories are initialize functions (e.g., `saNtfInitialize`). The methods are functions associated with an object type and which always take a handle as a first parameter. Object destruction is accomplished with finalize functions (e.g., `saNtfFinalize`). In C there is not, of course, automatic garbage collection. Details of how factories and life-cycles are mapped in Java are covered in the section below.

In Java, the object reference parameter is not needed.

These service objects tend to be large and often contain (internal) references to system resources and cooperating processes (e.g., area servers). In typical usage, only a few such objects are allocated per client process: often one per service. Because of the potential internal references, these service objects retain the explicit destruction methods.

One will not find the service objects defined using Java classes. Instead, they are defined using interfaces. These results in a Java mapping – expressed in Java source code – that is largely devoid of implementation. The notable exception to this rule is the `FactoryImpl` and related factories.

There are, however, many other class definitions in the Java mapping. The use of class definitions is generally reserved for defining data structures. That is, these classes have no methods; not even 'getters' and 'setters.' Objects of these classes are used only as parameters.

12.3.5   Types

The C language specifications define a fairly high number of special data types that are used primarily to pass information between the underlying middleware and the client code. These data types are mapped using the following conventions:

Primitive data types are mapped to their Java counterparts of the same bit-size. For example, a type defined in the C language specification as `SaInt16T` is mapped to a short type in the Java mapping. Unsigned integer types are mapped to the signed Java primitive type of the same bit size. Although this could in theory cause interpretation problems if, for example, arithmetic or relational operations were used carelessly. In practice, most unsigned integer types of the C language specification are used for special purposes, so this is not a problem: unsigned values used as IDs only need an equality check and never computation; flags are defined by their bit positions and the Java bitwise operations operate as expected on signed values. Still, the client programmer needs to aware of this use of signed primitive types. The alternative of defining or using more complex types for unsigned values was not adopted as experience showed that problems did not appear in practice.

When a particular parameter or return value can be one of many different numeric types, `java.lang.Number` is used in the mapping.

Strings are represented in the C language specifications in several ways including null-terminated strings, character arrays, and structures containing array and length fields. All of these types are mapped using the `java.lang.String` class. This means that all length information is unneeded in the Java mapping: it is provided by the `length()` method. As well, all the other String methods are available 'for free.'

In a similar manner, whenever the C specification defines a `struct` that encapsulates a buffer pointer and its length, a Java array is used instead as these always have a length property.

Enumerated types (enum) are used to define most constant values. Java enums define a set of IDs without the need to explicitly assign implementation values to the IDs. However, in the Java mapping enums are mapped to the numeric value defined by the corresponding C language definition. Java clients never need this mapping for internal use. The reason for the mapping is that the Java client may have to forward the numeric value to a non-Java, but SA Forum aware entity, through some other channel than the Java API. For example, the value may be sent in a message or written in a log record.

The mapping is achieved using the standard techniques associated with Java enums (i.e., defining a private constructor and value field) plus implementing `getValue()` defined in `org.saforum.ais.EnumValue`.

The Java mapping uses `public static` final fields of the appropriate primitive type for constants that cannot be defined reasonably using enum's. This is the case when the constant is a numeric value used in calculations and not simply an instance name. As well, some constants are not defined as enum's simply due to historical accident. Constants are used for bitmaps, strings, and (meaningful) numeric values.

C structs are mapped to public classes with public fields. For the sake of simplicity, there is no data hiding or encapsulation (e.g., no 'setters' or 'getters'). These classes are completely free of code. The reasons for this decision are: First, these classes are used only as data transfer objects, that is, they are only meant to transmit data between the client code and the service. Secondly, the likelihood of changing the existing fields of these structures is marginal, since they are likewise exposed in the C API specifications and changing them in newer AIS versions would break the compatibility with older AIS versions and client code. Finally, mapped in this way, the correspondence between the Java and C is evident.

One potential function of such setters and getters would be to provide validity checks for the fields of these objects. However, the AIS servers do the same validity checks when the objects are actually passed to them. It is not clear that the client library code would even be able to perform such checking.

C unions are mapped using an enum discriminator in the 'root' type that indicates the actual underlying type or sub-type. The use of this technique is not always consistent. For example, in org.saforum.ais.imm.AttrValues, one uses the attrValueType field to select which cast to apply to the java.lang.Objects in the attrValues array. In org.saforum.ais.ntf.Notification, the getEventType method returns an Event-Type used to select a cast of the entire Notification object to a subtype. Programmers need to pay particular attention when using these union types!

C arrays are mapped to Java arrays in a straightforward manner. Again, an associated length variable is not needed in Java. In the C language specification, arrays often have a maximum or allocated size plus the 'used' size. This complication is largely eliminated in the Java mapping: arrays are allocated to the size needed. This implies that reuse of allocated variable size data structures is not well supported or encouraged in the Java mappings.

Even more significantly, the responsibility for allocating memory is often different between the C specification and the Java mapping. The C specification style is for the client to allocate (and often deallocate) memory for shared use by the client and the client library. In the Java mapping, the style is that the producer of the data allocates memory and the memory is eventually garbage collected.

C function pointers – used only for callbacks in the C specifications – are mapped to interfaces with a single method defining the signature of the callback. The client implements the callback function by defining a class that implements the callback interface. The callback is provided to the library by providing an object of this class. This implies that callback methods may not be static.

12.3.6 Parameters, Exceptions, and Method Signatures

Functions of the C language specifications are mapped to methods in Java. These methods are defined in interfaces. For the most part, there is one-to-one correspondence in functionality between the C function and the Java method. As explained above, typically the naming has been simplified in the Java mapping in such a manner that the equivalent C function name is obvious. Noteworthy exceptions to the naming of such Java methods is the initialization and finalization of handles:

1. In the C specifications, the library handle lifecycle is controlled by sa<Area>Initialize() and sa<Area>Finalize() functions. Applying the default naming conventions would lead to a finalize() method. To avoid confusion with the well-known method with the same name inherited from java.lang.Object, the method name finalizeHandle() is used instead. For consistency, initializeHandle() is used instead of simply initialize().
2. For handles other than those referring to a service library instance, the pair of create <Entity>()/destroy() method names is used for life-cycle control.

There are cases where a single C function is re-factored to several Java methods with different parameter lists. These C functions use certain parameters to govern what the function actually should do while in the Java mapping this information is encapsulated in the method name. This is an exception to the normal naming conventions. The documentation of the Java methods contains a reference to the equivalent C functions, including information on the specific parameter values that belong to a 'refactored' Java method.

In Java, memory is never freed explicitly. Therefore, the C API functions used to free memory do not have their counterparts in Java. As noted above, the service objects do retain their 'destruction' methods so that system resources can be freed.

As already described, the C 'handles' are never passed as parameters in the Java methods.

The C language specifications employ a typical C error handling convention: the return value of each function indicates success or failure along with an encoding of the error cause. The enum type SaAisErrorT defines these errors in the C specifications. SA_AIS_OK indicates successful execution, whereas SA_AIS_ERR_<SOME_ERROR> values represent possible errors.

In the Java mapping, errors are handled using exceptions. The org.saforum.ais package defines corresponding exceptions for each SaAisErrorT value. So, Java methods may throw one of these exceptions and do not return a status value. The return value can then be used to return other data that in the C specifications is returned with an 'out' parameter.

There is a superclass, AisException, from which all the other AIS exceptions are subtyped. The API never throws this exception: explicit exception subtypes are always defined for each method. The superclass only serves to factor out common functionality.

The enumerated type AisStatus corresponds to the C SaAisErrorT values (including SA_AIS_OK – but there is no 'AisOk' exception). These C language enum values are available for each exception in the cause field. The status value is used to indicate the status of operations executed asynchronously in the Java mapping; that is, the status of an operation other than the one initiated by the actual method call. See, for example, org.saforum.ais.imm.om.AdminOwner. adminOperation.Invoke.

The replacement of handles with object references, the use of exceptions instead of a status return, the use of the (now available) return value for other data, and the elimination of descriptive parameters such as allocated and used array sizes results in the Java methods having far fewer parameters: approximately one half, on average compared to the C functions. Also, the elimination of the status return and producer allocated memory result in the Java signatures having very few 'out' parameters. The naming conventions shorten and simplify parameter, type, and method names. In total the Java signatures are quite different while their C counterparts are usually readily recognizable.

12.3.7 Factories, Callbacks, and Life-cycles

The prototypical C language AIS client library instance life-cycle is based upon the following model:

1. The client initializes a dynamic entity – an instance of the service client library – representing and implementing an association between the client and the service area server.
2. The client code communicating with the area server via the entity.
3. The client code shutting down the association when it is no longer needed.

In the Java mapping this generic model is translated to the following:

1. The client code obtaining an <Area>Handle object (an instance of the service client library).
2. The client code communicating with the local library and area server by invoking methods on the library handle or on other closely associated objects always obtained from this 'root' or 'core' object. As well, the area server initiates communication to the client via the callback mechanism.
3. The client code shutting down the library instance by invoking its finalizeHandle() method. The client should not retain any references to this now defunct object so that it can be garbage collected.

The client is free to create and use more than one <Area>Handle object. The only exception applies to the AMF service, which allows only a single library handle to carry out operations that require registration.

The two life-cycle operations of initialization and the finalization are done rather differently: finalization is a method invoked on the existing object; initialization via the generic factory framework is significantly more complicated and is explained here.

The factory framework consists of:

- The *factory method* `initializeHandle` defined by the org.saforum.ais.Factory interface. All the specific service factory classes implement this interface. This is a generic interface parameterized by the service interface 'S' representing the service library handle type; the callback class 'C' representing the set of callback objects used by the libraries to invoke client code.
- *Service handle types* (interfaces). These define the core objects for each service. The initialize methods return objects of this type.
- *Callback classes* defined for each service. These classes are nested within the handle interfaces. These classes contain a field for each callback function.
- A `<Area>HandleFactory` *class* for each service. These classes are derived from, and share the implementation from, the `org.saforum.ais.FactoryImpl` class. This class is the only one in the Java mapping that contains any significant code: everything else is either data or interface definitions.

Because of the use of interfaces, the factory methods cannot be static. Hence, it is required to create a factory object using the default constructor prior to calling the factory method. Details of the steps client code must take are covered below in the example.

Once the root object is obtained from the factory, the client may invoke methods on it, including methods that are used to obtain/create related objects representing functionality and data subsets. The life-cycles for these related objects follow one of the following patterns:

1. 'Getter' methods return references to objects that have a one-to-one association with the library handle: that is, all invocations of the getter method will return the same object. These objects simply bundle the methods providing part of the service. For example, AmfHandle defines many getter functions: getComponentRegistry(), getCsiManager(), and so on. These objects do not require any special method for clean-up after use as their life-cycles are associated with that of the core object.
2. 'Creator' methods – following the naming convention of create<Entity> – return references to <Entity> objects that are newly created on each method invocation and may also allocate system resources. For example, the Consumer interface defines a createReader() method that creates and returns a new Reader object upon each invocation. These created objects will usually have a method named destroy() that must be invoked on the object when it is no longer needed so that system resources can be cleanly deallocated. As in the case for core objects, the client should drop all references to such objects after invoking `destroy()` so that the garbage collector can recover their memory.

12.3.8   Callbacks and the Selection Object in Java

The library handle provides a mechanism for asynchronous communication between the client and the associated area server. This mechanism is shared by all the services, and hence, is factored into the `ais` package in the Java mapping.

There are three elements:

1. **The set of callback objects that are invoked when a requested asynchronous operation is ready.** The callback objects are provided when the library handle is initialized (see above). They are permanently associated with the library handle until the library handle is finalized. The client code is free to provide null for any callback object if the client does not use the operation associated with that particular callback. However, if an operation would require a callback and it is requested without a callback defined, the AisInitException will be thrown.

2. **Methods to control the dispatching of callbacks.** Following the model defined in the C language specifications, the client library never invokes a callback using a thread not provided by the client itself. This allows the client to be truly single threaded if necessary. Pending callbacks are always invoked in the context of one of the dispatch() methods defined in org.saforum.ais.Handle.

3. **Methods to detect pending callbacks.** The library handle provides several forms of the dispatch() method. Some will return immediately if there is no pending callback, some will block until a callback becomes pending, and some will specify a time-out before returning. Furthermore, the library handle provides methods for checking the availability of callbacks without actually dispatching them (see the hasPendingCallback() methods). Last but not least, the library handle supports multiplexed callback selection by integrating this API with the New I/O (NIO) [104] framework: the getSelectableChannel() method will return a SelectableChannel object that can be used by NIO selectors. This is the closest to the semantics of the C language selection object.

12.4   Using the Java Mappings

12.4.1   Integrating AIS Services with Java Applications

In this section, we will present a simple client code example that both illustrates the Java API use as well as highlights some fine points and practical concerns when the mapping specification is put to use.

The inspiration for this example is Appendix C of the NTF specification [39]. The appendix contains C code that sends and receives security alarm notifications. We don't attempt to translate the example in all the details but the basic functionality is the same.

When one compares the C and Java code for this example, a key difference is immediately apparent: most of the C code is occupied with obtaining and using storage; most of the Java code is the instantiation of objects – with the Java code being substantially shorter. Both versions of the example devote (or should devote) significant effort to error/exception recovery.

The Java example is cast in the light of two components: one that checks access attempts to various 'objects.' If the access is denied (always in this example), a security alarm notification is emitted. This component is called the 'GateKeeper' and we have defined a class with this name.

The second component, named the 'GateWatcher' waits for security alarm notifications of a certain (elevated) severity level emitted by components from the GateWatcher's domain (i.e., the same vendor). The GateWatcher does nothing more than print out a line of information for this example.

This example, as well as the C language version in the specification appendix, is not meant to – and does not – illustrate best practices of modularity, robustness, or other elements of politically correct style.

12.4.1.1   Finding the SA Forum Service

For a client to use an SA Forum service, several prerequisites must be met: the service implementation must be installed according to the service vendor's guidelines, the Java mapping classes must be available to the client Java code, and via some appropriate configuration the mapping library must be able to locate and load the Java service library implementation classes.

Different vendors and different services will have varying approaches to installation. We won't cover these here. In any event, the client software design is independent of the particular approach.

The Java mapping classes must be available. In this example of the NTF, we will need two jar files containing four packages: `ais.jar` and `ntf.jar`. They need to be available to the client's classloader(s), which often means they can be found on the classpath.

The third prerequisite is the trickiest: how to locate – in fact even how to name – the class supplied by the service vendor that implements (for the NTF in our example) `public static NtfHandle` initializeHandle(NtfHandle.Callbacks, Version). The SA Forum Java mapping does not specify vendor class or package names for the implementation. The only requirement is that the vendor must implement this initialization method. Without the classname, we can't even use the classpath to search for this method.

What the Java mapping does specify are two property keys (or, 'property names') for each service: one that is used to find the name of the class implementing initializeHandle, a second that is used to find the URL that a URLClassloader can use to find the named class. When the client allocates a factory for the service, the System properties are searched using these keys. One finds the definition of the keys for the NTF in the NtfHandleFactory class.

These properties might be set on behalf of the client when the client is launched, perhaps on the command line. That is, the client inherits these properties. The client *might* also be able to set these properties programmatically. We say 'might' because the client might not have permission to set these properties. The means for the client to set these properties was included in the specification for testing purposes: it is not a mechanism suggested for production use.

For our example, we will actually set the properties using client code:

```
/**
 * Change this as needed to adopt to vendors
 * and local install environment.
 */
static final String SAF_IMPL_PROVIDER_URL =
"http://localhost:8080/saforum/providers/com.provider/impl/classes/";

/**
 * Class name of implementation of factory
 */
static final String SAF_IMPL_NTF_CLASSNAME =
"com.provider.ais.ntf.ntfImpl.NtfHandleFactoryImpl";
/**
 * Set up implementation configuration information for
 * SA Forum services. This needs to be done before
 * instantiating any services. Note that the execution
 * environment, including security
 * settings applicable to system properties, may override
 * this configuration.
 */
static protected void safSetup() {

System.setProperty(NtfHandleFactory.IMPL_CLASSNAME_KEY,
      SAF_IMPL_NTF_CLASSNAME);
  System.setProperty(NtfHandleFactory.IMPL_URL_KEY,
      SAF_IMPL_PROVIDER_URL);
}
```

12.4.1.2 Instantiating a Service Instance

With the configuration set, we can now obtain a NTF instance.

In our example, each GateKeeper object has its own NTF instance. An alternative approach would be that different functional areas of the client code would share a single service. It is mostly a matter of client architecture choice: 'mostly' because creating a service instance can be rather heavyweight in terms of communication between servers and in the maintenance of shared state between servers. This might, in practice, constrain architectural choice.

Both the `GateKeeper` and the `GateWatcher` extend an abstract class called `Notification-Wrapper` that holds shared data and logic. This class also suggests a common lifecycle of initialize, use, and then shut down. The handle field is inherited from `NotificationWrapper`.

In the `GateKeeper` example, the initialize method instantiates the NTF as well as the associated `producer`. Data that will be sent with each notification is also initialized:

```
public void initialize() {

  Version version = new Version('A', (short) 2, (short) 1);

  NtfHandle.Callbacks callbacks = new NtfHandle.Callbacks();

  NtfHandleFactory factory = new NtfHandleFactory();
  try {
   handle = factory.initializeHandle(callbacks, version);
   producer = handle.getProducer();
  } catch (AisException e) {
        e.printStackTrace(); // acceptable for an example
  }

          classId = new ClassId();
          classId.vendorId = 33333;
          classId.majorId = 995;
          classId.minorId = 1;

      serviceProvider =
          new ServiceUser();
      serviceProvider.value =
    new ValueStringImpl("switch configurator");
    securityAlarmDetector =
          new SecurityAlarmDetector();
  securityAlarmDetector.value =
    new ValueStringImpl(
      "com.example.notifications.GateKeeper.checkAccess"
      );
}
```

Of particular note:

1. The GateKeeper is requesting version A.02.02 even though later versions are available.
2. A Callbacks structure is allocated, but no callbacks are supplied here since the GateKeeper is never going to request an operation that would require them.
3. Once the NTF instance is obtained (i.e., handle is initialized), we immediately request the associated Producer object. The handle will be used again only to shut down the service.
4. Marshaling data and requesting service.

The `GateKeeper`'s main task is to check access and return 'ok' or not. If access is denied, a security alarm is raised and a `SecurityAlarmNotification` sent:

```
public boolean checkAccess(String user, String role, String object,
      String operation) {
 boolean ok = false;    // for this example, we're not
      // letting anyone do anything

 if (! ok) {
          ServiceUser serviceUser = new ServiceUser();
          serviceUser.value = new ValueStringImpl(user);
          Date now = new Date();

          try {
           SecurityAlarmNotification alarm =
   producer.createSecurityAlarmNotification(
           EventType.OPERATION_VIOLATION,
           object,
           classId,
           now.getTime() * 1000 * 1000, // SAF uses nanoseconds,
           // not milliseconds
           ProbableCause.AUTHENTICATION_FAILURE,
           Severity.SEVERITY_MAJOR,
           securityAlarmDetector,
           serviceUser,
           serviceProvider);
   alarm.setAdditionalText("Access Denied!");
   alarm.send();
 } catch (AisException e) {
          e.printStackTrace(); // acceptable for an example,
                // but not for production
 }
 }
```

Some of the data sent is specific to each access check and some is common and reused in every notification. The SA Forum has defined time stamps in terms of nanoseconds while Java only provides milliseconds. This requires us to make a conversion. The mapping specification *could* have defined an automatic conversion, but this would have violated the principle of following the semantics of the original C language specification.

12.4.1.3 Modularity and Exceptions

Up to this point, we have not handled exceptions very reasonably. We will see better examples below.

One design choice that we have made is to insulate code using the GateKeeper and Gate-Watcher classes from the NTF: no SA Forum types or exceptions are visible from the outside. Presumably, a completely different type of alarm notification scheme could replace the SA Forum defined service. But, how do we report problems when they inevitably occur? We could define our own exception classes and map the AisExceptions to these. Or, we could use the AisExceptions even if we do not use the SA Forum service. Still another approach would be to map the appropriate AisExceptions to Java unchecked exceptions. Similar design choices pertain to SA Forum type definitions in general.

The design space is large and the choice is largely driven by the client application, not the mapping specification. In the following section covering containerized applications, we will see yet another approach: mapping between the NTF and JMX.

12.4.1.4 Being Called Back

While the `GateKeeper` sends notifications in response to access violations, the `GateWatcher` receives notifications when the NTF delivers them. The `GateWatcher`'s initialization involves more than obtaining the consumer object: a filter defining notifications of interest must be constructed and then used to subscribe to notifications. As well, a callback method must be provided to NTF so that it can be invoked when a message matching the filter arrives. Here is the initialization code:

```
public void initialize() {
  Version version = new Version('A', (short) 4, (short) 1);

  NtfHandle.Callbacks callbacks = new NtfHandle.Callbacks();
  callbacks.notificationCallback = new OnNotificationCallback();

  ClassId classIds[] = new ClassId[] { new ClassId() };
  classIds[0].vendorId = 33333;
  classIds[0].majorId = 995;
  classIds[0].minorId = 1;

  Severity severities[] = new Severity[] { Severity.SEVERITY_MAJOR,
  Severity.SEVERITY_CRITICAL };

  NtfHandleFactory factory = new NtfHandleFactory();
  try {
   handle = factory.initializeHandle(callbacks, version);
   Consumer consumer = handle.getConsumer();
   NotificationFilters filters = new NotificationFilters();
   filters.securityAlarmFilter =
   consumer.createSecurityAlarmNotificationFilter(
       null,
       (FilterName []) null,
       null,
       classIds,
       null,
       severities,
       null,
       null,
       null);
   consumer.createSubscription(filters); // we drop the return value
  } catch (AisException e) {
       e.printStackTrace();
             // acceptable for an example, but not for production
  }
 }
```

Of note:

1. The GateWatcher requests version A.04.01 that includes changes to how filters are specified (Java mapping in progress).
2. We filter on a single ClassId: that of our partner, the GateKeeper. We don't want to receive (here) other security alarms.
3. We provide a callback for security alarms only. We won't get notifications for other alarms, state changes, and so on.

4. The cast of null to a FilterName[] is required to disambiguate between a deprecated version of createSecurityAlarmNotificationFilter. In the C language API, the name of the function changes with each change made in its specification. In Java, we use features in the language to reduce such renaming.

5. We don't plan to ever unsubscribe, so we don't save the return value from createSecurityAlarm-NotificationFilter.

Here is what the callback looks like:

```
class OnNotificationCallback implements NotificationCallback {

  @Override
  public void notificationCallback(
    Subscription subscription,
    ConsumedNotification notification) {
   try {
   SecurityAlarmNotification alarm =
        (SecurityAlarmNotification) notification;
   // and if the cast fails?
   Date eventTime = new Date(alarm.getEventTime() / (1000*1000));
   // convert to milliseconds
   System.out.println(
                "Security Alarm at " +
                eventTime.toString() +
                ", with severity " +
                alarm.getSeverity() +
                ", type: " +
                alarm.getEventType() +
                ", probable cause: " +
                alarm.getProbableCause() +
                " (" + alarm.getAdditionalText() + " "
                + alarm.getLocalizedMessage() + ")");
  } catch (AisException e) {
   e.printStackTrace();
    // acceptable for an example, but not for production
  }
 }
}
```

When called, a few fields of the notification are printed out for example purposes. What would it mean if the cast to `SecurityAlarmNotification` failed? It would mean an error in the service implementation since it should only provide security alarm notifications to security alarm notification callback methods. This looks like a place to improve the mapping specification: the typing could be tighter so that no cast would be needed.

In SA Forum services, the client is responsible for providing the threads used to invoke callbacks. They are supplied with the dispatch method (and its variants). Everything above was initialization prior to 'use.' The use phase of the `GateWatcher`'s lifecycle is the `listen()` method:

```
public void listen() {
  try {
        handle.dispatchBlocking();
  } catch (AisException e) {
```

```
      e.printStackTrace();
            // acceptable for an example, but not for production
 }
 return;
 }
```

What is not shown here is the creation of the thread that (eventually) calls listen. The way this example is written, `listen` is going to repeatedly block waiting for notifications to arrive, invoke the callback for each, and then block again. Only when the service is shut down does this thread return from `dispatchBlocking`, and hence, from `listen`.

12.4.1.5 Cleaning Up

When it is time to stop checking access or stop receiving notifications, we can simply shut down, or 'finalize,' the NTF instantiations. The code to do this is identical for both consumers and producers and is contained in the `NotificationWrapper` class:

```
public void shutdown() {
  if (handle == null) return;   // already shutdown, or tried to

  long timeout = 500; // sleep timeout milli's if library
        is not responding
  int attempts = 3; // max tries to shutdown, double timeout
       each attempt
  do {
   try {
    handle.finalizeHandle();
   } catch (AisLibraryException e) {
    break;    // library dead, can we report this?, give up here
   } catch (AisTimeoutException e) {
    try {
     Thread.sleep(timeout);
    } catch (InterruptedException e1) { // ignore this, just try again
    }
    timeout >>= 2;    // wait long next time if we try again
    continue;
   } catch (AisTryAgainException e) {
    continue;     // might work next time?
   } catch (AisBadHandleException e) {
    // don't bury this, it's a bug! re-throw!
    throw new IllegalStateException("
    Expected the notification handle to always be valid",e);
   }
  } while (--attempts > 0);

  handle = null;
   // indicates we're done, and let's the garbage collector go to work
 }
```

The basic code is pretty simple: if we haven't already shut down (i.e., the handle is **null**), then call `finalizeHandle()`. What we will illustrate here is handling and attempting to recover from exceptions. This is the reason for the complexity.

Two of the exceptions deal with (hopefully) transient situations: 'Timeout' and 'TryAgain.' In either case, attempting the operation again might result is success. We keep track of these retries with the attempts counter which counts down from the initial value. With timeouts, we sleep between tries (and increase the time we sleep on each attempt).

The `AisLibraryException` indicates an internal error in the service (it is not our fault ... well, maybe, if we're coding in C with run-away pointers!). We can't recover from this and might not even be able to create a new service instance. In many situations, we should report this and perhaps initiate an application restart. Here, since we are shutting down anyway, we just 'swallow' it.

The `AisBadHandleException` indicates that, most likely, there is an error in our code. For instance, this exception is thrown when the handle has already been finalized. But we checked for the handle being null! Maybe somebody might change the code in the future and reporting this exception – we can't actually recover here – will help find a bug. Maybe there is a bug in the code already? What if two threads were to call shutdown at the same time? Might there not be a race to shut down the service first? The reader is invited to design a thread-safe version of shutdown without making assumptions about `finalizeHandle`'s implementation.

Even if we never successfully execute `finalizeHandle`, we set the handle to **null** (except for bad handles, where it might be useful to not clobber the handle reference). Alternatively, we might have decided to throw an exception if we couldn't shut down in the given number of attempts, but for an example it is probably OK since our entire application is going to terminate. A SA Forum service must be able to handle service instances terminating ungracefully in any event.

12.4.1.6   The Big Picture

Normally, the producers and consumers of notifications would be in separate processes. For this example, we run them together. Here is the code that runs each through their respective life-cycles:

```
public static void main(String[] args) {
    NtfImpl.safSetup();

    GateWatcher watcher = new GateWatcher();
    watcher.initialize();
    GateKeeper gate = new GateKeeper();
    gate.initialize();

    new Thread(watcher).start();
    gate.checkAccess("Robert","Administrator", "Switch042", "disable");

    watcher.shutdown();
    gate.shutdown();
}
```

12.4.2   Integrating AIS Services with Containerized Java Applications

When an existing – and commonly used/supported – interface for Java already exists, and when this overlaps with the functionality provided by an SA Forum service; a key idea is to continue to expose the already available Java standards to the client applications. These interfaces will then be integrated and implemented with SA Forum services using the Java mapping APIs.

These existing standards are generally part of Java EE. Java EE provides an execution container to Java code – a container composed of various services, some of which address cluster-wide and distributed execution.

The integration between the JMX and the SA Forum NTF is one example. The diagram of Figure 12.2 illustrates how a Java application can use JMX to send notifications. The Java application

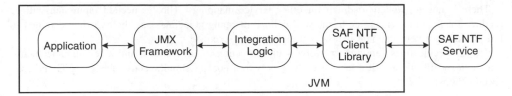

Figure 12.2   Integration of JMX and NTF.

is only using the JMX framework and design patterns. Integration logic is plugged-in between the JMX framework and the notification service. When producing notifications, this logic transforms the JMX notifications to NTF notifications and then forwards them to the SA Forum service. Similar transformation happens in the reverse direction. In this way all notifications in a mixed environment will be connected using the same service for sending and receiving notifications.

12.4.3   AIS Services in Mixed Language and Mixed Implementation Environments

It is unlikely that a single process is going to mix the usage of Java and C library API's (e.g., through the Java Native Interface). On the other hand and as described above, a single application composed of C as well as Java components (processes) is quite likely to use both interfaces. The likelihood is increased further if one considers a set of applications in a cluster.

When using the Java mapping directly or when implementing integration logic, what needs to be considered in this mixed environment? We have already discussed the provisions made to obtain the underlying C encoding for Java `enum` values for use when processes need to communicate outside of the SA Forum service channels. Other Java primitive values are clearly identified with the corresponding C types and could be likewise communicated. The communication of complex types has not been addressed in the mapping specifications although some tentative work is underway to define a SA Forum 'external representation.'

Another data consideration is *endian*-ness. However, this is a concern even in the C-only world with different machine architectures. Within the SA Forum services, this is handled by the service implementations insuring that data is properly converted as needed (or otherwise well defined for opaque client data). This is below, and is not a concern for, the Java mapping level.

Because the C and Java specifications are at the same level; because the mixed environment scenario was considered early in the mapping; and finally because we expected some Java service APIs to be implemented on top of C language implementations (and vice-versa!) the mapping specifications were written to be lossless in both directions: Java to C and C to Java. This avoids most of the problems that one might encounter in a mixed application language environment.

12.5   Going Further

12.5.1   The Java Mapping Roadmap

As of this writing, five SA Forum services have Java mapping specifications (plus the global mapping package). These have been developed by contributors on an 'as needed' basis. The advantage of this is that those working on the Java mappings usually have some direct and early feedback from real-world implementation experience. As well, the services that have Java mappings are, in some ways, the most used.

There is not an explicit roadmap for other service mappings. The 'as needed' approach will be retained for the time being. Due to the well-defined mapping guidelines and the experience gained so far, producing a new mapping is now a straightforward exercise.

The C specifications continue to evolve with corrections, clarifications, and a few enhancements. The Java specifications will track these changes with a goal of simultaneous releases of C and Java specifications.

12.5.2 Related Java Standards and Other References

As there did not exist any standardized way to handle HA in Java, we started a new JSR (Java Standardization Request) for this [100]. The purpose of the new JSR is to enable external availability managers, such as SA Forum AMF, to supervise and to control Java runtime entities. Thereby HA can be achieved in a standardized way. Availability Management for Java is not limited to SA Forum AMF but could be integrated with other HA frameworks as well.

Availability Management for Java has the following features and requirements:

- It provides the means by which an availability manager can control and supervise distributed Java processes and runtime entities, called availability units, within these Java processes. The availability manager controls the lifecycle of the availability units and of the Java processes and it can supervise their health. In case any error is detected, the availability manager may be configured to execute recovery actions.
- It supports different availability managers. There may be different availability manager with different capabilities. They may, for example, provide advanced error recovery actions such as recovery escalation, and they may support switchovers, when an availability unit in one Java process takes over the work of an availability unit in another Java process.
- It may be used for Java EE application servers and applications executing on Java EE 5.0 or later. The concepts are also possible to use in a Java Standard Edition environment, if complemented with definitions of the availability units to be controlled.
- It supports both Java EE applications that are not at all aware of the control of the availability manager and those that are availability aware. Most of the specification is implemented by the application server and current Java EE applications can be supported without any change.

Availability Management for Java can be integrated with SA Forum AMF by implementing a specialized agent. Thereby the application server instances will be mapped to AMF container components and the Java EE deployment units (applications or stand-alone modules) executing in an application server instance can be mapped to AMF contained components. It can be decided at deployment time if a deployment unit shall be mapped to a contained component or if it should be considered as an integrated part of the server instance and thereby included in the container component.

Availability Management for Java has a simpler state model than AMF due to the need to fit other availability managers as well. Basically, the server instances and the deployment units can be instantiated, activated, deactivated, and terminated. When setting the HA state of a component for a component service instance, this will map either to activate or deactivate of the server instance or deployment unit. Setting HA state active will map to activate and setting any of the other HA states or removing the HA state will map to deactivate. The attributes and the transition information (mapped to activation reason or deactivation reason) given when setting the HA state of the component will be forwarded to the deployment unit. The simplified state model implies that to each component it is meaningful to assign only one component service instance.

Container components and contained components must not be collocated in the same service unit and service units containing container service units and service units containing contained components must belong to different service groups. The only redundancy model supported by AMF for service groups containing container components, that is, server instances, is the N-way-active redundancy model. A service group containing contained components, that is, Java EE deployment units, can be associated with any of the redundancy models defined by AMF, except the N-way redundancy model which requires a component capability model not supported by Availability Management for Java.

Framework-invoked health checks are supported if optional health check methods are implemented for the individual deployment units.

13

SA Forum Middleware Implementations

Mario Angelic[1] and Ulrich Kleber[2]

[1]Ericsson, Stockholm, Sweden
[2]Huawei Technologies, Munich, Germany

13.1 Introduction

This chapter presents the two leading open source implementations of the Service Availability (SA) Forum Service Availability Interface specifications:

- OpenHPI, which implements the Hardware Platform Interface (HPI) [35]; and
- OpenSAF which delivers the services of the SA Forum Application Interface Specification (AIS).

Both projects were initiated and are driven by leading telecom and computing companies, many of whom contributed to the specifications themselves. These implementations can be deployed by themselves or together to provide a complete SA Forum solution.

13.1.1 OpenHPI

The OpenHPI an open source project [71, 105] aims at implementing the SA Forum HPI specification. The project started in 2004 and quickly provided its first HPI implementation. With the release of OpenHPI 2.0 the project implemented the HPI B.01.01 version of the interface specification in the same year as it was published by the SA Forum. The project remained well on track implementing the new versions of the specification. At the time of writing it provides the HPI interface according to B.03.02 version of the SA Forum standard [35]. The only exceptions are the diagnostics initiator management instrument (DIMI) and the firmware upgrade management instrument (FUMI) functions introduced by the B.02.01 version of the specification, which are not provided yet by some plugins.

OpenHPI adopted a design that makes it easily adaptable to different hardware architectures. The project also goes a few steps further than just implementing the HPI specification. HPI users can take advantage of various additional support programs the project provides:

- There is an extensive set of client programs – examples that show the use of the different HPI functions. These same examples can also be used for testing and/or monitoring the hardware.
- An hpi-shell provides an interactive interface to the hardware using HPI functions; and
- Different graphical user interfaces allow for browsing the hardware.

These support programs helped enormously the adoption of the HPI specification by different hardware manufacturers and implementers of hardware management software.

13.1.2   OpenSAF

The OpenSAF [91] open source project was launched in mid-2007. It is the most complete and most up-to-date implementation of the SA Forum AIS specifications that can be deployed even in carrier-grade and mission critical systems. OpenSAF is an open-source high-availability middleware consistent with SA Forum AIS developed within the OpenSAF Project. Anyone can contribute to the codebase and it is freely available under the LGPLv2.1 license [106].

OpenSAF has a modular build, packaging, and runtime architecture, which makes it adaptable to different deployment cases whether there full or only a subset of the services is required. Besides the standard C Application Programming Interfaces (APIs), it also contains the standard Java bindings for a number of services as well as Python bindings for all the implemented AIS services, which also facilitates a wide-range of application models.

In the rest of the chapter we take a more detailed look at each of these open source projects. We start with a closer look at the OpenHPI project. We continue with the discussion of the OpenSAF project: The evolution of the middleware from the start of the project till version 4.1 – the latest stable version at the time of writing of the book. We also describe the architecture and the main implementation aspects of OpenSAF, as well as management aspects of OpenSAF and process illustrating deployment of OpenSAF on target environments.

13.2   The OpenHPI Project

OpenHPI is one of the first implementations of SA Forum's HPI specification discussed in Chapter 5 and it is a leading product in this area. It has also become part of many Linux distributions.

This section describes the architecture and codebase of this open source project.

13.2.1   Overview of the OpenHPI Solution

The idea of an open source implementation of HPI immediately faces the problem that an interface to hardware must be hardware-specific. At the same time, an open source implementation has the goal to provide a common implementation of the interface.

OpenHPI addressed this problem by introducing the plugin-concept. Plugins contain the hardware specific part and use a common interface toward the OpenHPI infrastructure.

This approach created a new de-facto standard of the plugin application binary interface (ABI).

The client programs provided within the OpenHPI project can be used to exercise OpenHPI as well as third party HPI implementations. This demonstrates the portability of client applications that use the standardized HPI interfaces and the advantage of standard interfaces in general.

OpenHPI uses the client-server architecture which defines an interface between the server (OpenHPI daemon) and the client (the library linked to the HPI user process). OpenHPI being an open source project allows other implementations of the HPI specification to take advantage of this architecture by using the same interfaces and by that it also created a new de-facto standard. This creates a desired flexibility for the industry.

The OpenHPI client-server architecture is illustrated in Figure 13.1. This solution also provides remote management access to the hardware. As shown in the figure OpenHPI has introduced the hardware specific plugin concept on the server side. This makes the solution easily adaptable to different hardware architectures.

Some commercial HPI implementations use the interface between the OpenHPI base library and the OpenHPI daemon as defined by the OpenHPI client-server architecture. This allows for a combination of OpenHPI components with their own implementation. Nevertheless in most cases hardware vendors not yet provide HPI implementations or OpenHPI plugins in their deliveries automatically.

13.2.1.1 Base Library

The OpenHPI base library provides the user interface with all HPI interfaces as defined in the SA Forum specification. It opens the sessions to the HPI domains by connecting to the correct OpenHPI daemon.

The initial configuration of the base library is given in a configuration file, which defines the Internet Protocol (IP) address and port at which the library finds the daemon for each specific domain. This configuration can be modified at runtime by dynamically adding new domains to the library configuration or deleting existing ones from it.

13.2.1.2 Socket Interface

The message interface between base library and OpenHPI daemon uses a special marshaling, which supports different endianism and data alignment. Thus, the client together with the base library may run on a hardware architecture, which is different from that of hosting the OpenHPI daemon.

13.2.1.3 OpenHPI Daemon

Every OpenHPI daemon implements one HPI domain. This includes the data structures as defined by the HPI specification; namely, the domain reference table (DRT), the resource presence table (RPT), and the domain event log. It is also responsible for the uniqueness of the HPI entity paths.

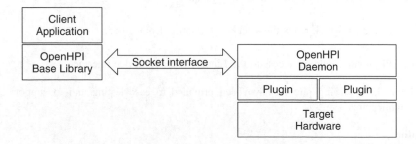

Figure 13.1 OpenHPI architecture.

The OpenHPI daemon is capable of hosting multiple plugins to access different hardware. Having multiple plugins loaded at the same time is especially important in cases where the hardware architecture provides a separate management controller like it is the case with the Advanced Telecommunication Computing Architecture (ATCA) shelf manager. Thus, a single daemon can manage multiple shelves using multiple instances of the ATCA `ipmidirect` plugin and at the same time host a specialized plugin to manage the local hardware, which cannot be accessed via the ATCA `ipmi` bus (see an example configuration in Section 13.2.2.4). This principle applies of course also for all other bladed or non-bladed architectures where there is a separate management controller with out-of-band management.

The OpenHPI daemon is configured with an initial configuration file, which specifies the plugins to be loaded at startup and their plugin specific parameters. As in case of the base library, plugins can be loaded or unloaded dynamically at runtime as necessary.

13.2.1.4 OpenHPI Plugins

Plugins are at the heart of OpenHPI as they serve as proxies to the different hardware management interfaces and protocols.

The plugin ABI allows that a plugin provides only a subset of the functions. This way plugins can also share the management tasks for some hardware such as have a specialized plugin for diagnosis via DIMI.

The open source project provides the following plugins:

- **ipmidirect**
  An Intelligent Platform Management Interface (IPMI) plugin designed specifically for ATCA chassis. It implements IPMI commands directly within the plugin.[1]

- **snmp_bc**
  An Simple Network Management Protocol (SNMP) based plugin that can communicate with the IBM BladeCenter (International Business Machines), as well as the IBM xSeries servers with (Remote Supervisor Adapter) RSA 1 adapters.

- **ilo2_ribcl**
  An OpenHPI plugin supporting the (Hewlett-Packard) HP ProLiant Rack Mount Servers. This plugin connects to the iLO2 on the HP ProLiant Rack Mount Server using a Secure Socket Layer (SSL) connection and exchanges information via the Remote Insight Board Command Language (RIBCL).

- **oa_soap**
  An OpenHPI plugin supporting HP BladeSystems c-Class. This plugin connects to the Onboard Administrator (OA) of a c-Class chassis using an SSL connection and manages the system using an (eXtensible Markup Language) XML-encoded simple object access protocol (SOAP) interface.

- **rtas**
  The Run-Time Abstraction Services (RTASs) plugin.

- **sysfs**
  An OpenHPI plugin, which reads the system information from `sysfs`.

- **watchdog**
  An OpenHPI plugin providing access to the Linux watchdog device interface.

In addition the following plugins have been provided to ease testing and to support complex hardware architectures:

- **simulator**
  An OpenHPI plugin, which reports fake hardware used for testing the core library.

---

[1] This plugin replaces the 'ipmi' plugin that used the open source code OpenIPMI to generate the commands.

- **dynamic_simulator**
  An OpenHPI plugin, which reports fake hardware defined in the `simulation.data` file and which is used to test the core library.
- **slave**
  An OpenHPI plugin, which allows the aggregation of resources from different domains (slave domains) and provides these aggregated resources as part of the one domain (master domain).

13.2.2   User Perspective

The OpenHPI architecture is very flexible in supporting different distributed system architectures because it is the user who gets to choose the components on which the OpenHPI daemons should run, and the ways the necessary plugins are deployed.

In this section we demonstrate this flexibility on a few ATCA-based examples. The principles that we present apply also for other hardware architectures that use separate management controllers.

13.2.2.1   Typical OpenHPI Deployment in ATCA

The diagram of Figure 13.2 shows the OpenHPI deployment in a typical ATCA configuration with two shelves. The management application, that is, the HPI user runs on a blade within one of shelves – Shelf 1 in this example. This blade also hosts the HPI daemon with two `ipmidirect` plugin instances talking to the shelf managers of both shelves.

Please note that in this deployment example the two shelves are managed as a single HPI domain.

13.2.2.2   OpenHPI Deployment in ATCA with Multiple Domains

This next example of Figure 13.3 shows that the OpenHPI daemon can run also on the shelf manager, a separate daemon in each shelf manager. This configuration can manage the same hardware as in the previous example, but the HPI user, that is, the management application in shelf 1 must work with two HPI domains.

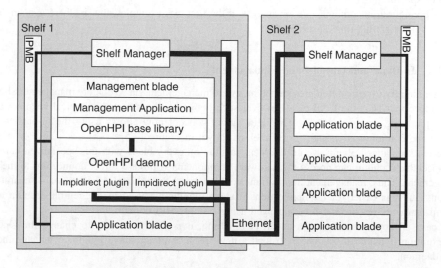

Figure 13.2   Typical OpenHPI deployment in ATCA.

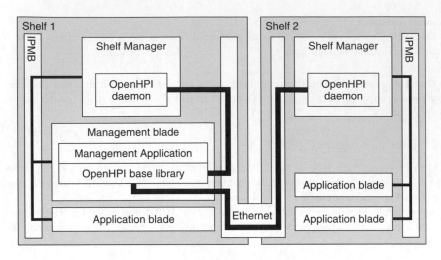

Figure 13.3 OpenHPI deployment in ATCA with multiple domains.

The OpenHPI daemon on the shelf manager may use the `ipmidirect` plugin or a shelf manager specific plugin.

This configuration provides better start-up times than the first one, because the discovery is performed on the shelf manager, and there the communication is faster.

13.2.2.3 OpenHPI Deployment Using the Slave Plugin

The configuration in Figure 13.4 is again very similar to the previous one. But it avoids creating different HPI domains by using the slave plugin with an additional instance of the OpenHPI daemon running on a blade of the first shelf. It is configured with two instances of the slave plugin, each talking to an OpenHPI daemon on a shelf manager.

Again the OpenHPI daemon on the shelf manager may use the `ipmidirect` plugin or a shelf manager specific plugin.

This configuration combines the advantages of the first two examples, that is, there is a single HPI domain yet the startup time is still good.

13.2.2.4 In-Band and Out-of-Band Management

High availability systems often provide the hardware management via a separate management bus like for instance in ATCA the IPMI bus. However such out-of band management typically cannot access all the hardware components. The OpenHPI plugin concept now allows using the in-band and the out-of-band communication paths in parallel, as shown in Figure 13.5.

For the specialized plugin, we need a local daemon on every blade with this specialized hardware. The specialized plugin can access the hardware components without a direct connection to the IPMI management. Thus, management applications could connect to the local daemons via a separate HPI session. But it is easier for the applications if slave plugins are used in the central daemon as this way all hardware resources can be managed using a single HPI domain.

In case an application blade is added at run time, it is also necessary to change the OpenHPI configuration dynamically: It needs to load a new instance of the slave plugin to manage the new local daemon.

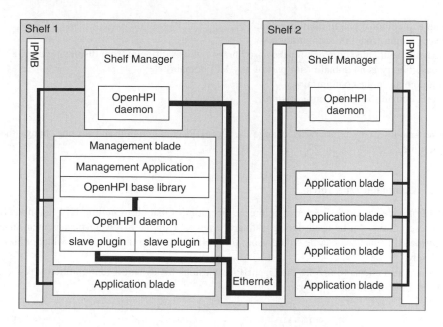

Figure 13.4 OpenHPI deployment in ATCA using slave plugin.

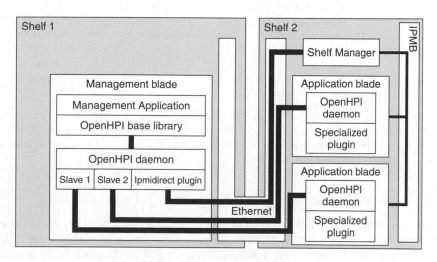

Figure 13.5 In-band and out-of-band management.

13.2.3 *OpenHPI Tools*

13.2.3.1 Dynamic Configuration Functions

OpenHPI provides a few functions additionally to the HPI specification, which are mainly needed for the dynamic configuration. These functions allow us to:

- display the configured plugins;
- find the plugin, which manages a particular resource;

- create an additional plugin configuration;
- delete a plugin from the configuration;
- display the configured domains for the base library – this is different from displaying the DRT since the base library is no domain (not part of any HPI domain);
- add a domain to the configuration;
- delete a domain from the configuration.

These functions only temporarily change the OpenHPI configuration and do not change the configuration files.

13.2.3.2 Command Line Tools

OpenHPI provides an extensive set of command line utilities that allow calling HPI functions from a shell. These utilities can be used during testing or even as a simple hardware management interface. Most of the HPI functions can be called using these clients or from the hpi_shell, which works more interactively.

There are also clients to invoke the OpenHPI extensions for dynamic configuration.

Here is a list of the client programs:

```
hpialarms        hpidomain          hpiel           hpievents
hpifan           hpigensimdata      hpiinv          hpionIBMblade
hpipower         hpireset           hpisensor       hpisettime
hpithres         hpitop             hpitree         hpiwdt
hpixml           ohdomainlist       ohhandler       ohparam
hpi_shell
```

13.2.4   Open Issues and Recommendations

As any open source project, OpenHPI depends on contributions by companies or by individuals. Therefore sometimes features are prepared in the architecture, but they are never completely implemented. Similarly, the HPI standard may include a feature, but if none of the OpenHPI contributors needs this feature, it will not be implemented.

13.2.4.1 Related Domains and Peer Domains

Related domains and peer domains as defined in the HPI specification are not really used by OpenHPI. OpenHPI implements all the functions for exploring the DRT, but there is no possibility for configuring related domains or peer domains. Thus, strictly speaking, OpenHPI is compliant to the standard even with respect to the domain discovery as described in the specification by SA Forum. But there were no users for these features, and thus the implementation is incomplete.

13.2.4.2 Diagnostics and Firmware Upgrade

HPI B.02 introduced new management instruments for diagnostics and firmware upgrade. This was a major step, since hardware vendors provide different solutions for this.

OpenHPI in its main parts implements the necessary interfaces for both the DIMI and the FUMI. But unfortunately, the functions are not supported by the major plugins provided by the project. Only the dynamic simulator plugin and the slave plugin support DIMI and FUMI.

The full support of DIMI and FUMI will only be possible when more implementations of the hardware specific plugins become available.

On the positive side, OpenHPI being an open source project provides an easy access to an implementation of the SA Forum HPI specification. It is also easy to use OpenHPI for different hardware architectures and the supporting tools provided within the project improve the usability at runtime as well as during application development.

13.3 The OpenSAF Project

In the rest of the chapter we give a high level overview of the OpenSAF project and middleware. First we provide some background information on how the project came about and the evolution of the middleware implementations up until latest stable version, which is 4.1 at the time of writing this book. Subsequently we describe the main architectural, implementation, and management aspects of OpenSAF. Finally we illustrate the process of deploying OpenSAF on different target environments.

13.3.1 Background

13.3.1.1 How It All Started

The OpenSAF open source project was launched in mid 2007 by an informal group and subsequently lead to the establishment of the OpenSAF Foundation [64] on 22 January 2008 with Emerson Network Power, Ericsson, HP, Nokia Siemens Networks, and SUN Microsystems as founding members. Soon after GoAhead Software, Huawei, IP Infusion, Montavista, Oracle, Rancore Technologies, Tail-f, and Wind River have also joined the foundation.

The initial codebase of the project was contributed by Motorola Embedded Communication and Computing business by open-sourcing their Netplane Core Services, an high availability (HA) middleware software. This initial contribution included some SA Forum compliant APIs, but mostly legacy services. Since the project launch the OpenSAF architecture has been streamlined and completely aligned with SA Forum AIS specifications.

To achieve portability across different Linux distributions, OpenSAF is consistent with the Linux Standard Base (LSB) [88] APIs. This way OpenSAF uses only those operating system (OS) APIs that are common in all LSB compliant Linux distributions. Ports to other, non-Linux, OS is still easy and has been done by some OpenSAF users since OpenSAF uses Portable Operating System Interface for Unix (POSIX) [73] as its OS abstraction layer.

13.3.1.2 OpenSAF 2.0 and 3.0: Pre-4.0 Era

The first release of the project, OpenSAF 2.0, was in August 2008. Its main highlights were the addition of the SA Forum Log service (LOG) implementation and the 64-bit adaptation of the entire codebase. While working on this first release the development processes and infrastructure has been settled: OpenSAF uses the Mercurial distributed version control system and Trac for the web-based project management and wiki.

The implementations of the SA Forum Information Model Management (IMM) and Notification services (NTFs) were added in OpenSAF 3.0 released in June 2009. At the same time most of the existing SA Forum services were also lifted up to match the latest specification versions. In addition the complete build system was overhauled to be based on the GNU autotools.

13.3.1.3 OpenSAF 4.0 Aka 'Architecture' Release

OpenSAF 4.0 was released in July 2010 and has been, so far, the biggest and most significant release earning the nickname the 'Architecture' release. It addressed three key areas: closing the last functional gaps with the SA Forum services, streamlining architecture, and improving the modularity.

The functional gaps with the SA Forum services were closed by the contribution of

- the Software Management Framework (SMF), which enabled in-service upgrades with minimal service disruption; and
- the Platform Management service (PLM) allowing the IMM-based management of hardware and low-level software.

Streamlining the architecture meant refactoring of the OpenSAF internals so that the OpenSAF services themselves became aligned with the specification and used the SA Forum service implementations instead of the legacy services. This allowed the removal of eight of the legacy services and despite the increased functionality (i.e., the addition of SMF and PLM) resulted in the reduction of the total code volume by almost two-thirds.

In this release modularity was a strong focus as well:

- The established modular architecture facilitates the evolution of OpenSAF without compromising the support for users requiring a minimum set of services.
- In the fully redesigned packaging each service is packaged in its own set of RPM Package Managers (RPMs) to further support the wide-range of deployment cases that different OpenSAF users might have.

13.3.1.4   OpenSAF 4.1: Post-4.0 Era

OpenSAF 4.1, which was released in March 2011 focused mainly on the functional improvements of the existing services. These included the addition of the rollback support in SMF; adding Transmission Control Protocol (TCP) as an alternative communication mechanism while still keeping Transparent Inter-Process Communication (TIPC) as the default transport protocol; some substantial improvements of IMM in regard to handling of large IMM contents and schema upgrades; as well as improved handling of large Availability Management Framework (AMF) models.

Additionally the release delivers a Java Agent compliant with the JSR 319 (Java Specification Request) specification, which makes OpenSAF ready for integration with any Java Application Server compliant with JSR 319 [100].

13.3.2   OpenSAF Architecture

We look at the OpenSAF architecture from the process, deployment, and logical views. In the process and deployment views we describe the architectural patterns used by the implemented OpenSAF services. In the logical view we expand further on the different services and how they use the corresponding architectural patterns.

13.3.2.1   Process View

Within OpenSAF the different service implementations use one of the following architectural patterns.

Three-Tier Architecture
The three-tier architecture decomposes the service functionality into three parts as show in Figure 13.6: the director, the node director, and the agent.

Director
A director for a particular service implementation manages and coordinates the key data among the other distributed parts of this service implementation. It communicates with one or more node directors.

Directors are implemented as daemons and use the Message Base Checkpoint service (MBCsv) discussed in Section 13.3.4.2 to replicate the service state between the active and the standby servers running on the system controller nodes.

Node Director
Node directors handle all the activities within the scope of a node such as interacting with the central director and with the local service agent. They are implemented as daemons.

Agent
A service agent makes the service functionality available to the service users such as client applications, by means of shared linkable libraries, which expose the SA Forum APIs. Agents execute within the context of the application process using the service interface.

The main advantage of the three-tier architecture is that by offloading the node scoped activities from the singleton directors to the node directors it offers better scalability.

Two-Tier Architecture
Not surprisingly the two-tier architecture decomposes the service functionality into two parts: the server and the agent shown in Figure 13.7.

Servers
The server provides the central intelligence for a particular service combining the roles of the director and the node directors of the three-tier architecture. The server communicates directly with service

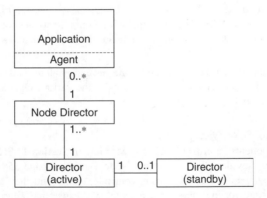

Figure 13.6   The three-tier architecture.

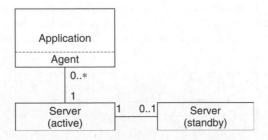

Figure 13.7   The two-tier architecture.

agents. Servers are implemented as daemons. Servers also use MBCSv discussed in Section 13.3.4.2 to replicate the service state between the active and the standby server instances running on the system controller nodes.

Agents
As in case of the three-tier architecture, agents expose the SAF APIs toward the client applications making for them the service functionality available in the form of shared linkable libraries. Agents execute in the context of the application processes.

This architecture is simple and suitable for services, which are not load-intensive. Using this architecture for load intensive services would limit their scalability.

One-Tier Architecture
In this architecture a service is implemented as pure-library, an agent only shown in Figure 13.8. It is the least used architecture, only some OpenSAF Infrastructure services are using it.

13.3.2.2 Deployment View

Node Types
For the easier development of distributed services including SA Forum services themselves, OpenSAF defines two categories of nodes within an OpenSAF cluster:

- **System controller node**
  There are at most two designated nodes within the OpenSAF cluster that host the centralized functions of the various OpenSAF services. Application components may or may not be configured on a system controller node depending on the application's requirements.
- **Payload node**
  All nodes which are not system controller nodes are termed payload nodes. From the perspective of the middleware, these nodes contain only functions of the various OpenSAF services, which are limited in scope to the node.

Mapping of Processes to Node Types
The processes and components described in the process view in Section 13.3.2.1 are mapped to node types of the deployment view described in section 'Node Types' on page 358 as follows:
    Directors as well as servers run on the system controller nodes and they are configured with the 2N redundancy model. An error in a director or a server will cause the failure of the hosting system controller node and the fail-over of all of the services of the node as a whole. There is one director component per OpenSAF service per system controller node.
    Node directors run on all cluster nodes (i.e., system controller and payload nodes), and are configured with the no-redundancy redundancy model. On each node there is one node director component for each OpenSAF service implemented according to the three-tier architecture.

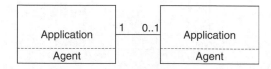

Figure 13.8   The one-tier architecture.

Agents are available on all cluster nodes and since they are pure libraries their redundancy model or the exact location where they execute are fully determined by the actual components using them.

Figure 13.9 illustrates the process mapping of a service implemented according to the three-tier architecture (e.g., IMM) to the nodes of a cluster. It also indicates some application processes using this three-tier service. In this particular case the active director runs on Node 2.

Figure 13.10 shows the process mapping in case of a two-tier service such as NTF to the cluster nodes, as well as, some application processes using the service. As in the previous example, the active Server runs on Node 2.

13.3.2.3 Logical View

The logical view of the architecture distinguishes between two groups of services: The OpenSAF services implementing the SA Forum specifications, that is, the SA Forum compliant services and the OpenSAF supporting services providing the infrastructure.

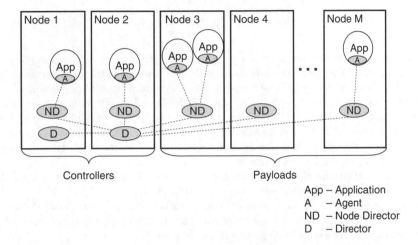

App – Application
A   – Agent
ND – Node Director
D   – Director

Figure 13.9 Mapping of the process view to the deployment view in case of a three-tier service.

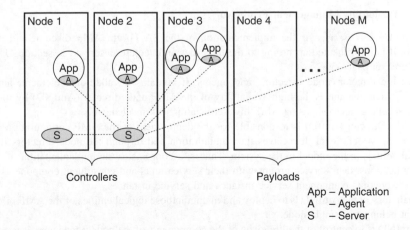

App – Application
A   – Agent
S   – Server

Figure 13.10 Mapping of the process view to the deployment view in case of a two-tier service.

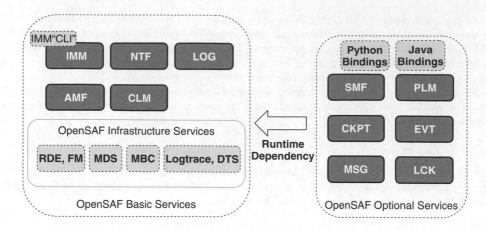

Figure 13.11 The OpenSAF services.

Only the first group of services is intended to be used by applications. The second group supports the implementation of the OpenSAF middleware itself, and is not intended for application. These are nonstandard services, on which OpenSAF provides no backward compatibility guarantees.

A subset of the SA Forum compliant services and the OpenSAF infrastructure services together compose the core of the OpenSAF implementation and referred as the OpenSAF Basic Services in Figure 13.11. They need to be deployed in any system using OpenSAF. The remaining SA Forum compliant services are optional and can be deployed as needed on the target system. They have a runtime dependency on the Basic Services.

This modular architecture allows the growth of the OpenSAF project without compromising the minimum system size for those who need only the basic services.

In the following section we take a closer look at each of the existing OpenSAF services and highlight some implementation specifics without going into details on the functional aspects for which we refer the reader to the corresponding chapters of the book.

13.3.3 SA Forum Compliant Services

13.3.3.1 Availability Management Framework

The *OpenSAF AMF service* is the implementation of the SA Forum AMF discussed in details in Chapter 6. It is implemented according to the three-tier architecture described in section 'Three-Tier Architecture' on page 356.

Besides the director, node directors, and agents, it contains an additional element, an internal watchdog, which guarantees the high availability of the AMF node director (AmfND) by monitoring its health at regular intervals using AMF invoked healthcheck mechanism.

The AMF Director (AmfD) is responsible for the deployment of the AMF configuration specified in IMM at the system level. It receives the configuration and through the node directors it deploys the logical entities of the model in the system. That is, it coordinates the implementation of the AMF nodes, applications, their service groups with their service units and constituent components, as well as the corresponding component service instance and service instances.

The main task of the AmfND is to deploy and maintain those logical entities of the AMF information model that belong to a single node.

The AmfND is in control of the life-cycle of the components of the node it manages. It is capable of disengaging, restarting, and destroying any component within its scope. This may occur in accordance with the instructions of the AmfD, as a result of an administrative action, or automatically based on the applicable configured policies.

The AmfND performs fault detection, isolation, and repairs of components on the node and coordinates these actions with the AmfD.

The AmfND interacts with the component through the AMF Agent (AmfA) as necessary; it maintains the states of the components and the encapsulating service units. It keeps the AmfD informed about the current status and any changes.

In the OpenSAF 4.1 release the AMF API and the functional scope are compliant to SAI-AIS-AMF-B.01.01 version of the specification while the AMF information model is consistent with SAI-AIS-AMF-B.04.01 specification.

13.3.3.2   Cluster Membership Service

The *OpenSAF CLM service* implements the SA Forum CLM presented in Section 5.4. It is implemented using two-tier architecture as discussed in section 'Two-Tier Architecture' on page 357. It also includes an additional element, the CLM Node Agent, which runs on each node and allows the node to join the cluster only if it is a configured node.

The CLM Server (ClmS) maintains the database of all the nodes in cluster as well as all application processes, which have registered for cluster tracking. When the CLM is deployed together with the PLM the ClmS subscribes with PLM for the readiness track callbacks to track the status of the underlying execution environments for all cluster nodes.

The ClmS considers a CLM node as a node qualifying for CLM when all of the following conditions are met:

- the CLM node is configured in IMM;
- the CLM node is not locked administratively;
- there is connectivity between the ClmS and CLM node over the Message Distribution Service (MDS) (see Section 13.3.4.1); and
- the readiness state of the execution environment hosting the CLM node in not out-of-service, provided the ClmS is a subscriber to the PLM track API.

The OpenSAF 4.1 CLM implementation supports two versions of the API specification: SAI-AIS-CLM-B.04.01 and SAI-AIS-CLM-B.01.01.

13.3.3.3   Information Model Management Service

The *OpenSAF IMM service* implements the SA Forum IMM which we presented in Section 8.4. The service is implemented according to three-tier architecture of section 'Three-Tier Architecture' on page 356.

IMM is implemented as in-memory database, which replicates its data on all nodes. This replication pattern is suitable for configuration data as configuration data are frequently read but seldom updated. Thus reading the data is fast and it does not require inter-node communication, while writing the data is more consuming since the change is propagated to all nodes in the cluster.

Although it is in-memory database, IMM can use SQLite as its persistent backend, in which case IMM also writes any committed transaction to the persistent storage. In this case none of the committed transactions is lost even in case of system restart.

Each IMM Node Director (ImmND) process contains the IMM repository (with the complete SA Forum information model) and all ImmNDs, which are in sync, are identical. The ImmNDs replicate their data using the reliable multicast service provided by the IMM Director (ImmD). The ImmD also elects one of the ImmNDs as a coordinator for driving any sync process required if any of the

ImmNDs goes out of sync, for example, because its node has been restarted, or if the node just joining cluster.

While such synchronization is in progress, the IMM repository (i.e., the SA Forum information model) is not writable, however, it will serve read requests.

In the OpenSAF 4.1 release the IMM API and functional scope are compliant with the SAI-AIS-IMM-A.02.01 version of the specification.

13.3.3.4 Notification Service

The *OpenSAF NTF service* implements the SA Forum NTF described in Section 8.3 according to the two-tier architecture presented in section 'Two-Tier Architecture' on page 357.

The NTF Server (NtfS) maintains the list of subscribers in the cluster together with their subscriptions. Based on this information the NtfS compares each incoming NTF with the subscriptions and if there is a match it forwards the NTF to the corresponding subscribers. The NtfS also caches the last 1000 alarms published in the cluster. These can be read by using the NTF reader API.

The OpenSAF 4.1 NTF service API and functional scope are compliant with the SAI-AIS-NTF-A.01.01 version of the specification.

13.3.3.5 Log Service

The *OpenSAF LOG service* implements the SA Forum LOG presented also in Section 8.2 according to the two-tier architecture (see section 'Two-Tier Architecture' on page 357).

It is intended for logging high-level information of cluster-wide significance and suited to troubleshoot issues related, for example, to misconfigurations, network disconnects, or unavailable resources.

The LOG Server (LogS) maintains a database with all the log streams currently open in the cluster. Using this database it writes each incoming log record to the file corresponding to the log stream indicated as the destination of the log record.

The LogS also implements the file management required by the specification: It creates a configuration file for each new log stream. It also handles the file rotation for the log streams according to their configuration.

In the OpenSAF 4.1 release the LOG implementation is fully compliant to the SAI-AIS-LOG-A.02.01 specification.

13.3.3.6 Software Management Framework

The SA Forum SMF presented in Chapter 9 is implemented by *OpenSAF SMF service* according to the three-tier architecture (see section 'Three-Tier Architecture' on page 356).

The SMF Director (SmfD) is responsible for parsing the upgrade campaign specification XML file and controlling its execution. The servers and directors of the OpenSAF service implementations discussed so far use the MBCSv (see Section 13.3.4.2) to maintain a 'hot' standby state in their redundant peer. The SmfD instead stores its runtime state in IMM to achieve a similar 'hot' standby behavior.

An SMF Node Director (SmfND) performs all the upgrade actions which need to be executed locally on the nodes it is responsible for such as the installation of software bundles. The SmfND only acts by direct order from the SmfD.

The SMF service implementation of the OpenSAF 4.1 release is compliant to the SAF-AIS-SMF-A.01.02 specification with respect to the API and functionality.

However it extends the scope of the specification, which applies only to AMF entities, to some PLM entities. In particular it introduces an additional set of upgrade steps to cover the scenarios of low-level software (OS and middleware itself) upgrades and to support systems with RAM disk based file systems.

13.3.3.7 Checkpoint Service

The *OpenSAF CKPT service* is the implementation of the SA Forum CKPT specification discussed in Section 7.4. It uses the three-tier architecture pattern described in section 'Three-Tier Architecture' on page 356.

The CKPT provides a facility for user processes to record their checkpoint data incrementally, for example, to protect the application against failures.

The OpenSAF 4.1 CKPT implementation is compliant to the SAI-AIS-CKPT-B.02.02 specification with respect to its API and functionality. However it provides additional non-SAF APIs to facilitate the implementation of a hot-standby component. Namely, it allows the tracking of changes in checkpoints and receiving callbacks whenever the content is updated.

The CKPT Director (CkptD) maintains a centralized repository of some control information for all the checkpoints created in the cluster. It also maintains the location information of the active replicas for all checkpoints opened in the cluster. In case of a non-collocated checkpoint, the CkptD designates a particular node to manage an active replica for that checkpoint and also decides on the number of replicas to be created. If a non-collocated checkpoint is created by an opener residing on a controller node then there will be two replicas, one on each system controller node. If the creator-opener resides on a payload node an additional replica is created on that payload node.

To reduce the amount of data copied between an application process and the CKPT Node Director, the OpenSAF CKPT uses the shared memory for storing checkpoint replicas. This choice also improves significantly the read performance.

13.3.3.8 Message Service

The SA Forum MSG presented in Section 7.3 is implemented by *OpenSAF MSG service* according to the three-tier architecture pattern discussed in section 'Three-Tier Architecture' on page 356.

The MSG provides a multipoint-to-multipoint communication mechanism for processes residing on the same or on different nodes. Application processes using this service send messages to queues and not to the receiver application processes themselves. This means, if the receiver process dies, the message stays in the queue and can be retrieved by the restarted process or another one on the same or a different node.

The MSG Director maintains a database for the message queues and queue groups existing in the system. It contains the location, state, and some other information for each of them.

A message queues itself is implemented by a MSG Node Director. It uses Linux message queues to preserve the messages irrespective of the application process life-cycle.

In the OpenSAF 4.1 release the MSG API and functional scope is compliant to the SAI-AIS-MSG-B.03.01 specification.

13.3.3.9 Event Service

The SA Forum EVT also discussed in Section 7.2 is implemented according to two-tier architecture by the *OpenSAF EVT service*.

The EVT Server (EvtS) is a process, which distributes the published events based on subscriptions and associated filtering criteria. It maintains the publisher and subscriber information. If an event is published with a nonzero retention time, the EvtS also retains the event for this retention time period. From this repository the EvtS will deliver to new subscribers any event that matches their subscription.

The OpenSAF 4.1 EVT implementation is compliant to the SAI-AIS-EVT-B.03.01 specification.

13.3.3.10   Lock Service

The *OpenSAF LCK service* implements the SA Forum LCK using the three-tier architecture presented in section 'Three-Tier Architecture' on page 356. The LCK is a distributed lock service, which allows application processes running on a multitude of nodes to coordinate their access to some shared resources.

The LCK Director (LckD) maintains the details about the different resources opened by each node represented by a LCK Node Director (LckND), and about the different nodes from which a particular resource has been opened. The LckD elects a master LckND for each particular resource responsible for controlling the access to this resource.

In OpenSAF 4.1 the LCK service API and functional implementation is compliant to SAI-AIS-LCK-B.01.01 specification.

13.3.3.11   Platform Management Service

The *OpenSAF PLM service* implements the SA Forum PLM discussed in Section 5.3. The service is implemented according to two-tier architecture (see section 'Two-Tier Architecture' on page 357). The architecture includes an extra element called the PLM Coordinator, which executes on each node.

The PLM provides a logical view of the hardware and low-level software of the system.

The PLM Server (PlmS) is the object implementer for the objects of the PLM information model and it is responsible for mapping the hardware elements of the model to the hardware entities reported by the HPI implementation.

The PlmS also maintains the PLM state machines for the PLM entities, manages the readiness tracking and the associated entity groups. It also maps HPI events to PLM notifications.

The PLM Coordinator coordinates the administrative operations and in particular the validation and start phases of their execution. It performs its task in an OS agnostic way.

The OpenSAF 4.1 PLM service implementation is compliant to the SAI-AIS-PLM-A.01.02 specification with respect to the provided functionality and APIs.

13.3.4   OpenSAF Infrastructure Services

These services are used within the OpenSAF middleware as support services, they are not intended for applications use. Of course since OpenSAF is an open source implementation one can find and access the APIs these services provide, however they are not SA Forum compliant. Moreover OpenSAF does not provide any backward compatibility guarantees with respect to these services.

13.3.4.1   Message Distribution Service

The OpenSAF MDS is a non-standard service providing the inter-process communication infrastructure within OpenSAF. By default it relies on the TIPC [107] protocol as the underlying transport mechanism. It is implemented using one-tier architecture discussed in section 'One-Tier Architecture' on page 358.

By changing its default configuration MDS can also use TCP [108] instead of TIPC. This is convenient:

- for OSs which do not have a TIPC port; or
- when layer three connectivity is required between the nodes of the cluster, for example, because the cluster is geographically distributed.

MDS provides both synchronous and asynchronous APIs and supports data marshaling. An MDS client can subscribe to track the state changes of other MDS clients. The MDS then informs the subscriber about the arrival, exit, and state change of other clients.

13.3.4.2 Message Based Checkpoint Service

The OpenSAF Message Based Checkpointing Service (MBCSv) is a non-standard service that provides a lightweight checkpointing infrastructure within OpenSAF.

MBCSv uses also the one-tier architecture presented in section 'One-Tier Architecture' on page 358.

MBCSv provides a facility that processes can use to exchange checkpoint data between the active and the standby entities to protect their service against failures. The checkpoint data helps the standby entity turning into active to resume the execution from the state recorded at the moment the active entity failed.

For its communication needs MBCSv relies on the MDS discussed in Section 13.3.4.1. Among its clients MBCSv dynamically discovers the peer entities and establishes a session between one active and possibly multiple standby entities. Once the peers are established MBCSv also drives the synchronization of the checkpoint data maintained at the active and standby entities.

13.3.4.3 Logtrace

Logtrace is a simple OpenSAF-internal infrastructure service offering logging and tracing APIs. It maps logging to the `syslog` service provided by the OS, while tracing, which is disabled by default, writes to a file.

It is implemented using the one-tier architecture discussed in section 'One-Tier Architecture' on page 358.

13.3.4.4 Role Determination Engine

The Role Determination Engine (RDE) service determines the HA role for the controller node on which it runs based on a platform specific logic. It informs AMF about the initial role of controller node and also checks controller state in case of switchover. In its decision-making process it relies on TIPC.

13.3.5 Managing OpenSAF

The OpenSAF management architecture is aligned with SA Forum principles. That is, the configuration of OpenSAF is managed via the IMM service; for fault management the middleware service use the NTF; and the SMF is used to orchestrate in-service upgrades of the middleware itself as well as its applications. Note that even the upgrades and SMF itself are controlled via IMM since it manages the SMF information model.

Management architecture is illustrated in Figure 13.12.

IMM and NTF expose their standard C APIs respectively. The OpenSAF project has made the deliberate architectural choice not to limit itself on a single management and/or transport protocol such a Netconf, SNMP, CORBA (Common Object Request Broker Architecture), HTTP (HyperText Transmission Protocol), SOAP, and so on. This enables users of OpenSAF to use management access agents best suited to their needs and preferences.

However to improve its usability (e.g., at evaluation, testing, etc.), OpenSAF offers a simple management client in the form of a few Linux shell commands that allow the manipulation of the IMM content in a user friendly way, as well as generation and subscription to notifications. It also includes a few high-level shell commands for typical AMF and SMF use cases.

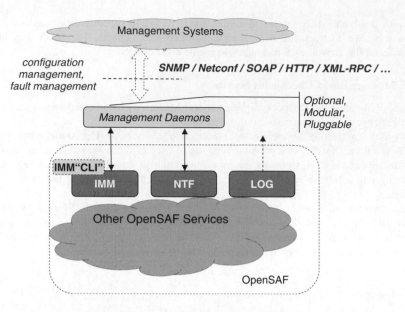

Figure 13.12 The OpenSAF management architecture.

13.3.5.1 IMM Shell Commands

These are the commands of the simple Object Management client:

- `immcfg`: It creates, modifies, or deletes an IMM object. It can also load IMM classes and/or objects from a file.
- `immadm`: It performs an administrative operation on an IMM object.
- `immlist`: It prints the attributes of IMM objects.
- `immfind`: It finds one or more IMM object satisfying the search criteria.
- `immdump`: It dump the IMM content to a file.

13.3.5.2 NTF Shell Commands

These two commands are implemented by the simple NTF client:

- `ntfsubscribe`: This command subscribes for all incoming notifications or notifications of a certain type.
- `ntfsend`: This command does not qualify as management command; but rather can be used by application or more likely test programs and scripts to send notifications.

13.3.5.3 AMF Shell Commands

The AMF shell commands are implemented based on the IMM shell commands. While each of the operations, which can be performed with the AMF shell commands, can be performed directly by IMM shell commands, these commands were defined with syntax and semantic suited to the domain of AMF. Thus, they simplify the management operations on AMF information model.

The commands are:

- `amf-adm`: It executes different administrative operations on objects of AMF information model.
- `amf-find`: It searches for objects in the AMF information model.
- `amf-state`: It displays different state information of objects of the AMF information model.

13.3.5.4 SMF Shell Commands

Similarly to the AMF commands described in previous section, the SMF shell commands are also implemented using based on the IMM shell commands and

- `smf-adm`: It executes different administrative operations on an upgrade campaign object.
- `smf-find`: It searches for objects in the SMF information model.
- `smf-state`: It displays different state information of objects the SMF information model.

13.3.6 Deploying OpenSAF

13.3.6.1 Real or Virtualized Target System

In this section we go through the steps involved with the initial deployment of OpenSAF on a target system. As target system we consider a set of computers (nodes) interconnected at layer-two such as by a switched Ethernet subnet.

1. **Getting OpenSAF:**
   The OpenSAF tarball can be fetched from OpenSAF Project we-page [91]. Alternatively it can be cloned from the mercurial repository.
2. **Bootstraping OpenSAF:**
   To bootstrap the build system execute:
   `$./bootstrap.sh`
   Note: OpenSAF has package dependencies, therefore it is recommended to check the README file in the OpenSAF source tree to find out about them.
3. **Configuring the OpenSAF build with the default options:**
   `$./configure`
4. **Building the OpenSAF RPMs:**
   `$ make rpm`
5. **Installing OpenSAF:**
   To simplify the installation OpenSAF provides the 'opensaf-controller' and the 'opensaf-payload' meta-packages whose dependencies toward other packages have been set-up so that it is clear which packages need to be installed in the controller nodes and which on the payload nodes. This way if OpenSAF is installed from a supported yum server then it will resolve dependency.
   To install OpenSAF on the controller nodes:
   `$ yum install opensaf-controller`
   To install OpenSAF on the payload nodes:
   `$ yum install opensaf-payload`
6. **Configuring OpenSAF:**
   OpenSAF provides the support needed to generate the initial AMF information model for OpenSAF itself for the specific target. This includes:
   Adjusting the model to the actual size of the cluster
   `$ immxml-clustersize -s 2 -p 6`
   The example above prepares a configuration for an eight-node cluster consisting of two system controller nodes and six payload nodes and stores it in the `nodes.cfg` file.
   From this file
   `$ immxml-configure`
   command generates the actual OpenSAF configuration which is then copied to the `/etc/opensaf/imm.xml` folder

These commands also make sure that on each node of the cluster the `/etc/opensaf/` `node_name` matches the value in the `nodes.cfg` file in the row corresponding to that node.

7. **Start OpenSAF:**
   OpenSAF is started after rebooting node, or by executing as root the following command:
   `# /etc/init.d/opensafd start`
   That's it! With these seven easy steps OpenSAF is up and running.

13.3.6.2 User-Mode Linux Environment

OpenSAF can run in a very convenient User-Mode Linux (UmL) environment included in the distribution. This environment is primarily used by the OpenSAF developers themselves, but users can use to get some quick hands-on experience with OpenSAF.

The first three steps of getting OpenSAF, bootstrapping, and configuring it are the same as for a real target system presented in the previous Section 13.3.6.1. So here we start with:

1. Build OpenSAF
   `$ make`
2. Move to the OpenSAF UmL cluster tool root directory
   `$ cd tool/cluster_sim_uml`
3. Set up the UmL environment
   `$ ./build_uml`
4. Adjust the configuration to the wanted cluster size (e.g., 4 nodes)
   `$ ./build_uml generate_immxml 4`
5. Star the four-node cluster running OpenSAF
   `$ ./opensaf start 4`
   Now we have a cluster consisting of four nodes running OpenSAF.
   Stop the cluster:
   `$ ./opensaf stop.`

13.4 Conclusion

In this chapter we review the two leading implementations of the SA Forum specifications OpenHPI and OpenSAF. Both are developed within open source projects with contributions from companies leading in the HA and service availability domain that were also the main contributors of the SA Forum specifications themselves.

With the OpenHPI project there is an easy access to an implementation of the SA Forum HPI specification. This fact facilitated the acceptance of the specification by vendors enormously. As a result OpenHPI has become a key factor and plays significant role in the industry, in the domain of carrier grade and high availability systems. Its position is also promoted by its architectural solution, which makes OpenHPI easy to use for many different hardware architectures. The supporting tools provided within the project improve the usability of the HPI specification at deployment time as well as during the development of hardware management applications.

OpenSAF has had an impressive development and evolution since the project launch. To reach maturity the major focus has been on streamlining its architecture and aligning it with SA Forum architecture and the latest releases of the specifications. This makes OpenSAF an open-source middleware which is the most consistent with the SA Forum AIS.

Besides improving the architecture significant effort has been made to improve its robustness, scalability, maintainability, and modularity. OpenSAF also includes Python bindings for all its services making it an attractive solution for quick system integration and prototyping. Together with the existing standard C and Java support OpenSAF is able to address the needs of a wide range of applications.

All the above and the emerging ecosystem of components and supporting tools ranging from commercial management clients to tools for automatic generation of AMF configurations as well as SMF upgrade campaign specifications make OpenSAF the number one choice for applications, which do not want to compromise service availability and which do not want to be locked into proprietary solutions.

14

Integration of the VideoLAN Client with OpenSAF: An Example

Anik Mishra[1] and Ali Kanso[2]

[1]*Ericsson, Town of Mount Royal, Quebec, Canada*
[2]*Concordia University, Montreal, Quebec, Canada*

14.1 Introduction

The Availability Management Framework (AMF) manages the high availability (HA) of the services provided by an application through dynamically assigning the workload to the application's components and controlling their life-cycle. To achieve the highest level of availability, the application's components typically need to interface with the AMF [48] and possibly with other services. This design decision is usually made in the development process of the application's components. However legacy applications do not implement this interface that would allow them to interact with AMF. In order to improve the availability of such applications, different levels of integration are possible: They range from the nonproxied-non-SA-aware integration that leaves the application's code intact, through SA-aware integration in which the application is modified allowing more interaction with AMF, to the integrations with additional services of the Service Availability (SA) Forum middleware (e.g., Checkpoint).

In any case, for AMF to manage the availability of an application, it requires a configuration describing the application.

In this chapter we illustrate how these different levels of integration offered by the SA Forum middleware can be used to improve the availability of a legacy application.

More specifically we focus of the steps and the efforts required in achieving various levels of integration and demonstrate them on the example of integrating the VLC (VideoLAN Client) application [109] with the OpenSAF implementation of the SA Forum services [91]. We discuss the achieved availability of the application services versus the complexity associated with implementing each of these levels.

Service Availability: Principles and Practice, First Edition. Edited by Maria Toeroe and Francis Tam.
© 2012 John Wiley & Sons, Ltd. Published 2012 by John Wiley & Sons, Ltd.

We chose VLC as our example for this exercise because with the increasing growth of internet bandwidth and the number of users, video streaming is gaining more and more interest from both users and suppliers.

HA is an important factor in the quality of service of the delivered stream. This availability is reflected in two main features: (i) the availability of the stream upon demand and (ii) the continuity of the stream during the transmission.

As a result a streaming application like VLC is ideal for examining the effectiveness of HA solutions, because it is a real-time application and the service outage is visually experienced, so the end user can easily appreciate fault tolerance.

In addition VLC is an open source video streaming application. It can be used as a streaming server, or a client receiving the video. It has a modular architecture and it is a product intended to be used by both developers and consumers, in the sense that it offers developers Application Programming Interfaces (APIs) which they can use to add VLC functionalities to their own applications. The application is also reasonably documented.

14.2   Going Under the Hood: The VLC Workflow

VLC's code is structured into cohesive functional modules shown in Figure 14.1. We can divide them into two major categories: stream modules (shown in the rounded rectangle) and management modules.

- Stream modules take input from various sources, process this input and produce the required output. Examples of this processing are multiplexing/de-multiplexing, encoding/decoding, and so on.
- The management modules are necessary to coordinate and manage the stream modules, one such management module is the VideoLAN Manager (VLM) module.

For instance, to broadcast a video file, the Control Module informs VLM of the location of the file, and instructs it how this file needs to be streamed (whether it is a broadcast or video-on-demand (VoD), the broadcast address, etc.). VLM in turn will request (i) the I/O module to open the file and (ii) the other streaming modules to process the file as needed (e.g., convert its format if needed). Finally, the Real-time Transport Protocol (RTP) module will stream the file and send it to the network.

Video streaming can be configured to function in one of two modes: broadcast or VoD:

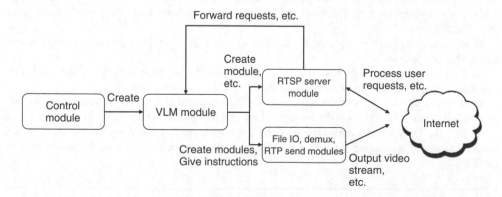

Figure 14.1   VLC workflow.

- Broadcast simply sends the video stream to a configured address. This is typically used with a multicast destination address. Within this dedicated range of Internet protocol (IP) addresses, special handling is defined for end users to be able to subscribe to an address instead of establishing a connection. This allows multiple clients to subscribe to the same feed without consuming additional resources on the video server. Of course, this entails some drawbacks as individual clients cannot personalize their experience. They simply receive what is on the network. As such, pause and seek commands would affect all clients and there is no option to resend data in case of packet drop.
- By contrast, VoD is client request driven. A user contacts an Real-Time Streaming Protocol (RTSP) service to set up a video stream. This stream starts at the beginning and can be paused/resumed as requested by the user without interfering with other users' feeds. When configured for VoD, VLM starts an RTSP server module to accept the streaming requests. Each requesting user then gets its own instance of the video stream to control.

VLM wraps almost all the functionality needed for a streaming service; however, it does not start by itself and it also requires some input to properly perform its task. This is provided by a Control Module.

VLC offers several Control Modules: For clients, VLC offers a graphical user interface (GUI) Control Module that makes it into a full featured video player application. On the server side, among others Telnet and HyperText Transmission Protocol (HTTP) Control Modules are offered.

For our exercise, we stripped down VLC of certain functionalities: We only support the broadcast aspect of VLM, and we implemented our own Control Module that we explain in more details in the next section.

14.3   Integrating VLC with OpenSAF

There is no standard (or single) way of integrating legacy applications with the SA Forum middleware. The method we present is based on our experience with the specifications. We start with the nonproxied-non-SA-aware integration of VLC with OpenSAF. Then we move onto the SA-aware version. Finally we present the addition of checkpointing to this SA-aware version.

In all three cases we followed these generic steps:

1. Selecting an integration method.
2. Defining the component types and their component service types (CSTs).
3. Implementing the life-cycle commands.
   a. For nonproxied-non-SA-aware: Implementing the health monitoring.
4. Integrating with the middleware APIs.
   a. Selection of services to integrate with.
   b. For SA-aware: Implementing the AMF API.
   c. If with checkpointing: implementing the Checkpoint service (CKPT) API [42].
5. Designing the AMF configuration.
6. Deploying and testing the implementation.

The steps that are in common for all integrations may in fact differ slightly in their details as we will see subsequently.

The application integrator must make a choice of which integration technique to adopt. Normally this decision is driven by the specificities of the application itself and the implementation efforts to be invested in the integration.

In terms of the implementation efforts the nonproxied-non-SA-aware integration is the least demanding. We present first this approach.

14.3.1   Nonproxied-Non-SA-Aware Integration

14.3.1.1   Component and CS Types

AMF manages the availability of the services of a nonproxied-non-SA-aware component by controlling its life-cycle. The assumption is that the component starts to provide its service – the single component service instance (CSI) it can provide – at the moment of its instantiation. Obviously when the component is terminated it stops providing the CSI.

All VLC modules discussed in Section 14.2 run as a single process, so without code modifications we need to equate the AMF component – at least – to such a single process. This determines the VLC-component component type.[1]

When the VLC process starts, it reads a configuration file, which is a list of command lines. It defines the mode in which different medias – different streams – are enabled and disabled and all the required attributes for them. Among these attributes each media is associated with one or more inputs composing a playlist the VLC process should stream. In broadcast mode each media stream is associated with a broadcast IP address. This means that a client can access the broadcast by subscribing to this broadcast IP address.

This behavior determines for our VLC-components the CST (Video-CST): it is a request to broadcast a preconfigured playlist to the configured IP address. Accordingly, different configuration files represent different CSIs.

Since the VLC process implementing the VLC-component reads its configuration at its start, there is no need for additional environment variables for assigning the CSI. It also means that different configuration files are needed if we need to run more than one component providing different CSIs on a node. At this time we limited our solution to a single component per node.

Note that even though a VLC-component could support VoD (since we did not change the code) we do not consider and do not enable this mode in the configuration.

From the clients' perspective each stream is identified by two IP addresses: The destination broadcast IP address mentioned above and the source IP address of the streaming server. This means that when, for example, the node goes down and the VLC service is failed over to another node it would result in the change of the source IP address (i.e., the IP address of the originally standby node). The clients would perceive this as a different stream due to the different source IP address and would not play it until the original stream times-out.

In this case and for any service where a client should expect continued communications from the same system, or needs to perform a request, the IP address of the service must be preserved; and therefore in case of a failure, when the service is failed over to a redundant component deployed on a different node, the IP address must be migrated to this node in order to mask the failure. This is applicable to both the broadcast and the VoD – not covered here – services offered by VLC.

This can be done in one of several ways. Many vendors have prepackaged solutions; however, we will use the simplest one: binding the IP address, when needed, to the node with the active video service.

To make this transition automatic we decided to use the life-cycle management provided by AMF for components. Therefore we have created an additional nonproxied-non-SA-aware component type – namely the IP-component – that binds and unbinds the IP address from the node. AMF is able to perform the migration for us simply by specifying a second CST (IP-CST).

This IP-CSI is required for the proper streaming represented by the Video-CSI as it needs to be initiated at this preserved IP address. In other words we also defined a dependency between the two CSIs.

---

[1] For short and easier read we will refer to the VLC-component component type as VLC-component and components of its type as VLC-component instances.

14.3.1.2   Implementing the Life-Cycle Commands

AMF requires three CLC-CLI (Component Life-Cycle – Command Line Interface) commands to be implemented for a nonproxied-non-SA-aware component: the INSTANTIATE, the TERMINATE, and the CLEANUP.

OpenSAF is implemented on the Linux operating system therefore we implemented the CLC-CLI commands as BASH (Born Again Shell) scripts.

To perform error recovery AMF also needs to detect component failures therefore the CLC-CLI commands include also the optional AM_START and AM_STOP commands to start and stop external active monitoring.

We start our discussion of the CLC-CLIs with the issue of health monitoring.

VLC-Component Health Monitoring

AMF can only monitor the health of nonproxied-non-SA-aware components through passive or external active monitoring, because these types of monitoring can be implemented without modifications to the component itself.

External active monitoring involves defining some entity external to the component (referred to as the active monitor) that assesses the health of the component and that reports back to AMF when it detects a component error using the AMF API.

On the other hand passive monitoring uses mostly operating system features to assess the health of the component therefore in our nonproxied-non-SA-aware integration we opt for the later one.

This solution still requires the implementation of the API that instructs AMF to start the passive monitoring namely saAmfPmStart_3(), which our nonproxied-non-SA-aware VLC does not implement obviously. One way of doing this is through the instantiate command where the instantiation script will not only start VLC but also passive monitoring.

The INSTANTIATE command is implemented as shell script that cannot invoke the passive monitoring function of AMF; therefore we implemented a small program in C, which performs this task:

```
#include <saAmf.h>
...
SaVersionTver = {.releaseCode = 'B', .majorVersion = 0x01,
        .minorVersion = 0x01};
SaAisErrorTrc;
...
// initialize a handle
rc = saAmfInitialize(&amf_hdl, &reg_callback_set, &ver);
if (rc!= SA_AIS_OK)
    {
    fprintf(stderr, "cannot get handle to AMF - %u\n", rc);
    return 1;
    }
/* call the passive monitoring function, where comp_name and
      argv[2] would be the component name and process ID
      that were passed as arguments.*/
rc = saAmfPmStart(amf_hdl, &comp_name, atoi(argv[2]),0,
    SA_AMF_PM_NON_ZERO_EXIT|SA_AMF_PM_ZERO_EXIT,
    SA_AMF_NO_RECOMMENDATION
    );
if (rc!= SA_AIS_OK)
    {
fprintf(stderr, "saAmfPmStart FAILED - %u\n", rc );
return 2;
    {
...
```

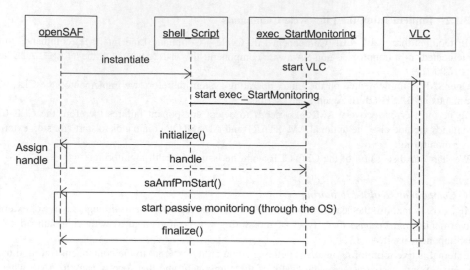

Figure 14.2 OpenSAF interactions during the instantiation of VLC.

The INSTANTIATE command will run the executable file (we refer to as exec_ StartMonitoring) of this code, as illustrated in Figure 14.2.

It is important to note that when the INSTANTIATE shell script runs the exec_ StartMonitoring, it will pass as arguments the process ID and the component name assigned to VLC, as this information is part of the parameter list of the passive monitoring API invoked later. The code snippet above shows the saAmfInitialize() and the saAmfPm- Start() function calls made to start the passive monitoring. These calls constitute the bulk of the C code. In the code above if the passive monitoring fails, we give no recommendation to AMF, while in fact we could have specified a recovery action to take place, for example, a component restart.

Figure 14.2 illustrates the interactions performed to instantiate and start the passive monitoring on the nonproxied-non-SA-aware VLC implementation.

VLC-Component CLC-CLIs
Instantiate

```
cvlc --daemon --pidfile /var/run/vlc/vlc.pid
exec_StartMonitoring
```

The cvlc command invokes VLC.

We use the first argument (--daemon) for two reasons:

- It serves as a notification toward AMF as it returns a zero integer value when the command is executed successfully satisfying the AMF requirement on CLC-CLI commands to have a zero exit status in case of success.
- It also instructs VLC to detach itself from the controlling terminal and run in the background.

The next pair of arguments instructs VLC to create a file containing its process ID (--pidfile) and store it under the name defined by the following argument – /var/run/vlc/vlc.pid in our case. The value stored in this file will be used by the other two CLC-CLI commands.

The `exec_StartMonitoring` initiates the AMF passive monitoring as presented in section 'VLC Component Health Monitoring' on page 375.

Cleanup

```
kill -9 "$(< /var/run/vlc/vlc.pid)"
rm /var/run/vlc/vlc.pid
```

The CLEANUP command will read the process ID (`pid`) defined in the (`/var/run/vlc/vlc.pid`) file and pass it to the kill CLI command to terminate the process. The `SIGKILL` (9) signal cannot be caught or ignored by a process and therefore it is used for immediate termination.

For a complete cleanup we also need to remove the file containing the process ID using the remove command (`rm`).

Terminate

```
pid= "$(< /var/run/vlc/vlc.pid)"
kill $pid
if [[ -n "$pid"] ]; then
  i=0
  while [ -n "$(ps -ef | grep "$pid")"] ; do
    i=$i+1
    if [ $i -gt 5] ; then
     exit 1
    fi
    sleep 1
  done
fi
rm "/var/run/vlc/vlc.pid"
exit 0
```

The TERMINATE command will also read the process ID (`pid`) defined in the (`/var/run/vlc/vlc.pid`) file and pass it to the `kill` CLI command to terminate the process. This `kill` command sends the default signal (`SIGTERM`), which is interpreted by the application as a request to terminate itself.

After issuing the termination command we check to see if a process with this `pid` still exists, if that is the case then we wait for 1 second, and repeat the loop. Otherwise we remove (`rm`) the file storing the process ID and return success (`exit 0`).

When the loop is repeated five times, that is, 5 seconds have passed by and the process is still alive, then we exit with an error (`exit 1`), which notifies AMF that the termination was unsuccessful.

IP-Component CLC-CLI

For the nonproxied-non-SA-aware IP-component we also need to implement the same three life-cycle commands. For this component type the TERMINATE and CLEANUP implementations are identical, because we are simply unbinding an IP address; there is no process to kill or resources to de-allocate.

Instantiate

```
ip addr add $ip dev $dev
arping -U -c 1 -I $dev $ip
```

The INSTANTIATE command script consists of two commands: The first adds the IP address (held in `$ip`) to the selected device (given by `$dev`).

The second command performs an ARP (Address Resolution Protocol) takeover of the address to indicate to any attached network switch that the address now belongs to this server. This is done by indicating to the `arping` command to send an unsolicited ARP reply to the networking equipment.

Terminate, Cleanup

```
ip addr del$ip/32 dev$dev
```

This command removes the IP address (given by `$ip`) from the device (given by `$dev`).

14.3.1.3 The AMF Configuration

For AMF to manage any application, it requires the configuration of this application. Using this configuration AMF selects the components to instantiate and the workload to assign to them.

Figure 14.3 illustrates the AMF entities of our VLC configuration:[2] we have two redundant service units (SUs) that form a service group (SG); each SU includes two components – one instance of IP-component and one instance of VLC-component. As discussed in Section 14.3.1.1 a CSI is defined for each component: the IP-CSI of type IP-CST and the Video-CSI of type Video-CST.

The AMF configuration is specified in an XML (eXtensible Markup Language) file compliant to the IMM (Information Model Management) XML schema [110] which is loaded by IMM at cluster start.

There are certain attributes in the configuration that require more considerations:

As mentioned in Section 14.3.1.1, the IP-CSI must be assigned first (before streaming any video); this we capture through the CSI dependency attribute in the CSI class, where the Video-CSI is configured to depend on the IP-CSI.

For the SG of our nonproxied-non-SA-aware VLC we select the no-redundancy redundancy model, which is captured in the redundancy model attribute of the SG type.

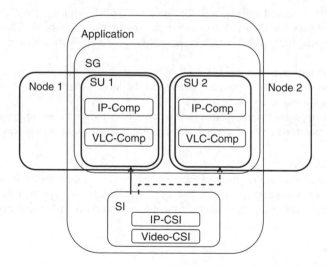

Figure 14.3 The AMF configuration structure for VLC.

---

[2] The same entities are used for all three integrations we implemented, but some their configuration attributes differ as we will discuss subsequently.

There are several timer attributes that are typically set based on experimenting with the application. Among these attributes are the timeouts for the life-cycle commands.

Another attribute which is configured based on the relation of the measured timings is the recovery on error attribute for the components (e.g., whether to set the recommended recovery is to restart or to failover). Typically the choice is based on which recovery takes less time to complete. For the nonproxied-non-SA-aware integration of VLC the time needed to recover the service is the same for both recoveries. However we favor the failover, because it starts the service on a different node and therefore platform related faults that cannot be recovered by restarting the component are also covered by this recovery.

We set the appropriate BASH script for each of the component types as the CLC-CLI commands. While in the class associating the software bundles with nodes we indicate the location of the executables as required by the deployment environment.

14.3.1.4 Test Deployment

To test our application, we used a test environment available in OpenSAF's development sources, under the `tools/cluster_sim_uml/` directory. It is a preconfigured User-mode Linux (UmL) cluster that allows one to easily start and stop an entire cluster.

UmL is a lightweight virtualization solution in that it simply uses the current real Linux kernel to provide actual functionality. However, applications running within it see a complete Linux environment, with its own independent root user. Furthermore, to simply run a UmL instance, root access is not necessary.

To integrate with this test system, we used the XML file containing our AMF configuration for the VLC application and merged it with the existing `imm.xml` describing other parts of the system. We included this file in building the test UmL software (consisting of the Linux kernel and the system image to be loaded at the virtual cluster startup). We then installed our VLC software into the common mapped folder, which is in the `root_fs/` directory. The merged IMM configuration that we made indicated that the executables resided in that directory. That means that if we want to test a new version, we simply need to replace the software and restart the test cluster.

Since our application is network oriented, it is useful for the host system to be able to talk to the test cluster through standard networking interface. This is needed so that the broadcasted video can be seen by the hosting system, and thereafter forwarded to the system's external network interface if needed (i.e., if the end user receiving the stream is on a different machine). To do this we created a tunnel using the following steps:

- Using the `tunctl` program we created the tunnel.
- We then assigned an IP address to the interface within the same subnet as the test cluster and brought the interface up.
- We started the cluster with the tap environment variable set to – `tap tap0`. This indicates the cluster startup script to use the `tap0` interface for networking.

On the end user side, she would have to start the VLC client application, and request VLC to start playing the current broadcasted stream by specifying the broadcast IP address and port, where the stream is being broadcast, that is, as defined in the configuration file of the VLC-component instances running the server side.

14.3.2 SA-Aware VLC Integration

The main difference between the SA-aware and the nonproxied-non-SA-aware integrations is the addition of the AMF API implementation to the VLC application so that it can interact with the AMF implementation of the OpenSAF middleware. This means that we need to change the application code in order to allow dynamic work load assignments and other required features.

This adds more complexity to the integration and requires a deeper understanding of the application's workflow, compared to the superficial knowledge needed for the nonproxied-non-SA-aware integration.

14.3.2.1 Component and CS Types

VLC is a highly threaded application: it consists of various modules, each of which runs in its own thread. These threads are tightly coupled; therefore it would be a tedious task to separate each thread into its own process with its own independent life-cycle.

An alternative is to consider the different threads as contained components; however, their fault isolation is still a problem and they would fail together anyway.

Consequently, and instead of making substantial modifications to the application, we decided again to represent the VLC process as a single component. We call this component type again the VLC-component.

In contrast to nonproxied-non-SA-aware components that start to provide their CSIs the moment they are instantiated, SA-aware components are assigned the CSIs any time after their instantiation. Therefore we also implement the Video-CST of our SA-aware VLC-component differently.

The workload represented by a VLM-configuration file, which contains information such as the media to broadcast and its properties is no longer associated with the instantiate command. Instead it is passed as an attribute in the assignment of a CSI, when this CSI is assigned to a VLC-component. We again refer to our modified CST as Video-CST.

By the same reason as presented in Section 14.3.1.1 we need the IP-component and the related CST. They can be reused as-is.

14.3.2.2 Implementing the Life-Cycle Commands

As the implementation of the IP-component remains the same its CLC-CLI commands remain as well.

For the VLC-component the `INSTANTIATE` and `CLEANUP` CLC CLI command implementations remain the same as defined in Section 14.3.1.2 for the nonproxied-non-SA-aware implementation.

We do not need the `TERMINATE` CLC-CLI command any more as the termination is implemented as a callback function for SA-aware components. We will discuss the implementation of this callback in the next section.

Regarding health monitoring we can use the same passive monitoring as discussed for the nonproxied-non-SA-aware solution, which is started by the `INSTANTIATE` command. However since the application code is now linked with the AMF library implementation, AMF can use its own tools and the passive monitoring is not essential.

In either case the health monitoring can be enhances with the use of health-checks, however we will not cover that in this chapter.

14.3.2.3 Integrating with the AMF API

In its original form VLC is capable of being idle – when no VLM-configuration file is provided yet. Once the configuration file has been loaded VLC starts to broadcast the requested media, that is, it assumes the HA active state for the CSI described in the VLM-configuration file that has been loaded.

For our SA-aware integration we would like VLC also to be able to assume the HA standby state: In this first approach this would mean that when a CSI is assigned to a VLC-component, it loads the configuration file, but does not start the broadcast.

As discussed in Section 14.2, in VLC it is the Control Module, which instructs VLM and the other modules what to do in terms of setting up and controlling streams by loading a configuration, then starting and stopping them. Considering that the primary goal of the AMF API is to control of the workload of the components this seems to be a suitable integration point with AMF. Therefore our intention is to create a new Control Module with the interface providing the interaction with AMF, thus implement the SA-awareness in the VLC-component type.

Component Life-Cycle API

Internally, any VLC module provides three functions: `Open`, `Close`, and `Run` (or equivalents) that implement the VLC module life-cycle API. These functions respectively are responsible for initializing, terminating, and performing the assigned tasks of the module.

This means that it is within these functions that we can map the AMF life-cycle instructions (CLC CLI commands and callbacks) and forward them to other VLC modules.

When an SA-aware VLC-component is instantiated by AMF (such as the AMF implementation of OpenSAF), our Control Module invokes first the Open function on VLM (in VLM it is called `vlm_New`), then it would initialize a handle and register this instance of the VLC-component with AMF.

Thereafter a selection object is obtained by calling the `saAmfSelectionObjectGet()` function. This allows the SA-aware VLC-component to discover AMF callbacks and dispatch them without continual polling. Figure 14.4 illustrates these interactions.

For each of the above calls, we must of course verify that the invocation was successful. When it is not the case, we close any resources we have successfully opened and we tell VLC to exit.

The Control Module registers with AMF the following three callbacks: `saAmfCSISet-Callback()`, `saAmfCSIRemoveCallback()`, and `saAmfComponentTerminateCall-back()`. AMF can use the first two for managing the component's workload:

- `saAmfCSISetCallback`: When AMF desires the VLC-component instance to take an HA state for a CSI it calls this function. As arguments it provides the name and attributes of the CSI and the desired HA state as well as some information about the status of other assignments for the same CSI.
- `saAmfCSIRemoveCallback`: When AMF desires the VLC-component to drop a CSI assignment, it calls this function indicating which CSI is concerned.
- `saAmfComponentTerminateCallback`: SA-aware components are terminated by invoking this callback function as opposed to the `TERMINATE` CLC CLI command, which is used in the case of nonproxied-non-SA-aware components. This callback invokes the Close function implemented in the Control Module. As the reader might expect, we free allocated resources and make VLC quit when this function is called.

By calling `saAmfDispatch` (Figure 14.4) we allow AMF to perform callbacks to the API implemented in the Control Module.

The continuous operation tasks of the VLC-component are performed within the Run function of the Control Module: It handles the AMF workload related requests as they arrive. We look at these details next.

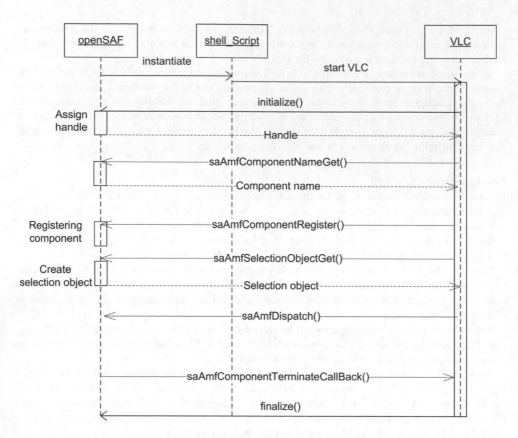

Figure 14.4   The main (SA-aware) VLC interactions upon instantiation.

CSI Management Callbacks

As we mentioned, when the saAmfDispatch() function is invoked, three functions may be called. Here we elaborate further on the two callbacks AMF uses to manage the CSI assignments.

The first one is saAmfCSISetCallback(), which is the most complicated. This function is responsible for assigning an HA state on behalf of some CSI. The HA states are the already mentioned active and standby, as well as the quiescing and quiesced states. The function has three arguments of note: the component name, the desired state and a CSI descriptor. In our case, in each component instance the component name is constant since each component registers only itself. So we are only interested in the desired state and the CSI descriptor.

This descriptor is a structure that has four properties: The first is csiFlags. This indicates if this invocation applies to one CSI or all currently assigned CSIs. In our implementation a component takes only a single CSI assignment, so the csiName property indicates the name of the CSI the callback is invoked for. CsiStateDescriptor holds additional information about the state transition. Since it does not affect the behavior of our solution we ignore it. Finally, the csiAttr property may contain a list of the CSI attributes for the CSI if an active or standby assignment is being given. In our case it contains for the active and standby HA state assignments the name of the VLM configuration file describing the workload represented by the CSI for which the HA state is being assigned.

The most important question is what to do when assigned one of those four HA states for a CSI:

- **Active:** This means that the component should be providing the service for the CSI indicated in the csiName. The csiAttr provides the name of the VLM configuration file to be loaded if necessary. Since the original VLC implementation does not include and the AMF API by itself also does not provide any functionality that helps synchronizing the stream position between different VLC instances, the component receiving this assignment starts to broadcast from the beginning the media indicated in the VLM configuration file after loading it if necessary (e.g., initial CSI assignment). For this purpose VLM may open and run additional modules.
- **Standby:** It means another component is tasked with providing the service; and the component receiving this standby assignment needs to prepare to be ready to take over the active assignment if needed. As we set out at the beginning of the section, in this integration the only preparation we want the standby component to make is to load the VLM-configuration file.
- **Quiesced:** This state is assigned when AMF wishes to switch-over active assignments from the active component to another component. In this integration the switch from active to quiesced state is done instantly by stopping all the broadcast streams the VLC-component instance is providing.
- **Quiescing:** This state is assigned when AMF wishes for the active assignment to terminate gracefully. That is to say that it should finish servicing its current clients before giving up its assignment. When the component is assigned this state, it continues streaming the current video in the playlist until its completion and then calls the saAmfQuiescingComplete() once to indicate to AMF that the task is done.

The second function is saAmfCSIRemoveCallback(). This is called when AMF wishes to remove an assignment. That is when this callback is invoked on a VLC-component instance we offload the configuration file by cleaning up any objects and releasing any resources created and allocated in association with the configuration file. The Control Module closes all other modules that are not needed for the idle state.

14.3.2.4 AMF Configuration

The entities of the configuration and their arrangement of the SA-aware VLC application remain the same as the one presented in Figure 14.3. However, we need to adjust certain attribute values and add some objects:

- the VLC-component type category: we set it to SA_AMF_COMP_SA_AWARE instead of SA_AMF_COMP_LOCAL specified for nonproxied-non-SA-aware local components);
- the redundancy model of the SG type changes to SA_AMF_2N_REDUNDANCY_MODEL instead of SA_AMF_NO_REDUNDANCY_MODEL;
- the Video-CST now includes a CSI attribute, so we specify the attribute name, and accordingly;
- we add to the Video-CSI configuration object an association object specifying the VLC configuration file.

14.3.2.5 Test Deployment

The deployment and testing remains the same, as described in Section 14.3.2.5. As noted there for deployment we need to install the new version of the VLC application into the root_fs/ common mapped folder.

14.3.3   SA-Aware VLC with Service Continuity

Service continuity is a feature that any highly available streaming application must acquire as it would be frustrating for the end user to have to watch the same video from the beginning if a failure occurs on the streaming server side. Therefore the purpose of this third integration variant is to solve the service continuity by using additional middleware services.

When the application provides a state-full service to ensure service continuity the state information needs be communicated to the standby component so that in case of a failure it can resume the work from the state where the now failed previously active component left of the execution. There are several ways of doing this synchronization with the different AIS utility services as discussed in Chapter 7. There are two factors to consider:

The first one is the data required to describe the state so that another component can take over the service provisioning. The important point here is to make sure that with the state information we do not communicate the fault that caused the failure of the active component. In case of our streaming application the state of the streaming is described by the contents of the VLM configuration file (e.g., playlists for the different streams, their associated broadcast address, etc.) and the position of the broadcast for each configured media stream. This data needs to be externalized from the application, and potentially duplicated on other nodes. Note that only the position information changes over time but the amount of data required describing it remains constant.

Secondly, we need to consider the frequency at which the state updates need to be propagated to the standby. This is usually a tradeoff among several factors: the amount of data required to transfer, the conceptual effort to segment the work and produce this data, the resources needed to recover the proper state from the communicated data and most importantly the experience of the user and the guarantees promised to the user.

We need to compare these needs with the functionality of the different services. The Event or Messaging services can be used to propagate the stream position; however the initial configuration would need to be repeated in case of a standby failure. On the other hand all the information can be stored easily in a checkpoint, which can be read by any standby or if there is no standby even the restarted active can use it for picking up its previous state.

We decided to use the Checkpoint Service as it meets all our needs. This service allows us to atomically update the checkpoint, guaranteeing that either all or none of the changes were performed. Furthermore, the service is responsible for all of the logic with regards to duplicating the data on other node(s). Additionally using collocated checkpoint we can improve the performance of the checkpoint update operations.

Accordingly there is a need for some changes in our SA-aware VLC-component. Namely, we need to add the Checkpoint API implementations and we need to incorporate them in the handling of the CSI assignment callbacks. In our discussion we will focus on the traits where the difference exists.

In particular we will not discuss the component type and CST definitions as they remain the same as for the basic SA-aware solution as well as the CLC-CLI commands.

We also need to migrate the source IP address to the new standby component the same way as we presented for the nonproxied-non-SA-aware component in section 'IP-Component CLC-CLI' on page 377. Therefore our IP-component and related IP-CST need to remain part of our application and they require no change.

14.3.3.1   Integrating the Middleware APIs

The objective is to store the current stream position in a checkpoint so that when another VLC-component instance need to continue the broadcast currently broadcasted by the active instance it can do so from the stored position; that is we obtain service continuity.

As a result the `Run` function of the Control Module will have more tasks to perform: When a component is active it will also periodically update the video position within the checkpoint. While at take-over this checkpoint needs to be read to obtain the position.

To do this in addition to the initialization of the AMF library described in section 'Component Life-Cycle API' on page 381 and shown in Figure 14.4, the component also needs to initialize the CKPT by calling `saCkptInitialize()` and obtain a handle and in turn a selection object for the CKPT as well.

CSI Management Callbacks

To obtain service continuity the workload assignment callback (i.e., `saAmfCSISetCallback`) is implemented differently for the different HA states to incorporate checkpointing:

- **Active:** This means that in addition to providing the service for the CSI the component needs to periodically checkpoint its state information (i.e., the current position of the stream being broadcasted). When the component is assigned the active state, it first opens the checkpoint associated with the CSI and tries to read it. If the checkpoint is not empty then so this means that this is not an initial assignment and if the component was the standby for the CSI it already obtained the VLM configuration contents and only needs to read the current position from the checkpoint to be able to resume the stream from the position specified in the checkpoint.

  If the component was not the standby (i.e., the active which is being restarted) it needs to read the configuration information and the position information.

  If the checkpoint is empty, then it was created with the open call and this is an initial assignment; therefore the component needs to read the configuration information from the VLM configuration file, store the information in the checkpoint and to start to broadcast all streams from the beginning. In any case the component also needs to update the checkpoint periodically with the current streaming position. To allow this we need to modify the way the component detects callbacks. Rather then continuously waiting now it should wait for a limited amount of time. If an event comes up before the timeout, it is dispatched by calling the `saAmfDispatch()` function and this time it should dispatch one callback at a time so that it does not miss the checkpointing time. After the dispatch returns, the component checks if it is time to perform a checkpoint and, if that is the case, it does so. Then it resumes the timed wait on the selection object.

- **Standby:** The component needs to open the checkpoint first. Opening the checkpoint indicates to the Checkpoint service to ensure that there is a replica of the checkpoint on the local node. With the change of storing the configuration information in the checkpoint itself taking the standby assignment for a CSI means that the component needs to obtain the configuration information from the checkpoint rather than from the configuration file itself. (With that the configuration file now can be modified without impacting ongoing stream.) The OpenSAF implementation of the service also allows for tracking checkpoint updates to keep the standby up to date. We do not use this extension as in our case the state information is not incremental.

- **Quiesced:** This state assignment is given to change the current active one so that another component – the standby can take over the active assignment. Accordingly the component reacts to the assignment by stopping all broadcasts and saving their position in the checkpoint so that the component taking over resuming the broadcast can do so from the stored position. It also closes the checkpoint.

- **Quiescing:** This state assignment is also given to the component currently active for the CSI. In this state, the component continues to broadcast as well as checkpoint all the streams, but only till the end of the current item on the playlist. When the end of the current item is reached in each stream the quiescing is complete and the quiesced state is reached by the component, which confirms this change with AMF by calling the `saAmfCSIQuiescingComplete` function. It also closes and unlinks the checkpoint as going through the quiescing state indicates that the service is being removed gracefully.

We also need to modify the processing of the `saAmfCSIRemoveCallback` for the case when the service is abruptly stopped, that is if the CSI is being removed from the component when it has the active assignment. In this case the component unlinks the checkpoint as part of the removal procedure, so that the checkpoint can be removed completely from the system as soon as all components close it.

14.3.3.2 Implementing the Checkpointing

The components assigned the active and the standby HA states need to open the checkpoint for the CSI of concern.

For this they use the `saCkptCheckpointOpen()` call. It takes as arguments:

- The handle to the Checkpoint service obtained at initialization.
- The distinguished name of the checkpoint, we use the CSI name for this purpose.
- The checkpoint attributes, which are set only by the component with the active assignment:
  - The checkpoint creation flags. We chose to use a collocated checkpoint to maximize the performance of the active component by needing to write to the active replica only and have it as a local replica. Therefore we set the `SA_CKPT_WR_ACTIVE_REPLICA` and the `SA_CKPT_CHECKPOINT_COLLOCATED` flags.
  - The maximum checkpoint size: 10 000 000 (i.e., 10 MB).
  - The retention duration of the checkpoint, that is, until what time should it be kept around. We gave SA_TIME_MAX. This is significant only for the time when no component keeps the checkpoint open, that is, when the service is not being provided. We do not keep the state for the CSI if it has been stopped gracefully or abruptly.
  - The maximum number of section: 3 (only two of them are used).
  - The maximum size of each section: 10 000 000 (i.e., 10 MB).
  - The maximum length of each section Id: 2.
- For the opening mode for the checkpoint. The component with the standby assignment sets the `SA_CKPT_CHECKPOINT_READ` flag only. The component with the active assignment sets the `SA_CKPT_CHECKPOINT_READ`, the `SA_CKPT_CHECKPOINT_WRITE` and the `SA_CKPT_CHECKPOINT_CREATE` flags.
- A pointer to a checkpoint handle instance, so that the open can return it to the calling function.

Opening the checkpoint ensures that it is being replicated on the local node of the component opening it.

Following that, if the component is assigned the active HA state it also needs to tell the Checkpoint service to make the local checkpoint replica the active one by calling `saCkptActiveReplicaSet()`.

If the checkpoint was just created (it is empty) then the sections need to be created using the `saCkptSectionCreate()` function. It takes as arguments the checkpoint handle, the creation attributes, which include the section name and the section expiration, a pointer to initial data to put in the checkpoint, as well as the size of this data.

We use two sections. The first is to host a copy of the content of the VLM configuration file. This guarantees that the active configuration stays consistent, even if the version on disk has been modified. The second section maintains the current position of the media streams, so that we can resume at the same position whenever is needed.

Next to write the initial data to the checkpoint, `saCkptCheckpointWrite()` is used. It takes as arguments:

- The checkpoint handle the we obtained at the opening of the checkpoint.
- A pointer to an array of vectors. These vectors contain the section ID, the offset to write at, the amount of data to write and finally a pointer to the data that needs to be written.
- The number of vectors in the array, which is two for the two sections.
- A pointer to a `SaUint32T`. This number is modified on error. It gives the index of the vector that caused the error.

For subsequent updates the component actively broadcasting only updates the second section with the current positions for this it overwrites the section using `saCkptSectionOverwrite()` function, which requires the checkpoint handle, the section ID, the pointer to the data, and the data size.

To read from a checkpoint, an `saCkptCheckpointRead()` function exists which is analogous to `saCkptCheckpointWrite()`, it has the same arguments.

To close and unlink a checkpoint the `saCkptCheckpointClose()` and the `saCkptCheckpointUnlink()` need to be called respectively. They take as argument the handle to the checkpoint.

14.4 Summary and Conclusion

In this chapter we have presented three different levels of integration of the VLC application with the OpenSAF middleware:

In the first one we left the application code intact and had AMF manage the application as a nonproxied-non-SA-aware component. This is the minimum level of integration where the interaction between AMF and the application is limited to life-cycle management, and therefore the least effort is required to implement it. The integration simply consisted of implementing the three life-cycle commands and the passive monitoring code – approximately 70 lines of C code – that instructed AMF to start monitoring the health of the process implementing the component and providing AMF with the configuration based on which it could perform the management.

With this level of integration AMF was already capable of detecting the failure of the VLC process and restart it on the other node within a couple of seconds that would result in an availability measure of approximately 4×9's (with a mean time to fail of roughly every 8 hours). The main shortcoming of this integration was that it did not offer the service continuity needed for streaming: In case of a failure the streams restarted from the beginning.

In the second version we rendered VLC a SA-aware application. This was done by creating our own Control Module for VLC, which implemented the APIs required for the interaction with AMF. This new Control Module is roughly 400 lines of C code which also required about 30 lines for the build system. In this case only two CLC CLI were implemented. In terms of effort we used a similar configuration as in the first version.

In this SA-aware VLC implementation the standby was better prepared as it did not need to read the configuration file, so the failover became faster, below half a second on average, which would allow for 5×9's of availability (again with the same failure rate). But this solution would still not provide the service continuity expected from a streaming application as the streams would still be restarted. However the introduction of the standby was a necessary step toward the next level of integration.

In order to ensure service continuity, in the third integration besides AMF we also used the Check-point service. This addition was handled within the Control Module that now also opened, read and updated a checkpoint with the current stream positions. When the active component failed, the redundant standby VLC-component instance would fetch the stream positions from the checkpoint and would continue the streaming from the same positions. Adding check-pointing increased the effort needed for the integration: We added approximately 390 lines of C code to the 400 lines of our Control Module. We also had to add 150 lines of C code to the VLM and RTP modules of VLC. Again we used a similar AMF configuration as in the previous cases.

With these changes the time required for the recovery remained similar as in the previous case but with this added effort we finally achieved the service continuity experience that the end user watching the stream expects.

Table 14.1 summarizes the efforts and the benefits of the different integration levels for the VLC application.

In conclusion we can say that there is more than one method to integrate legacy applications with an SA Forum middleware implementation. Depending on the application little effort may go a long way, for example, if the application does not have state information or already uses some solution, such as a database to store it. With more efforts better integration is possible, which is still far from a complete application rewrite or implementing all the availability concepts from scratch. For applications that do support some of these concepts and most importantly the concept of the standby, additional options such as the use of proxies or containers are also possible that we did not experiment with. We have also seen that the SA Forum concepts allow not only for achieving HA, but also providing service continuity.

The appropriate level of integration is really influenced by (i) the application itself and the type of service(s) it provide and whether it already implements some availability concepts, (ii) the availability level expected after the integration, and (iii) the amount of effort to be spent on implementing the integration.

In our experience the integration challenge was more on the side of understanding the application features and mapping them with the middleware concepts. But at the simplest level for the nonproxied-non-SA-ware integration even this was not necessary. It was straightforward and did not require any deep understanding or changes/additions to the application's code. On the other hand such a solution may not offer a good user experience for state-full services as we have seen for our media streaming application.

Table 14.1 Comparison of the different integration levels of VLC

Integration level	Coding requirements	Recovery time	Achievable availability	Missing feature
Non-proxied-non-SA-aware	70 lines of C 6 CLC-CLI scripts	~2 seconds	~99.99 ...	Service continuity (streams are restarted)
SA-aware	400 lines of C 5 CLC-CLI scripts 30 line build	<0.5 second	~99.999 ...	Service continuity (streams are restarted)
SA-aware with checkpointing	940 lines of C 5 CLC-CLI scripts 30 line build	<0.5 second	~99.999 ...	–

At the other extreme the SA-aware integration with checkpointing has offered the expected availability level and user experience, but required a deeper understanding of the application and its structure as it required some modifications to the code.

Finally while the API integration of the application is major part of the story, it is not the complete one. AMF requires a configuration of the application without which it cannot perform its task and its settings do affect the availability at runtime. It is important that this configuration is carefully defined, and populated with the proper values.

15

Migration Paths for Legacy Applications

Mario Angelic
Ericsson, Stockholm, Sweden

15.1 Introduction

Systems that are required to provide highly available services typically have long life-cycles. Once deployed, they are in service easily for decades during which they go through periodic updates and upgrades of their hardware and software. Furthermore, during this time new technologies also emerge and for the system owners it becomes a challenge not just to understand these technologies but to decide whether any of them should be deployed in their existing system, and if the answer is yes then which ones.

As a result, we cannot speak about the updates in the traditional sense which would update some old hardware and/or software to a newer version of the same kind. The changes are more substantial. Some examples one may think of are: Replacing an object-oriented database with a relational one; migrating from one operating system to another; migrating from a legacy high availability (HA) middleware to the Service Availability (SA) Forum middleware. In this chapter we focus on this last case: On the migration of applications from a legacy HA middleware to the SA Forum Application Interface Specification (AIS) middleware.

We first look at the drivers for the migration, and then we continue with the review of the different aspects of system integration to define its levels with respect to the SA Forum specifications. Using this scale we set our ambition level for the migration, which is followed by a detailed discussion on how to achieve it. We present the applicable approaches and techniques that can be used for the different services and most importantly for the Availability Management Framework (AMF) [48].

Throughout this chapter we will use the term SA Forum middleware referring to a middleware implementation compliant to the SA Forum specifications.

Service Availability: Principles and Practice, First Edition. Edited by Maria Toeroe and Francis Tam.
© 2012 John Wiley & Sons, Ltd. Published 2012 by John Wiley & Sons, Ltd.

15.2   Reasons for Migration

There are various benefits for migrating legacy software to a standard-based system, such as one based on the SA Forum specifications. They range from business benefits to technical benefits. In this section we focus on those relevant to system owners/operators and to software vendors.

15.2.1   Benefits for System Owners

System owners or operators who want to provide some application services in a highly available manner may achieve this in different ways.

The first option is that they develop themselves the functions that support the availability and management needs of the target application system. Since there are different application components and most of them need these functions, it makes sense to provide them as middleware services.

Sometimes this alternative seems attractive since when building a new application system the designers often perceive that they need only a subset of the functions typically found in an HA middleware. This choice however comes with the associated risk that as the application system evolves it will require more and more such features. In the long run this leads to a full-fledged proprietary middleware solution with the full maintenance responsibility as well as with significant integration efforts any time a third party product (3PP) component needs to be brought into the system.

Therefore a better alternative is to use an already available middleware. This avoids the efforts and expenses associated with the development and maintenance of the new proprietary middleware. It also allows the application system designers to focus on the logic of their application.

To build a complete application system, a number of application components may be needed such as a database, different protocols stacks, a web server, an Internet protocol (IP) load balancer, and so on. The number of components depends on the nature of application being built. An enterprise application will usually contain more application components while a low-end embedded system will have less. Using an existing middleware allows the designers to leverage the existing ecosystem of 3PP components resulting in a shorter time-to-market without compromising the overall system characteristics including availability.

There is a clear advantage of selecting a middleware with interfaces compliant to some standard. Such a choice increases the size of the ecosystem, that is, the number of already available application components. It also diminishes the chances to become locked-in with a proprietary middleware solution.

Up to date the only standard based solution is a middleware implementation compliant to the SA Forum specifications. At the same time there exist already a large number of application components, both commercial-of-the-shelf (COTS) and proprietary, that have been developed for different proprietary middleware solutions. Many of these have years and even decades of development investment. It is a no surprise that system owners would like to leverage these existing components and focus on building new added-value.

This is possible only by migrating these existing application components to the standard based middleware, that is, an SA Forum middleware.

15.2.2   Benefits for ISVs

From the discussion of Section 15.2.1 follows that the most important benefit for an independent software vendor (ISV) to produce COTS components, which have been integrated with an SA Forum middleware, is the portability of their product. In other words, the reduction of the integration efforts needed by their customers, owners of an SA Forum middleware based systems.

Thus, from the perspective of owners of SA Forum compliant systems ISVs who offer components already integrated with the SA Forum middleware would typically fall into the preferred category among providers of COTS components for their system.

In this chapter we discuss the efforts required to integrate applications with the availability management as well as with system management. These integration efforts are hardly ever negligible and far too often underestimated. We would like to point out also that the efforts needed to integrate an application with the system management are often at same level or even exceed the efforts required for the availability integration management.

15.3   Integration Criteria

15.3.1   Main Factors

When integrating an application with the SA Forum middleware and analyzing its different aspects a rather common question emerges:

What it means to integrate an application with SA Forum middleware?

More specifically, which SA Forum services the application component needs to use to qualify as an integrated component?

As we have concluded in Chapter 10, the standard is completely open here, so we need to come up with some criteria ourselves. For this we analyze the needs of the system owners operating the deployed systems as well as those relevant to vendors delivering end-solution, that is, complete systems. Note that in some cases the vendor and the customer are the same entity.

The two factors most important to operators or end-customers owning the deployed systems are: The service availability (SA) provided by the system, which is also referred as in-service-performance (ISP), and the cost of ownership, which the customer would like to have as low as possible.

- **The ISP of the system:** The system owner obviously would like to have the best possible availability figures as well as maintaining service continuity since any problem in this area directly impacts the user experience, which in turn impacts the system owner's business. It is not so easy to sell the services of an unreliable system.
  To reduce downtime and have a good system ISP, it is important to have a system that is easy to manage, which has a streamlined architecture and which runs high quality code.
- **Cost of ownership:** HA systems are typically nonstop embedded systems with a long life-cycle, which accordingly, have a significant operational cost. Thus the system operator is interested in having these ownership costs as low as possible. Interestingly the criteria mentioned in the previous bullet apply here as well. The system needs to be easily manageable as handling problems impacting the ISP will decrease the revenue and increase cost; it needs to have a streamlined architecture so there is a common way of handling of the entire system; and it needs to run high quality code creating fewer problems and needing less troubleshooting.

The factors important for (end-)vendors are:

- **Easy maintenance:** This includes capabilities like easy localization of the source of problems when they occur; the capability of exposing events that have the potential of escalating into severer problems in the system; and a streamlined architecture.
- **ISP:** As we have seen bad ISP figures have a direct business impact on the system owners, that is, the vendors' customers, thus, they impact (in)-directly the (end-)vendor.

We can see that the *easy management*, the *streamlined architecture*, and the *code quality* are reoccurring factors that are important for both the system owners (end-customers) and the (end-) vendors. In following sections we take a deeper look at each of these factors to understand how they relate to the integration of applications with SA Forum middleware.

15.3.2 Easy Management

By easy management we mean that throughout the product life-cycle the system operator has an easy way of executing the typical management tasks. In other words, while managing the system the operator is able to achieve the desired results, deploy the desired changes in the system in a relatively simple way, and with a minimal risk of causing unwanted side-effects.

To determine this we consider the typical management use cases of configuration management, software management, and fault management.

We elaborate on each of these areas in following sections.

15.3.2.1 Configuration Management

During the prolonged operation of a system it is inevitable that an operator will need to change some of the configuration parameters as a result of changes in the operator's network, adjustments to service parameters, and so on. Most importantly the operator would like to perform these operations in a *transactional* fashion to ensure that the related configuration changes are either all applied to the system or none of them are applied.

It is more difficult to manage a system that does not support transactional handling of configuration changes because it allows for the partial deployment of the changes that may be inconsistent. Consequently it is easier to end up with a miss-configured node or a miss-configured application, which in the worst case may impact ISP and in turn the revenue.

By the ease of configuration management we also mean a *uniform* way of handling of the different parts of the system; that the configuration management of the whole system is done in a common way. An operator does not need to be aware of the internal structure of the system and its potential complexity. To reduce mistakes he or she should not be required to configure the different parts of the system in different ways.

With the definition of the Information Model Management service (IMM) discussed in Chapter 8 [38] the SA Forum offers a basic configuration infrastructure capable of changing configuration data in a transactional manner. It allows for application specific validation of configuration data represented as configuration objects and attributes as discussed in Chapter 4. It also provides a representation for the state and other runtime properties of the system via runtime attributes and objects.

By modeling in IMM any configurable property of the system (or application) that requires exposure to the operator, vendors can achieve an easy and uniform way of handling of configurations, which also results in a streamlined architecture for configuration management. Such a solution additionally offers:

- easier implementation options of the configuration data backup and restore procedures;
- the possibility of spanning transactions across the complete configuration space; and therefore
- a global consistency of configuration data across the entire system.

We need to mention another possible, but less attractive alternative of achieving the ease of configuration management: the aggregation of the different configuration management services of the different subsystems can be done closer to the operator.

While this approach seemingly achieves the same result as the integration with a single configuration management service (like IMM) there are severe drawbacks:

The most important drawback is that the consistency cannot be guaranteed across the complete configuration space since, typically, the different composing configuration services are not able to guarantee the transactional semantics for the changes spanning across multiple configuration services. Additionally, the internal architecture of such a solution is also not streamlined, which negatively impacts maintainability as among others it becomes more complicated to create a backup of the entire system configuration and to restore it.

15.3.2.2 Software Management

During the life-cycle of a system an operator will also need to upgrade the system numerous times let it be due to the periodic correction and functional releases provided by the software vendors, or because of the need to adapt the structure and size of the system to the new functional and performance requirements.

Similarly to the configuration management discussed in Section 15.3.2.1; by the ease of software management we mean that the same mechanism can be used no matter which part of the system is being upgraded. Whether the change is in some kernel modules of the operating system, in the middleware itself, or in an application running in a Portable Operating System Interface for Unix (POSIX) or a Java environment, it is beneficial for the system operator if all software upgrades are done in a uniform way in the system.

Considering HA systems and SA an even more critical factor is whether in-service upgrades are possible, that is, whether the software management solution has the capability of upgrading the system without any downtime and with no or minimum service disruption.

The SA Forum has specified the Software Management Framework (SMF) [49] to support the upgrade of systems from one deployment configuration to another while they are in-service. It defines the upgrade process in such a way that no manual steps are required during the upgrade. An important characteristic of the SMF is its close interaction with AMF, which ensures that the impact on the provided services is kept at minimum. SMF has been presented in Chapter 9.

To achieve a seamless software management, it should be possible to upgrade all components of the system via SMF. SMF is suitable for this task since it has been specified as a framework, which – among others – is open toward the different software packaging concepts.

The unappealing alternative to this uniform software management is a system with a different upgrade mechanism for each part of the system. For example, the operating system level functionality is upgraded using the operating system's software management functionality, while some application software running in the Java Application Server is upgraded via the upgrade mechanism specific to that particular Application Server. When such different mechanisms are directly exposed to the operator, the operator becomes aware of the heterogeneity of the system and the task of software management is not so easy any more. In such a system it is almost impossible to coordinate the simultaneous upgrade of multiple interdependent functionalities.

For SA Forum compliant systems the ease of software management can be achieved by ensuring that the upgrade of the different system components can be driven by SMF. As a result this enables operators to upgrade any part of the system in a seamless and transparent way.

15.3.2.3 Fault Management

Fault management refers to the capability of the system to notify the operator that some event happened in the system that requires attention and possibly the intervention of the operator. There are two main classes of such events:

- Alarms, which are events that require operator action to resolve the problem; and
- Notifications that are sent for informational purposes.

Similarly to the discussions on the configuration and software management we consider fault management easy if it is done in a common way using the same infrastructure service.

The SA Forum has defined the Notification service (NTF) [39] as such a service that provides support for fault management in SA Forum compliant clusters. The NTF allows for the classification of events as alarms and notifications and setting delivery guarantee policies, filtering, suppressions criteria. The service is described in more details in Chapter 8.

System components that generate these types of events need to be integrated with the NTF Producer application programming interface (API) so that they can report the events via the NTF.

Again the unattractive counterpart is a system, which internally has multiple different ways of reporting errors and problems. This complicates the fault management, since the internal complexity and heterogeneity of the system is exposed to the operator and it becomes hard to correlate the information about errors coming from different sources if is possible at all.

15.3.3 Streamlined Architecture

We have mentioned streamlined architecture several times, but have not defined what we meant by it. The best way to explain is through examples that make an architecture not streamlined.

In the context of our migration discussions we do not consider an architecture streamlined when there is a duplication of functionality in the system, and especially if these are functionalities of system-wide services that are required by many system components at all system levels such as multiple logging services, multiple configuration services, different availability managers, different ways of upgrading different parts of system. Multiplication of such services leads to a nonstreamlined architecture.

There can be multiple causes for the duplication of functionalities such as combining certain type 3PP software with (legacy) application components developed in-house, or when different parts of the system consists of components developed in different programming languages.

In an effort to make their product more adaptable to different environments and for faster portability some ISVs embed some basic availability and manageability support in their product. This is in contrast with the architectural pattern where the 3PP components contain only the core functional parts, but using a portability interface they are prepared to be fitted to different availability and management services already existing in the different environments they are being integrated with.

While embedding some HA functionality initially looks appealing since it is faster to get the 3PP up-and-running in the different environments, it is typically harder to achieve a good integration of such components. This approach duplicates the availability and management functionalities leading to a nonstreamlined architecture.

The cornerstone functions in any HA middleware are the group membership service and the availability management service. In the SA Forum architecture these are provided by the Cluster Membership service (CLM) [37] and the AMF [48].

Aligning the system in such a way that it uses a single instance of these services significantly helps in streamlining the availability management of the system. It results in better and more predictable availability characteristics, easier troubleshooting, and less service disruption during system upgrades. As an example we can see that in the SA Forum architecture SMF avoids duplication by interacting with AMF whenever it needs to ensure minimal service disruption during upgrades.

In the area of system management we have already identified that configuration management, fault management, and software management are the services that have direct impact on manageability. By applying the appropriate services across all the levels of the system we can streamline the management architecture as well.

In the SA Forum architecture it is the IMM that provides support for configuration management, the NTF provides support for fault management, and the SMF for software management.

During troubleshooting and root cause analysis of the problems occurred in the system it is important to be able to reach a conclusion about what went wrong and localize the fault based on the available logs. Having multiple logging services with different rules and settings for logging data, some being cluster-wide while others are node-local, with different rotation policies makes the troubleshooting task rather challenging. The SA Forum has also standardized a cluster-wide Log service (LOG) [40], which can be used for logging cluster-significant events. Adapting the functionality across the system and using the LOG makes the troubleshooting work easier.

Of course, we should not forget that in case of logging the largest responsibility lies always on the shoulders of the application designer – the user of the logging service – to ensure that the appropriate information is logged and with the appropriate severity level. It is an art by itself to make the right tradeoff between not logging too much information so that the logs indicating real problems are not rotated out too quickly, and at the same time also not to log too little so that based on the information present in the logs fault localization becomes impossible.

While the task of generating logging information can only be done by the application and components designers, the LOG provides the mechanisms that guarantees that such log records produced anywhere in the cluster are stored in a reliable and persistent way in the files associated with each centralized log stream.

The LOG has a few preconfigured log streams, but if it is desired to separate the logged information, application can configure and create their own additional log streams dynamically via the LOG API. Chapter 8 has more details on the LOG.

In some migration cases it could be rather challenging and costly to adapt 3PPs to use the LOG if they were not designed to use it, or if they do not support any portability interface to hook in different log implementations. Nevertheless efforts should be made at least to minimize the number of logging services used in the system.

15.3.4   Code Quality

Code quality depends on the quality assurance level a vendor is using in its development process. The SA Forum indirectly helps to achieve this by the facts that it has standardized a set of APIs, and that such standard APIs stimulate the growth of the ecosystem of components reusable even across company borders. This reuse results in improving quality since each time users reuse such a component they contribute to the quality by testing the component in their own deployment context.

While code quality is an important factor for both the vendors and the system owners, it cannot be associated with any specific service or services of the SA Forum architecture.

15.3.5   Integration Levels

Now that we defined them it is time to see how we can use the criteria of easy management and streamlined architecture to understand which services an application component needs to use to be qualified as integrated with the SA Forum middleware.

From our discussions we have seen that a streamlined architecture is an important prerequisite to ease system management. We have also seen that the AIS management services have the features that make system management easy, which makes them desirable integration points.

Additionally we identified for availability management CLM and AMF as crucial services therefore to streamline the architecture we need to integrate with them as well. This integration also indirectly eases software management using SMF, which collaborates with AMF during upgrade to minimize service disruption.

Accordingly we define the following integration levels for applications:

- **Service Availability Forum (SAF)-Availability-integrated:** For an application component to state that it is *availability-integrated* with the SA Forum middleware implies that its availability is managed by AMF. Note that according to the SA Forum architecture we do not need to require explicit integration with the CLM since the AMF itself is a user of CLM. Therefore applications managed by AMF do not need to use CLM directly.
- **SAF-Manageability-integrated:** For an application component to state that it is *manageability-integrated* with the SA Forum middleware implies that
  - if the component requires any configuration data its configuration management uses the IMM;
  - if the component generates any relevant notification it needs to do so via the NTF to facilitate fault management;
  - for the purpose of software management it is upgradeable by the SMF; and
  - optionally: for logging it uses the LOG.

Note that it is possible that a component uses only a subset of the listed management services. For example, it may use NTF and is upgradeable by SMF, but it is not configured via IMM, we consider such a component as *partially-manageability-integrated*.

- **Well-SAF-integrated:** To state about an application component that it is *well-SAF-integrated* with the SA Forum middleware implies that it is:
  - SAF-Availability-integrated; and
  - SAF-Manageability-integrated at the same time.
- **Fully-SAF-integrated:** For full integration we consider all AIS services therefore to state about an application component that it is *fully-SAF-integrated* with the SA Forum middleware implies that:
  - it is well-SAF-integrated; and
  - if the application component requires any functionality provided by some other SA Forum services – such as the AIS utility services – it uses those services.
- **Using-SAF:** Finally the statement that an application component uses the SA Forum middleware implies that:
  - the component is not SAF-availability-integrated and is not SAF-manageability integrated (neither partially manageability integrated); but
  - the component uses some of the SA Forum services that are not considered in availability and manageability domain. Examples of such services are the Checkpoint service, the Message service, and so on.

It is obvious that fully-integrated components and systems offer the highest benefit for streamlining the architecture and easing system management. Well-integrated components do offer similar benefits at least from the perspective of easing system management. Therefore for the purpose of migrating legacy HA applications we consider well-integration as the optimal goal to achieve. This is the ambition level we focus in the rest of this chapter.

Does this mean that the use of other services like the SA Forum utility services, which includes the Messaging, Checkpoint, Event services, is not needed? The answer is definitely no. Though such services are not directly contributing to the easy handling of the system, they do help in streamlining the architecture by avoiding duplication of the respective functionality.

15.4 How to Migrate

15.4.1 Availability Integration

There is a strategic choice that an application designer has to make when integrating some legacy software with the AMF. He or she has to decide whether to keep the legacy code intact or to modify it so that it interacts with AMF natively.

Here we give a short overview of the different possibilities and provide some general recommendation when some of these approaches are most appropriate.

One of the first questions to ask is whether the legacy software has the capability of being started without immediately providing its service. This means, for example, that a process can be started and it will not provide any service immediately but wait until it is instructed by an external entity to do so. In AMF terminology this is called pre-instantiable component. If a legacy software has the (hot-)standby capability that would typically indicate that it is capable of being started and remain idle until it needs to take either the active or the standby role for a specific task (service).

On the other hand if the legacy software starts to provide its service as soon as it is started, in AMF terminology it is called a non-pre-instantiable component. Components of the non-pre-instantiable category are never SA-aware.

The options in front of the application designer are to integrate the software natively with AMF, or to use one of the techniques offered by AMF for easy porting. These techniques include the wrapper process (WrapP) technique, the proxy-proxied technique and the non-SA-aware nonproxied technique. We describe these techniques further in subsequent sections.

15.4.1.1 Native Integration

The native integration approach is feasible for legacy software which is or could be turned into a component of the pre-instantiable category.

In this approach the legacy software code is adapted to directly interface with AMF, effectively transforming the non-SA-aware software to an SA-aware. While this approach seems to be intrusive at first glance, it is very common that only a minor part of legacy functionality needs to be adapted: Only those parts directly related to the control of the process life-cycle, the health monitoring, and the assignment of service responsibilities need to be considered. In most cases the core functional parts related to the service that the process provides, the traffic handling, and so on, do not need to be touched.

The biggest obstacle is that such modifications require access to and the knowledge of the source code, which is not the case with most 3PP software.

15.4.1.2 Wrapper Approach

The wrapper approach is a design pattern to make some 3PP functionality integrated with AMF so that, form the perspective of AMF, it behaves and it is treated as any other SA-aware component interacting directly with AMF. The essence of this approach is to make the 3PP, which natively does not use the AMF interfaces (thus by itself it is a non-SA-aware component) look like to AMF as if it was a normal SA-aware component.

The approach consists of developing a wrapper that encapsulates the legacy 3PP into an SA-aware component. The wrapper itself is typically a single process (though it can be several processes as well) which on one side interfaces directly with AMF and on the other it interacts with the legacy software. The task of the wrapper is to map the interactions with the AMF to the interface specific to the legacy software for the handling of its life-cycle, health monitoring, and workload assignments.

Mediating the life-cycle control and workload assignments is rather straightforward therefore here we focus on the health monitoring issues as error detection is key for availability management that the wrapper needs to enable.

Accordingly the wrapper may initiate one or more healthchecks, and evaluate the health of the 3PP process in a 3PP specific way within the context of these healthcheck invocations then report the result to AMF. The 3PP may already have some proprietary interface to check its health that the wrapper can use. Alternatively the WrapP can make its assessment by executing some application requests toward the 3PP component: If the 3PP is a web server, the WrapP could invoke HyperText Transmission Protocol (HTTP) requests for a test web page. If the 3PP is a database it can execute some database operation on some known content. Whenever the wrapper detects an error, it reports this error to AMF which then triggers a recovery procedure according to the recommendation provided by the WrapP in the error report or if no recommendation is given according to the configured recovery policy. It also applies any necessary escalation policy.

Alternatively or additionally to the healthcheck, the WrapP may register the processes of the 3PP with AMF for passive monitoring. In this case the AMF implementation will utilize an appropriate mechanism of the underlying operating system to detect if any of the monitored processes crashes.

It is important to note that for the AMF the legacy software integrated this way will look like any other SA-aware component. AMF can not see the difference between the legacy software that interacts with AMF through a WrapP and the software that was designed to interact with AMF natively.

Figure 15.1 illustrates at a high level an example of using the wrapper approach. It shows a subset of the AMF entities in the AMF information model (for readability we omitted the application entity). In the figure the shapes labeled 3PP represent instances of the legacy software which are hooked to the AMF via the wrapper processes – the shapes labeled WrapP.

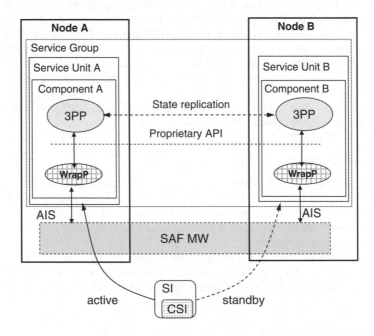

Figure 15.1   Integration through a wrapper process.

The WrapP and 3PP processes are encapsulated in a single AMF component. To achieve this there is not a single-line of code modified in the 3PP process. An AMF component containing a WrapP and a 3PP process is contained in each service unit. There are service units allocated to Nodes A and B. These service units are grouped into a service group which is defined to run in 2N redundancy to protect one service instance representing workload performed by the 3PP process.

Accordingly, at runtime AMF will assign one service unit active for the protected this service instance while the second service unit will be standby for the same service instance. Figure 15.1 illustrates also that the 3PP may use some internal mechanism for the state replication needed to achieve the hot-standby capability.

For more detailed descriptions on the AMF concepts like the AMF information model, and its associated entities like components, service units, service groups, service instances, healthcheck, and passive monitoring we refer the reader to Chapter 6.

15.4.1.3 Proxy-Proxied Approach

In the wrapper approach, the legacy software processes are wrapped by a new process so that the legacy software behaves and is represented in AMF information model as a standard SA-aware component. From implementation perspective the proxy-proxied approach is similar in that sense that there is a dedicated piece of software, which is SA-aware and which mediates the interaction between the AMF and legacy software.

The main difference is that in the proxy-proxied approach the mediation function and the legacy software are decoupled into separate components of the AMF information model. AMF is aware of the mediation software as the proxy component and the legacy software as the proxied component, as well as the workload (service instance and its component service instance (CSI)) representing mediation tasks of the proxy components; and workload served by legacy software, the proxied components.

This leads to an important characteristic of this approach. Namely that the proxy component and the proxied component are different entities with separate life-cycles. If the active proxy component fails AMF can request another component typically acting as the standby to the failed proxy component to resume the mediation task. This new proxy component then assumes the mediation task of the failed component and registers the proxied component again without affecting the service provided by the proxied component. Thus, the failure of the proxy component does not indicate a failure of the proxied components. Similarly, the AMF does not consider the failure of the proxied component to be a failure of the proxy component.

As in case of the wrapper approach the proxy component is the primary source of information for AMF for error detection on the proxied components. It may use similar techniques as discussed for the wrapper.

Since the proxy and the proxied components are different entities, the redundancy model of the proxy components can be different from that of the proxied components.

Note that the proxied component can be pre-instantiable or non-pre-instantiable. In case of non-pre-instantiable component the proxy component will start the proxied component only when it gets the active CSI assignment for this proxied component.

For more information on proxy and proxied components and the related concepts please see Chapter 6.

In general, the proxy-proxied solution is appropriate when one of the following is true:

- The redundancy model of the proxied component (the legacy software or hardware) needs to be different from the redundancy model of the proxy component. The proxy component usually requires a very simple redundancy model such as 2N, whereas the legacy component may need a more complex redundancy model such as N+M or N-way active.

- The failure semantics and the fault zone of the proxied component are different from the ones for the proxy component. For example, the proxied component may be running outside of the cluster, whereas the proxy component is always located on a node within the cluster.

Figure 15.2 illustrates an example where the processes mediating the requests (Proxy) from the AMF toward the legacy software processes (3PP) is allocated on two nodes (Node 1 and 2); and the two service units containing the proxy components (P1 and P2 each encapsulating a Proxy process) are allocated to a service group configured with the 2N redundancy model. This means that at runtime one of the proxy components (e.g., P2) will receive the active assignment for Mediation workload; while the other proxy component (P1) will have the standby assignment for the same Mediation workload.

In this specific example the legacy software is modeled with the N-way-active redundancy model and it is configured in such a way that the number of preferred assignments for the service instance is equal to number of service units in the service groups (i.e., three).

From the AMF configuration the AMF knows that the service instance (mediation) is specific for proxying the components L1, L2, and L3. From the moment the proxy component P2 receives the active assignment for the Mediation service any AMF interaction related to any of these components will go via component P2. Examples of such interactions are the requests for instantiating the proxied components (if they are pre-instantiable components), the assignment of the CSI, as well as any healthcheck.

15.4.1.4 Nonproxied-Non-SA-Aware Approach

In this approach the role of the AMF is limited to the management of the component life-cycle (CLC). AMF instantiates the component when it needs to provide its service and AMF terminates the component when it must stop providing the service. The interaction with the component is only via Component Life-Cycle Command Line Interface (CLC-CLI); more specifically via the INSTANTIATE, TERMINATE, and CLEANUP commands.

This is the approach that requires the least integration efforts since for a basic integration not a single line of code is needed. The only things needed are the CLC-CLI scripts for instantiating, terminating, and cleaning up the component.

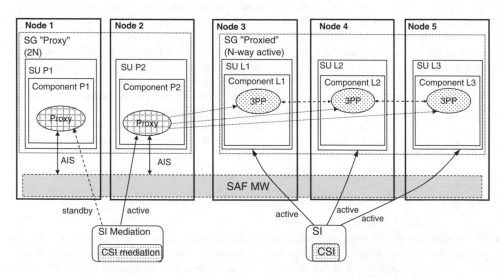

Figure 15.2 Example of the proxy-proxied integration approach.

To be able to assess the health of the component it is common to combine this approach with the external active monitoring concept offered by AMF. The external active monitoring is supported with two CLC-CLI commands, namely the AM_START, which starts a monitoring process for the associated component and the AM_STOP, which stops the monitoring process for the component.

The monitor process may assess the health of the component by submitting some service requests to the component and checking that the service is provided in a timely fashion. The monitor process uses the AMF interface to report any error it detects. It can build on the AMF healthcheck facility to trigger the monitoring activity toward the legacy software but also to monitor its own health.

Additionally if the monitoring process (or another process started at the component instantiation) can find out the process identifier(s) associated with the legacy software it can start the AMF passive monitoring for the faster detection of process crashes or unexpected process exits.

15.4.1.5 Recommendations: Which Approach to Use

In the previous sections we went through several integration approaches and in this section we try to give some suggestions and recommendations on which approach is most appropriate for which case.

We start with the classification of legacy software that we use to provide a recommendation on the approach, which in our view is most suitable for the class.

Figure 15.3 presents the following decision points:

- **Distributed:** Is the legacy software developed to run as distributed application or it runs within single node? In the later case the software belongs to Class 1, otherwise we continue with the next question.
- **Built-in HA:** Does the distributed legacy software have a built-in HA functionality? For example, does it maintain a group membership status of the different parts of its own functionality running on different nodes? Distributed software with no HA functionality forms Class 2.
- **Execution environment (EE) Control:** Does the distributed legacy software with built-in HA functionality also control the life-cycle of the encapsulating EE? Here by the term execution environment we refer to the entity defined in the Platform Management service discussed in Chapter 5. More specifically, what we are interested in is if as part of the error escalation and recovery procedures the software would restart its encapsulating EE such as the instance of the operating system.

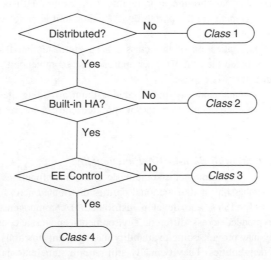

Figure 15.3 Decision points for the classification of legacy software.

Class 1: Nondistributed Legacy Software

Examples of this category are non-distributed software like an Simple Network Management Protocol (SNMP) daemon, a domain name server (DNS), a nondistributed Web server, an network time protocol (NTP) server. This class of software by itself is not capable of providing high-available services.

This type of software usually starts to provide its service as soon as it is started so from AMF's perspective it maps to a non-pre-instantiable component. When integrating this class of software with AMF, the nonproxied-non-SA-aware approach is used most often and that is what we recommend as well.

When integrating with AMF this class of software can be configured in a service group with the no-redundancy redundancy model, which includes one or more additional service units as spares. It can also be used in an N-way-active redundancy model when multiple service units share the workload associated with the same service.

When an instance of such a component needs to be assigned the active role for its service (its CSI), AMF instantiates it and the component starts to provide the service immediately. Other component instances with no active assignment stay uninstantiated spares. They will be started only when they need to provide service. In particular, in case of a switchover or a failover AMF will respectively terminate or clean up the component instance that previously provided service first then will instantiate one of the spare components.

Note that such a component can be combined with other components in other redundancy models as well. In this case AMF assigns the standby state only to those components that are capable of taking such an assignment. AMF will not attempt to assign the standby state to the non-pre-instantiable component running the software of this class.

Alternatively one may use the wrapper approach to create an SA-aware component that can take only active assignments. In this case when AMF instantiates the component only the WrapP is instantiated and it becomes responsible to handle the life-cycle of the processes of the legacy software. When AMF assigns a CSI in the active state the wrapper instantiates the processes and tears them down when the assignment is removed.

Class 2: Distributed Software with No Built-in HA

This category represents legacy software, which is designed to run in a distributed environment and to use of the availability services through a portability API. Some of these applications may also abstract the workload management via the portability API, though it is more common that they do not expose this.

If the components of an application of this class can be pre-instantiated, that is, it is capable of remaining idle when instantiated then the wrapper approach is recommended. This would be the case when the workload management is exposed.

If the application components cannot be instantiated without providing their services, that is, they map to non-pre-instantiable components then the nonproxied-non-SA-aware approach is usually the simplest choice.

Class 3: Distributed Software with Built-in HA Functionality

The legacy software of this class implements some basic availability functions most often to maintain the state of the group membership of its functional entities distributed across the cluster nodes. This is a rather common case for ISVs who develop distributed HA components. They try to increase the portability of their product across different environments and as a consequence embed some clustering support. They may provide some availability management functions as well for monitoring and restarting some of their entities. They typically still require some external middleware functions to start the base processes.

Examples of this class are some of the distributed protocol stacks, the distributed databases, some of the application servers like the Java 2 Enterprise Environment products.

If the legacy software has a centralized software entity, which manages rest of the legacy application then the proxy-proxied approach is preferred as it minimizes the integration efforts. Otherwise again we recommend the wrapper approach.

The nonproxied-non-SA-aware approach is also possible to use, but in our experience it is worthwhile to investment into a tighter integration in case of such complex applications. This relatively small initial integration cost using the wrapper or the proxy-proxied approaches pays off over time as they result in more flexible solution than the nonproxied-non-SA-aware approach. Particularly the proxy-proxied approach is easily adaptable for both pre-instantiable and non-pre-instantiable categories of components.

Class 4: Distributed Software with EE Control

The legacy software of this class is similar to Class 3 but has the major distinction that the built-in HA capabilities control more then their own functionality.

This is not common for 3PP components developed by ISVs, but it is relatively common for proprietary legacy applications. When a complete vertical solution was developed and there was no clear cut between the application and the middleware functions.

For this class of legacy software systems there is no simple approach to migrate them to an SA Forum based system or, for that matter, to any other middleware. The most feasible approach is to identify the parts of such a legacy software system that belong to Class 2 or at least Class 3; and then integrate them according those categories.

15.4.1.6 Integration Examples

Shared-Storage Database Management System

This is an example of the typical shared-storage database architecture. Such a database management system (DBMS) consists of a Database Connector implemented as a shared library delivering the DBMS API to applications, which for the application designers also abstracts the complexity of the rest of the DBMS architecture.

The Database Connector communicates with the core of the DBMS system. The DBMS server is typically distributed over a number of nodes for availability and capacity reasons. Each instance of the DBMS server is connected to a shared-storage system via the storage area network and has access to the complete persistently stored DBMS content.

In this architecture each instance of the DBMS server can serve application requests from the complete DBMS content. For better load balancing the Database Connector distributes the application requests among the instances of the DBMS servers.

Among the redundancy models defined by the AMF this can be best represented by the N-way-active redundancy model. In this redundancy model the service units of the service group protecting a given service instance are all assigned active for this service instance.

In this DBMS solution if any of the service units is locked or a node becomes unavailable, the remaining DBMS server instances can still provide the full service as it is the case with the N-way-active redundancy model. Thus, we map each DBMS server instance to a service unit of a service group configured for the N-way-active redundancy. In the AMF information model we also configure the service instance served by the DBMS servers in such way its preferred number of active assignments is equal to the number of service units in the service group of the DBMS server.

Application software components of this type usually have a very light built-in support for availability management. It mainly serves the needs of the Database Connector to detect the availability the DBMS server instances so that it can distribute among them the application requests. Therefore according to our classification method we classify this software as Class 2, where the portability API consists of the interface to manage the life-cycle of the DBMS server instances and the interface to assess the health of a DBMS server instance.

For legacy software of Class 2 we recommended the wrapper approach, which is shown in the following Figure 15.4.

A WrapP encapsulates each DBMS server instance into an SA-aware component. In this example we assume that the legacy software is pre-instantiable (able to remain idle); thus it can be instantiated as part of wrapper instantiation.

If it could not stay idle and therefore it was non-pre-instantiable, then the WrapP would instantiate it as part of the procedure of accepting the active assignment for CSI representing DBMS workload. Note that in this later case the nonproxied-non-SA-aware approach is also a possibility.

As previously stated, the appropriate redundancy model is the N-way-active for the service group containing the DBMS server instance.

The Database Connectors are shared libraries and they will execute in the context of the application process using the DBMS. These application processes will be part of components grouped into service units and service groups with redundancy models suiting the application's needs.

Shared-Nothing DBMS

This is an example of the shared-nothing database architecture where neither the disk nor the memory is shared among the cluster nodes that host the DBMS.

The content of the DBMS is replicated in the memory of the nodes hosting the DBMS and it is periodically backed-up on a persistent file system. Such an in-memory DBMS system also has some disk log where each transaction performed since the last complete backup of the database is persisted. This way even in the case of a total outage such as a cluster restart these systems can operate without any data loss by restoring the data from the last backed up state, to which they apply all the transaction persisted from the log to recreate the last state before the outage.

Again the DBMS consists of a Database Connector implemented as a shared library, which delivers the DBMS API to application and abstracts the complexity of the DBMS architecture. It communicates with DBMS server, which is distributed on a set of nodes to improve the availability, processing, and memory capacity.

In this case the data is stored in-memory: it is partitioned to optimize the memory consumption and it is also replicated for availability. Each DBMS server instance has

- its own partition of data on which it can work without the risk of contention; and
- one or more partition of data replicated from other DBMS server instances for HA purposes.

Figure 15.5 illustrates an example of the partitioning and replication of the database content. In this example the data has been divided into four partitions, that is, P1, P2, P3, and P4. The DBMS system has been configured to create one replica for each database partition (P1', P2', P3', and P4'). In our example the DBMS system groups the partitions to create partition groups in such a way that the DBMS server instances, which are members of the same partition group, will have same content. For example, Partition Group A in Figure 15.5 has the content (P1 + P2) of the database. The partition group is also a replication domain, that is, an update of the content in one DBMS server will require the update of other DBMS servers within same partition group.

Note that there is a specific configuration case where the number of replicas is the same as number of DBMS servers resulting in a single partition group, and in the total replication of all the data. This use case is feasible when the amount of data is relatively small, and when the data is seldom written but often read such as in the case of configurations. In this case each DBMS server has a full copy of database. After the failure of any DBMS server the remaining ones will be able to serve the traffic since they have a full copy of database.

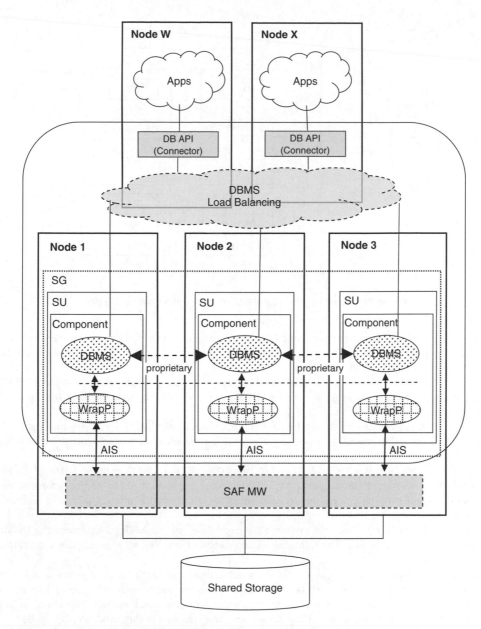

Figure 15.4 DBMS integration using the wrapper approach.

From the perspective of its integration with AMF this specific case becomes the same as our previous example with the shared-disk solution discussed in section 'Shared-Storage Database Management System' on page 405. Therefore we will not consider it any further in this section; the conclusions for the shared-disk example are relevant for this specific case as well.

In this example we will assume that there exists a specific part of the DBMS software, the DBMS management daemon which controls the DBMS servers. This is a valid assumption as one of the most popular clustered in-memory databases use this approach.

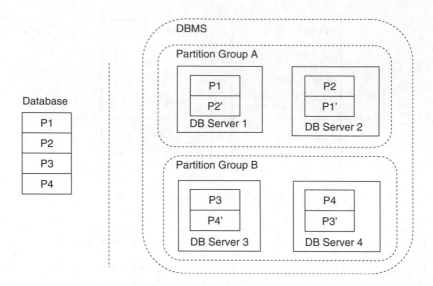

Figure 15.5 Data partitioning and replication of a share-nothing DBMS.

Since the communication controlling the DBMS servers goes via the DBMS management daemon, we can apply the wrapper approach to the DBMS management daemons to make this component SA-aware and interact directly with AMF.

Subsequently we also apply the proxy-proxied approach and use the now SA-aware DBMS management daemons as proxies for the DBMS servers, which now become the proxied components.

This integration is relatively straightforward and fits naturally the legacy DBMS architecture.

The AMF modeling aspect is less straightforward and depends on the amount of built-in HA functionality.

Let first assume that there is no built-in HA capability. Such a DBMS maps to Class 2. All the HA support is abstracted via the portability API including the control of the number of partitions, the definitions of partition groups, the decision about the owners of the primary partitions and their replicas, and so on.

For the integration it is desirable to utilize the capabilities of the AMF to reproduce the DBMS behavior described previously. The following example shows the modeling approach that can achieve this behavior.

Each data partition is modeled as a specific service instance, and each partition group as a service group. That is, for our example we have four service instances and two service groups. Two of the service instances P1 and P2 are protected by the service group mapped to Protection Group A and the other two P3 and P4 are protected by the service group mapped to the Protection Group B.

Figure 15.6 illustrates AMF instance model for this specific case.

Now let us look at our example assuming that the legacy DBMS is of Class 3.

This means that it includes some built-in HA capabilities, for example, that based on the number of replicas configured the DBMS internally makes the decision about the number of partition groups and where within a partition group the primary and the secondary replicas of a given partition will reside.

This means that looking from outside the DBMS behaves like a 'black box.' A certain set of DBMS servers will provide the service and the Database connector will ensure that each request is forwarded to the appropriate instance of the DBMS server. Any internal decision of the DBMS to reshuffle the partitions is not visible from outside the DBMS system.

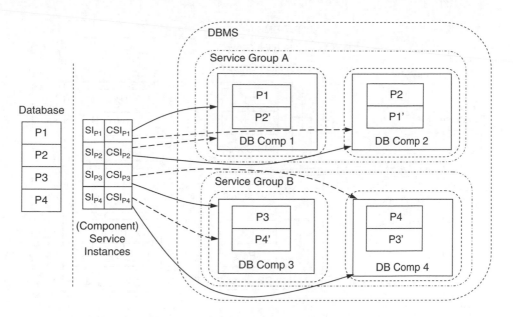

Figure 15.6 The AMF entities reflecting the data partitioning and replication example.

As shown in Figure 15.7, this solution can be modeled with the N-way-active redundancy model since to the outside user and therefore to the AMF as well it looks like that all the DBMS server instances are providing same service. The difference compared to true N-way-active redundancy model is that if all DBMS servers of the same partition group suddenly go out-of-service – which of course should be an extremely rare occasion – then the DBMS cannot function anymore since the full DBMS content is not reachable anymore. This case needs to be escalated, for example, by the proxy component. It becomes aware of the situation using the proprietary interface of DBMS and reports to the AMF the error with the appropriate recovery recommendation.

Although this integration example results in a light integration with AMF (due to the built-in HA capabilities of the DBMS) there are still benefits of performing it, and they are related to software management. It allows a smoother execution of software upgrades using the SMF since SMF inter-acts with AMF during upgrade to handle workload switchovers and to impose the least possible service impact.

15.4.2 Manageability Integration

In this section we present the approaches that can be used to integrate the system management aspects of legacy software with the SA Forum middleware. According to our definition of the manageability-integration we need to use for configuration management the IMM, for fault management the NTF, and for software management the SMF.

15.4.2.1 Configuration Management Using IMM

The process of integration of legacy software with IMM consists of the tasks of identifying the dynamic configuration space, designing the application's information model and encoding it, and implementing the Object Implementer (OI) interface of IMM, that is, the code which handles the validation of configuration changes as well as propagating them toward the legacy software.

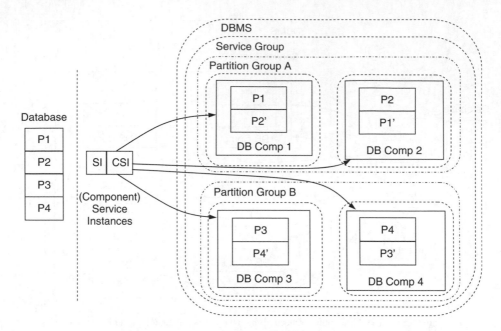

Figure 15.7   Class 3 share-nothing DBMS integration example.

Dynamic Configuration Space

In this step we need to identify those properties within the configuration of the legacy application that need to be configurable at runtime. These belong to the dynamic configuration space and need to be exposed as writable properties in IMM.

Other properties of the legacy application that are configured only at the installation time do not belong to the dynamic configuration space; they do not need to be exposed via IMM. (If desirable these can be represented as read-only properties.)

Identifying the dynamic configuration space is usually challenging for new (i.e., greenfield) applications, but it is quite trivial for legacy application since they already provide some configuration properties through some configuration framework. Whatever schema the legacy software uses to describe its configuration model, it normally contains sufficient information to understand the dynamic configuration space of the application and to derive model compliant to the SA Forum information model.

Designing the Information Model

After identifying the dynamic configuration space the next step is to design an information model, which reflects this configuration space. For this we need to identify the different relationships between the different configuration properties of the system we selected in the previous step.

In IMM terms this means that we need to map the properties as configuration attributes, group them into classes, and define relationship between those classes to create an information model. As discussed in Chapters 4 and 8, in the IMM meta-model there are no specific modeling concepts to specify the relationship among classes, except parent-child relatinonship, and this task is left to the OIs. They need to enforce all the needed constraints.

To document these relations among the different model elements it is advisable to use a high-level modeling language capable of describing the application's information model and the related constraints. The SA Forum itself uses UML (Unified Modeling Language) to describe the information model of AMF and other services that can be used as examples.

Note that there are also tools that support automatic code generation from UML and some other popular high-level modeling languages. So using these languages to describe the information model and its constraints allow the automatic generation of a skeleton code for the OI interface that would enforce the constraints described in the modeling language.

Typically the migration efforts will depend on the modeling language used to represent the information model of the legacy software. If the meta-model of this modeling language is in line with the main principles of the SA Forum information model and the associated IMM meta-model, then it may even be possible to use the information model of the legacy software directly or with few modifications by defining the transformation rules. This allows the automatic transformation of the instance models used with the legacy software to their IMM representations. This significantly simplifies the migration to IMM. The main principles of the SA Forum information model that need to be aligned to simplify the transformation are the tree-based structure, where each object in the model has no more then single parent, and objects are accessed using single key.

The classes and objects of the information model can be created in IMM programmatically using the IMM Object Management API or by providing a initial configuration according to the IMM XML (eXtensible Markup Language) schema as defined in [109].

For more information on the SA Forum information model and IMM see Chapters 4 and 8 respectively.

OI Interface Implementation

Finally, the legacy application needs to interact with IMM to receive its configuration changes. For this the application code that implements the related IMM API needs to be written. More specifically IMM OI API needs to be implementation to register the legacy application with IMM as the implementer of the classes and/or objects of the information model defined for the legacy application in section 'Designing the Information Model' on page 410.

This piece of code receives the configuration changes proposed on any of these objects, validates the changes, and subsequently it propagates them to the legacy application itself for deployment when IMM indicates that the validation has passed by all involved OIs.

The amount of the work depends on the complexity of the information model; more specifically on the complexity of the constraints that need to be enforced. As indicated in the previous section 'Designing the Information Model' it may be possible to automate this process to some degree by generating the OI skeleton code based the constraints defined in the information model.

15.4.2.2 Fault Management Using NTF

The process of integrating some legacy software with NTF consists of the steps: identifying the events that are relevant to fault management and among them those that need to be sent to the system administrator; the classification of these events into the notification types as defined by NTF; and adapting the application code to the NTF API, more specifically, to the NTF producer API used to send notifications presented in Chapter 8.

Identify Notifications

The task is first to identify the events that are relevant to fault management and among them those that need to be exposed to the system administration and then to identify to which NTF notification type each of the events should be mapped. NTF defines six notification types: alarms, security alarms, state change notifications, object create/delete notifications, attribute change notifications, and miscellaneous notifications.

A legacy application might already use some other fault management mechanism for which the events have already been identified and classified into different categories. Those categories may not match exactly the notification types defined by NTF nevertheless the mapping between the categories used by the legacy software and the NTF notification types are usually straightforward. The reasons for this is that legacy HA applications are often compliant to the ITU X.73X recommendations (International Telecommunication Union) [85], which lies at the basis of the NTF specification as well.

Sending Notifications Using the NTF API

Compared to the previously discussed services, the migration to the NTF requires much less efforts and it is usually straightforward. The easiest approach is to map the API used by the legacy software to the producer API of the NTF. This approach then minimizes the impact on the application code.

15.4.2.3 Software Management Using SMF

The process of integrating legacy software with SMF consists of the following steps: describing the software in an entity types file (ETF), packaging of the software, and creating an upgrade campaign specification (UCS). For the actually deployment only the last two are needed, but ETF helps to automate these and other integration tasks.

The biggest value of using the SMF comes when the upgrade would impact entities controlled by AMF. In this case, SMF can indeed reduce the impact on the availability of their services during upgrades. However SMF can orchestrate upgrades regardless whether the software is under the control of AMF or not.

For more information on the SMF refer to Chapter 9 and [49].

Creating an Entity Types File

As described in Chapter 9, in the ETF the software vendor describes the accompanying software bundle. The software bundle description provides the information on how to install and uninstall the software in question, the CLI commands provided by the vendor for this purpose and if applicable the scope of impact they may have (e.g., if the node is rebooted). This information serves as input to the upgrade campaign designer and it may need adjustments for the particular deployment.

Additionally, for AMF integrated software, the vendor also specifies the AMF prototypes from which the AMF types and their entities can be derived for the AMF configuration. This is an input for the availability integration, for the AMF model design mentioned in Section 15.4.1.

Note that ETFs may or may not directly be consumed by an SMF implementation as the information is related to the AMF configuration and the software repository, which is not standardized by the first version of the SMF specification.

Software Packaging

The SMF has been specified as a framework and it does not define any particular packaging mechanism; it leaves this aspect to the SMF implementation. As we have seen in case of the AMF CLC-CLIs, SMF also interacts with different software bundles – often referred as software packages – via CLI commands specified in the information model. As a result SMF is flexible enough to install and uninstall (i.e., upgrade) almost any type of software. Typically an SMF implementation supports the packaging technologies established by the operating systems (like rpm, deb, jar, and so on) it is able to execute on.

Each software bundle in the software repository is represented in IMM as an object in the SMF information model. The attributes of this object contain the CLI commands as they apply to the actual deployment. They are derived from the CLI commands provided by the vendor in the ETF. The software bundle object is created with this information usually when the software bundle is imported to the system's software repository.

Upgrade Campaign Design

The software is deployed in a SA Forum compliant system through an upgrade campaign executed by SMF. Hence in this step a campaign designer need to create an UCS compliant to the XML schema defined by SMF specification to instruct SMF of the appropriate procedures.

The UCS is based on the desired configuration of the legacy software created using the description contained in its ETF and the current configuration of the system. It is desirable that the campaign designer uses a tool to generate the UCS as upgrades are probably the riskiest operations in HA systems. If no tool is available, we prefer to develop that rather than use manual design. It is too error prone. Using a tool for UCS generation simplifies the integration of legacy software with SMF as well as guarantees the creation of reliable upgrade campaigns.

15.5 Open Issues

From the perspective of manageability the most important open issue is the lack of a standard system-wide trace service, which could provide a common tracing mechanism across the complete system. The SA Forum has not yet specified such a service even though it has been on the roadmap for some time. When it comes to the integration of system management, this has been identified as a gap toward the specifications.

Until such a service specification becomes available the solution is left to the middleware implementers to offer a system-wide tracing solution and if that is also not available then the system integrators need to find their own solution, which might be challenging.

15.6 Conclusion

In this chapter we described the main drivers for migrating legacy application to the SA Forum middleware. We listed the benefits of such a migration and defined a set of integration levels that can be used as reference when discussing software with respect to the SA Forum middleware. Finally we made some recommendations to help application designers during the migration process.

Our strong belief is that the SA Forum architecture offers a number of mechanisms that can ease the migration path of legacy applications and systems to the SA Forum middleware. These range between techniques explicitly supported by the specific services such as the techniques enabled by the AMF that we discussed in Section 15.4.1 and the fact that the SA Forum architecture is very tolerant in supporting a stepwise migration. By that we mean that the architecture does not couple the services for the service users. For example one does not need to create an AMF component to use the IMM or any other AIS services; nor is it required the use of IMM API to configure an AMF component. Basically, an application or component may start with using a single AIS service and migrate its other functionalities gradually. Of course, the full benefit from the use of the middleware is reached once the application is well-integrated with SA Forum services.

When it comes to migrating to a standardized, open middleware combined with the business aspects, the SA Forum architecture is the system of choice for applications that do not have the luxury to compromise their SA.

16

Overcoming Complexity: Formal Modeling Techniques at the Rescue

Maria Toeroe[1] and Ferhat Khendek[2]

[1]*Ericsson, Town of Mount Royal, Quebec, Canada*
[2]*Concordia University, Montreal, Quebec, Canada*

16.1 Introduction

As we have seen in Chapter 6 the Availability Management Framework (AMF) [48] performs its task based on an information model, and in particular the configuration information contained in this model. This model defines the scope within which AMF manages the availability of the services. That is the guarantees AMF can provide with respect to service availability (SA) is built into the model.

The model describes each entity AMF should manage as objects of the classes we touched upon in Chapter 6. Among them the configuration object class describing an AMF component defines over 20 configuration attributes all of which need to be considered when constructing the configuration for AMF. It is easy to see how the task of designing a configuration, which is not only syntactically and semantically correct – that is compliant to AMF's expectations – but that also guarantees the availability and maybe other characteristics of the system becomes a designer's nightmare if it needs to be done manually.

The Platform Management service (PLM) [36] presented in Chapter 5 requires a similar model, but it poses somewhat different challenges. Here the main issue is to make sure that when PLM collects the platform information from the Hardware Platform Interface (HPI), it needs to be able to map the detected configuration into the provided configuration model in a deterministic way. For example, one may have two similar hardware elements that are expected to run different execution environments. If the execution environment software is installed in such a way that it can boot only on one particular hardware element then the configuration needs to contain enough information for PLM that it can distinguish the hardware elements and does not try to force the wrong one to boot the execution environment.

As for the Software Management Framework (SMF) [49] discussed in Chapter 9 it needs to manipulate the entire information model to express the configuration changes for all the effected services. That is the combination of the above models and it expects this information as an input in the form of the upgrade campaign. Just like AMF, SMF also provides guarantees only within the scope of the upgrade campaign and the actual configuration it manipulates.

With all that we wanted to illustrate that designing a correct configuration for AMF or PLM and maintaining the system through upgrade campaigns is a challenge and requires significant brain power.

We can describe the problem in a different way: Building and maintaining a system that can provide highly available services on a continuous basis requires some efforts from the system and application designers and developers as well as from the system maintainers, the system administration. We can look at the SA Forum services – primarily the two frameworks AMF and SMF – as ways to shift the complexity of the task from the system maintainers and administration side toward the designers and developers because the design and development can be performed offline while maintenance and administration in high availability systems always mean online manipulation that may jeopardize the system. Simplifying these tasks in an inherently complex system is essential. However with shifting the problem from one desk to another we did not solve it. We still need a solution but because we want to solve it offline we can deploy tools of a wider range. We can use formal modeling techniques typically not considered – at least for the time being – for live systems.

We also look at them in a somewhat different context. Traditionally these techniques were designed and developed to support the software development process, that is, to ensure fault free software that satisfies the requirements. Accordingly the design process started with the requirements specification, followed by the design of the models, which could be validated and refined until they were ready for deriving the software of higher quality.

We believe that such methods and techniques are relevant to software management at configuration design time as well as at the design of upgrade campaigns, which deploy new configurations.

In this chapter we first present a high level overview of the relevant formal modeling techniques and the prerequisites the SA Forum specifications offer to enable their application. Subsequently we discuss in more details the approaches one may want to take to resolve the issues at hand. Throughout the chapter we reference work that has been done in the MAGIC[1] (Modeling and Automatic Generation of Information for Configuration and upgrade campaigns for SA) project that we invite the reader to study if interested in further details.

16.2 Background

16.2.1 The Model-Based Approach

The Unified Modeling Language (UML) [59] – as it name indicates – is a modeling language standardized by the Object Management Group (OMG). It is widely accepted by the academia as well as by the industry. It defines a set of notations including class diagrams, sequence diagrams, state charts, and so on, to describe different aspects of systems such as its structure and behavior. It uses primarily graphical elements to facilitate intuitive understanding and it does not impose strict formalism.

To achieve well-formedness, and improve the formalism, UML is complemented with a textual formal language, the Object Constraints Language (OCL) [111] to express constraints on elements of a UML model and which can also be used as a query language to extract elements from a model. OCL is based on first order logic.

---

[1] A Collaborative Research and Development project between Ericsson Software Research and Concordia University partially funded by the Natural Sciences and Engineering Research Council of Canada.

OCL is also used together with the Meta Object Facility (MOF) [112], OMG's standard for defining meta-models such as the meta-model of UML itself.

OMG has defined also different extension (or specialization) mechanisms for UML to tailor it for a particular domain and define what is called a UML profile. Through the specialization, UML profiles become more concise and formal than UML itself.

Model Driven Engineering (MDE) [113] reduces complexity by raising the level of abstraction. The idea is to start out with a high level overall picture of the target system and refine it with more and more details step-wise until a low level description is obtained that then can be transformed even automatically into the target code. Throughout the cycles of refinement validation and verification techniques can be used to guarantee that the refined model still satisfies the requirements set out at the beginning as well as the overall description is consistent and obeys the rules of the particular domain.

UML has been designed to support MDE and accordingly OMG defined its own approach referred as Model Driven Architecture™ (MDA™) [114, 115]. Its primary goals are portability, interoperability, and reusability through architectural separation.

UML is not well-suited for all the areas where formal methods and techniques have been used. For example, there exist notations and techniques better suited for the analysis of nonfunctional characteristics such as performance or our main concern availability. Techniques based on Petri-nets or Markov chains have been in use for decades [116] to determine these system characteristics. To apply them to a system described in UML, the model needs to be transformed into these formalisms.

16.2.2   Starting Points in the Specifications

MDE and MDA have been deployed mostly for code generation to raise the level of abstraction and improve quality of (prevent faults in) software. This is a valid goal in case of applications developed for SA Forum clusters and they need to follow the MDE/MDA concepts. In this chapter however we want to explore whether the same paradigm can be applied to other artifacts the SA Forum cluster needs.

AMF and some other services perform their task based on a configuration that describes the different entities they need to manage. The problem with such configuration information is that it is a significant amount of interconnected data. Any piece of it has dependencies and relations to other pieces and they should change in a consistent manner. It is like a spider net when it is touched at one point, the whole web reacts.

As we pointed out in Chapter 4 this configuration information is part of the SA Forum information model managed by the Information Model Management service [38] discussed in Chapter 8. It is described in UML, which is used in a somewhat specific way as described in Chapter 4. But the actual semantics, the behavior that defines the interdependencies between the different model elements are described in the text of the relevant service specifications. That is, they are not formalized.

Nevertheless the UML model itself is a good starting point for the formalization. As we mentioned OCL can be used to add the constraints to the UML model elements to formalize the textual descriptions of the specification.

A formalized configuration model describes the target for a site designer who needs to develop the site configuration. To create a configuration the designer has some requirements:

- the target system needs to provide some services, which may be characterized by functional and also nonfunctional requirements;
- there is a set of software that may be used to provide these services; and
- there is a cluster, a set of nodes to run that selected software.

We have presented one piece of this puzzle in the SMF in Chapter 9 at the discussion of the information flow envisioned by SMF.

The SMF specification includes the definition of the entity types file (ETF) XML (eXtensible Markup Language) schema [90], which is the means intended for software vendors to describe in a formalized way their software with respect to the SA Forum information model and in particular with respect to the different service entities. At the time of writing these are the AMF entities.

The SA Forum specifications describe the service configuration and the related semantics and the idea behind ETF is to describe the software intended for these services in similar terms. This description implies a range of deployment options so that different configurations could be derived to satisfy different systems and user requirements. That is, the ETF describes the prototypes for the entity types that can be deployed in particular systems.

From the perspective of formalization we again have a semi-formal description in the form of the XML schema the semantics for which is given in the text of the relevant specifications or even worse, implied through the entity types to be derived.

The ETF contents is the starting points for the configuration development therefore its model is an input for an automatic configuration generation process.

Unfortunately generating configurations does not solve all our headaches with the maintenance of a system designed for SA. Such a system has a long life-cycle during which it goes through numerous upgrades and reconfigurations.

The SA Forum specification dealing with these system aspects is again the SMF. As presented in Chapter 9, besides the ETF XML schema it also defines an XML schema to describe upgrade campaigns [90].

An upgrade campaign specifies the process, which migrates a system from one deployment configuration to a new desired one. One may consider this upgrade campaign specification similar to some interpreted code which guides the SMF implementation: SMF applies the upgrade campaign to the current system configuration and executes the resulting interpretation.

If we have a running system deploying a configuration to provide some given services and we have also generated a new configuration for the new (modified) requirements, the comparison of the two would identify what we need to change in the deployed configuration, that is, the target for the upgrade campaign we need to run.

If the new configuration is generated then the naming of the entities in the running system may not be consistent with the naming in the new one. We cannot use the entity features either for the mapping, since the whole purpose of the upgrade is to change features of the software entities. Therefore the mapping of the current and the desired configuration becomes a challenge.

Of course once we have identified the differences we still need to figure out the proper grouping and ordering of the upgrade procedures to be able to specify the upgrade campaign that SMF can apply to take the running system to the new desired configuration. As long as the comparison shows that only the software version changes, but not the arrangement of the software entities the upgrade campaign is relatively straightforward. However to truly take advantage of the new features of the software the arrangement of the entities such as the service group redundancy model may need to be changed as well. There might be none or more than one possible upgrade processes that maintains SA and we may need to evaluate which one suits best the constraints of the system. Besides avoiding potential service outages we need to consider dependencies and compatibility issues that do not occur in a stable running system, but characteristics of such transitioning periods.

We have to say up front that we cannot answer all these questions and it is not our goal to do so in this chapter. Nevertheless we would like to offer some ideas and discuss the experience learnt in the course of the MAGIC project mentioned in the introduction. The results of this project have been published in numerous papers that we are going to reference so that interested readers can investigate the subject beyond our high level overview.

16.3 Model-Based Software Management

16.3.1 Configuration Model

To apply any type of formal method we need to have a formal description of our problem domain, that is, a description free of any ambiguity. We need a model that we can manipulate according to some logic that finds a solution to our problem or determines that there is no solution.

The SA Forum information model is described in UML, but as we pointed out in Chapter 4 the primary goal of this model is to reflect the system status for the system administration and also to support the SA Forum services in performing their task. This is not that same task as reasoning about the correctness or constructing a valid configuration, which requires more information – information which is provided only informally in the specification text.

For reasoning about the correctness of the configuration we need to extend the UML model defined by the specifications with the constraints that describe the interrelations between the different model elements and their attributes. Most of these are currently hidden in the text of the relevant specification.

We use the word hidden on purpose because the specifications define the system runtime behavior. For example, in the AMF specification one of the main emphases is on the (component) service instance (SI) assignment distribution among the service units (SUs) and their components. This is only performed at runtime and it is not part of the configuration information. At the same time it has certain implications that need to be satisfied at configuration time so that the specified behavior can be performed at runtime. For example, if the redundancy model indicates that five assignments need to be given out for a SI then at least five SUs need to be in the protecting service groups.

The extent to which one is able to identify these rules in the specification determines the powerfulness of the model. Since the information model is described in UML, OCL could be a suitable language to describe the related constraints.

When trying to formalize these constraints one realizes quickly that, for example, the AMF information model is defined in such a way that it emphasizes similarity to suit the management needs and accordingly few classes are defined with mostly optional attributes. The mandatory attributes define the conditions when the optional attributes are used, which in turn have some relation with other attributes.

This conditional nature of attributes complicates the constraints definition of this already complex domain. A way to deal with complexity is to refine the standard model following the classification presented in the standard and define a class hierarchy where each class contains only the attributes appropriate for the subclass, following the basic object oriented principles.

Another nuance that one has to pay attention is that the standard information model defines the multiplicities in such a way that the Information Model Management service managing the model can accept incomplete configurations so that the model can be changed at runtime, for example, during a system upgrade. But when we want to determine whether a configuration is correct or not from the perspective of stable operation, it needs to be complete.

This shows that one may have different versions of the completed model depending on the system's mode of operation. For example, the verification that the system will transition through valid configurations during an upgrade implies a different notion of validity than when it is checked for normal operation mode.

Within the MAGIC project the work has been completed for the AMF information model and the results reported in [117, 118]. It encompasses an AMF domain model and a UML profile for AMF that can be plugged into an appropriate UML CASE (Computer-Aided Software Engineering) tool to create and validate AMF configurations.

It turns out that the general question whether a service group can protect all the SIs assigned to it is an nondeterministic polynomial-time hard (NP-hard) problem for most of the redundancy models as shown in [119] and therefore we cannot guarantee the answer in all cases within a polynomial time even using these techniques. However for SA this is a very important question (an important property of the configuration) and because of its NP-hardness we resort to heuristics to resolve it as proposed in [120].

The work referenced so far used as its starting point the UML model defined in the AMF specification and defined a new UML profile for AMF.

However others may approach the problem differently and base their work on results of the work done in the field of real-time embedded systems. Several UML profiles have been defined and among them Modeling and Analysis of Real-Time Embedded systems (MARTE) [121] for real-time embedded systems, which is the basis of the Dependability Analysis Modeling (DAM) [122] profile – an extension to cover dependability analysis. Depending of the ultimate goal of the model these can serve as starting points as well.

16.3.2   Configuration Generation

After we formalized our target – the system configuration, we are able to reason and construct valid configurations. UML CASE tools, for example, can take a UML profile and guide the user in crafting a syntactically valid configuration. This is however still a manual process and may involve a lot of guessing work on the site designer's side until he or she gets it right so that the tool throws no more errors. Meanwhile his or her focus is divided between the errors and the requirements the configuration would need to satisfy.

Knowing the dependencies built into the profile one can come up also with a systematic approach to construct valid configurations, which then can be turned into a deployable implementation.

16.3.2.1   Inputs for the Configuration Generation

As we mentioned in Section 16.2.2 to design a configuration one needs to start with the services the system needs to provide, the description of the software that can be used to provide these services, and the cluster on which the software will be deployed to provide these services.

Within the SA Forum information model the application services are reflected only in the AMF information model. They are the SIs provided and protected by service groups. The actual assignments are performed at the component service instance (CSI) level as AMF selects the components most appropriate for the job.

Since only the site designer knows the services that the target system needs to provide, he or she needs to define the SIs and their CSIs. If the target site exists, that is, its reconfiguration or upgrade is the goal then the information can be collected from its current configuration together with the cluster information.

This also means that the AMF model itself can be used to represent these inputs.

With that we can slightly reformulate our goal: We want to generate the configuration of service provider entities – components, SUs, service groups – that are capable of providing and protecting the SIs and their CSIs the system needs to provide and distribute them on the cluster nodes; all in a way that satisfies the constraints imposed by AMF.

A software component is the execution of some software, which is installed in the system. This software typically is developed by a software vendor who describes its product in terms prototypes in an ETF. The ETF needs to include at least the prototypes for the components that can be instantiated from the software and the CSIs these components can provide. Other prototypes need to be present only if there are limitations on how these basic prototypes can be combined.

Here we need to take a short detour as the format of the ETFs is defined by an XML schema and the semantics is described in text. So the situation is similar to the AMF configuration except that it is not provided initially in UML. The reason for that is that ETF is expected to be packaged and delivered together with the software for which XML is much more suitable.

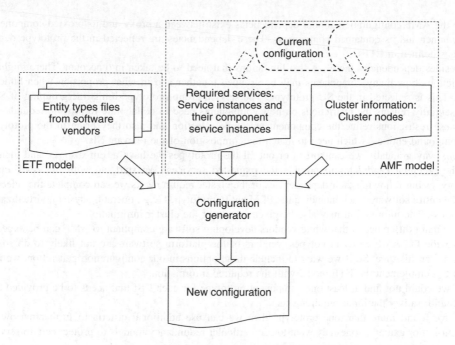

Figure 16.1   Inputs for the configuration generation.

Looking at the actual contents of the schema it is easy to see that it maps well into a UML model therefore the formalization process we described in Section 16.3.1 applies easily to ETF as well and we can create an ETF domain model and profile.

Figure 16.1 summarizes the inputs of the configuration generation.

16.3.2.2   Prototype Selection and Dependency Handling

In the AMF information model the SIs and their CSIs that our target configuration need to provide all reference their types, the service types and the component service types.

Among them the component service types at least were derived from some component service prototypes. This is the link we are looking for as in ETF we find the information on the component service prototypes and which component prototypes can provide them.

With respect to the derivation while theoretically it is possible to derive different component service types from the same prototype, on practice there might not be a reason for this. Because of its link to the component prototypes, it may be better reasons to equate the component service type defined in AMF to the component service prototype defined in ETF.

Based on this information we can select from our ETF model the component prototypes that are potential candidates to provide the required CSIs composing the SIs.

When making this selection we also need to take into account the vendor defined limitations, restrictions, and dependencies.

A component prototype may require another component prototype to be able to provide some component service prototype. This means that components of both prototypes need to be included in the configuration grouped together in the same SU, and to activate them the relevant CSIs need to compose a SI.

In the AMF configuration similar dependencies exist between a proxy and its proxied components, a container and its contained components. These dependencies are reflected in the prototype dependencies defined in ETFs.

Besides dependencies there are also limitations that need to be taken into account. There might be restrictions how many components may be grouped together and in what proportion when building SUs. This is reflected in the SU prototype. The service group prototype limits the grouping of SUs. It essentially reflects the restrictions on how components need to collaborate to provide redundancy.

These restrictions refine the component prototype selection and also they bring in the prototypes of compound entities, which need to match the composition of the relevant SIs.

This way gradually we can sort in or out all the prototypes we have at our disposal and identify all the potential candidates to provide the requested services. We start out with the AMF entity prototypes but following their dependencies or the cluster requirements we can complete the selection for the entire software stack including the PLM entity prototypes (e.g., operating system, virtualization facilities). The information may also be given as part of the cluster information.

The main difference is that while vendors developing software compliant to AMF can be expected to provide the AMF entity prototypes, vendors of the platform software are not likely to do so – at least not at this time. So if we want to include these options in our configuration generation, we may need to complement the ETF model with the required information.

If we could not find at least one component prototype for each CSI that needs to be provided then we cannot satisfy the input requirements.

If we found more than one prototype then we can use additional criteria to further narrow the selection. For example, one may want to use particular redundancy models to protect certain services or use particular versions of the software.

16.3.2.3 Generating Entities with Their Types

When we look at the prototype candidates we have selected we may realize that they may not make up complete stacks. For example, we may have component prototypes, but no SU prototypes grouping them.

Also when we made our selection we might have selected a prototype because it was in the range we needed, however for the configuration to work properly we need to narrow the choice to a single or a set of valid options only. For example, a SI may indicate that it contains three CSIs of a given component service type, but the matching SU may be configured in the range of 1–10.

This means that we need to go through our selected prototypes and derive from them the entity prototypes (e.g., in case of AMF the component type, SU type, etc. as defined by the AMF specification) that match exactly the input requirements. Depending on the attribute this may mean defining a single option or a valid range. When doing so we again need to look at the input requirements to find the best match for all the attributes of the relevant service configuration.

For the configuration entity types for which we have no selected prototypes we need to create the types required by the configuration. These are typically compound entity prototypes and the fact that we have no prototypes for them means that there is no limitation or restriction on the way we can construct them. Therefore we can use the input requirements to construct them the way they match best the needs. A typical case would be that for the set of CSIs of a SI we found the component prototypes capable of providing them, but no SU prototype. Since the CSIs are all part of the same SI, the component types derived from selected component prototypes need to be included in the same SU type, which need to contain enough components so that the SI can be provided and protected.

So we need to look at the capabilities of the component types with respect to CSIs and make sure that the SU incorporates enough of them to provide the capacity required by the SI.

This leads us to the next step: the generation of the entities themselves. Again we need to look at the defined entity types and the services they need to provide and protect according to the input requirements. We need to generate the model representing the individual entities and includes enough of them so that the capacities required for active and standby assignments are met or – if desired – even exceeded. In this case the provisioning and protection of the SIs are guaranteed by construction, contrarily to the NP-hardness issue we run into at the validation of configurations.

Once we have all the entities generated we need to distribute them on the cluster nodes. Here the main criteria is to ensure that redundant entities protecting the same services are distributed on different cluster nodes so that when a node fails it cannot take out both the active and the standby assignments.

We also need to take into account dependencies. As we mentioned earlier it is possible that a component type has dependencies. This is the case with contained components which need to be configured together with the container components. Similar dependency may exist toward an execution environment or components that include hardware elements.

For some entities the SA Forum information model is very specific and the exact mapping needs to be provided in the configuration information. In other cases a pool of hosting nodes can be defined that can host one or more entities and the relevant service implementation will need to make the decision at runtime. An example of the former case is the AMF node – CLM node (Cluster Membership Service) – execution environment mapping, which is given as 1 : 1 mapping in the configuration. In contrast, AMF allows for more flexible configurations with respect to the nodes hosting the SUs of a service group. These can be given in a 1 : 1 mapping or as a node group within which any node may host any SU.

Figure 16.2 illustrates the different steps in the configuration generation process with their respective input and output.

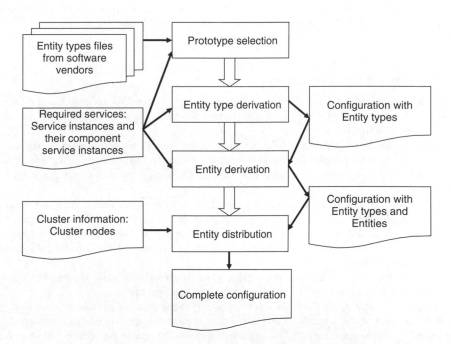

Figure 16.2 Overview of the configuration generation process.

16.3.2.4   Implementation Options

The general principles of the configuration generation have been implemented within the MAGIC project in different ways. Both cover the AMF portion of the information model.

The first approach is algorithmic and it has been presented in more details in [123, 124], while specifics on SU ranking attributes settings is discussed in [80, 125].

The proposed solutions were implemented as Eclipse [126] plugins.

The basic principle in this solution is that given the input requirements the implementation searches through the ETF model to find the appropriate prototypes. Once a prototype has been found for the required CSI or SI he search moves on to the next until there is a selection for each of them.

A completely different approach is reported in [127], which is based on UML profiles defined for ETF and AMF and which fully exploits the MDA paradigm using the model transformation techniques. Rather then defining a selection algorithm, it defines the transformation rules that gradually mutate the model instances provided as input to the configuration generation to an instance of the output model.

The transformation rules are implementations of the relations between prototypes, types and entities and among the elements of each of these groups as we presented in the previous sections. They create links among the model elements if they satisfy the rule and prune the rest of the model. Applying the rules one by one to the input models eventually we reach the output model, which is constructed only of elements that satisfied all the applicable rules.

This declarative method has been implemented within Eclipse using ATLAS Transformation Language (ATL) [128].

16.3.3   Upgrade Campaign Generation

When it comes to upgrades we can talk about two main approaches: replacing the running entities with their new versions and migrating the system to a newly generated configuration.

In this section we look at each of these in more details.

16.3.3.1   Upgrading the Running Entities

This first approach is quite straightforward and used in most systems today. It is based on the fact that we usually want to upgrade the system or some parts of it when new versions of the running software become available.

Accordingly, the goal is to identify the entities running the old version of each of the software for which a new version is ready for deployment; and replace the old software with the new one by installing it and adjusting the configuration attributes then restarting the entity so that it starts to run the new software.

In Section 16.3.2 we have seen that entity types are derived from the entity prototypes describing the software implementation, that is, the entity type represents a particular software implementation. At the discussion of the AMF information model in Chapter 6 we also saw that entity types are grouped into entity base types as the reflection of different versions of the software delivering the same basic functionality.

This relationship puts in relation a new version of the software with those running in the system. Thus, it helps identifying the entities in the system which are upgradeable.

Of course, becoming upgradeable does not necessarily mean that we have to or want to upgrade the entities. Moreover, the upgrade of one entity may require the upgrade of other entities. Related entities need to remain consistent after an upgrade. Therefore the choice may depend on other factors and we may want to make a selection.

Once we identified the entities we would like to upgrade we can create a new configuration by replacing the entity types of the selected entities with the new entity types. It is essential that we validate this newly created configuration, for example, using UML CASE tools and the appropriate profiles because when we change the type of an entity its characteristics change, so may its dependencies and limitations. Many of these dependencies and restrictions are reflected in the relations of the different configuration attributes. While probably one can spot such a dependency change when the new version of a proxied component type becomes SA-aware. But when the change is as subtle as the modification of the component capabilities or the number of entities in a compound entity type, it may be difficult to find the difference. Both cases may render the modified configuration invalid without further modification of the configuration.

If the new target configuration evaluates as valid we can start the upgrade campaign generation. Existing entities that are part of a redundancy schema typically can be upgraded by a rolling upgrade procedure without causing any service outage. Therefore we can generate a rolling upgrade procedure for each service group whose SUs or some components within being upgraded.

To deal with the problem of the proxy becoming unnecessary by changing the proxied component type to an SA-aware type, we also need to be able to remove entities from the configuration. Symmetrically we would like to be able to add entities to the system.

The SMF specification suggests the use of single-step upgrade method in both cases. This suggestion is valid from the perspective of the services being added and removed. Unfortunately the operation is not performed in the vacuum, software installation, and removal, the dependencies between the entities and/or their grouping may impact other entities in the system beyond those being targeted by the upgrade campaign. In such cases we may want to generate a series of single-step upgrade procedures based on the redundancy of the impacted entities so we can make sure that the scope of each single-steps is not service impacting.

Figure 16.3 summarizes the upgrade campaign generation process.

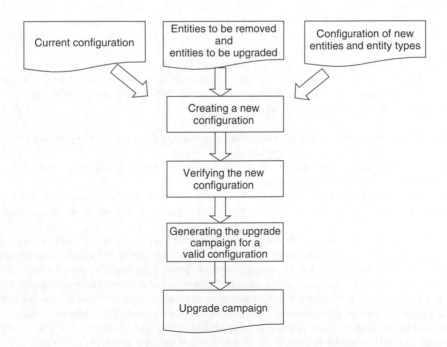

Figure 16.3   Upgrade campaign generation for adding new, removing, and upgrading existing entities.

To reduce the overall execution time of the upgrade campaign the different procedures may be further grouped so that operations on the same node, for example, are collected together and the entire node is upgraded in a single step rather than locking and unlocking it multiple times.

An implementation of this semi-automatic upgrade campaign generation method composes the core of the work presented in [129], which has been implemented as an Eclipse plugin.

16.3.3.2 Upgrade to New Generated Configuration

We have seen the problem of the type replacement method: the new version of the type may not fit properly into the old configuration. Additional changes may be required, which may or may not be straightforward.

Also, since the running configuration was generated with the features of the old entity prototypes it possible could not take into account any new features of the newly delivered one, while the whole idea of the upgrade usually focuses on deploying those new features. A better idea is to generate a new configuration with the updated input requirement and upgrade the system to this new configuration.

Unfortunately it is easier said than done. The problem is that when we generate a configuration we generate the entity types and the entities including their names. Since the upgrade is about to change the characteristics of the entities, we cannot rely on them either to identify which entity is new in the system, which one was upgraded and which ones were removed. This mapping between the configuration of the running system and the newly generated system is the key element from the perspective of maintaining the continuity of the services.

This issue normally does not come up in systems that we can shut down and bring up with the new configuration. To maintain the continuity of the service we need to know exactly which entity participates in the redundancy for the provisioning of which service so that this redundancy can be used to avoid service outage.

The continuity of the service seems to be the problem, but it also provides us with the solution: as we have seen in Section 16.3.2.1 one of the input requirements for configuration generation is the description of the SIs and their component service instances. All those service instances existing in the current configuration that need to be provided continuously throughout the upgrade and afterwards need to be part of the input requirements for the new configuration generation. This means that they put in relation the entities providing these service instances in the existing and the new configuration therefore we can resolve the mapping between the running configuration and the new one.

The mapping identifies three groups of entities:

- **Entities to be removed:** These are the entities that exist in the current configuration but do not map into any entity of the new configuration
- **Entities to be added:** These are the entities that exist in the new configuration but do not map into any entity of the current configuration
- **Potentially upgraded entities:** These are the entities that map between the current and the new configuration. These need to be analyzed further to find out whether they indeed have been upgraded.

Entities of the first two groups are always part of the upgrade target, while from the third group only those entities that changed some of their properties become part of the target. This information resembles the input for the upgrade campaign generation method presented in the previous section.

The differences are in the nuances. For example, it may occur that two service instances that are protected by the same service group in the current configuration, may be assigned to two different service groups in the new configuration. While in the previous process there was no way to specify such a change, the configuration generation method may create such differences. It may also select a different redundancy model to exploit the new features of the new software, which may not easily transform into an upgrade procedure.

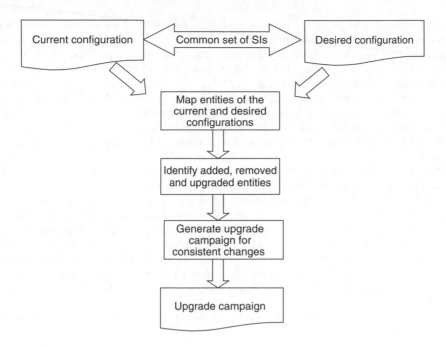

Figure 16.4   Upgrade campaign generation to deploy a new configuration.

Figure 16.4 summarizes the generation of an upgrade campaign that deploys a new configuration. Compared to Figure 16.3, there is no need for configuration validation since one can assume that the desired target configuration is valid. As not all changes may be consistently deployable while preserving service continuity we need to analyze the changes and may want to generate the campaign only if all the changes are consistent.

This approach while it is more generic requires further investigation. The results achieved so far as part of the MAGIC project are reported in [130] including a prototype implementation as an Eclipse plugin.

16.3.4   Analytical Models and How They Can Help

In this chapter so far we have looked at the ways satisfying the requirements posed by the SA Forum specifications at automatic generation of valid configurations, upgrade campaigns and at the validation of third party configurations.

These specification requirements are related to the techniques offered by the SA Forum to increase the availability of the different services through the different fault tolerance mechanisms.

While the specifications pose many requirements in regarding the correct use of the specified techniques none of them are non-functional requirements and therefore do not characterize SA, which is the main subject of our book.

So the question is do we know what availability guaranties we achieve when deploying these techniques? This is the realm of fault forecasting as introduced in Chapter 1. We need to pose this question mainly for efficiency reasons: Since redundancy means extra cost we want to deploy just enough of it to meet our targeted availability since that is the level for which the customer is willing to pay.

Without attempting to give an exhaustive overview of the state of the art, we can say that the quantitative analysis of the availability of systems is based on the state models which describe the transitions of the system between healthy and faulty states. Failures that move the system to a faulty state may occur at different rates, while the repairs that move it back to a healthy state take some time and therefore characterized by durations.

To deal with the erratic nature of failures, stochastic models such as the Markov models [131, 132] have been developed early on and used for availability analysis. The Markov reward model is an extension of the Markov model, which allows the association of a reward rate with each state of a Markov chain, thus capturing nonfunctional aspects of the system such as performance and dependability.

Markov models are capable of representing complex real time system, however the number of states required doing so may be forbiddingly high [131] for manual processing.

An alternative model is the Stochastic Reward Net (SRN) [133], which is an extension to the Stochastic Petri Nets. It can represent the system in a still manageable size while it can be transformed automatically, for example, into Markov reward models which in turn can be solved analytically with available tools like stochastic Petri net package (SPNP) [134].

More recent work focuses on even higher level representation of the system using, for example, the already mentioned DAM [122] profile. In turn these higher level representations are also transformed into stochastic models for analysis, into Deterministic and Stochastic Petri Nets (DSPNs) in case of DAM.

Some work has been done with respect to the SA Forum specification as well: [135] presents the use of SRNs for SA computation.

All this would suggest that we should be able to answer the question what availability guarantees an SA Forum compliant system can provide. It turns out that this is not quite the case.

The AMF chooses a recovery action based on three factors among which the configuration, that is, the model – similar to those used by the discussed methods – is only one. In addition AMF takes into account the recommendations of the processes reporting the error and it also correlates the error with previous errors. All these may escalate the actual recovery recommended for a component to a recovery on the fault zone of a higher level.

This dynamic adaptation of the recoveries makes it hard to evaluate the availability of the AMF managed services. Hence the failure rates and the repair durations typically measured for the components in some test environment need to be re-interpreted for each new configuration, each new context the component is placed in. This means that components running the same software but on different nodes of the cluster may undergo different recovery actions and at different rates due to the impact of their different environment. The most tangible example for this is the software running on an overheating board.

All that is to say that a high level model discussed in Section 16.3.1 is a starting point for the evaluation of the availability and other non-functional characteristics, however the specifics of the transformation of this model into stochastic analytical models requires some further research.

When these results become available, they can be integrated into a configuration generation method as well as into the upgrade campaign generation to ensure that the outcome of the generation meets targeted non-functional requirements as well as compliant to the specifications.

16.4   Conclusion

This chapter addressed the missing pieces necessary to operate a system compliant to the SA Forum specifications: Namely, the creation of system configurations and upgrade campaigns deploying them. The AMF and the SMF cannot perform their respective task without these artifacts, but creating them for complex systems is a challenging task without tool support. To be able to provide tool support at least some formalization is necessary.

That is, we need a formal model to represent configurations. We looked at the SA Forum information model and in particular at its AMF portion to determine that in the form it is specified in the specification is suitable for runtime management, but for analysis one may benefit from further refinements, extensions, and formalization. One of the techniques suggested is the definition of a UML profile, which then can be used in UML CASE tools to support site designers in defining valid configurations and validating existing ones before their deployment.

One may go even further and relieve designers all together from the details of the AMF configurations by defining rigorous methods and techniques for the automatic generation of such configurations. We presented its principles and most important steps and pointed to two different approaches that have been implemented for this purpose.

The enabler for these techniques was the SMF, which requires software vendors to describe their software formally, so that this description is machine processable.

SMF also formalizes the specification of upgrade campaigns to be used to deploy a new configuration in a running system while maintaining SA. We have looked at the main challenges and the steps of coming up with an upgrade campaign. Since the goal is to maintain SA throughout the execution of the upgrade campaign we suggested the use of these services as the basis for the mapping of generated configurations to be able to identify the upgrade target. The upgrade target, in turn, determines the applicable upgrade methods based on which the upgrade campaign can be constructed.

Finally we looked at whether we could predict the SA achievable by a given configuration. For decades different stochastic models have been used for this purpose, but such analysis models capture different perspectives than the UML-based models used nowadays for system specification. Nevertheless there are solutions to transform these UML-based specifications into stochastic models to bridge the gap and allow for the evaluation of availability of the systems under consideration.

The SA Forum specifications include some features that require modifications/extensions to the existing transformation techniques, but there is no reason to assume this would be impossible.

It is desirable to go even further and based on the results of the analysis study to refine the configuration generation techniques to incorporate the lesson learned in the generation process itself so that the generated configuration meets an availability (and potentially other) nonfunctional criteria as well as the functional criteria imposed by the SA Forum specifications.

Throughout our discussion we referenced work that has been done in the area and we invite the interested reader to explore further the subject as here we only had the chance to give a taste to this work.

Complex systems such as grids and clouds cannot be operated without adequate tool support. One of the biggest challenges of clouds is already SA, which in turn requires proper design for manageability and maintenance. It is unimaginable to perform these tasks without appropriate tools that can provide provable guarantees of the design.

17

Conclusion

17.1 Summary

In Part One of the book, we introduced the notion of service availability and presented a set of relevant principles and concepts that are associated with dependability, for the purpose of achieving service continuity even in the presence of some failures in the underlying system. We went on to give a brief description of the circumstances, the vision of developing open standards for service availability and how the Service Availability (SA) Forum was founded.

Part Two discussed some of the key services and frameworks specified by the SA Forum for delivering service availability. The main emphasis was on the reasons behind the design choices for the standard as much of this information was not documented in the specification at all. Therefore, this part can be viewed as the main principles behind the SA Forum services and frameworks providing the readers with valuable insight into the specifications and therefore enable them to use the specifications as they were intended.

We started this part of the book with an overview of the architecture of a SA Forum system, and discussed the dependencies and interrelations among these services. In the information model chapter, we explained the background and considerations that were taken into account as the different part of SA Forum Information Model were developed; and application designers need to follow these same considerations whenever they extend the information model for their application.

In the chapter that explained the various platform services we provided the rationale behind the designs of the Hardware Platform Interface and the Cluster Membership service reflecting respectively the physical and the logical platform views of clustered systems. One is being discovery driven, while the other is based on configuration these two worlds could be perceived as antagonistic therefore as part of this discussion we described how the Platform Management service bridged the gap between the corresponding hardware level and the level where membership in a cluster was handled.

In the Availability Management Framework (AMF) chapter, we presented some of the key dependability concepts that were introduced in Part One and placed them into the context of a SA Forum system. Through this we gave a comprehensive introduction to AMF showing how service availability could be maintained. The AMF specification is often perceived as too complex without clear distinction of what should be addressed by the developers and what is of concern of configuration designers and administrators. Therefore we divided our discussion around the tasks of component development, the configuration design and the system administration.

We trust that the main takeaway from the chapter is that AMF can manage the service availability at different levels of sophistication for a wide variety of applications. Its capabilities range from simple life-cycle management of application components implementing none of the application

Service Availability: Principles and Practice, First Edition. Edited by Maria Toeroe and Francis Tam.
© 2012 John Wiley & Sons, Ltd. Published 2012 by John Wiley & Sons, Ltd.

programming interface (API) to the orchestration of sophisticated switch-over and failover scenarios among components implementing all the bells and whistles possible.

The subsequent chapter on communication and synchronization utilities developed further the support offered to applications developers by discussing the Event, Message, and Checkpoint services. They provide the active and standby entities with facilities for state replication and communication, which are appropriate for the clustered environment. Namely, that they decouple the interacting entities in a location agnostic way.

The system management chapter discussed the essential functions for fault management that included the Log, Notification, and Information Model Management services. One might consider these services less important as they typically do not contribute to the functionality provided by the application to its users. However, when it comes to service availability all revolves round management and fault management is of primary concern as we presented it through these services.

Next we explained the approach taken by the Software Management Framework for upgrading a SA Forum system with minimum service disruption – another aspect essential for service availability. In particular, we focused on the notions of software inventory and upgrade campaign in the chapter. The former geared toward software vendors responsible for supplying the new or modified software as the expectation was/is that they need to describe their product developed for SA Forum compliant systems in terms of the SA Forum specifications. The latter is concerned with the deployment of new software or configuration, including the failure handling mechanism during an upgrade.

We concluded this part by giving hints on how to combine the SA Forum services and frameworks, whether, when and by whom they should be used together, and what to be avoided, in order to achieve the desired service availability goal.

Part Three of the book contained a collection of topics that had a more practical nature, addressing issues such as the programming model, the implementation of the SA Forum middleware, simple examples, the migration paths for non-SA Forum applications, and the use of formal techniques. The aim was to take the readers one step closer to the daily software design and development tasks so that a link between the discussed principles and practice could be established.

We started this part with the chapter on the programming model and API conventions in the C programming language, the definitive specification of the SA Forum interfaces. Apart from the general discussions on the consistent usage pattern across all the SA Forum services and frameworks, we have also included a number of topics and issues that are frequently considered by the developers in practice. These included the interaction with Portable Operating System Interface for Unix (POSIX), memory management, pointers handling, finding out the implementation limits, the availability of area service libraries, and the issue of backward compatibility. The Java mappings chapter continued with the discussions on the history, rationale, and mapping idioms used in the SA Forum Java language mappings, together with its usage and experience to date. The SA Forum middleware implementations chapter described OpenHPI and OpenSAF, the two most complete and up-to-date open source implementations of the SA Forum specifications.

The chapter on integrating the VideoLAN Client with OpenSAF demonstrated an example of using the SA Forum specifications in practice. We presented the different levels of integration with OpenSAF and discussed the relative implications of each approach on the overall availability achievement and complexity of the implementation. The migration paths for legacy applications chapter explained the benefits of migrating non-SA Forum applications to use the SA Forum middleware. We presented the various integration aspects and gave recommendations and guidelines to developers for their choices of integration approaches.

In the last chapter of this part, we showed how formal techniques could be used as a basis for tool support for the generation of SA Forum system configurations and upgrade campaigns for the Software Management Framework. We discussed the challenges and the most promising techniques and platforms for delivering tool support to the tasks of site design and maintenance.

In terms of the specific SA Forum services, we have omitted the Security, Naming, Lock, and Timer services in this book. As pointed out in the beginning, we left security out because it is a topic that crosscuts all the services and it could have easily filled the book of this size. Due to the nature of the other three services, we feel that their omissions do not compromise the overall treatment of the subject on service availability. Following our philosophy of making this book as pragmatic as possible, we would have liked to tackle topics such as testing systems after they have been implemented. That would have included the discussions of techniques such as failures and events simulation, fault injection and processing, and so on. Again, this would have caused a significant increase in the number of pages needed in the book.

We believe the presentation in this book has achieved the goal we set out in the beginning, and has indeed captured the undocumented reasoning behind the many design decisions of the SA Forum specifications. We feel that this book has succeeded in linking the principles with practice of service availability.

17.2   The Future

At the time of writing, the SA Forum specifications still remain to be the only open standard with a rather comprehensive support for service availability and a proven track record in the telecommunications domain. Coupled with the ongoing open source implementations and the continuous feedback loop from the implementation experience to rectify and improve the specifications, the anticipation is that the quality will continue to increase.

Although this book has been written to address specifically the SA Forum specifications, the principles, and concepts used in arriving at the various services and frameworks, which are the results of many years of accumulated experience of the many contributors to the SA Forum, are applicable to other types of systems and new technologies as well. One example is the much publicized cloud computing, which has evolved recently into a main trend in the IT and enterprise sectors.

As with any new buzzword coming into town, there is no overall agreement on the definition of cloud computing. However, the term cloud computing is generally characterized as having access to IT infrastructures, platforms, and applications over the Internet, and the users are charged based on the usage of the requested services. The main offerings come from a diverse range of companies including Amazon's Elastic Compute Cloud (EC2) [136], Microsoft's Windows Azure [137], VMware's vCloud® [138], and Google's App Engine [139]. On the horizon there are also telecom grade cloud infrastructures [15] that are suitable for the network equipment providers and operators to offer better communication solutions to their customers.

One of the most cited advantages and at the same time top obstacles of adopting cloud computing has been high availability (HA) [140]. It is evident from recent reports of service outage of cloud computing providers such as Amazon Web Services [4] and Microsoft [141], the HA issue needs to be addressed. Even when this is resolved at the cloud computing provider level, a cloud computing user, typically via platform/infrastructure as a service, still needs to deal with HA issues at the application level.

As an example, given the choice of VMware HA and VMware FT (fault tolerant) [142], an application designer must decide whether loss of data and service for the period of time during which a failover is performed by VMware HA is acceptable or not. If not, then VMware FT should be used instead, which provides a replica running lock-step with the primary so that there is no loss of data nor service even when there is a host failure. Similarly, a developer using Windows Azure must also decide whether loss of work during a virtual machine restart [143] is acceptable or not. If not, the developer must introduce his or her own solution. Yet another example comes from the recent Amazon EC2 service disruption report [5] in which the problem was reportedly caused by an erroneous step in an upgrade. The report has further identified that some EC2 customer applications were impacted in this event because they only used a single Availability Zone. This was because either they did not have access to multiple Availability Zones due to their types of requested services or they did not understand the consequence of using a single Availability Zone was basically the cause of a single point of failure. As one imagine, all of these issues must be thought through by developers at the application level, regardless of whether they are using cloud computing or the SA Forum services and frameworks.

It is important, however, to put these cases into perspective: Virtualization as used in cloud computing addresses typically host, that is, hardware failures whether there is a standby virtual machine running in lock-step or not. The fact is that failures occur not only in hardware, but even more frequently in software whether it is the operating system, the hypervisor, the middleware, or the different applications that include the offending piece of code, or maybe it is the interaction among them causes the failure. In this latter case it is extremely difficult, if not impossible, to detect and debug the software at development time. Unfortunately this does not reduce the impact of the failure and may not even reduce its frequency in the live system.

Furthermore systems designed for service availability also have to consider and protect at least to some extent against administrative and human errors. From this perspective upgrades are the most critical operations as they are 'part of life' of the systems of such continuous operation.

It is unlikely that any single technology or method is able and will provide a solution for all these cases. Experience shows that the different techniques need to be used in unison to provide the best possible/feasible protection at the lowest possible/feasible overhead. Indeed, we did not emphasize this aspect throughout our book as it is quite obvious that all these techniques introduce some overhead in the system and accordingly they also imply economical and performance consequence which could be studied and evaluated. That in itself is again a whole area of research.

To remain at our subject, one needs to approach the problem systematically by the analysis of potential faults at all levels of the system contributing to the delivery of the services, their impact on other parts of the system and the system's services in general. These need to be matched with the applicable methods of protection. Only with such global systematic view one is prepared to make the decision on the design the system and the techniques deployed for the protection of the different services. Note that when looking at availability at the service level the needs may vary based on the different service level agreements regulating each one of them.

Ultimately, in spite of the relatively long history of research and investigation into dependable systems with considerable success on the technology and solution fronts, the fact remains that system failures causing unavailability of services are occurring far too frequently. More often than not, postmortem reports on these system failures suggested that the causes of the failures were more to do with how the systems were designed, rather than the lack of viable solutions. This casual observation is also reinforced by the fact that the subject of designing dependable systems is not in the core of computing education, and that computing professionals with skills and experience in HA remain to be a very small minority of specialists in the field.

This very last point brings out an issue related to educating the next generation of computing professionals. We advocate the teaching of dependable systems design in the core of the Computer Science [144], Information Technology [145], and Information Systems [146] curriculums, instead of treating it as an advanced, specialized, and optional topic. Since we are depending more and more on computer-based systems, it is extremely important to bring this issue upfront to the computer-based system designers, rather than trying to fix the implementations afterwards. Contrary to the popular belief that this is a difficult topic, the subject can even be taught with no complex mathematics, but by showing the concepts, principles, design, and practice, as we have demonstrated in this book. If we have our own way, the days of collecting dossier on unavailability of services and their consequences could very well be numbered.

References

1. YLE (2008) Widespread Disturbance in Elisa's GSM Network Fixed. YLE Uutiset, http://yle.fi/news/id90683.html (accessed 14 May 2008).
2. The Local (2008) Telia Customers Lose Network Coverage, http://www.thelocal.se/11828/ (accessed 17 May 2008).
3. Helsingin Sanomat (2008) Lengthy Blackout Stalls Tuesday Share Trading Across Nordic Region. International Edition, http://www.hs.fi/english/article/Lengthy+blackout+stalls+Tuesday+share+trading+across+Nordic+region+/1135236901403 (accessed 4 June 2008).
4. BBC (2011) Amazon Apologises for Cloud Fault One Week on, http://www.bbc.co.uk/news/business-13242782 (accessed 4 May 2011).
5. Amazon Web Services (2011) Summary of the Amazon EC2 and Amazon RDS Service Disruption in the US East Region, http://aws.amazon.com/message/65648/ (accessed 12 December 2011).
6. Narasimhan, P. (2010) Downtime Incidents, http://www.cs.cmu.edu/~priya/downtime.html (accessed 5 December 2010).
7. Service Availability Forum (2011) Service Outage Examples, http://saforum.org/Service-Outage-Examples~310991~16627.htm (accessed 17 July 2011).
8. International Federation of Information Processing (2011) IFIP Working Group 10.4 on Dependable Computing and Fault Tolerance, http://www.dependability.org/wg10.4/ (accessed 17 July 2011).
9. Laprie, J.C. (1985) Dependable computing and fault tolerance: concepts and terminology, *Proceedings of the 15th IEEE International Symposium on Fault Tolerant Computing*, IEEE Computer Society, pp. 2–11.
10. Avižienis, A., Laprie, J-C., Randell, B., and Landwehr, C. (2004) Basic concepts and taxonomy of dependable and secure computing. *IEEE Transactions on Dependable and Secure Computing*, **1** (1), 11–33.
11. IEEE (1990) Std 610.12-1990. *IEEE Standard Glossary of Software Engineering Terminology*, Institute of Electrical and Electronics Engineers.
12. Department of Defense (2011) Federal Acquisition Regulation (FAR), https://www.acquisition.gov/comp/far/index.html (accessed 11 July 2011).
13. Oberndorf, P.A. (1997) Facilitating component-based software engineering: COTS and open systems, in *5th International Symposium of Assessment on Software Tools and Technologies* (ed. E. Nahouraii), IEEE Computer Society, Pittsburgh, PA, pp. 143–148.
14. Anderson, T. and Lee, P. (1981) *Fault Tolerance: Principles and Practice*, Prentice Hall, London.
15. SCOPE Alliance (2011) Telecom Grade Cloud Computing, Version 1.0, 2011-05-03.
16. Gray, J. (1986) Why do computers stop and what can be done about it? *5th Symposium on Reliability and Distributed Software and Database Systems*, IEEE Computer Society Press, Los Angeles, CA, pp. 3–12.
17. Geffroy, J.C. and Motet, G. (2002) *Design of Dependable Computing Systems*, Kluwer Academic Publishers, Dordrecht.
18. Birman, K.P. (2005) *Reliable Distributed Systems Technologies, Web Services, and Applications*, Springer, New York.
19. Pullum, L. (2001) *Software Fault Tolerance Techniques and Implementation*, Artech House, Norwood.
20. Schmidt, K. (2006) *High Availability and Disaster Recovery: Concepts, Design, Implementation*, Springer-Verlag, Berlin.

Service Availability: Principles and Practice, First Edition. Edited by Maria Toeroe and Francis Tam.
© 2012 John Wiley & Sons, Ltd. Published 2012 by John Wiley & Sons, Ltd.

21. High Availability Forum (2001) Providing Open Architecture High Availability Solutions. Revision 1.0, http://www.lynuxworks.com/products/whitepapers/ha-solutions.pdf and http://www.saforum.org/ (accessed May 2011).

22. Gimpelson, T. (2001) New Forum Seeks to Set Service Availability Specs. Computerworld, http://www.computerworld.com/s/article/66555/New_forum_seeks_to_set_service_availability_specs (accessed 27 September 2011).

23. Mannion, P. (2001) Forum Aims to Heighten Packet Network Reliability. News & Analysis, EE Times, http://www.eetimes.com/electronics-news/4042863/Forum-aims-to-heighten-packet-network-reliability (accessed 27 September 2011).

24. Alliance for Telecommunications Industry Solutions (2011) ATIS Telecom Glossary 2011, http://www.atis.org/glossary/ (accessed 27 September 2011).

25. Wikipedia (2011) World Wide Web. http://en.wikipedia.org/wiki/World_Wide_Web (accessed 28 September 2011).

26. Bernstein, P.A. (1996) Middleware: a model for distributed system services. *Communications of the ACM*, **39** (2), 86–98.

27. Object Management Group (2011) The OMG's CORBA Website, http://www.corba.org/ (accessed 30 September 2011).

28. Ferraro-Esparza, V., Gudmandsen, M., and Olsson, K. (2002) Ericsson telecom server platform 4. *Ericsson Review*, **3**, 104–113, http://www.ericsson.com/res/thecompany/docs/publications/ericsson_review/2002/2002032.pdf (accessed 26 September 2011).

29. Oracle (2011) Sun Netra Carrier-Grade Servers, http://www.oracle.com/us/products/servers-storage/servers/netra-carrier-grade/index.html (accessed 28 September 2011).

30. GoAhead Software Inc. (2011) SelfReliant, http://www.goahead.com/products/selfreliant/high-availability-middleware.aspx (accessed 28 September 2011).

31. Wikipedia (2011) Dot-com Bubble, http://en.wikipedia.org/wiki/Dot-com_bubble (accessed 30 September 2011).

32. Bass, L., Clements, P., and Kazman, R. (1998) *Software Architecture in Practice*, SEI Series in Software Engineering, Addison-Wesley, Reading, MA.

33. Object Management Group (2001) CORBA 2.5 – Chapter 25 – Fault Tolerant CORBA, http://www.omg.org/cgi-bin/doc?formal/01-09-29.pdf (accessed May 2011).

34. Object Management Group (2007) XML Metadata Interchange (XMI®), http://www.omg.org/spec/XMI/ (accessed 8 August 2011).

35. Service Availability™ Forum (2011) Hardware Platform Interface Specification. SAI-HPI-B.03.02, http://www.saforum.org (accessed 30 September 2011).

36. Service Availability™ Forum (2011) Application Interface Specification, Platform Management Service. SAI-AIS-PLM-A.01.02, http://www.saforum.org (accessed 30 September 2011).

37. Service Availability™ Forum (2011) Application Interface Specification, Cluster Membership Service. SAI-AIS-CLM-B.04.01, http://www.saforum.org (accessed 30 September 2011).

38. Service Availability™ Forum (2011) Application Interface Specification, Information Model Management Service. SAI-AIS-IMM-A.03.01, http://www.saforum.org (accessed 30 September 2011).

39. Service Availability™ Forum (2011) Application Interface Specification, Notification Service. SAI-AIS-NTF-A.04.01, http://www.saforum.org (accessed 30 September 2011).

40. Service Availability™ Forum (2011) Application Interface Specification, Log Service. SAI-AIS-LOG-A.02.01, http://www.saforum.org (accessed 30 September 2011).

41. Service Availability™ Forum (2011) Application Interface Specification, Security Service. SAI-AIS-SEC-A.01.01, http://www.saforum.org (accessed 30 September 2011).

42. Service Availability™ Forum (2011) Application Interface Specification, Checkpoint Service. SAI-AIS-CKPT-B.02.02, http://www.saforum.org (accessed 30 September 2011).

43. Service Availability™ Forum (2011) Application Interface Specification, Event Service. SAI-AIS-EVT-B.03.01, http://www.saforum.org (accessed 30 September 2011).

44. Service Availability™ Forum (2011) Application Interface Specification, Message Service. SAI-AIS-MSG-B.02.01, http://www.saforum.org (accessed 30 September 2011).

45. Service Availability™Forum (2011) Application Interface Specification, Lock Service. SAI-AIS-LCK-B.03.01, http://www.saforum.org (accessed 30 September 2011).

46. Service Availability™ Forum (2011) Application Interface Specification, Naming Service. SAI-AIS-NAM-A.01.01, http://www.saforum.org (accessed 30 September 2011).
47. Service Availability™ Forum (2011) Application Interface Specification, Timer Service. SAI-AIS-TMR-A.01.01, http://www.saforum.org (accessed 30 September 2011).
48. Service Availability™ Forum (2011) Application Interface Specification, Availability Management Framework. SAI-AIS-AMF-B.04.01, http://www.saforum.org (accessed 30 September 2011).
49. Service Availability™ Forum (2011) Application Interface Specification, Software Management Framework. SAI-AIS-SMF-A.01.02, http://www.saforum.org (accessed 30 September 2011).
50. Distributed Management Task Force, Inc. (2009) Systems Management Architecture for Server Hardware (SMASH), http://www.dmtf.org/standards/smash (accessed May 2011).
51. Internet Engineering Task Force (1990) Management Information Base for Network Management of TCP/IP-Based Internets. RFC 1156, http://tools.ietf.org/html/rfc1156 (accesssed 23 September 2011).
52. Internet Engineering Task Force (1991) Management Information Base for Network Management of TCP/IP-Based Internets: MIB-II. RFC 1213, http://tools.ietf.org/html/rfc1213 (accessed 23 September 2011).
53. Internet Engineering Task Force (2002) An Architecture for Describing Simple Network Management Protocol (SNMP) Management Frameworks. RFC 3411, http://tools.ietf.org/html/rfc3411 (accessed 23 September 2011).
54. International Telecommunication Union (2002) Information Technologies – Abstract Syntax Notation One (ASN.1): Specification of Basic Notation. ITU-T Recommendation X.680.
55. Internet Engineering Task Force (1999) Structure of Management Information Version 2 (SMIv2). RFC 2578, http://tools.ietf.org/html/rfc2578 (accessed 23 September 2011).
56. Internet Engineering Task Force (2006) NETCONF Configuration Protocol. RFC 4741, http://tools.ietf.org/html/rfc4741 (accessed 27 April 2011).
57. Internet Engineering Task Force (2010) YANG - A Data Modeling Language for the Network Configuration Protocol (NETCONF). RFC 6020, http://tools.ietf.org/html/rfc6020 (accessed 27 April 2011).
58. Distributed Management Task Force, Inc. (2005) Common Information Model (CIM) Infrastructure Specification. DSP0004, Version 2.3, http://dmtf.org/standards/cim (accessed 23 September 2011).
59. Object Management Group (2011) Unified Modeling Language (UML) Version 2.4, http://www.omg.org/spec/UML/2.4/ (accessed 27 July 2011).
60. Internet Engineering Task Force (2006) Lightweight Directory Access Protocol (LDAP): Technical Specification Road Map. RFC 4510, http://tools.ietf.org/html/rfc4510 (accessed 5 May 2011).
61. Service Availability™ Forum (2011) Service Availability Interface, Overview Document. SAI-Overview-B.05.03, http://www.saforum.org (accessed 30 September 2011).
62. Service Availability™ Forum (2012) Errata for the Application Interface Specification, Information Model. SAI-IM-XMI-A.04.02.errata, http://www.saforum.org (accessed 31 January 2012).
63. Service Availability™ Forum (2011) Service Availability Interface, C Programming Model. SAI-AIS-CPROG-B.05.02, http://www.saforum.org (accessed 30 September 2011).
64. OpenSAF™ (2011) The Open Service Availability Framework, http://www.opensaf.org/ (accessed 2 August 2011).
65. PCI Industrial Computers Manufacturers Group (PICMG) (2011) CompactPCI®Specification, PICMG 2.0 R3.0, Short Form, http://www.compactpci.org/index.php/compactpci/compactpci-specification-overview.html (accessed 23 September 2011).
66. PCI Industrial Computers Manufacturers Group (PICMG) (2003) AdvancedTCA PICMG 3.0®Short Form Specification, http://www.picmg.org/v2internal/resourcepage2.cfm?id=2 (Retrieved 23 September 2011).
67. PCI Industrial Computers Manufacturers Group (PICMG) (2006) Micro Telecommunications Computer Architecture PICMG® MTCA.0, Short Form Specification, http://www.picmg.org/v2internal/resourcepage2.cfm?id=5 (accessed 23 September 2011).
68. Intel, Hewlett-Packard, NEC, & Dell (2009) IPMI – Intelligent Platform Management Interface Specification, Second Generation v2.0, http://www.intel.com/design/servers/ipmi/ (accessed 23 September 2011).
69. Service Availability™ Forum (2004) Specification Distributed Systems Management for HPI-SNMP. SAI-HPI-SNMP-B.01.01, http://www.saforum.org (accessed 26 August 2011).
70. Service Availability™ Forum (2011) Specification for HPI Mapping to AdvancedTCA and MicroTCA. SAIM-HPI-B.03.02-xTCA, http://www.saforum.org (accessed 30 September 2011).
71. OpenHPI (2011) OpenHPI Project Home Page, http://www.openhpi.org/Home (accessed 2 August 2011).

72. International Telecommunication Union (1992) Information Technology – Open Systems Interconnection – System Management: State Management Function. ITU-T Recommendation X.731.

73. IEEE (2008) IEEE Std 1003.1 TM -2008. *Standard for Information Technology – Portable Operating System Interface (for Unix). (POSIX®)*, Institute of Electrical and Electronics Engineers.

74. virtlib: The Virtualization API (2011) http://libvirt.org/ (accessed 13 December 2011).

75. Internet Engineering Task Force (2006) XDR: External Data Representation Standard. RFC 4506, http://tools.ietf.org/html/rfc4506 (accessed 23 September 2011).

76. Distributed Management Task Force, Inc. (2009–2010) Virtualization Management, http://dmtf.org/standards/vman (accessed 23 September 2011).

77. Chandra, T.D. , Hadzilacos, V., Toueg, S., and Charron-Bost, B. (1996) On the impossibility of group membership. The Proceedings of the 15th ACM Symposium on Principles of Distributed Computing, Philadelphia, PA, pp. 322–330, http://www.cs.utoronto.ca/~vassos/research/publications/CHTCB96/paper.ps.gz (accessed 23 September 2011).

78. Fischer, M.J., Lynch, N.A., and Paterson, M.S. (1985) Impossibility of distributed consensus with one faulty process. *Journal of the ACM* , **32** (2), 374–383.

79. Heller, J. (1961) *Catch-22* , Simon&Schuster, New York.

80. Kanso, A., Khendek, F., Hamou-Lhadj, A., and Toeroe, M. (2010) Ranking service units for providing and protecting highly available services with load balancing, in *10th Annual International Conference on New Technologies of Distributed Systems(NOTERE 2010)* (eds K. Drira, A.H. Kacem, and M. Jamiel), IEEE Computer Society, Tozeur, Tunisia, pp. 33–40.

81. World Wide Web Consortium (W3C) (2008) Extensible Markup Language (XML) 1.0 (Fifth Edition), http://www.w3.org/TR/2008/REC-xml-20081126/ (accessed 24 September 2011).

82. Service Availability™ Forum (2010) The Software Management Framework: Basic Concepts Explained. Whitepaper, http://www.saforum.org (accessed 3 September 2011).

83. Wikipedia (2011) Publish/Subscribe, http://en.wikipedia.org/wiki/Publish/subscribe (accessed 13 June 2011).

84. Wikipedia (2011) Two-Phase Commit Protocol, http://en.wikipedia.org/wiki/Two-phase_commit_protocol (accessed 12 July 2011).

85. International Telecommunication Union (1992) Information Technology – Open Systems Interconnection – System Management: Object Management Function. ITU-T Recommendation X.730.

86. International Telecommunication Union (1992) Information Technology – Open Systems Interconnection – System Management: Alarm Reporting Function. ITU-T Recommendation X.733.

87. International Telecommunication Union (1992) Information Technology – Open Systems Interconnection – System Management: Security Alarm Reporting Function. ITU-T Recommendation X.736.

88. Linux Foundation (2009) The LSB 4 Specification, Linux Standard Base Specification Archive, http://refspecs.freestandards.org/lsb.shtml (accessed 23 August 2011).

89. World Wide Web Consortium (W3C) (2004) XML Schema, Second Edition, http://www.w3.org/standards/techs/xmlschema#w3c_all (accessed 24 September 2011).

90. Service Availability™ Forum (2011) Application Interface Specification, SMF Entity Types and Upgrade Campaign XML Schemas. SAI-AIS-SMF-XSD-A.01.02, http://www.saforum.org (accessed 30 September 2011).

91. OpenSAF™ (2011) OpenSAF™ Project Home Page, http://devel.opensaf.org/ (accessed 2 August 2011).

92. ISO (1999) ISO/IEC 9899:1999. *Programming Languages – C* , International Organization for Standardization.

93. Service Availability™ Forum (2011) Application Interface Specification Mapping Javadoc for Release 6. SAIM-AIS-R6-JD-A.01.01, http://www.saforum.org (accessed 30 September 2011).

94. Distributed Management Task Force, Inc. (2010) Web-Based Enterprise Management, http://dmtf.org/standards/wbem (accessed 25 August 2010).

95. Service Availability™ Forum (2008) Java Usage of the SA Forum Notification Service, http://www.saforum.org (accessed 20 October 2010).

96. Service Availability™ Forum (2007) Java Usage of the SAF Log Service, http://www.saforum.org (accessed 20 October 2010).

97. Oracle (2010) Java Enterprise Edition (Java EE), Technical Documentation, http://download.oracle.com/javaee/ (accessed 6 April 2011).

98. Oracle (2000) Java Management Extensions (JMX) Technology. JSR 3, Java Community Process, http://jcp.org/en/jsr/detail?id=3 and http://www.oracle.com/technetwork/java/javase/tech/javamanagement-140525.html (accessed 6 April 2011).

99. Oracle (1995, 2011) JavaBeans™, http://download.oracle.com/javase/tutorial/javabeans/ (accessed 6 April 2011).

100. Ericsson, A.B. (2010) Availability Management for Java, JSR 319. Java Community Process, http://jcp.org/en/jsr/summary?id=319 (accessed 26 August 2010).

101. Sun Microsystems (2004, 2010) Javadoc Tool Documentation, http://www.oracle.com/technetwork/java/javase/documentation/index-jsp-135444.html (accessed 6 April 2011).

102. Oracle (1993, 2010) Java Native Interface, http://download.oracle.com/javase/6/docs/technotes/guides/jni/index.html (accessed 6 April 2011).

103. Sun Microsystems (1995–1999) Java Coding Conventions, http://www.oracle.com/technetwork/java/codeconventions-135099.html (accessed 25 August 2010).

104. Sun Microsystems (2002) New I/O APIs for the Java™ Platform. JSR 51, Java Community Process, http://www.jcp.org/en/jsr/detail?id=51 (accessed 6 April 2011).

105. OpenHPI (2011) OpenHPI Code Base and Man Pages, http://sourceforge.net/projects/openhpi/ (accessed 2 August 2011).

106. Free Software Foundation, Inc. (1991, 1999) GNU Lesser General Public License, Version 2.1, http://www.gnu.org/licenses/lgpl-2.1.html (accessed 19 August 2011).

107. TIPC (2011) TIPC Home Page, http://tipc.sourceforge.net/ (accessed 25 August 2011).

108. Internet Engineering Task Force (1981) Transmission Control Protocol, DARPA Internet Program, Protocol Specification. RFC 793, http://tools.ietf.org/html/rfc793 (accessed 25 August 2011).

109. VideoLAN Organization VideoLAN, http://wiki.videolan.org/ and http://www.videolan.org/developers/ (accessed 2 August 2011).

110. Service Availability™ Forum (2011) Application Interface Specification, IMM Initial XML Schema. SAI-AIS-IMM-XSD-A.01.02, http://www.saforum.org (accessed 30 September 2011).

111. Object Management Group (2011) Object Constraint Language (OCL) Version 2.3, http://www.omg.org/spec/OCL/2.3/Beta2/ (accessed 27 July 2011).

112. Object Management Group (2011) Meta Object Facility (MOF) Version 2.4, http://www.omg.org/spec/MOF/2.4/Beta2/ (accessed 27 July 2011).

113. Schmidt, D.C. (2006) Special issue on model driven engineering. *IEEE Computer Magazine*, **39**, 2.

114. Object Management Group (2003) MDA Guide Version 1.0.1. (eds A. Miller and J. Mukerji), http://www.omg.org/cgi-bin/doc?omg/03-06-01 (accessed 27 July 2011).

115. Kleppe, A., Warmer, J., and Bast, W. (2003) *MDA Explained: The Model Driven Architecture: Practice and Promise*, Addison-Wesley Professional, Boston, MA.

116. Trivedi, K. (2001) *Probability and Statistics with Reliability, Queuing, and Computer Science Applications*, 2nd edn, John Wiley & Sons, Ltd, Chichester.

117. Gherbi, A., Salehi, P., Khendek, F., and Hamou-Lhadj, A. (2009) Capturing and formalizing SAF availability management framework configuration requirements. Paper presented at the Domain Engineering Workshop @ CAiSE'2009, Amsterdam, Netherlands.

118. Salehi, P., Hamoud-Lhadj, A., Colombo, P. *et al.* (2010) A UML-based domain specific modeling language for the availability management framework, *In the 12th IEEE International High Assurance Systems Engineering Symposium, HASE 2010*, IEEE Computer Society, San Jose, CA, pp. 35–44.

119. Salehi, P., Khendek, F., Toeroe, M., *et al.* (2009) Checking for service instance protection for AMF configurations, *In the Third IEEE International Conference on Secure Software Integration and Reliability Improvement, SSIRI 2009*, IEEE Computer Society, Singapore, pp. 269–274.

120. Salehi, P., Khendek, F., Hamou-Lhadj, A., *et al.* (2011) AMF configurations: checking for service protection using heuristics, *In the 7th International Conference on Network and Service Management, CNSM 2011*. IEEE Computer Society, Paris.

121. Object Management Group (2009) UML Profile for MARTE: Modeling and Analysis of Real-Time Embedded Systems, Version 1.0, http://www.omg.org/spec/MARTE/1.0/ (accessed 28 July 2011).

122. Bernardi, S., Merseguer, J., and Petriu, D. (2011) A dependability profile within MARTE. *Software and Systems Modeling*, **10** (3), 313–336.

123. Kanso, A., Toeroe, M., Khendek, F., and Hamou-Lhadj, A. (2008) Automatic generation of AMF compliant configurations, in *The 5th International Symposium on Service Availability*, Lecture Notes in Computer Science, Vol. **5017** (eds T. Nanya *et al.*), Springer, Berlin, pp. 155–170.

124. Kanso, A., Khendek, F., Toeroe, M., and Hamou-Lhadj, A. (2009) Generating AMF configurations from software vendor constraints and user requirements, *In International Conference on Availability, Reliability and Security, ARES 2009*, IEEE Computer Society, pp. 454–461.

125. Kanso, A., Khendek, F., Toeroe, M., and Hamou-Lhadj, A. (2010) Ranking service units for providing and protecting highly available services with load balancing, in *Proceedings of the 10th Annual International Conference on New Technologies of Distributed Systems, NOTERE 2010* (ed. K. Drira *et al.*), IEEE Computer Society, pp. 33–40.

126. The Eclipse Fundation (2011) Eclipse, http://eclipse.org/ (accessed 29 July 2011).

127. Salehi, P., Colombo, A., Hamou-Lhadj, A., and Khendek, F. (2010) A model driven approach for AMF configuration generation, in *System Analysis and Modeling: About Models. 6th International Workshop, SAM 2010*, Lecture Notes in Computer Science, Vol. **6598** (eds F.A. Kraemer and P. Herrmann), Springer, Berlin, pp. 124–143.

128. Jouault, F., Allilaire, F., Bézivin, J., and Kurtev, I. (2008) ATL: a model transformation tool. Special Issue on Second issue of experimental software and toolkits (EST), in *Science of Computer Programming*, vol. **72** (1–2) (ed. M.G.J. vandenBrand), Elsevier, Amsterdam, pp. 31–39.

129. Kohzadi, S. (2009) Automatic generation of upgrade campaign specifications. MASc thesis. CSE Department, Montreal, QC: Concordia University.

130. Mishra, A. (2010) Automated AMF configuration difference generation. MASc thesis. ECE Department, Montreal, QC: Concordia University.

131. Goyal, A. and Lavenberg, S.S. (1987) Modeling and analysis of computer system availability. *IBM Journal of Research and Development*, **31** (6), 651–664.

132. Wood, A. (1994) Availability modeling. *IEEE Circuits and Devices Magazine*, **10** (3), 22–27.

133. Muppala, J., Ciardo, G., and Trivedi, K.S. (1994) Stochastic reward nets for reliability prediction. *Communications in Reliability, Maintainability and Serviceability*, **1** (2), 9–20.

134. Ciardo, G., Muppala, J., Trivedi, K., *et al.* (1989) SPNP: stochastic petri net package, *Proceedings of the 3rd International Workshop on Petri Nets and Performance Models, PNPM89*, IEEE Computer Society, pp. 142–150.

135. Wang, D. and Trivedi, K. (2005) Modeling user-perceived service availability, in *2nd International Service Availability Symposion*, Lecture Notes in Computer Science, Vol. **3694** (eds M. Malek, E. Nett, and N. Suri), Springer, Berlin, pp. 107–122.

136. Amazon Web Services (2011) Amazon Elastic Compute Cloud (EC2), http://aws.amazon.com/ec2/ (accessed 25 September 2011).

137. Microsoft (2011) Windows Azure, http://www.microsoft.com/windowsazure/ (accessed 25 September 2011).

138. VMware Inc. (2011) VMware Cloud Computing, http://www.vmware.com/solutions/cloud-computing/index.html (accessed 25 September 2011).

139. Google (2011) Google App Engine, http://code.google.com/appengine/docs/whatisgoogleappengine.html (accessed 25 September 2011).

140. Armbrust, M., Fox, A., Griffith, R. *et al.* (2010) A view of cloud computing. *Communications of the ACM*, **53** (4), 50–58.

141. BBC (2011) Microsoft Online Services Hit by Major Failure, http://www.bbc.co.uk/news/technology-14851455 (accessed 13 September 2011).

142. VMware Inc. (2009) Protecting Mission-Critical Workloads with VMware Fault Tolerance. VMware White Paper, http://www.vmware.com/files/pdf/resources/ft_virtualization_wp.pdf (accessed 25 September 2011).

143. Khalidi, Y.A. (2011) Building a cloud computing platform for new possibilities. *IEEE Computer*, **44** (3), 29–34.

144. IEEE Computer Society and Association for Computing Machinery (2008) Computer Science Curriculum 2008: An Interim Revision of CS 2001, http://www.computer.org/portal/c/document_library/get_file?p_l_id=2814020&folderId=3111026&name=DLFE-57604.pdf (accessed 25 September 2011).

145. Association for Computing Machinery and IEEE Computer Society (2008) Information Technology 2008 Curriculum Guidelines for Undergraduate Degree Programs in Information Technology, http://www.computer.org/portal/c/document_library/get_file?p_l_id=2814020&folderId=3111026&name=DLFE-57605.pdf (accessed 25 September 2011).

146. Association for Computing Machinery and Association for Information Systems (2010) IS 2010 Curriculum Guidelines for Undergraduate Degree Programs in Information Systems, http://www.computer.org/portal/c/document_library/get_file?p_l_id=2814020&folderId=3111026&name=DLFE-57608.pdf (accessed 25 September 2011).

Further Reading

Gray, J. and Reuter, A. (1993) *Transaction Processing: Concepts and Techniques*, Morgan Kaufmann Publishers Inc., San Francisco, CA.

Sahner, R., Trivedi, K.S., and Puliafito, A. (1996) *Performance and Reliability Analysis of Computer Systems*, Kluwer Academic Publishers, Dordrecht.

Utas, G. (2005) *Robust Communications Software: Extreme Availability, Reliability and Scalability for Carrier-Grade Systems*, John Wiley & Sons, Ltd, Chichester.

Index

3G, 24

Activate, 9, 113–14, 185, 243–4, 344
Activation unit, 58, 275–7, 279–80, 285, 290
Active
  element, 15, 18, 40, 154
  entity, 365
  process, 137
  role, 15, 17–19, 54, 141, 144, 404
Administrative attention, 228
Administrative operation, 68, 80–82, 201, 210, 220, 315
  AMF, 53, 148, 159, 177–87, 188–92, 366
  CLM, 126–9
  HPI, 94–5
  IMM, 38, 228, 247–52, 255–60, 366
  LOG, 231, 233
  NTF, 243
  PLM, 48, 98, 101, 104–5, 107, 110–120
  SMF, 274, 276, 278–80, 285–7, 289–91, 367
Administrative ownership, 249, 251–2, 255, 261, 290
Administrative state, 101, 104–5, 113–16, 127–9, 177–81, 185–6, 276
  locked-inactive, 101, 105, 113–14
  locked-instantiation, 104–5, 112, 114, 178, 186
  shutting down, 101, 104–5, 115–16, 127–9, 179–81, 186
  unlocked, 101, 104, 127, 179–81, 185–6, 276
Admin-operation-pending, 105, 113–14

Agent, 248–9, 344, 356–61, 365
AIS *see* Application Interface Specification
  Services, 46–60, 81, 258, 298–300, 306, 328
Alarm, 46, 50–51, 195, 299–300, 338, 395
  clock, 136
  HPI, 87–8, 91, 94, 96
  LOG, 230, 234–6
  NTF, 228, 238–9, 242–6, 247
  PLM, 98, 111–12, 118–19
  security, 50–51, 118, 238–9, 244–6, 263, 335–40, 411
AMF *see* Availability Management Framework
  cluster, 157–9, 179, 181–2, 186–7
  configuration, 52, 190, 360, 378–9, 383, 412, 419–20
  domain model, 419
  node, 52–3, 155–9, 178–82, 187, 360, 423
AMF configuration validation, 419, 425
Analytical model, 427–8
API, 34, 60, 68–70, 80–83, 305
  HPI, 87–8, 91, 93–4, 96
  track, 107–10, 114–16, 126–31, 209–10, 311–12
Application, 25, 36, 81, 156–7, 297–9, 371–3, 392–400
Application Interface Specification (AIS), 43–4, 46–7, 347
Application programming interface *see* API
Architecture, 25, 36–40, 43–59, 105, 328, 348–51, 356–8
Area library, 306

Service Availability: Principles and Practice, First Edition. Edited by Maria Toeroe and Francis Tam.
© 2012 John Wiley & Sons, Ltd. Published 2012 by John Wiley & Sons, Ltd.